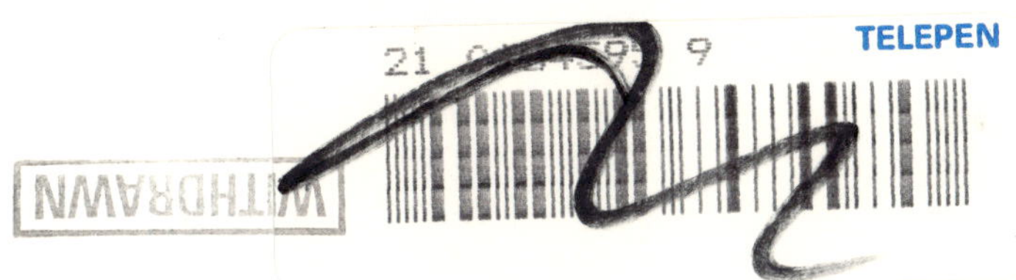

SYMPOSIA OF THE
SOCIETY FOR EXPERIMENTAL BIOLOGY

NUMBER XXXIV

SYMPOSIA OF THE SOCIETY FOR EXPERIMENTAL BIOLOGY

I Nucleic Acid
II Growth, Differentiation and Morphogenesis
III Selective Toxicity and Antibiotics
IV Physiological Mechanisms in Animal Behaviour
V Fixation of Carbon Dioxide
VI Structural Aspects of Cell Physiology
VII Evolution
VIII Active Transport and Secretion
IX Fibrous Proteins and their Biological Significance
X Mitochondria and other Cytoplasmic Inclusions
XI Biological Action of Growth Substances
XII The Biological Replication of Macromolecules
XIII Utilization of Nitrogen and its Compounds by Plants
XIV Models and Analogues in Biology
XV Mechanisms in Biological Competition
XVI Biological Receptor Mechanisms
XVII Cell Differentiation
XVIII Homeostasis and Feedback Mechanisms
XIX The State and Movement of Water in Living Organisms
XX Nervous and Hormonal Mechanisms of Integration
XXI Aspects of the Biology of Ageing
XXII Aspects of Cell Motility
XXIII Dormancy and Survival
XXIV Control of Organelle Development
XXV Control Mechanisms of Growth and Differentiation
XXVI The Effects of Pressure on Organisms
XXVII Rate Control of Biological Processes
XXVIII Transport at the Cellular Level
XXIX Symbiosis
XXX Calcium in Biological Systems
XXXI Integration of Activity in the Higher Plant
XXXII Cell–Cell Recognition
XXXIII Secretory Mechanisms

The Journal of Experimental Botany
is published by the Oxford University Press
for the Society for Experimental Biology

SYMPOSIA OF THE
SOCIETY FOR EXPERIMENTAL BIOLOGY

NUMBER XXXIV

THE MECHANICAL PROPERTIES OF BIOLOGICAL MATERIALS

Published for the Society for Experimental Biology

CAMBRIDGE UNIVERSITY PRESS

CAMBRIDGE

LONDON NEW YORK NEW ROCHELLE

MELBOURNE SYDNEY

Published by the Press Syndicate of the University of Cambridge
The Pitt Building, Trumpington Street, Cambridge CB2 1RP
32 East 57th Street, New York, NY 10022, USA
296 Beaconsfield Parade, Middle Park, Melbourne 3206, Australia

First published 1980

Text set in 10/13 pt Linotron 202 Times Roman, printed and bound
in Great Britain at The Pitman Press, Bath

British Library Cataloguing in Publication Data

Society for Experimental Biology. *Symposium, 34th, Leeds University, 1979*
The mechanical properties of biological materials.
1. Biomechanics
I. Vincent, J. F. V. II. Currey, John D.
574.1′912 QH505 80–40111

ISBN 0 521 23478 6

CONTENTS

Preface vii
J. F. V. VINCENT AND J. D. CURREY

Biomechanics: the last stronghold of vitalism 1
J. E. GORDON

Fracture 13
E. H. ANDREWS

The mechanical behaviour of composite materials 37
B. HARRIS

Mechanical properties of mollusc shell 75
J. D. CURREY

Some mechanical and physical properties of teeth 99
N. E. WATERS

The structure and biomechanics of bone 137
J. L. KATZ

Wood, one of nature's challenging composites 169
G. JERONIMIDIS

Insect cuticle: a paradigm for natural composites 183
J. F. V. VINCENT

Molecular structure and mechanical properties of keratins 211
R. D. B. FRASER AND T. P. MACRAE

Silks – their properties and functions 247
M. DENNY

Theories of rubber-like elasticity and the behaviour of filled rubber 273
L. MULLINS

The theory of viscoelasticity in biomaterials 289
K. L. DORRINGTON

The mechanical properties of plant cell walls 315
D. B. SELLEN

The elastic properties of rubber-like proteins and highly extensible tissues 331
J. M. GOSLINE

The viscoelasticity of mucus: a molecular model 359
R. H. PAIN

Articular cartilage 377
S. A. V. SWANSON

Deformation in tendon cartilage 397
J. KASTELIC AND E. BAER

Adaptive materials: a view from the organism 437
S. A. WAINWRIGHT

Posters displayed at the Symposium (4–6 September 1979) 455

Author Index 491

Subject Index 501

PREFACE

The thirty-fourth symposium of the Society for Experimental Biology was held at Leeds University from 4 to 6 September 1979. We are grateful to Professor Alexander for suggesting that the symposium be held at Leeds, and for making us feel so welcome. Dr D. B. Sellen worked hard and efficiently as local secretary, and all arrangements went smoothly and pleasantly.

Because the subject of the symposium is perhaps not quite so much in the mainstream of the Society's usual interests as usual the organisers made considerable efforts to publicise the meeting to all kinds of scientists through journals and other media, so the composition of the conference members is interesting. There were roughly equal numbers of biologists and of materials scientists in the broad sense. There were virtually no members whose main interest was clinical.

The discussions after the talks were very lively, and always had to be cut short by the chairman. The nature of the discussions was instructive: on one hand it was clear that biologists who work on the mechanical properties of materials need much more to call upon the materials scientists for guidance on what is known already about the phenomena they are studying; on the other hand it was equally clear that the materials scientists had little idea of the richness of biological diversity and of the pervasive ability of natural selection to provide the optimum solutions to very complex problems.

Biomechanics is a relatively recent animal: it feeds off a combination of engineering, materials science, polymer science, morphology and physiology. For the biologist its closest ancestor is the functional morphology of the turn of the century. At that time the aim was primarily the taxonomy of animals and of their bits and pieces. Morphology was mainly descriptive and a proof (in both senses) of evolution. In the early years of this century several people – of whom d'Arcy Thompson (1917) has been shown the most eminent – attempted to put morphology on to a more physical and mathematical basis but succeeded, mostly, only in drawing rather better parallels between biological and physical/engineering morphologies. This was useful in that it confirmed that biological morphology is just as subject to rational analysis as any other branch of science (see also Gordon, 1980) but the science of the mechanical properties of materials, more particularly of polymers, had to wait for the physicists and mathematicians (Dorrington, 1980). Meanwhile, developmental morphology of the d'Arcy Thompson variety flowered briefly in Huxley's hands (Huxley, 1932) then sank almost without trace. The side of morphology concerned with embryology flourished in a more physiological direction and d'Arcy Thompson was left as a beacon, used as a reference point by engineers, architects, embryologists and, later, materials scientists. Biomaterials science has had a far steadier following from the medical fraternity who have progressed from peg legs to plastic arteries. But their response to any problem has often, with good reason, been expedient rather than experimental.

So it is only in the last 15 or 20 years that biomechanics has come to have any obvious place in biology: before 1953 the number of papers on biomaterials published in the *Journal of Experimental Biology* averaged one every five years. Since that date it has averaged about two a year – a tenfold increase. The real increase is much greater since in the time many other journals have been founded specifically for the publication of papers on biomechanics. It seems likely that biomechanics in its various manifestations (this symposium covered only certain of the materials of animals and plants and was not concerned with structures) will come to replace a large part of what is now functional morphology and so take its place in what will be the triumverate of biology: ecology, physiology and biomechanics.

For any engineer or biologist finding biomechanics for the first time in this volume, the best introduction is Wainwright, Biggs, Currey & Gosline (1976). For biologists, additional background material is available from most first-year physics and engineering texts; at a more general level nothing surpasses Gordon (1976, 1978) who also gives a lead in to the more complex aspects.

J. F. V. VINCENT
J. D. CURREY

References

DORRINGTON, K. (1980). The theory of viscoelasticity in biomaterials. In *The Mechanical Properties of Biological Materials* (34th Symposium of the Society for Experimental Biology), ed. J. F. V. Vincent & J. D. Currey, pp. 289–314. Cambridge University Press.

GORDON, J. E. (1976). *The New Science of Strong Materials*, 2nd edn. Harmondsworth: Penguin.
GORDON, J. E. (1978). *Structures*. Harmondsworth: Penguin.
GORDON, J. E. (1980). Biomechanics – the last stronghold of vitalism. In *The Mechanical Properties of Biological Materials* (34th Symposium of the Society for Experimental Biology) ed. J. F. V. Vincent & J. D. Currey, pp. 1–11. Cambridge University Press.
HUXLEY, J. S. (1932). *Problems of Relative Growth*. London: Methuen.
THOMPSON, d'A. W. (1917). *On Growth and Form*, Cambridge University Press.
WAINWRIGHT, S. A., BIGGS, W. D., CURREY, J. D. & GOSLINE, J. M. (1976). *Mechanical Design in Organisms*. London: Edward Arnold.

BIOMECHANICS: THE LAST STRONGHOLD OF VITALISM

J. E. GORDON

Department of Engineering, University of Reading, Whiteknights, Reading RG6 2AJ, UK

Most people – and especially most biologists – have a special relationship with the science of elasticity; they don't want to know anything about it. It is too trivial, too difficult and too unimportant – above all boring. But in fact the mechanical problems of plants and animals are not trivial; they are of quite extraordinary difficulty and complexity and, of course, mechanical strength is absolutely vital to the existence of the organism. But, almost invariably, living things are so successful in solving their structural problems that we do not notice how they do it. For nothing attracts less curiosity than total success.

The theory of engineering structures is not easy, even at an elementary level: it is intellectually very difficult – as students are apt to discover. Nevertheless, ordinary engineering theory is not really adequate to cope with the majority of the problems in biomechanics. There is a real danger, on the one hand, of trying to force the mechanics of living structures into a simplistic engineering mould (at Friday Harbor Marine Laboratory some student had painted on a testing machine 'Procrustes was here') or, on the other hand, of regarding the whole subject as an arcane, vitalistic mystery to which the laws of science can hardly be said to apply.

But living creatures of course sustain their loads by making use of the ordinary physical laws; just as they make use of the ordinary laws of chemistry. However one cannot expect to understand biological metabolism by the simple, crude application of 'O' level school chemistry. Equally, the understanding of biomechanics takes one into regions of elasticity which are, for the most part, well beyond the level of the elementary text-books.

For one thing the engineer generally assumes Hooke's law; for another, most engineering structures work at small strains and deflections. This is partly because that is the way in which most engineering materials, such as steel and concrete, behave. It is also partly because these are the only conditions for which it is possible to do the mathematics. As a result, many engineers have come to think in this way and some of them have come to believe that high-strain, non-Hookean structures are somehow not quite respectable.

Biological materials like wood and bone do, of course, more or less conform to engineering preconceptions but soft tissues seldom obey Hooke's law and the strains in an animal are usually about a thousand times higher than those to which the engineer is accustomed. The engineer might however be reminded that, although the stresses in animals are sometimes low yet, because the strains are so high, the strain energies with which one is likely to be dealing in biomechanics are at least as high (and sometimes much higher) than those which have to be contained in artificial structures – thus the problems of fracture mechanics in living organisms are very severe.

However there is another, and perhaps a more important, difference between the structural philosophy of the engineer and that of Nature. This lies in what I may, perhaps, christen 'Catastrophe design'. In his design thinking the engineer applies 'factors of safety' and so on, seeking to postpone the collapse of his structure for as long as possible. In fact he generally tries to make the structure much too strong for its purpose and he hopes that it will never break. If it does collapse however, the broken remains are chiefly of interest to the accident inspectors and the lawyers – the engineer is not generally concerned that the broken structure should still be serviceable.

This kind of philosophy may be applicable, to a limited extent, to the trunks of trees and, sometimes, to bones; but it is certainly not the case with soft biological structures. A large number of living organisms are designed to cope with excessive loads by deflecting. Small plants, like grasses, stand up to routine loads, arising from the wind and their own weight much like the structures of the engineer. But small plants get trodden on and when this happens they simply buckle and deflect – and in due course recover. One cannot 'break' a carpet or a doormat. But this is a way of structural life which hardly enters into the thinking of the engineer. In fact its practical advantages are very great since it enables one to get rid of that expensive weight-eater the 'factor of safety' – which engineers are beginning to realise is largely rubbish anyway. This is a field in which engineers have a great deal to learn from biomechanics.

In his attempts to avoid or to postpone the dreaded 'collapse' of his structures the engineer has traditionally made use of the concept of the 'breaking stress' of materials. Generations of students have been taught that if, at any place in a structure, the working stress reaches the breaking stress of the material then the structure will fail. This is nonsense. For one thing it does not adequately take account of stress concentrations. At every crack and scratch and re-entrant in a stressed material the local stress will be raised – in accordance with Professor Inglis' well-known formula,

$$\sigma_{max} = \sigma_{mean}(1 + 2\sqrt{l/r})$$

where σ_{max} is the peak stress at the crack tip. σ_{mean} is the mean stress away from the crack. l is the crack length and r is the crack tip radius. A fine scratch is just as bad, from this point of view, as a great big hole since the concentration of stress depends only upon the ratio of l to r. Using Inglis' formula it is quite easy to show that if one were to scratch a girder of the Forth Bridge with an ordinary sharp pin the stress which must exist at the bottom of the scratch would be well above the 'breaking stress' of the steel and so, according to the traditional engineer's view, the bridge should break and fall into the sea. It is well known that this is not likely to happen. Nor are similar effects experienced by most plants and animals which are continually being scratched and punctured and abraded during the course of their lives.

Engineering structures are not, as a rule as good at surviving mechanical defects as plants and animals but, of course they do – and must – have a certain capacity for doing so; even though super-tankers break in two from time to time and distribute lavish supplies of free oil. In consequence engineers have been compelled to take notice of that subversive man A. A. Griffith. In fact the coming of fracture mechanics has blown a lot of fresh air into some pretty stuffy and hypocritical engineering cubby-holes.

And, of course, fracture mechanics applies in biology; but again I don't think that we can necessarily apply elementary fracture mechanics theory in the sort of way that engineers do. Accepted theory assumes that pretty well the whole of the strain energy which is released when a material is in the process of breaking is taken up as 'work of fracture' in the fracture zone. As a matter of fact I do not think that this assumption will stand up to sceptical examination, even with Hookean materials, and it may not be even approximately true for soft biological tissues.

As we have said, we are very frequently dealing with very large amounts of strain energy; in fact the strain energy in the tissues of one's cheek when one yawns is approximately equal, per unit volume of material, to that in mild steel at an ordinary engineering stress – though, as a rule, one's cheek is in no danger of splitting across, like a box-girder bridge, even in the widest yawn. In fact animal tissues are

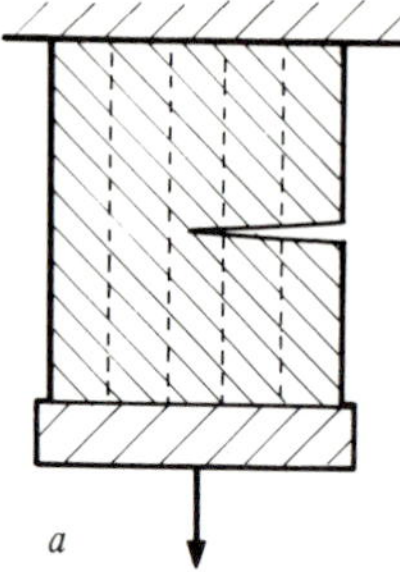

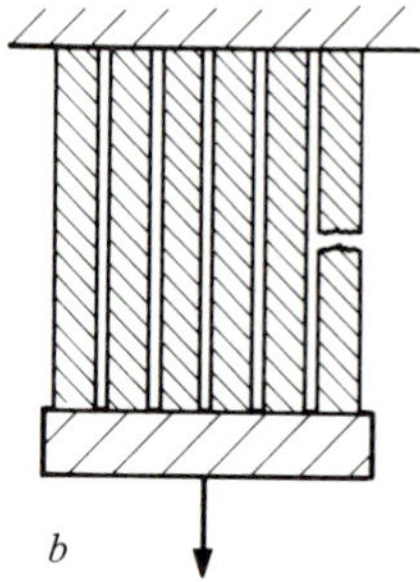

Fig. 1. Crack propagation in continuous and sub-divided material or structure. *a*. Elements joined (e.g. continuous material), crack propagates. *b*. Elements isolated (e.g. a rope), crack is stopped.

exceptionally tough – in terms of critical crack length at a given strain – almost certainly tougher than steel. As we have said, animals do not go pop, like a balloon, when you stick pins into them. All this seems to imply a high work of fracture. No doubt the work of fracture *is* high – but is it as high as all that?

I have suggested elsewhere (Gordon, 1979*a, b*) that fracture mechanics is not entirely about such matters as 'work of fracture'; it is about the communication of energy between the parts or elements of a structure – for, if the released strain energy cannot reach the fracture zone, then the material or the structure cannot break.

Tension structures

One way of restricting the transmission of energy through a structure is by sub-dividing it and this, of course, is most easily done with a tension structure (Fig. 1). It has been shown by H. L. Cox (1965) that the sub-division of a tension structure into many parallel members does not incur extra weight, indeed some weight is actually saved on the end attachments by sub-division. The multiplication of tension members will result in what is called a 'redundant' structure (Fig. 2). In engineering structures, especially those made from ductile metals, this can be highly dangerous. However the nature of animal structures and materials is generally such that this danger does not arise.

Even in engineering the sub-division of tension members in this way can be highly beneficial. Ordinary ropes are a case in point. Modern long-span suspension bridges are entirely dependent upon the use of many wires made from high-tensile steel. Because they are so strong these wires are almost dead brittle but, since the individual wires are entirely separated from each other elastically, no strain energy can be

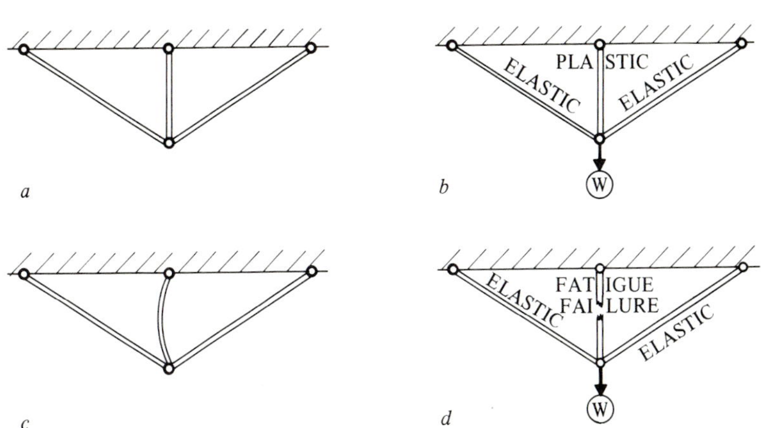

Fig. 2. The danger of fatigue failure in a redundant structure. *a*. Before loading. *b*. Initial loading. *c*. Load removed. *d*. Link broken by high-strain/low-cycle fatigue.

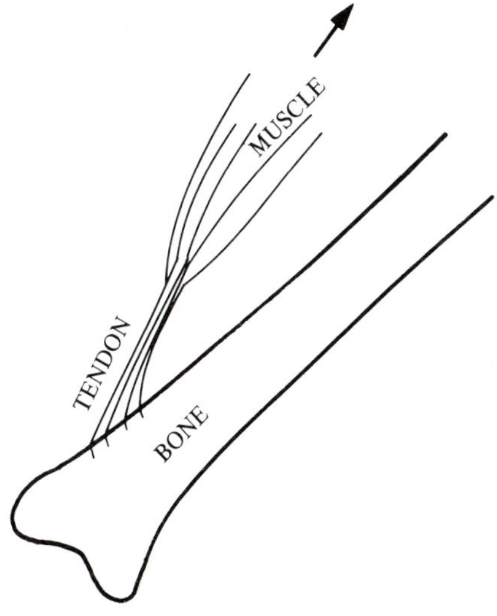

Fig. 3. Sub-division of a tension structure (a tendon).

transmitted between them – and so such cables have proved to be quite safe. In fact they are just about ten times as strong as the older cables which were made from tough, but weak, mild steel; and so suspension bridges can now be built having a span ten times as great as was formerly possible.

This principle is widely used in animals, especially in tendons (Fig. 3). It is also used in such highly sophisticated designs as H.M.S. *Victory* (Fig. 4) where it will be seen that the lower, topmast, and topgallant

Fig. 4. H.M.S. *Victory*

shrouds closely resemble animal tendons. Similar biological principles prevail throughout this brilliant piece of naval architecture – and it might be remembered that at Trafalgar, one of the most decisive naval actions in history, no ship on either side was sunk as a direct result of enemy action. Such structures could not, as a rule, be destroyed by the cutting action of cold shot.

Furthermore, it is interesting to notice how extraordinarily close is design analogy between square-rigged ships, such as the *Victory*, and vertebrate animals, such as man. The backbone, with its vertebrae, corresponds to a mast, the ribs to yards and so on. The immense complication of tension members – the standing and running rigging – is analogous to the tendons and muscles; the sails, I suppose, to skin and other membranes. These analogies are not fanciful, for like structural requirements are apt to produce like structural effects and it has to be emphasised, most strongly, that ships like *Victory* had achieved the very highest standards in safe structural design.

Bending structures

The principle whereby structural safety can be ensured by the elastic isolation of the elements or components can often be applied to bending structures. Fig. 5 shows the simplest kind of wooden bridge in which the

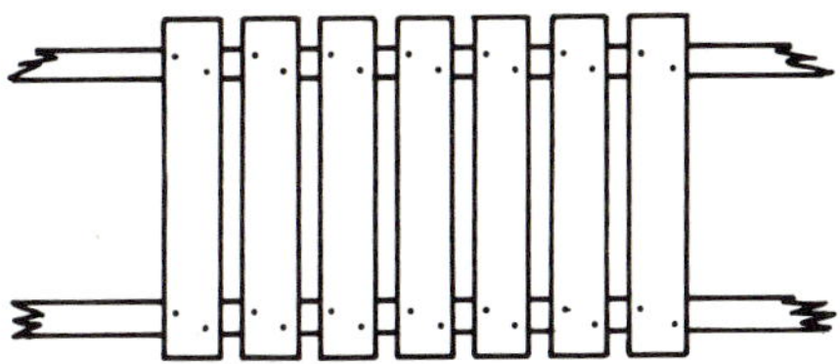

Fig. 5. The ordinary decking of simple wooden bridges is an example of 'isolated' structure in bending. If one plank breaks little harm is done and damage may not spread.

decking is constructed from separate slats or beams. If any individual slat breaks, the damage is unlikely to spread, which would not be the case if the decking were continuous. Very similar principles apply to an ordinary tiled roof, which can thus be made from brittle ceramic tiles without incurring the danger of any major failure. Thus birds, unlike aeroplanes, are not covered with a continuous layer of shiny aluminium plates but rather by a system of cantilevers, called feathers. Each feather is a separate cantilever and so is each of the barbs from which the feather is made up. All these cantilevers are only attached to each other in the weakest possible way. No doubt to the horror of aeronautical inspectors birds can and do fly around quite safely with one or more feathers missing. Moreover feathers seem to constitute an excellent protective armour. Birds of prey do not, as a rule, seek to penetrate the feathers of their victims, they generally try to break their necks by 'stooping' from a height. In the same sort of way 'primitive' people, like the mediaeval Japanese, have often used feather armour as a protection against swords and similar weapons. In modern times the protective armour of battleships has generally been arranged in a scale-like manner with little elastic connection between the armour plates which, as a rule, were not expected to contribute to the structural strength of the ship as a whole.

Compression structures

If a structure is entirely in compression, like masonry, then the problem of preventing cracks from spreading may not arise; but such examples are rare in biology. Much more often the material of the structure has to be rendered safe both in compression and in tension; wood and bone and teeth are cases in point. As Fig. 6 shows, if we attempt to sub-divide the structure after the fashion of Fig. 1, then the members will be liable to buckle separately in compression so that some form of lateral connection is essential. One way of promoting the safety of such

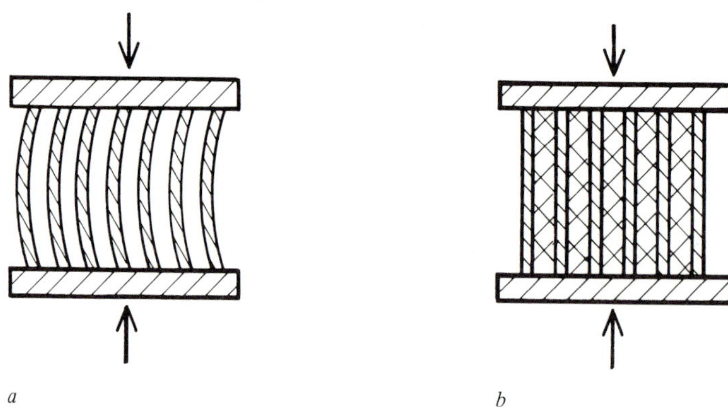

Fig. 6. Compressive behaviour of a sub-divided structure. *a*. Separate elements buckle separately at a low load (e.g. grass and doormats). *b*. Compressive strength is much improved by some lateral communication (e.g. timber).

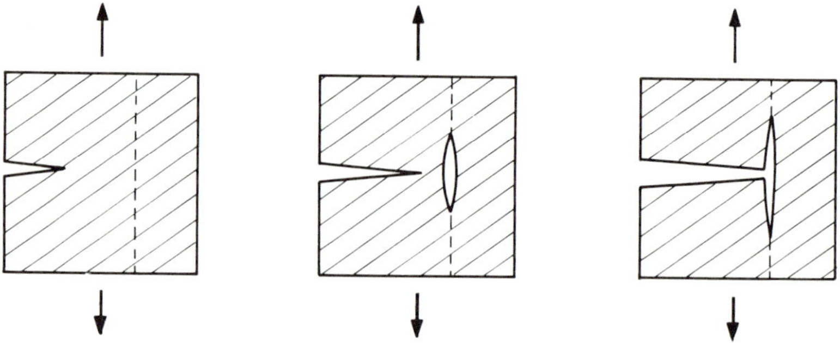

Fig. 7. Adhesion of interfaces is 'rigid' but weak. The interface will open in advance of the crack by the Cook–Gordon mechanism and will often stop the crack. (e.g. wood, fibreglass, teeth, mother-of-pearl, etc.)

arrangements in tension is to connect the elements to each other by means of 'weak interfaces'. If this is properly done then the spread of cracks may be inhibited (Fig. 7) by means of the Cook–Gordon mechanism (Cook & Gordon, 1964).

This is in fact what happens in wood, in mother-of-pearl, in teeth and probably in most bones. It is also what happens in artificial composites such as fibreglass.

The safety of continuous membranes

It is of course possible to employ the device of separating the elements when designing membranes which are intended to take tension and this

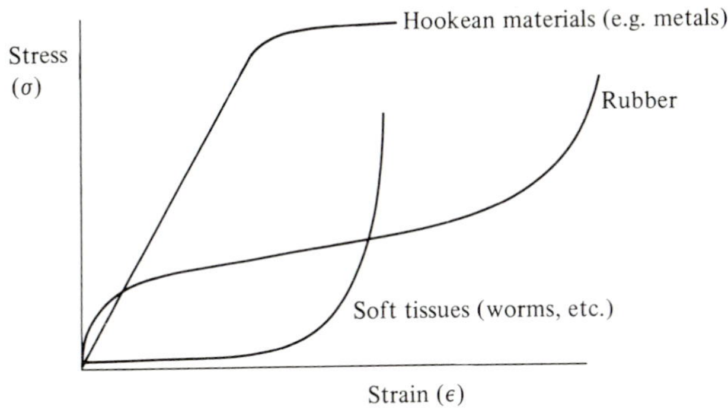

Fig. 8. Typical stress–strain curves (not to scale).

is why cloth is difficult to tear. However cloth is permeable to liquids and gases and attempts to 'proof' it by the application of polymeric films and adhesives sometimes cause embrittlement. Artificial membranes like cloth and paper often obey Hooke's law but it is very noticeable that animal membranes, such as skin and artery walls, do not do so. Generally they have a J-shaped elastic curve, like Fig. 8. It is also noticeable that membranes of this sort are generally very hard to tear. When we became aware of the extreme toughness of some of these membranes our reaction, in the Biomechanics Group at Reading, was to suppose that such materials must have an exceptionally high work of fracture. In fact my colleague, Peter Purslow, has measured the work of fracture of a considerable number of soft animal tissues; it is not exceptionally high, being usually in the range 10^3–10^4 $\mathrm{Jm^{-2}}$. Where rat-skin, or worm cuticle, or human arteries differ from metal sheet, or indeed from rubber sheet, is not in the work of fracture but in the shape of the stress–strain curve.

Nearly all tough soft tissues have this J-shaped curve and I am inclined to think that it is of great importance in ensuring the mechanical safety of the animal. It will be seen that the *initial* part of the curve is almost horizontal, very much like the surface tension of a liquid. Over this part of the curve, therefore, the material has virtually no shear modulus and so, at low and moderate strains, there can be virtually no communication of strain energy between the parts or elements of the material or the structure. Thus, whatever the work of fracture, Griffith is, so to speak, frustrated and the material is unable to tear. If this is the case then the shape of the stress–strain curve is of the greatest importance. It is noticeable that membranes which are intended to tear, such as amniotic membranes and egg-shell membranes, do not have the

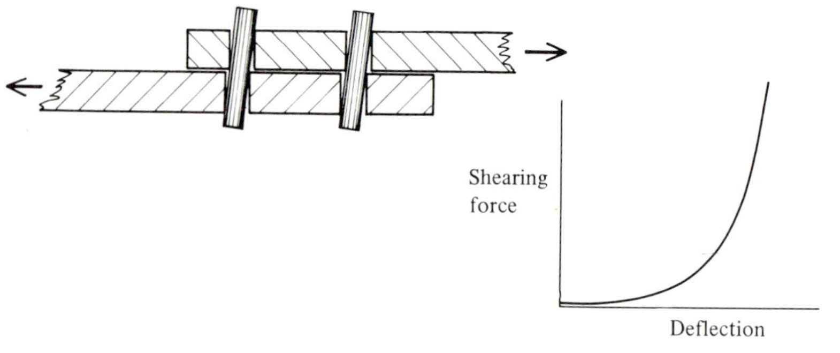

Fig. 9. In a 'wobbly' pegged or lashed joint (e.g. wooden ship fastened with 'tree nails') shear communication is non-linear.

J-shaped curve but more-or-less obey Hooke's law. Furthermore, ordinary soft rubber has a sigmoid stress–strain curve (Fig. 8) and, as is well known, if you stick a pin into a rubber balloon, it will immediately burst with a loud pop. It is difficult to reproduce this behaviour using a natural bladder.

Joints in 'rigid' structures

Traditional technological structures, such as farm carts and wooden ships, were generally held together with nails or bolts or wooden pegs (treenails). These joints soon became 'wobbly' and enabled the structure as a whole to 'work', and wooden ships invariably leaked. Until recently iron and steel ships were riveted together and the riveted joints 'worked' to a greater or less extent; the classical description of this behaviour is, of course, to be found in Rudyard Kipling's 'Ship that Found Herself', which is a valuable engineering text. Modern engineers are nearly always scornful of joints of this type and tend to teach the advantages of rigid, welded joints, which do not 'work'. This is true, but welded steel structures are frequently full of cracks and welded ships break in two much too often. Traditional ships and similar structures had all sorts of faults but they almost never broke in two, and I think that this was at least partly due to the wobbly characteristics of their joints. The characteristics of the traditional pegged or bolted or riveted joint when subjected to a shearing load is shown in Fig. 9. It will be seen that this is a J-shaped curve not at all unlike that of an animal membrane and it is probably 'safe' for much the same reasons. Joints between the 'rigid' parts of animals, such as bones, seem to be of this elastic type. The sutures in the human skull are a case in point – anatomical writers have supposed that these joints are not 'functional' in the grown animal, but I think it is very likely that they exist in skulls for excellent protective structural reasons. In fact the safety of living, and other, structures seems to me to depend largely upon an enormously sophisticated attention to what the casual observer might regard as 'details'.

References

Gordon, J. E. (1979*a*). Catastrophe Design – or how to behave like a worm. Paper given at Conference on N.D.E., La Jolla, California.

Gordon, J. E. (1979*b*). Designing with Brittle Materials. Paper given at ICM 3 Conference at Cambridge, UK.

Cox, H. L. (1965). *The Design of Structures of Least Weight*. Oxford: Pergamon.

Cook, J. & Gordon, J. E. (1964). A mechanism for the control of crack propagation in all-brittle systems. *Proceedings of the Royal Society of London, Series A*, **282**, 508–20.

FRACTURE

E. H. ANDREWS

Department of Materials, Queen Mary College, Mile End Road, London E1 4NS, UK

Introduction

The term 'fracture' can be used to embrace a wide variety of failure phenomena in materials under stress. Such phenomena include impact failure, fatigue failure, environmental cracking, tearing of soft materials, ductile failure, adhesive failure and so on. With biological materials the expression includes for example, impact and fatigue of bone, the rupture of soft tissue and adhesive failure between tissue of differing kinds. Although an enormous amount of information has been amassed on the subject of fracture in engineering solids, this knowledge and the theories which relate to it tend still to be fragmented into materials-related or phenomenon-related compartments. Thus the fracture of metals is described by a body of data distinct from that describing the fracture of ceramics, and theories of fracture *mechanics* are often unrelated to theories of fracture *mechanisms*. For some years the author and his colleagues have been developing a generalized theory of fracture which goes a long way towards unifying the treatment of fracture in all classes of solids and on all levels (e.g. microscopic and macroscopic) of observation. The theory has the added advantage of being physically simple in that it displays directly the role of the important physical properties governing the fracture resistance of the material.

The theory is outlined and its applications reviewed. The roles of surface energies and mechanical hysteresis (energy loss) are described and demonstrated from experimental results. It is suggested that the theory represents the best available starting point for the consideration of fracture and failure in biological systems.

An enormous volume of literature now exists on the fracture of engineering solids, and the prediction of fracture probabilities is part and parcel of modern engineering design. In spite of this, however, anyone approaching the subject of fracture in a biological context is likely to encounter serious problems as he attempts to employ existing theories and experimental findings. The main reasons for this are not difficult to discover.

First, the mechanical behaviour of biological materials is frequently different from that of typical engineering solids. Engineering 'fracture mechanics', for example, is based on the assumption that solids are either linearly elastic or elastic-plastic in their response to load. To be more explicit, the assumptions are as follows.

(i) That strain is proportional to stress (linearity).
(ii) That all deformations up to some limit are elastic i.e. reversible (no energy is lost in a stress–strain cycle). This also implies that all deformations are fully recovered.
(iii) The elastic strains are infinitesimally small.
(iv) That if the material does depart from linear elastic behaviour, it does so by undergoing fully irrecoverable plastic flow.
(v) The material is isotropic in its mechanical properties.

Needless to say, biological materials often behave very differently from this. Apart from compact bone, which can be considered linearly elastic to a first approximation, most living tissue exhibits large recoverable deformations in which significant energy loss (hysteresis) takes place in a loading/unloading cycle. Examples are wood, skin, muscle and tendon. True plastic flow, as observed in metallic alloys, is almost unknown in biological materials. But true linear elasticity is equally uncommon, even in relatively brittle solids such as bone. Finally, most biological materials are mechanically anisotropic.

Thus traditional engineering fracture theory fails to provide a suitable framework in which to consider the fracture of biological materials, simply because it was designed for classes of solids which exhibit different mechanical properties. There is, however, a second, more philosophical, problem in using conventional fracture theories.

The *purpose* of engineering 'fracture mechanics' is to allow us to predict the conditions of stress or deformation under which a material will fail in service. In contrast, a consideration of fracture in biological systems is concerned much more closely with the physical aspects of the materials. Typical questions might be the following. Can fracture phenomena throw light on the constitution of biological materials? Can they reveal the nature of the forces of adhesion or cohesion in bio-systems? What fracture-resistant mechanisms are employed in load-bearing tissue, and how do they originate? What causes pathological conditions such as *osteogenesis imperfecta* ('brittle bones') in which fracture resistance is severely reduced?

In order to address ourselves to these and similar questions we require a theory of fracture that contains explicit parameters describing the physical and molecular properties of the solid. Here conventional fracture mechanics fails us completely, for such theories simply characterize the material's fracture resistance in terms of an empirical parameter which has no explicit relationship to its physical properties.

In what follows we shall consider a new theory of fracture mechanics ('Generalized Fracture Mechanics' or GFM) which promises to overcome many of the limitations of conventional theories and which therefore lends itself more readily to the study of biological materials. In particular it is applicable to all materials regardless of their modes of deformation and, second, it contains two explicit physical parameters which relate directly to the molecular constitution of the medium concerned.

Fracture mechanics

So far we have employed two terms 'fracture theory' and 'fracture mechanics' which call for more precise definition. 'Fracture theory' embraces any theoretical model of the fracture process, whether concerned with mechanics or mechanisms. 'Fracture mechanics' (FM) on the other hand, is a theoretical formulation of the *conditions of deformation under which a crack will propagate in a continuous medium*. Fracture mechanics theory is thus expressed in terms of stresses, strains, elastic moduli, stored energies and geometric factors such as crack length. Fracture mechanisms, in contrast, relate to the physical or molecular processes which occur in the vicinity of a propagating crack. As we have seen, conventional FM fails to provide a framework within which both mechanics and mechanisms can be related, but GFM is capable of providing such a framework, as we shall see.

Fracture mechanics theory always assumes the existence of a flaw or crack in the solid which is capable of growth or propagation to cause failure. It then considers the conditions of stress, strain etc. under which propagation will occur. This requires the introduction of a criterion for propagation which can take one of two forms.

(i) The crack will propagate when the stress at its tip reaches some critical value which overcomes the forces of cohesion.

(ii) The crack will propagate when the energy released from the body by crack growth just exceeds the energy required for the creation of the new surfaces of the crack.

At first sight, criterion (i) would seem the better, since criterion (ii) is a necessary but not sufficient condition, i.e. criterion (ii) implicitly

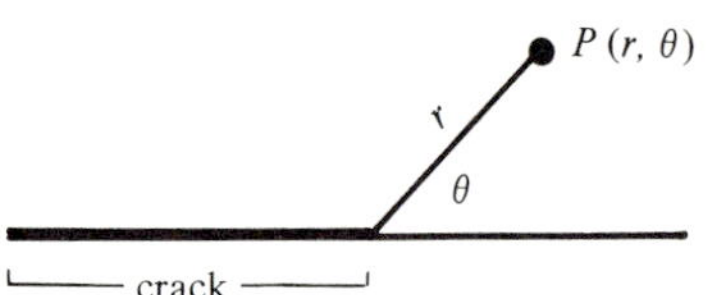

Fig. 1. Specifying the stress field around a crack in terms of the polar co-ordinates r, θ, of the point P.

requires criterion (i) to be satisfied simultaneously. In practice, however, this distinction proves to be rather academic and criterion (ii) is often the more useful. This is because the detailed stress pattern at the tip of a crack is almost impossible to ascertain.

Linear elastic fracture mechanics (LEFM) employs criterion (i) and overcomes the difficulty just mentioned by relating the applied and local stresses (see Fig. 1) by a stress field parameter K, the 'stress intensity factor'. Thus (see e.g. Knott, 1973)

$$\sigma_{ij}(\mathrm{P}) = K f_{ij}(\theta, r) \tag{1}$$

where $\sigma_{ij}(\mathrm{P})$ are the components of stress at a point P close to the crack and $f_{ij}(\theta, r)$ are known functions of the polar co-ordinates of point P. Typically, for a short crack in the edge of a wide sheet of material

$$K = \sigma_0 \sqrt{\pi c} \tag{2}$$

where σ_0 is the applied stress remote from the crack and c is the crack length. Regardless of the actual location of P, the intensity of stress at the point is controlled by K. The condition for crack propagation is then considered to be the attainment of some critical value K_c of K, which is hopefully a constant for the solid under given conditions of rate, temperature and environment. The parameter K_c can only be found empirically; it has no explicit relationship to the physical properties of the solid.

Turning to the second criterion we may consider the energy balance between the strained material, in which elastic energy $\mathscr{E}$ is stored, and the surface of the crack, which has a surface energy γ. If the crack surface grows by an amount ΔA, energy conservation requires that

$$-\Delta\mathscr{E} = \gamma \Delta A \tag{3}$$

provided, of course, that no energy is lost by other mechanisms. This provides the criterion, in the limit, that the crack will propagate when

$$-\frac{d\mathscr{E}}{dA} > \gamma \tag{4}$$

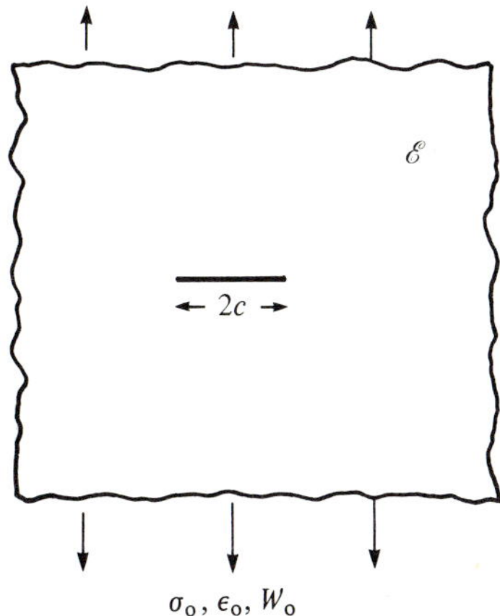

Fig. 2. Infinite lamina under uniform tensile stress containing a central crack of length $2c$. Stress, strain and energy density in the uniform region remote from the crack are denoted respectively σ_0, ε_0, W_0. The total strain energy of the system is $\mathscr{E}$.

For linear elastic materials it is possible to evaluate the left-hand side of this inequality in terms of stresses, moduli and geometrical factors. Thus for a centre crack of length $2c$ in an infinite thin sheet under a uniform tensile stress σ_0 (Griffith, 1920)

$$-\frac{d\mathscr{E}}{dA} = \frac{\pi\sigma_0^{\,2}c}{2E} \qquad (5)$$

Rivlin & Thomas (1953) showed that $(-d\mathscr{E}/dA)$ can also be deduced for particular geometries of test specimen when the material is a non-linear, highly extensible elastic solid like rubber. For an edge crack of length c in a semi-infinite sheet, for example, they found,

$$-\frac{d\mathscr{E}}{dA} = kcW_0 \qquad (6)$$

where W_0 is the elastic stored energy density remote from the crack and k an undetermined constant. Their treatment, however, still requires the material to be elastic i.e. does not allow mechanical hysteresis to occur.

Equations such as eqn (5) and eqn (6), when substituted in eqn (4), provide an experimentally testable criterion for crack propagation. The results from a wide variety of materials confirm the form of the predictions, e.g. in linear solids the product of the square of the applied stress required to cause propagation and the crack length, is constant.

The value derived experimentally for γ is, however, orders of magnitude too high for a genuine surface energy (Andrews, 1968).

Generalized fracture mechanics

GFM adopts the energy balance approach of Griffith and of Rivlin & Thomas but achieves great generality by adopting a dimensional analytical approach and relaxing the requirements of linearity and elasticity. Thus, consider an infinite sheet of material under a uniform tensile stress and containing a small crack of length $2c$ orientated perpendicular to the line of action of the tension (Fig. 2). The applied load is characterized by a stress σ_0, a strain ε_0 and an energy density W_0. However, W_0 is not an elastically stored energy, but merely the energy deposited in unit volume of material during monotonic deformation, regardless of whether or not it can be recovered. Dimensional analysis then relates W_0 to the energy density $W(\mathrm{P})$ at a point $\mathrm{P}(X, Y)$ in the sheet, thus (Andrews, 1974)

$$W(\mathrm{P}) = W_0 f\left(\frac{X}{c}, \frac{Y}{c}, \varepsilon_0\right) \tag{7}$$

where f is an unknown function. The 'reduced co-ordinates' X/c, Y/c are written x, y respectively. Clearly x, y are dimensionless co-ordinates. By differentiating eqn (7) with respect to crack length, and summing over all space, we immediately obtain an expression for $-d\mathscr{E}/dA$ (remembering that the crack surface area $A = 4ch$, where h is the sheet thickness). The result is (Andrews & Billington, 1976)

$$\begin{aligned} -d\mathscr{E}/dA &= k_1(\varepsilon_0)\, cW_0 \\ k_1 &= \frac{-p}{1-p}\, \Sigma g \delta x \delta y \end{aligned} \tag{8}$$

where $0 \leqslant p \leqslant 1$ and $p = 0.5$ for linear materials; g is a distribution function of energy density; k_1 is independent of c but depends on ε_0 and thus on W_0. The critical value of $(-d\mathscr{E}/dA)$ at which the crack propagates, is denoted $(-d\mathscr{E}/dA)_{\mathrm{crit}}$ or $\mathscr{T}$.

The critical energy release rate so defined is only the apparent energy needed to create unit area of interface. It includes the surface energy, of course, but it also includes all other energy losses *wherever* they occur in the specimen. There is therefore no particular reason why $\mathscr{T}$ should be constant, except in so far as the energy losses occur mainly close to the crack tip, where the strains are most intense. Under these conditions the creation of unit area of crack surface may be held to involve energy loss in a well-defined zone adjacent to the surface, and $\mathcal{T}$ may then be sensibly constant for given environmental conditions.

In practice it has been found that the expression $k_1 c W_{0(\mathrm{crit})}$ i.e. $\mathcal{T}$, is frequently constant for highly non-linear and inelastic solids as diverse as metal alloys, elastomers and plastic films (Andrews & Billington, 1976; Andrews & Fukahori, 1977). Thus the 'apparent critical energy release rate', $\mathcal{T}$, can be used, just as for linear elastic solids, to characterize fracture resistance in such materials. However, $\mathcal{T}$, is by no means a physical constant of the solid and depends on the temperature, propagation rate and overall strain level. It can also vary with sheet thickness and with the extent of crack growth. Although $\mathcal{T}$ may therefore characterize the fracture resistance of an inelastic material under some circumstances, it does no more to relate fracture resistance to physical and molecular properties than do the parameters of LEFM.

Physical parameters of fracture

In order to relate the apparent fracture resistance $\mathcal{T}$ to physical properties it is necessary to return to the original calculation of $-d\mathscr{E}/dA$. Implicit in this calculation (Andrews, 1974) was the assumption that a single spatial distribution function f relates the applied energy density W_0 to the energy density at any point P. This is true under monotonic loading until the crack begins to propagate, but *when* propagation occurs some regions of the stress field begin to *unload* i.e. the energy density at some points P begins to decrease. If the material is not perfectly elastic, the unloading will follow a different path in 'stress–strain space' to that followed in loading. This complex situation can be simply accounted for by stating that the energy density recovered in an element of material during unloading is only a fraction α of the energy stored during loading. Alternatively we can refer to the fractional energy *lost* during the loading/unloading cycle. This fraction, β, equals $(1 - \alpha)$ and is the 'hysteresis ratio'. Clearly $\beta = 0$ for a perfectly elastic material and $\beta = 1$ for a material (e.g. a flowing liquid) in which no elastic energy is stored by deformation.

If we now re-calculate $(-d\mathscr{E}/dA)$ *taking into account* this energy loss, we obtain the true energy available to create unit area of crack surface. This turns out to be (Andrews, 1974)

$$(-d\mathscr{E}/dA)_2 = \{k_1(\varepsilon_0) - \sum_{\mathrm{PU}} \beta g \delta x \delta y\} c W_0 \qquad (9)$$

where PU denotes summation over only the unloading regions of the stress field, and the suffix 2 denotes that the left-hand side is now the true energy available in contrast to $(-d\mathscr{E}/dA)_1$ previously calculated.

Suppose now that the crack propagates. The apparent energy available (per unit area of crack surface) is by definition, $\mathscr{T}$.

$$\mathscr{T} = (-d\mathscr{E}/dA)_{1\text{crit}} = k_1(\varepsilon_0)cW_{0\text{crit}} \tag{10}$$

In contrast, $(-d\mathscr{E}/dA)_{2\text{crit}}$, being the energy actually available for the creation of surfaces, must be the energy required for this purpose, which at first sight appears to be the surface energy γ. This is half the energy required to sever unit area of inter-atomic bonds across the fracture plane. For reasons which will emerge later we prefer not to press too closely this identity with surface energy, and therefore use a 'neutral' symbol $\mathscr{T}_0$ rather than γ. Thus we define

$$(-d\mathscr{E}/dA)_{2\text{crit}} = \mathscr{T}_0 = k_2(\varepsilon_0)cW_{0\text{crit}} \tag{11}$$

Dividing eqns (10) and (11) leads immediately to the result

$$\mathscr{T} = \mathscr{T}_0 \left\{ \frac{k_1(\varepsilon_0)}{k_1(\varepsilon_0) - \sum_{\text{PU}} \beta g \delta x \delta y} \right\} \tag{12}$$

or, for brevity,

$$\mathscr{T} = \mathscr{T}_0 \Phi(\varepsilon_0, \dot{c}, T) \tag{13}$$

where Φ is the 'loss function' and depends not only upon ε_0 but also upon any environmental condition, such as crack velocity, $\dot{c}$, and temperature, T, which affect the hysteresis ratio β of the solid. This equation thus tells us that the apparent energy required to form unit area of crack surface equals the actual energy so required ($\mathscr{T}_0$) multiplied by a loss function which varies from unity for perfectly elastic solids ($\beta = 0$) to infinity for solids with a sufficiently large mechanical hysteresis. The case $\Phi = 1$ corresponds to ideal brittle fracture. The case $\Phi = \infty$ corresponds to a situation where the material cannot sustain crack propagation (e.g. wet chewing gum, elastic liquids).

Eqn (13) immediately explains the fact that apparent fracture energies are usually far in excess of 'surface energies'. It also explains in principle the strong time and temperature dependence of fracture resistance observed in most materials. Most important, however, is the fact that eqn (12) contains two parameters which are properties of the physical continuum, namely $\mathscr{T}_0$ and β. It is therefore possible for the first time to relate explicitly the fracture resistance parameter $\mathscr{T}$ to the physical properties of the solid in question. This in turn enables us to

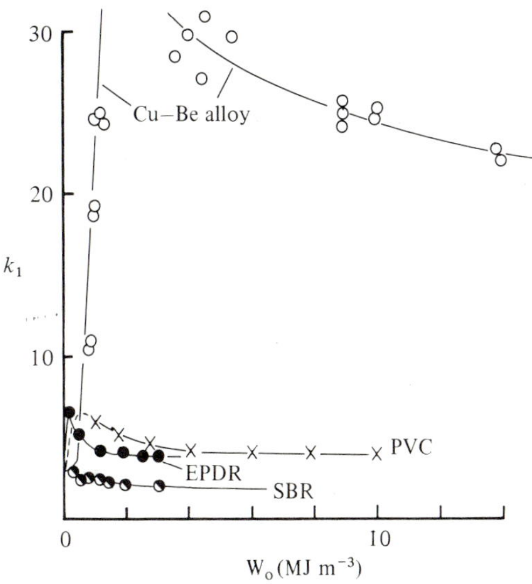

Fig. 3. The function $k_1(W_0)$ for various solids. SBR, styrene-butadiene rubber; EPDR, ethylene-propylenediene rubber; PVC, plasticized poly-vinylchloride.

discuss quantitatively how material structure, on a molecular and micro-morphological level, affects bulk strength. Let us now examine the various terms on the right-hand side of eqn (12).

The function $k_1(\varepsilon_0)$

Although in principle $k_1(\varepsilon_0)$ can be calculated using eqn (8), such a calculation requires a knowledge of the energy density distribution function g over the whole specimen and this is impossible to derive theoretically for any but perfectly elastic materials. However, $k_1(\varepsilon_0)$ may be measured experimentally by performing load–deformation tests on identical specimens containing edge cracks of different lengths. The method is fully described elsewhere (Andrews & Billington, 1976), and will not be elaborated here. The results obtained are, however, of interest and examples from some diverse solids are given in Fig. 3. For elastic-plastic solids such as metals and polycarbonate, $k_1(\varepsilon_0)$ (or $k_1(W_0)$ as it is more conveniently expressed) rises from its 'elastic' value of π at $W_0 = 0$ to a peak and then falls again to a relatively constant level, which may be as high as 20 but is more usually in the range 6–8. For elastomeric materials the peak is less noticeable and may be missing altogether, whilst 'lossy' materials such as plasticized poly-vinylchloride (PVC) and polyethylene film exhibit intermediate behaviour. The value of $k_1(W_0)$ to be used in calculating $\mathcal{T}$ or Φ is the value appropriate to

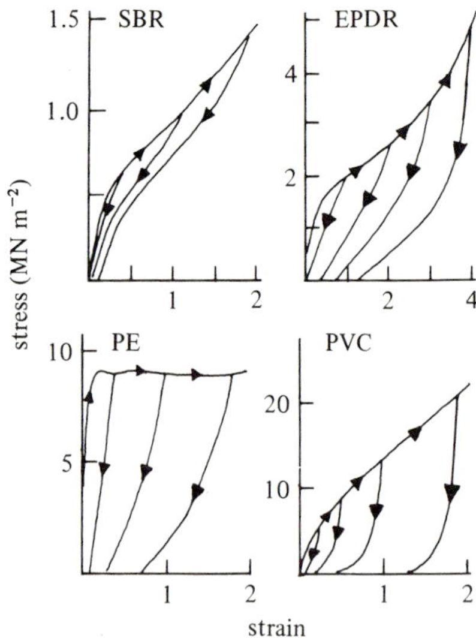

Fig. 4. Stress–strain loops for various extensible solids. SBR, styrene-butadiene rubber; EPDR, ethylene-propylenediene rubber; PE, low density polyethylene; PVC, plasticized poly-vinylchloride.

crack propagation conditions. This value usually lies to the right-hand side of the peaks in Fig. 3 in the region where $k_1(W_0)$ varies only slowly. This is not always the case, however, and constancy in $k_1(W_0)$ cannot be assumed under all conditions.

The hysteresis ratio β

Typical hysteresis data are shown in Fig. 4 for four materials, styrene-butadiene rubber (SBR), ethylene-propylenediene rubber (EPDR), plasticized poly-vinylchloride and polyethylene (Andrews & Fukahori, 1977). These diagrams show how the hysteresis (energy loss), which is proportional to the area enclosed by a complete load–deformation loop, varies with material and with strain. For a given material at a given strain-rate and temperature, hysteresis rises from a low value at small strain levels, to values which may approach unity at high strain levels (Fig. 5). Thus β varies from point to point in the strain field of a propagating crack. In materials which undergo plastic yielding, a reasonable approximation is to set $\beta = 1$ inside the plastic zone and $\beta = 0$ outside, thus greatly simplifying calculation of the summation term in eqn (12). In certain relatively brittle solids it may be acceptable to replace β by a strain independent quantity such as $2\pi \tan \delta$ (the

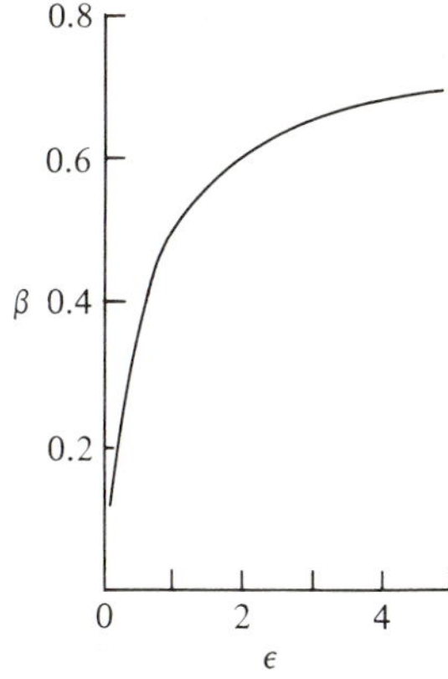

Fig. 5. The dependence of hysteresis ratio β upon strain for ethylene-propylenediene rubber.

quantity which measures energy loss in dynamic mechanical testing at small strains, see e.g. Ward, 1971).

The important point is that β reflects the internal friction of the solid and can often be attributed to specific molecular energy-absorbing processes (e.g. hindered rotational movement of molecular groups).

The effect of β upon fracture resistance is clear from eqn (12). At $\beta = 0$, $\mathcal{T} = \mathcal{T}_0$ and the only energy required to propagate the crack is that needed to form the new surfaces, namely $\mathcal{T}_0$. As β increases, however, $\mathcal{T}$ rises without limit, becoming infinite at values of β less than unity. (This does not of course mean that β cannot achieve the value of unity over a *restricted* region of the test-piece as e.g. a plastic zone.) One important consequence of this is that as β changes with, say, temperature or strain rate, the equation predicts the possibility of a brittle (finite $\mathcal{T}$) to ductile (infinite $\mathcal{T}$) transition in the response of the solid to tensile stress. Systematic variations of $\mathcal{T}$ with temperature, which reflects the temperature dependence of tan δ $(=\beta/2\pi$ for small $\beta)$, have been observed by J. G. Williams (personal communication) as shown in Fig. 6.

The time dependence of β in many polymeric solids gives rise to a time dependent $\mathcal{T}$ or, conversely, to a dependence of crack velocity on $\mathcal{T}$. Thus, for example, under a given applied constraint, slow crack growth may occur, giving rise to failure after a considerable time interval.

Finally, β may be dependent on the amount of crack growth which has occurred. This effect can arise because the effective hysteresis ratio in an element which does not unload *completely* is a function of the

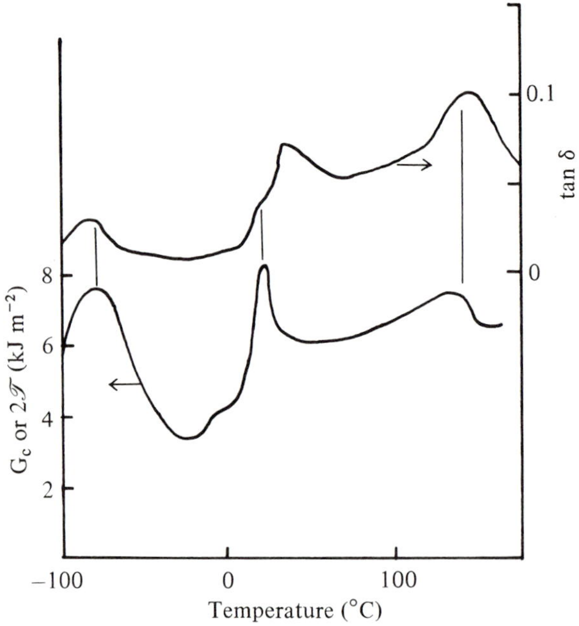

Fig. 6. Correlation between critical energy release rate (G_c) and tan δ for polytetrafluoroethylene (PTFE). (J. G. Williams, personal communication.)

amount of unloading. It is evident that for stress–strain loops such as those in Fig. 4, the first increments of unloading provide a greater relative return of energy than do the final stages of shape recovery. Thus β *increases* with Δc, the amount of crack growth, since Δc controls the amount of unloading in the vicinity of the crack tip. This effect gives rise to the well-known phenomenon of fatigue in which only a small increment of crack growth occurs on each loading of the specimen; Δc increases until the increase in β occasioned thereby is such that the right-hand side of eqn (12) exceeds the value of $\mathcal{T}$ appropriate to the applied load, at which point the equation is no longer satisfied and propagation ceases. Only increase of the applied load, or its removal and re-application, will then cause further growth. Fatigue fracture may thus be related to the bulk physical properties via the parameter β.

The 'surface energy' $\mathcal{T}_0$

Surface energy in solids is not an unambiguous concept. It can refer to the secondary inter-atomic forces which operate between the solid and a liquid deposited on its surface or, alternatively, to the energy required to separate one plane of atoms in the solid from another. Clearly the latter involves the fracture within the solid of cohesive bonds which are normally primary bonds. The situation is even more complicated, however, in materials such as polymers where the forces of cohesion may be both primary (e.g. the C–C bonds of a single polymer molecule) and secondary (the inter-molecular forces which resist disentanglement of molecules).

A second level of complexity arises because the fracture of inter-atomic bonds can be regarded as an activated process. More energy is required to initiate rupture (in raising the system to its activated state) than is represented by the equilibrium free-energy difference between broken and unbroken states. Clearly $\mathcal{T}_0$ must be taken to represent the energy required to initiate bond fracture (normalized to unit area) rather than the equilibrium free energy difference.

This conclusion leads to some important predictions, two of which we will consider here. First, it implies time and temperature dependence in $\mathcal{T}_0$. The rate of activated bond fracture is given approximately (Andrews, 1979) by

$$R = R_0 \exp(-[G - b\sigma]/kT) \tag{14}$$

where R_0 is the fundamental vibration frequency and is proportional to the thermodynamics temperature T, G is the activation free energy for thermal dissociation of the bond, σ is the tensile stress on the bond, b is the activation volume and k is Boltzmann's constant. The product $b\sigma$ is the mechanically supplied energy.

For instantaneous fracture (i.e. allowing no time for random thermal activation to cause bond severence), the entire activation energy G must be provided by the mechanical stress, giving

$$\mathcal{T}_0 \sim GN \tag{15}$$

where N is the number of bonds broken per unit area. If time is allowed, however, for thermal fluctuations to assist bond fracture, $\mathcal{T}_0$ is diminished so that

$$\mathcal{T}_0 \sim N\{G - kT \ln(R_0/R)\} \tag{16}$$

i.e. as R decreases below the fundamental vibration frequency, R_0, the value of $\mathcal{T}_0$ falls. The minimum possible value of $\mathcal{T}_0$, of course, is the equilibrium free-energy change from unbroken to broken state, again normalized to unit area (Andrews, 1979).

We have already seen that $\mathcal{T}$ is rate and temperature dependent on account of Φ. Eqn (16) now reveals a further rate and temperature dependence arising from $\mathcal{T}_0$ and superimposed upon the visco-elastic effects imposed by the loss function. These $\mathcal{T}_0$ effects are only likely to

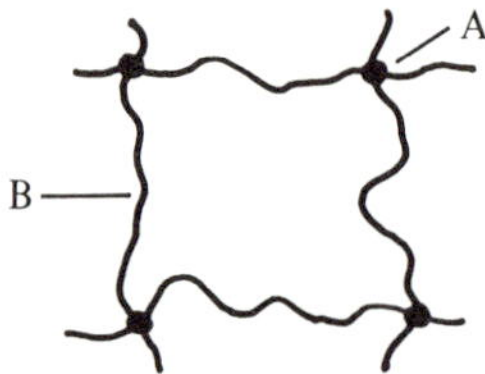

Fig. 7. Two-dimensional diagram of a molecular network showing cross-links A and network chains B.

be observed in practice under conditions of very low strain level, where $\beta \to 0$.

Our second example of the way 'activated bond fracture' modifies $\mathcal{T}_0$ occurs in network polymers such as chemically cross-linked rubber. Lake & Thomas (1967) pointed out that in order to break a single C–C bond in such materials, it is necessary to extend a long chain of such bonds to breaking point. This chain is the molecular length between cross-links, the so-called 'network chain' (see Fig. 7). In a typical vulcanized rubber such a chain may contain as many as 100 C–C bonds, so that the activation energy *for the entire chain* may be 100 times larger than for a single bond. If we take the surface energy for covalent bond fracture to be of the order of 1 J m^{-2} we might therefore expect $\mathcal{T}_0$ values for cross-linked rubbers to be as high as 100 J m^{-2}, taking into account the 'network effect' of Lake & Thomas (1967). In fact the cross-link density in such solids affects both the number of monomer units per chain, n, and the number of chains crossing unit area, so that the detailed theory of Lake & Thomas predicts

$$\mathcal{T}_0 = (3/32)^{1/2} \gamma \ell U N n^{3/2} \tag{17}$$

where γ is a factor related to the freedom of bond rotation, ℓ is the monomer unit length, N is the number of network chains per unit volume and U is the energy to break one C–C bond. The multiplicative effect of n is however still evident, and actual measured values of $\mathcal{T}_0$ are found to lie between 20 and 100 J m^{-2}, as predicted (see Andrews & Fukahori, 1977). These ideas have been extended successfully to semi-crystalline polymers such as polyethylene (Andrews & Fukahori, 1977) and to thermosetting polymers such as epoxy resins (King & Andrews, 1978).

The measurement of $\mathcal{T}_0$

In cross-linked elastomers it is often possible to carry out crack propagation studies at such low levels of stress and such low rates of

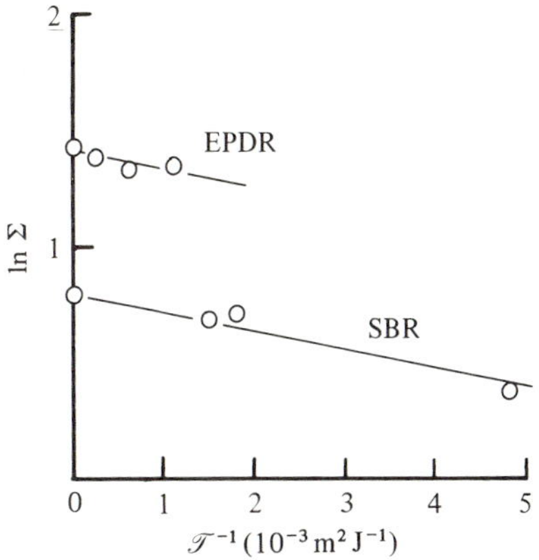

Fig. 8. Experimental plot of $\ln\Sigma$ versus $\mathscr{T}^{-1}$ for styrene-butadiene (SBR) and ethylene-propylenediene (EPDR) rubbers. $\mathscr{T}_0$ is obtained from the negative slope.

growth that β can be reduced to zero (Lake & Lindley, 1965). This is done by fatigue cycling at increasingly low levels of stress until a limiting stress level is reached below which no further growth occurs. Since $\dot{c}$ is reduced to zero, and the materials are ideally viscoelastic, $\beta \to 0$ and the observed lower limit to $\mathscr{T}$ can be taken as $\mathscr{T}_0$. This method is not applicable, however, if permanent set or non-viscoelastic hysteresis occurs.

Alternatively, we may use eqn (12) as a means of determining a value for $\mathscr{T}_0$. Re-arranging and using appropriate abbreviations

$$k_1(\mathscr{T}_0/\mathscr{T}) = k_1 - \Sigma \tag{18}$$

or alternatively

$$1 - (\mathscr{T}_0/\mathscr{T}) = \Sigma/k_1$$

$$-(\mathscr{T}_0/\mathscr{T}) \simeq \ln\Sigma - \ln k_1 \tag{19}$$

provided $\mathscr{T}_0 \ll \mathscr{T}$.

A graph of Σ against $\mathscr{T}^{-1}$ (or of $\ln\Sigma$ against $\mathscr{T}^{-1}$) should therefore produce a straight line with a negative slope of $k_1\mathscr{T}_0$ (or of $\mathscr{T}_0$). In either case, also, $k_1(\varepsilon_0)$ should appear as the value of Σ at which $\mathscr{T}^{-1} \to 0$.

It is possible to vary $\mathscr{T}$ for a given material in several ways. In viscoelastic solids $\mathscr{T}$ depends on rate and temperature. In elastic-plastic solids $\mathscr{T}$ also varies with the bluntness of the original crack tip. By recording the strain field around cracks propagating at different $\mathscr{T}$ values, Andrews & Fukahori (1977) were able to evaluate the summation term Σ for SBR, EPDM, PE and plasticized PVC. Different Σ

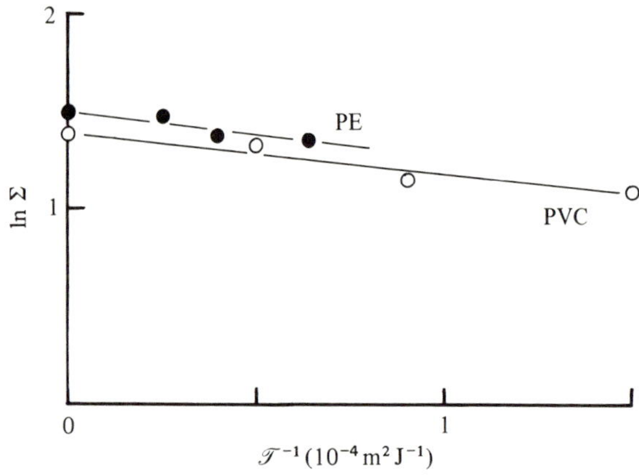

Fig. 9. Experimental plot of $\ln \Sigma$ versus $\mathcal{T}^{-1}$ for polyethylene (PE) and plasticized poly-vinylchloride (PVC).

Table 1. *Theoretical and experimental values for the 'surface energy'* $\mathcal{T}_0\,(\mathrm{J\,m^{-2}})$

Material	Theoretical	Experimental (Andrews & Fukahori, 1977)	Experimental (other work)
			(*in vacuo*)
NR	10 ~ 14	–	20
			30
			30
SBR	16	65 ± 15	45
EPDM	18	65 ± 15	–
p-PVC	15 ± 5	100 ± 50	–
PE	31 ± 6	200 ± 100	–

terms corresponded to different $\mathcal{T}$ values, and $k_1(\varepsilon_0)$ was measured in independent experiments. Figs. 8 and 9 show the data plotted according to eqn (19), giving the predicted linear relation with negative slope and the correct intercept $k_1(\varepsilon_0)$ at $\mathcal{T}^{-1} \rightarrow 0$. Table 1 shows the results deduced for $\mathcal{T}_0$ from these experiments, compared with data from limiting fatigue experiments and the theoretical result from Lake & Thomas' equation. It can be concluded that, in these materials, $\mathcal{T}_0$ has been properly identified as to its physical and molecular significance, and that its magnitude has been correctly ascertained within a factor of two relative to other experimental determinations.

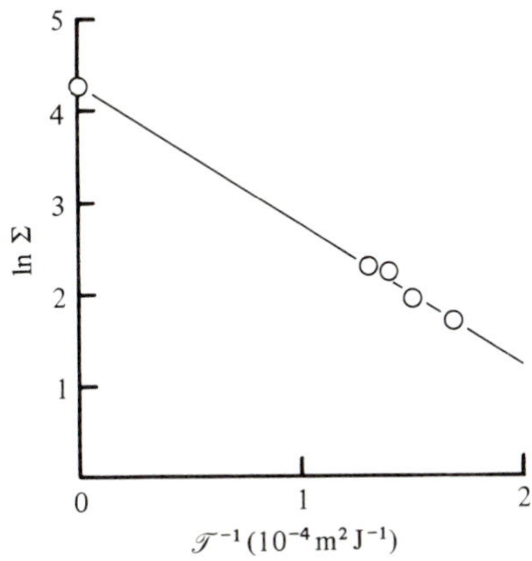

Fig. 10. Experimental plot of Σ versus $\mathcal{T}^{-1}$ for polycarbonate. $\mathcal{T}_0$ is equal to the negative slope.

The significance of $\mathcal{T}_0$ in a glassy polymer

Andrews & Barnes (personal communication) have carried out derivations of Σ and thence of $\mathcal{T}_0$ on a glassy plastic, polycarbonate of bisphenol A, which displays elastic-plastic deformation behaviour. Fig. 10 shows data from this material plotted according to eqn (18). Once again the linear behaviour and correct intercept are found, but the value deduced for $\mathcal{T}_0$ is even larger than for elastomeric materials, being of the order of 3.3 kJ m^{-2} (some 10^3 larger than for simple covalent bond fracture). Since this material is not cross-linked, the 'network effect' used to explain the large $\mathcal{T}_0$ for elastomers, cannot be appealed to for justification. It is known, however, that polycarbonate undergoes small-scale crazing in advance of the crack, as indeed do most glassy polymers. (A craze is a crack-like zone bridged by load-bearing filaments of deformed polymer.) Since this zone is microscopic in size it is not included in the summation term Σ, so that all energy used in the production of the craze appears within the term $\mathcal{T}_0$. Independent measurements of crazing energy in polycarbonate have been made by Fraser & Ward (1977), under conditions where normal plastic deformation does not occur, and they obtained a value of some 3.5 kJ m^{-2}. In glassy polymers, therefore, we conclude that $\mathcal{T}_0$ is likely to be the energy to form and fracture the craze which precedes the crack. Even here, however, when $\mathcal{T}_0$ fails to reflect the ultimate molecular fracture mechanisms, it does relate directly to the microscopic event of craze formation and thus to observable physical processes in the solid.

Adhesive failure

Although adhesion and adhesive failure are normally considered to be quite different phenomena from cohesion and cohesive failure, GFM Theory enables us to unify their treatment. Thus adhesive failure is considered as the propagation of a crack along (or close to) the adhesive interface. Just as $\mathcal{T}_0$ can be determined by the proper analysis of cohesive fracture data, so a corresponding quantity, θ_0, can be determined from adhesive failure experiments. More precisely we define, for adhesive failure,

$$\theta = -2(d\mathscr{E}/dA)_{1\,\mathrm{crit}} \tag{20}$$

$$\theta_0 = -2(d\mathscr{E}/dA)_{2\,\mathrm{crit}} \tag{21}$$

so that a formal correspondence exists between θ and $2\mathcal{T}$ on one hand and θ_0 and $2\mathcal{T}_0$ on the other. The factor 2 arises because $\mathcal{T}$ and $\mathcal{T}_0$ are referred to unit area of crack surface, A, whilst θ, θ_0 are referred to unit area of crack plane (two surfaces are formed from one crack plane). Otherwise θ, θ_0 and $2\mathcal{T}$, $2\mathcal{T}_0$ are really identical quantities and the alternative symbols are employed for convenience to differentiate at a glance adhesive and cohesive fracture parameters.

Adhesion is an important phenomenon in biology, with effects ranging from inter-cellular adhesion and the clotting of blood to the structural adhesion which exists between different kinds of tissue, e.g. tendon and bone. Furthermore, many organisms secrete and use adhesive substances, obvious examples being spiders and barnacles. The key value of GFM in the study of adhesion lies in its ability to isolate θ_0 from the other factors which affect adhesive strength and so identify the strength (and hence the character) of the atomic and molecular interactions at the adhesive interface.

The model to keep in mind is of a crack located at an interface between two semi-infinite sheets of material, the substrate and the adhesive respectively, and is shown in Fig. 11. Just as in the model for cohesive fracture (Fig. 2), the system is subject to a uniform tensile stress σ_0 applied perpendicular to the crack axis. Because the mechanical constants of the two media differ, the stress, strain and energy distribution above and below the crack are different. In order to simplify the analysis, therefore, we arrange for the substrate material to be effectively rigid (high modulus) in comparison with the adhesive phase. This ensures that no significant energy is stored or lost in the substrate phase. If we wish to study the adhesion between two non-rigid materials we can 'cheat' by applying one of the non-rigid media as a very

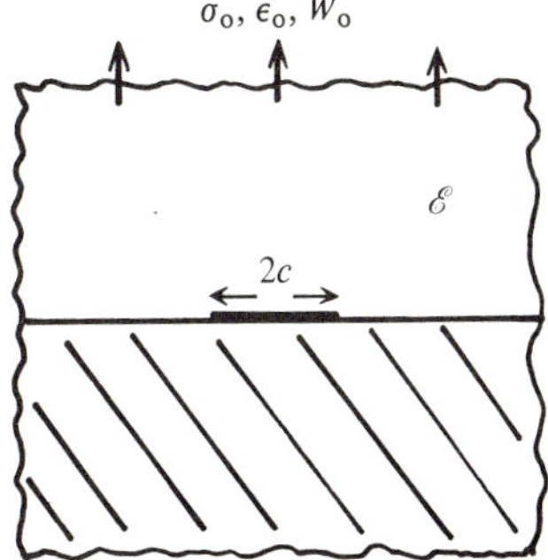

Fig. 11. Infinite lamina under uniform tensile stress containing a central crack of length $2c$, in an adhesive system. Stress, strain and energy density in the uniform region remote from the crack are denoted respectively σ_0, ε_0, W_0. The total strain energy of the system is $\mathscr{E}$.

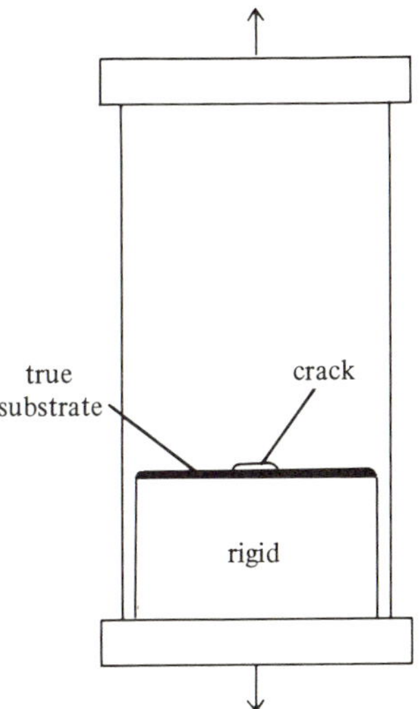

Fig. 12. Experimental arrangement for non-rigid substrate materials.

thin layer on top of a rigid (e.g. metal) block and then using this coated block as the substrate (see Fig. 12). Because the volume of the coating is so small, its effect can be ignored and the whole substrate treated as rigid.

With this arrangement, an identical analysis to that used for cohesive fracture leads to the following equations, characterizing propagation of the interfacial crack.

$$\theta = 2k_1(\varepsilon_0)cW_{0\text{crit}} \tag{22}$$

$$\theta = \theta_0\Phi(\varepsilon_0, \dot{c}, T) \tag{23}$$

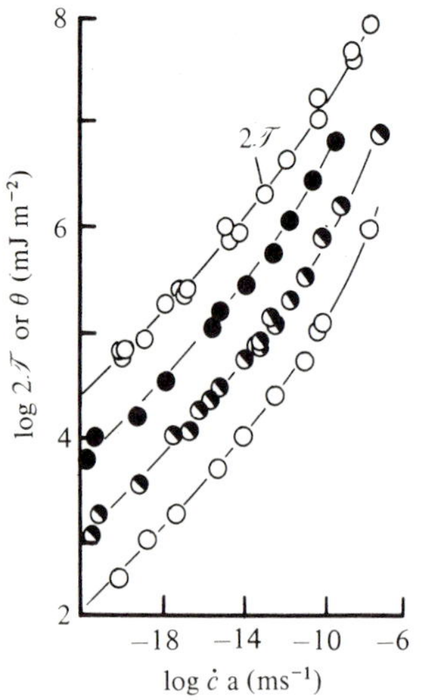

Fig. 13. Data for log $2\mathcal{T}$ and log θ plotted against 'reduced' crack velocity for SBR bonded to a variety of substrates. Note the parallelism with respect to a vertical displacement.

where W_0, ε_0 and Φ apply to the non-rigid phase. (For an infinitely rigid material, W_0, ε_0 are always zero and Φ is unity.) Thus the loss function Φ is *exactly the same* quantity as for cohesive fracture of the adhesive.

Taking logarithms of eqns (22) and (23)

$$\log \mathcal{T} = \log \mathcal{T}_0 + \log \Phi$$

$$\log \theta = \log \theta_0 + \log \Phi.$$

Thus if we plot experimentally measured values of log $\mathcal{T}$ and log θ against any variable on which Φ depends (say crack velocity, $\dot{c}$) we should obtain parallel curves which are displaced along the log θ, log $\mathcal{T}$ axis by an amount

$$\log \mathcal{T}_0 - \log \theta_0 = \log(\mathcal{T}_0/\theta_0) \tag{24}$$

If, further, we repeat the experiment for a range of substrates (1, 2, 3 . . .) we should obtain a family of parallel curves displaced from the cohesive failure curve by differing amounts, namely $\log(\mathcal{T}_0/\theta_{01})$ $\log(\mathcal{T}_0/\theta_{02})$ and so on. If $\mathcal{T}_0$ has been determined by one or other of the methods described earlier, values can immediately be deduced for θ_{01}, θ_{02} etc.

Fig. 13 shows data of Andrews & Kinloch (1973) for just such a series of experiments in which a SBR adhesive was bonded to a variety of thin plastic films, which in turn were mounted on a steel block. The theoretically predicted parallel curves were indeed obtained. (The

Table 2. *Values of θ_0 and thermodynamic work of adhesion, w_A, for styrene-butadiene rubber bonded to various unetched plastic substrates. Andrews & Kinloch, 1973*

Substrate	θ_0/mJ m^{-2}	w_A/mJ m^{-2}
Fluorinated ethylene propylene A	21.9	48.4
Polychlortrifluorethylene	74.9	62.5
Nylon 11	70.8	71.4
Polyethylene terephthalate	79.4	72.3
Fluorinated ethylene propylene P	68.5	56.8

Table 3. *Values of θ_0 and w_A for styrene-butadiene rubber bonded to fluorinated ethylene propylene film etched with sodium naphthalene*

Etching time (s)	θ_0/mJ m^{-2}	w_A/mJ m^{-2}
10	851	68.0
20	1170	70.2
60	1290	69.8
90	1620	71.1
120	1780	71.1
500	2420	72.7
1000	1990	71.8

variable $\dot{c}a_T$ employed is the 'reduced' or temperature compensated crack velocity which allows data obtained at different temperatures to be combined into a master curve. This only works because SBR is an ideally viscoelastic substance.)

Table 2 records the values obtained for θ_0 from the displacements of the curves, and compares these values with the independently determined 'thermodynamic works of adhesion', w_A, for the SBR/plastic film pairs. (See Andrews & Kinloch, 1973.)

The close correspondence between θ_0 and w_A is evident and, indeed, is exactly what is expected theoretically since w_A is simply the interfacial energy for an interface across which only secondary (van der Waals') inter-atomic forces operate. This correspondence between θ_0 and w_A does not apply if primary interfacial atomic bonds are present, nor if the crack propagates through the adhesive or substrate phases rather than interfacially. These effects account for the θ_0 values shown in Table 3 where the plastic films had been subject to etching before the SBR was applied. Now $\theta_0 \gg w_A$ because covalent bonding is present at the interface. However, θ_0 still has a significance and the $\log \theta$ against $\log \dot{c}a_T$ curves are still parallel. Andrews & Kinloch (1973) showed that these θ_0 values arose from mixed-mode failure, the crack propagating partly along the interface (I) and partly through the rubber (B) and substrate (S) phases. If the fractions of crack surface occurring in these three ways are, respectively, i, b and s ($i + b + s = 1$), then,

$$\theta_0 = iI_0 + bB_0 + sS_0 \tag{25}$$

where I_0 is the θ_0 value for the interface, B_0 is the value of $2\mathcal{T}_0$ for the rubber phase and S_0 is the value of $2\mathcal{T}_0$ for the substrate phase.

In subsequent studies (Andrews & King, 1976; King & Andrews, 1978; Andrews & Stevenson, 1979) it has been possible, using other adhesive systems, to characterize the interfacial atomic bonding as either primary or secondary, to identify the likely chemical character of the bond and to study quantitatively the hydrolysis of interfacial atomic bonds in aqueous environments.

Conclusion

The application of Generalized Fracture Mechanics enables us to characterize the fracture resistance of inelastic and highly deformable solids in a way that is otherwise impossible. Not only can a quantity $\mathcal{T}$ be measured that represents this fracture resistance, but it can also be broken down into two components, namely $\mathcal{T}_0$, the energy actually required to overcome the forces of cohesion, and Φ, a term describing the energy loss processes in the bulk solid; Φ can be further analysed in terms of the hysteresis ratio β, a physical property relatable to molecular energy loss processes in the material. Examination of $\mathcal{T}_0$ provides quantitative information on the molecular or microscopic mechanisms of cohesive failure in a wide variety of solids. Finally the entire theory can be carried over into the area of adhesive failure, providing direct measurement of the interfacial atomic bonding energy and illuminating the molecular mechanisms of adhesion.

References

Andrews, E. H. (1968). *Fracture in polymers*, p 121. London: Oliver & Boyd.

Andrews, E. H. (1974). A generalized theory of fracture mechanics. *Journal of Materials Science*, **9,** 887–94.

Andrews, E. H. (1979). Molecular structure and strength of polymers. In *Developments in polymer fracture, 1*, ed. E. H. Andrews, pp. 1–16. London: Applied Science Publishers.

ANDREWS, E. H. & BILLINGTON, E. W. (1976). Generalized fracture mechanics, II: Materials subject to general yielding. *Journal of Materials Science*, **11,** 1354–61.

ANDREWS, E. H. & FUKAHORI, Y. (1977). Generalized fracture mechanics, III: Prediction of fracture energies in highly extensible solids. *Journal of Materials Science*, **12,** 1307–19.

ANDREWS, E. H. & KING, N. E. (1976). Adhesion of epoxy resins to metals, I, *Journal of Materials Science*, **11,** 2004.

ANDREWS, E. H. & KINLOCH, A. J. (1973). Mechanics of adhesive failure, I & II. *Proceedings of the Royal Society of London, Series A*, **332,** 385–414.

ANDREWS, E. H. & STEVENSON, A. (1979). Adhesive failure energies of epoxy-titanium bonds. In *Adhesion, 3*, ed. K. Allen, p. 81. London: Applied Science Publishers.

FRASER, R. A. W. & WARD, I. M. (1977). The impact fracture behaviour of notched specimens of polycarbonate. *Journal of Materials Science*, **12,** 459–68.

GRIFFITH, A. A. (1920). The phenomena of rupture and flow in solids. *Philosophical Transactions of the Royal Society of London, Series A*, **221,** 163–98.

KING, N. E. & ANDREWS, E. H. (1978). Fracture energy of epoxy resins above T_g. *Journal of Materials Science*, **13,** 1291–1302.

KNOTT, J. (1973). *Fundamentals of fracture mechanics*, pp. 57–61. London: Butterworths.

LAKE, G. J. & LINDLEY, P. B. (1965). The mechanical fatigue limit for rubber. *Journal of Applied Polymer Science*, **9,** 1233–51.

LAKE, G. J. & THOMAS, A. G. (1967). The strength of highly elastic materials. *Proceedings of the Royal Society of London, Series A*, **300,** 108–19.

RIVLIN, R. S. & THOMAS, A. G. (1953), Rupture of rubber. I. Characteristic energy for tearing. *Journal of Polymer Science*, **10,** 291–318.

WARD, I. M. (1971). *Mechanical properties of solid polymers*, p. 96. London: Wiley Interscience.

THE MECHANICAL BEHAVIOUR OF COMPOSITE MATERIALS

BRYAN HARRIS

School of Materials Science, University of Bath, Claverton Down, Bath BA2 7AY, UK

Introduction

In engineering practice, as in nature, it is a common principle that two or more components may be profitably combined to form a composite material so as to make best use of the more favourable properties of the components while simultaneously mitigating the effects of some of their less desirable characteristics. Of the many composite materials available the most important are structural composites – materials used for their ability to sustain loads or resist deformations. These man-made composites have interesting natural analogues of strikingly similar character, although it goes without saying that nature's composites are far more successful than man's when considered in the most general terms.

The structural engineer traditionally requires materials that are very strong and rigid, and for applications where compressive buckling is likely to be a limiting factor it is also advantageous to choose materials with a high ratio of elastic modulus to density, E/ϱ. For the same reason, and also on account of the obvious economic factor relating to payloads, the aeronautical engineer also sets great store by low-density materials. The choice of materials having very high strength is frequently limited by the undesirable characteristic of brittle behaviour which leads to catastrophic failure with a concomitant high coefficient of variation of breaking strength. When designing with conventional materials therefore the engineer must always optimize strength and toughness, according to his requirements, and he will usually refrain from using in a tensile structure any material that has a strain to failure which is less than about 3%.

A search for materials that should theoretically have high strength and modulus leads directly to elements or compounds possessing a high density of either pure covalent or mixed covalent/ionic bonds (Kelly, 1973). Elements having these characteristics are carbon, boron and silicon; compounds that suggest themselves are ceramics like SiO_2, SiC,

BN, Si_3N_4. It is of course not coincidental that these are all very brittle solids of low density. Such materials are unpromising in the bulk form, and it is only when we convert them into fine fibres, eliminating as far as possible the strength-limiting defects normally present in brittle solids, that we obtain strong, rigid materials capable of being used as engineering materials. The most important fibres that are at all widely used in modern man-made composites are glass (SiO_2), carbon, boron, and highly drawn polymers like polypropylene, polyethylene terephthalate (Terylene) and aromatic polyamides such as Kevlar 49. In many of these fibres it is possible to obtain strengths, characteristic of the strong covalent bond, which reach a high proportion of the theoretical tensile failure stress, approximately $E/20$.

Apart from their use in ropes and cables in pure tensile structures fibres are of limited value in structural applications. For this reason, therefore, these strong filamentary materials must be combined with a matrix of metal, plastic, or (less frequently) ceramic so that a substantial proportion of their high tensile strength and rigidity can be utilized in stress systems more complex than simple tension. For composite matrices we choose weak, ductile metals like aluminium; weak, brittle plastics like the thermosetting epoxides and polyesters; weak, ductile thermoplastics like Nylon; or weak (in tension), brittle ceramics like concrete. The matrix serves several purposes. It is, of course, the agency through which stress is transferred, usually by shear, into the strong, reinforcing fibres and its inherent ability to resist shear and compressive forces permits the composite to sustain stresses other than pure tension. The matrix protects the fibres from the mechanical damage through abrasion and chemical attack from the environment that, together or separately, limit the durability of unbonded ropes and cables. It also serves the necessary function of keeping the fibres separated and in doing so prevents a crack in any one fibre from propagating rapidly through other fibres in contact with it. A tough metallic or polymeric film of matrix interposed between the brittle reinforcing filaments thus modifies the fracture behaviour or toughness of the material.

If we choose a low-density matrix for reinforcement by light, strong fibres we can therefore expect to produce strong, rigid composites that are also light and tough, thereby fulfilling, to a greater or lesser degree, all of the engineers' requirements for a modern structural material. The most successful of modern rigid composites are glass and carbon-fibre reinforced plastics (grp and cfrp), boron-fibre reinforced aluminium, Kevlar 49 reinforced resins, and polypropylene or steel-fibre reinforced cement (see, for example, Royal Society, 1980). The most recent developments relate to the combination of two types of reinforcing fibres in a single resin matrix to produce materials known as hybrid composites which carry the optimization principle even further. Important flexible composites, such as those used for tensile structures, are materials like PVC-coated Terylene cloth and PTFE-coated glass cloth. There are many similarities, in behaviour as well as in construction, between rigid natural composites like wood, bone and arthropod cuticle and synthetic composites like grp. Soft tissues like skin and cartilage are, likewise, not unlike the coated flexible membranes used for rubber boats, dirigibles and air-houses. In the following discussion we shall consider the behaviour of reinforced polymeric materials rather than composites in general since this is more appropriate to this particular symposium and gives more scope for the identification of analogies in natural systems.

Fibres and matrices

It is instructive at the outset to make an elementary comparison of the characteristics of the reinforcing elements and matrices of natural and artifical composites. The most-used synthetic reinforcing fibres in rigid structural composites are glass and carbon, which are both very strong and brittle, and the aromatic polyamide, Kevlar 49, which is strong but not brittle. These are continuous fibres and can be made into a variety of composite materials by rather complex fabrication techniques. The products are usually of high quality and the location and orientation of the fibres is very precise. The composite can therefore be tailored to cope with a given stress system by selecting the amount and type of fibre and the percentage of filaments arranged in any given direction. With such complex methods of manufacture, however, the range of shapes of articles that can be produced is limited and the cost is usually high. Ordinary versatile moulding techniques like injection moulding, extrusion, or transfer moulding can also be used, however, for forming short-fibre reinforced thermoplastics, and glass-filled Nylon and polyacetal are examples of common materials that are processed in this manner. Thermoset resins like polyesters are also used as a base for mouldable composites containing chopped glass fibres in compounds known as dough-moulding and sheet-moulding compounds. The glass fibres used in these much cheaper materials are chopped up from continuous strands, but it rarely pays to treat the more expensive carbon or boron fibres in this manner. An injection-moulded composite

Table 1. *Comparison of some properties of natural and artificial fibres*

Fibrous material	Young's modulus (GPa)	Tensile strength (MPa)	Breaking strain (%)
Cellulose (dry ramie)	80	900	2.3
Chitin	80	4000	4.2
Collagen (wet tendon)	2	100	8–10
E-Glass	70	1500–2000	2.0–3.0
Carbon	200–400	1900–2600	0.5–1.5
Boron	400	3400	0.5–1.0
Kevlar 49 (aromatic polyamide)	130	2700	2.0–3.5
Steel (high carbon)	210	2800	0.5

artefact would clearly possess poorer properties than, say, a filament-wound structure, partly because the fibres are shorter, partly because there are fewer of them, and partly because they are randomly organized (or nearly so). The pliant composites used in tensile structures are usually based on cloth woven from continuous glass or textile fibres, the commonest textile currently in use being Terylene, although Nylon and polypropylene are also used. The cloth supports the entire load and designs are based upon the principle that the coatings of PVC, Neoprene, PTFE, polyurethane, etc, serve only to protect the cloth from damage and degradation.

A very important structural fibre in the animal kingdom is the crystalline polymeric material collagen. It occurs as a component in pliant, connective tissues, both in vertebrates and in invertebrates, in tensile structures like tendon, and in passive structural elements like skin, cartilage and bone. Chitin, the second most widely distributed animal reinforcing fibre, also a crystalline polymer and much more rigid than collagen, is the fibrillar component in composites like arthropod cuticle. The most abundant of natural fibrous substances in the vegetable kingdom is cellulose. The basic molecule is organized in various ways in different vegetable structures, perhaps the most interesting being the complex, filament-wound, multi-layer, tubular elements that form the load-bearing tracheids of soft woods. The mechanical characteristics of these principal natural fibres may be compared with those of the commoner synthetic fibres in Table 1. In bone the collagen fibrils are given added support by needles or platelets of an imperfect form of the mineral hydroxyapatite.

The significance of the matrix material in natural composites depends on the nature of the composite. In tendon, for example (Kastelic &

Baer, 1980) the behaviour of the crimped collagen fibrils largely determines the instantaneous mechanical response of the composite, but the viscoelastic nature of the interfibrillar glycosaminoglycan introduces time-dependent effects. The viscoelastic mucopolysaccharides, which form the matrix of pliant composites, allow the material to stretch and bend as the animal moves and perhaps lubricate the movement of the more rigid fibrous component but contribute little to the loadbearing ability of the composite. In pliant vertebrate composites in which elastin fibres are combined with collagen in the glycosaminoglycan matrix the elastin, having a very low elastic modulus and high extensibility, confers long-range reversible extensibility upon the composite. A similar effect is obtained if the matrix polymer is cross-linked, like lightly vulcanized rubber, and the resultant long-range elasticity means that a restoring muscle is not needed to bring about a return to the undistorted shape. In cartilage (Swanson, 1980) the collagen, and perhaps also elastin, fibres are contained in a proteoglycan matrix, but the properties of the composite are different from those of skin because of osmotic swelling of the gel matrix against the elastic constraint of the collagen network, which gives rise to the variable rigidity characteristic of the material. In bone (Katz, 1980) the situation is reversed since the more pliant collagen serves as a matrix for the mineral crystals and as a result confers toughness and flexibility on the composite similar to that of mineral-filled plastics instead of the brittleness that would be characteristic of monolithic mineral solids. The reinforcing cellulose fibrils in wood are bound together by a resinous matrix consisting of amorphous lignin combined with crystalline hemi-cellulose material. The matrix separates the cells, but also permeates the cell walls, giving a more graded transition from fibre to matrix than can normally be achieved in man-made composites. The matrix is more rigid than in many animal composites, but is nevertheless sufficiently plastic (or viscoelastic) to confer time-dependent properties and toughness on the wood.

In synthetic and natural composites we meet a variety of fibre arrangements. Tendon, for example, contains parallel arrays of continuous collagen fibres which act in a purely tensile manner like the glass fibres in a structural tie of grp. Pliant animal composites, however, like the connective tissues of the sea anemone, contain parallel arrays of discontinuous fibres which result in higher extensibility. The body wall connective tissue and cuticles of worm-like animals and some vertebrates resemble laminated or filament-wound structures in which continuous fibres are arranged in layers at various angles to one another.

The deformability of such materials depends strongly on the winding or laminating angles, which in man-made structures are selected to satisfy the structural requirements of some particular design. A pipe, for example, filament wound with a high winding angle (hoop winding) will be very resistant to radial expansion but will have little resistance to deformation along the pipe axis. Structures of this kind are anisotropic, but less so than unidirectional composites and have little resistance to in-plane shear deformation. Coated fabrics for inflatable structures are materials of this type. Pliant tissues containing laminated arrays of discontinuous fibres are capable of large deformation, simultaneously, in the length and width, unlike continuous fibre structures, and in such tissues the matrix plays a much more significant role in determining mechanical properties. Uterine cervix, for example, contains a discontinuous fibre system in a polymer matrix which undergoes substantial reorganization during pregnancy. The cervix from a non-pregnant woman behaves almost like a continuous fibre composite, exhibiting very little ability to deform and properties which are substantially non-time-dependent. Indeed, deliberate dilatation beyond a limiting diameter (about 10 mm) is accompanied by irreversible deformation and damage which can cause incompetence of the cervix during subsequent pregnancy, resulting in miscarriage (Black, Melcher, Melville & Alderman, 1975). During pregnancy, however, the matrix polymer becomes capable of viscous deformation which permits large extensions at relatively low loads. This change, which is rapidly reversed after parturition, is explained in terms of changes in the cross-linking and breaking down of the continuity of the fibre network.

Felted systems of discontinuous fibres in which there is a more three-dimensional arrangement give moderately extensible material with more isotropic behaviour and better shear resistance. Such arrangements are found in the dermal connective tissue of birds and mammals. Skin, by contrast, is a three-dimensional felt of continuous collagen fibres, together with elastin fibres, in a protein–polysaccharide matrix, which possesses the characteristics of rubber elasticity. This is an example of a hybrid composite where at extensions up to 60% the behaviour of the material is determined by the elastin while at much greater strains the collagen network limits the extensibility of the composite and the modulus increases rapidly with strain.

This discussion on natural composite materials is largely based on the reviews given by Wainwright, Biggs, Currey & Gosline (1976).

Elastic behaviour of fibre composites

Because of the importance of being able to predict elastic response when designing structures there have been many sophisticated mathematical treatments of elastic properties of simple, uniaxial composites and of laminates. These are all, for the most part, more successful than models of composite strength and toughness, perhaps because the boundary conditions are easier to define for elastic behaviour.

The earliest treatments of particulate reinforcement of rubbers were based on Einstein's model for the viscosity of a Newtonian fluid containing infinitely rigid, spherical, non-interacting particles. Such a model leads to an expression for the modulus, E_c, of a composite containing a low volume fraction, V_f, such that

$$E_c = E_m(1 + 2.5\,V_f) \tag{1}$$

More generally, for $V_f > 0.05$

$$E_c = E_m(1 + 2.5\,V_f + 14.1\,V_f^2) \tag{2}$$

and further improvements include empirical shape factors for non-spherical particles. The filler modulus, E_f, is never infinite, of course, and deformation of the filler itself should be accounted for. The ultimate simplification is to imagine, as extreme cases, arrays of well-bonded lamellae having different moduli but identical Poisson's ratios, as shown in Fig. 1. For case *a* there is continuity of strain at the interfaces and we can treat the model as a parallel arrangement of springs while in the series arrangement of case *b* there is continuity of stress. Simple arithmetic gives the longitudinal elastic modulus for each model, and the parallel and series estimates are known as the Voigt and Reuss estimates:

$$E_c = E_m V_m + E_f V_f \text{ (Voigt)} \tag{3}$$

$$\frac{1}{E_c} = \frac{V_m}{E_m} + \frac{V_f}{E_f} \text{ (Reuss)} \tag{4}$$

where $V_m + V_f = 1$. These estimates are sketched in Fig. 2. The Voigt estimate is often called the 'rule of mixtures' and although apparently despised by theoreticians it is nonetheless an appropriate approximation, frequently found to be a perfect predictor, for the modulus of unidirectional fibre composites in the fibre direction. In a particulate or

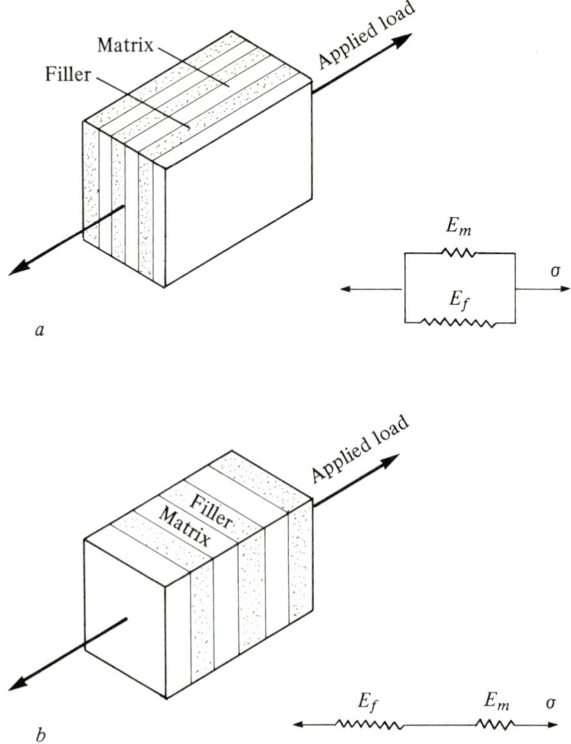

Fig. 1. Parallel and series arrangements of matrix and filler in a model composite. In each case $E_f > E_m$ and $\nu_f = \nu_m$. *a*, parallel model ($V_f = 0.5$); *b*, series model ($V_f = 0.5$).

discontinuous fibre composite, or in a continuous composite taken perpendicular to the fibres, the matrix phase is continuous but the filler is not and the actual composite modulus will be within the bounds given by the Voigt and Reuss estimates. For the transverse modulus of fibre composites good theoretical treatments are available, but a much simpler approach is to use an empirical 'contiguity' factor (Tsai, 1964). The transverse composite modulus is then given, usually to an acceptable degree of accuracy, by the weighted sum of the two estimates

$$E_c = E_{\mathrm{REUSS}} + C(E_{\mathrm{VOIGT}} - E_{\mathrm{REUSS}}) \tag{5}$$

C is the contiguity factor for which a value of 0.2 is appropriate for a uniform, unidirectional fibre composite.

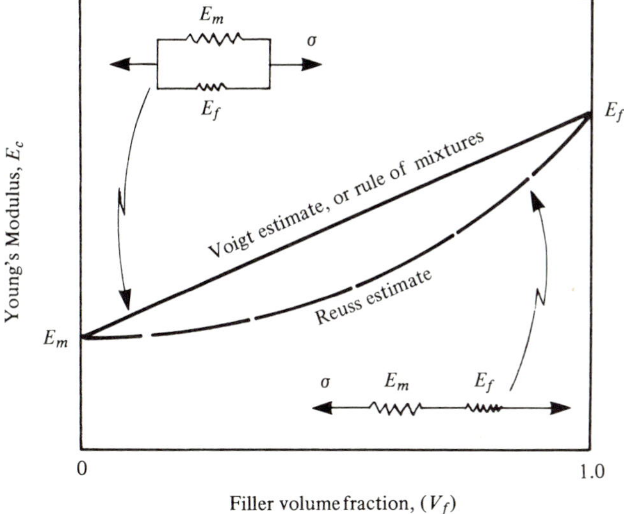

Fig. 2. Estimates of composite Young's modulus, E_c, as a function of reinforcement volume fraction, V_f, for the parallel and series combinations of Fig. 1.

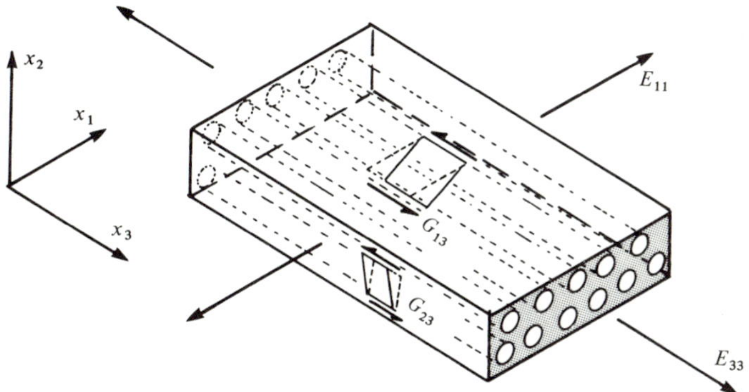

Fig. 3. Elastic properties of a composite material referred to Cartesian co-ordinates. In a unidirectional composite of this type the shear moduli, G_{13} and G_{23}, are approximately equal and less than G_{12}, and $E_{22} \approx E_{11}$.

When the Poisson's ratios of the two phases are dissimilar additional elastic constraints are present and the mixture rule (eqn (3)) is no longer a valid approximation for the modulus of an aligned composite. It becomes instead a lower bound for E_c. If we use the nomenclature of Fig. 3 to define the elastic properties of an anisotropic lamina, we must therefore write

$$E_{33} > E_f V_f + E_m(1 - V_f) \quad . \ . \ . \quad \nu_f \neq \nu_m \tag{6}$$

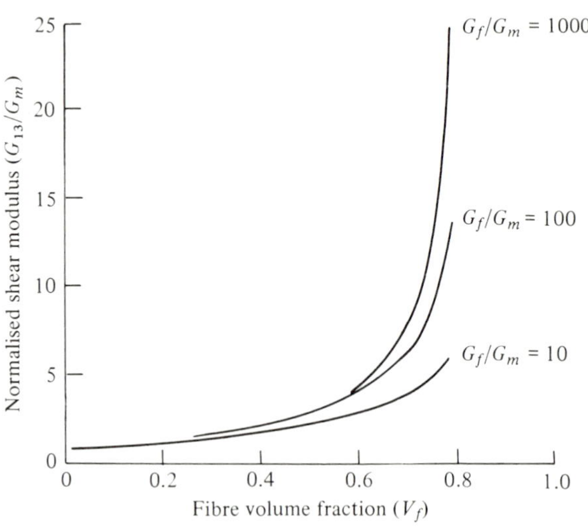

Fig. 4. Composite shear modulus, plotted as the reinforcement factor, G_{13}/G_m, as a function of the fibre volume fraction for various values of the stiffness ratio, G_f/G_m. After Tsai *et al.*, 1966.

where ν is Poisson's ratio, and Hill (1964) has shown that for a fibre of circular section in a cylindrical shell of matrix

$$E_{33} = E_m V_m + E_f V_f + \frac{4V_m V_f(\nu_m - \nu_f)^2}{\frac{V_m}{K_f} + \frac{V_f}{K_m} + \frac{1}{G_f}} \tag{7}$$

where K_f and K_m are the fibre and matrix bulk moduli. This reduces to the mixture rule when $\nu_m = \nu_f$.

The extensional Poisson's ratio of a unidirectional composite is also approximated by a mixture rule

$$\nu_{31} \approx \nu_f V_f + \nu_m(1 - V_f) \tag{8}$$

but the shear modulus, G_{13}, for shear along the fibres is a complex function of the stiffness ratio, G_f/G_m. Results of calculations by Tsai, Adams & Doner (1966) based on a square array of fibres, are plotted in Fig. 4. Their results for the transverse modulus, E_{11}, of a similar array are shown in Fig. 5.

A unidirectional composite is highly anisotropic and the modulus, E_θ, at any angle, θ, to the aligned fibres can be predicted accurately by transforming the compliance tensor. For example, assuming that the composite has a single axis of anisotropy (x_3) and is transversely

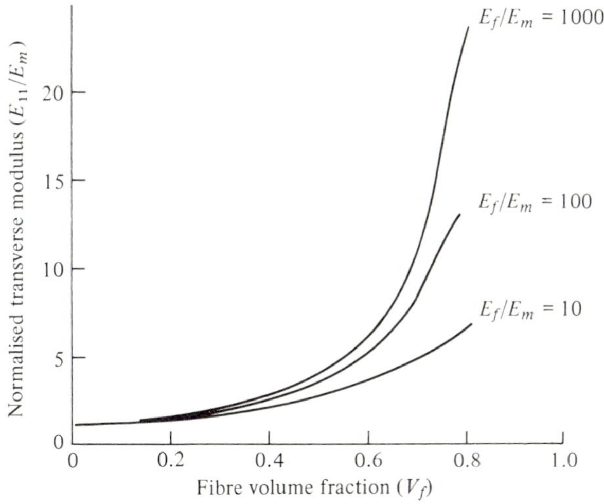

Fig. 5. Transverse Young's modulus, plotted as the reinforcement factor, E_{11}/E_m, as a function of V_f for various values of the component stiffness ratio, E_f/E_m. After Tsai *et al.*, 1966.

isotropic, like wood or drawn polymer fibres, the compliance matrix, with five independent elastic constants, is

$$\begin{matrix} S_{11} & S_{12} & S_{13} & - & - & - \\ S_{12} & S_{11} & S_{13} & - & - & - \\ S_{13} & S_{13} & S_{33} & - & - & - \\ - & - & - & S_{44} & - & - \\ - & - & - & - & S_{44} & - \\ - & - & - & - & - & 2(S_{11}-S_{12}) \end{matrix}$$

For the special case of a thin lamina of an orthotropic material, like grp, we can assume a state of plane stress, and Hooke's law for the composite becomes

$$\begin{bmatrix} \varepsilon_{11} \\ \varepsilon_{33} \\ \varepsilon_{13} \end{bmatrix} = \begin{bmatrix} S_{11} & S_{13} & 0 \\ S_{13} & S_{33} & 0 \\ 0 & 0 & S_{66} \end{bmatrix} \times \begin{bmatrix} \sigma_{11} \\ \sigma_{33} \\ \sigma_{13} \end{bmatrix} \tag{9}$$

in which $S_{11} = 1/E_{11}$, $S_{33} = 1/E_{33}$, $S_{66} = 1/2G_{13}$, and $S_{13} = -\nu_{13}/E_{33} = -\nu_{13}/E_{11}$. Transformation of the full tensor in the x_1x_3 plane gives the Young's modulus, E_θ, as a function of orientation

$$\frac{1}{E_\theta} = \frac{\cos^4\theta}{E_{33}} + \frac{\sin^4\theta}{E_{11}} + \left[\frac{1}{G_{13}} - \frac{2\nu_{31}}{E_{33}}\right] \sin^2\theta \cos^2\theta \tag{10}$$

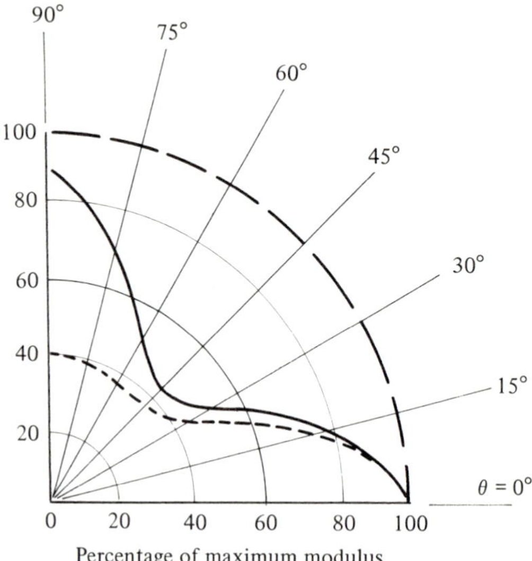

Fig. 6. Orientation dependence of the elastic modulus of typical grp laminates given by eqn (10). The curves represent E_θ as a percentage of the modulus E_{33} (for $\theta = 0°$). The three curves are for an isotropic laminate (interrupted line), a 0°/90° laminate, not perfectly balanced since $E_{33} > E_{11}$ (full line), and a unidirectional laminate (dotted line).

The theoretical variation of E_θ with orientation for three characteristic types of glass-reinforced epoxy laminates is shown in Fig. 6. Experimental values of E_θ are almost always in excellent agreement with predictions of this model.

The high degree of anisotropy in unidirectional laminates is unacceptable for many engineering purposes since practical stress systems are seldom unidirectional. The degree of anisotropy is therefore controlled by laminating together plies of different orientation or by the filament winding of pipes. By varying the cross-ply angle and the relative proportions of fibre lying in different orientations the ratio E_{33}/E_{11} and the values of G_{13} can be tailored to suit the application (Fig. 7). Calculation of the anisotropic elastic constants is then more complicated, but a well-developed classical theory of thin laminates is available for the purpose which effectively sums the contributions to stiffness of the separate laminae and accounts for the additional contributions from interlaminar shear coupling constraints (Jones, 1975). For the estimation of orthogonal moduli of simple 0°/90° balanced laminates, however, a mixture-rule summation of the volume fraction weighted contributions from the longitudinal and transverse plies is often adequate

$$E(0°) = E_{33}V_f(0°) + E_{11}V_f(90°)$$

where E_{33} and E_{11} are obtained from eqns (3) and (5).

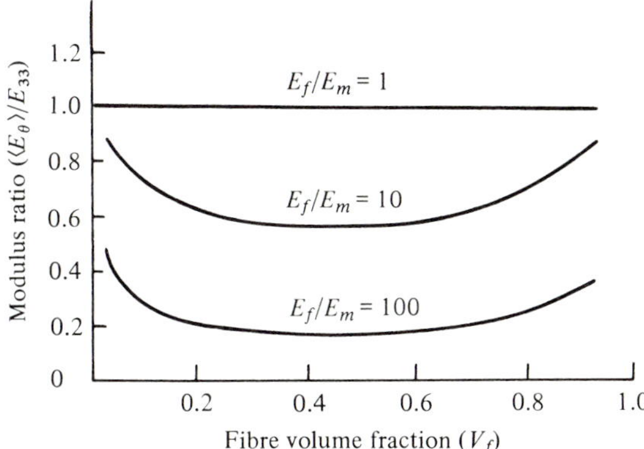

Fig. 7. The effect of randomizing the fibre orientation on the composite modulus, plotted as a fraction of the modulus E_{33} of a unidirectional composite, as a function of fibre volume fraction. The separate curves represent various fibre/matrix modulus ratios, E_f/E_m. After Nielsen & Chen, 1968.

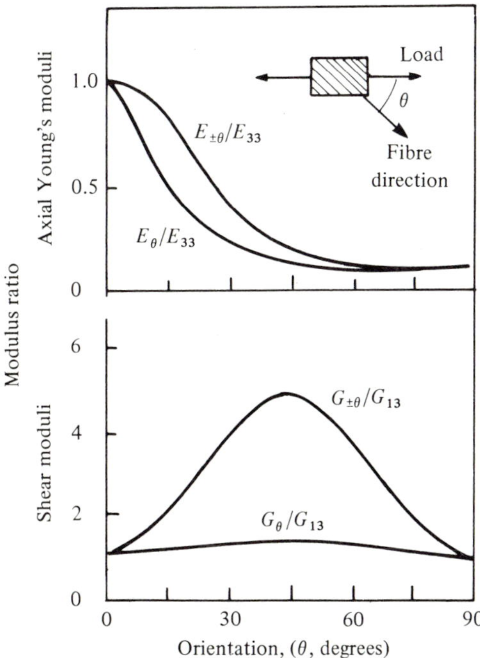

Fig. 8. The effect of cross-laminating on the axial Young's moduli and shear moduli of carbon-fibre composites, relative to the modulus, E_{33} and G_{13} of unidirectional composite. After Goatham, 1970.

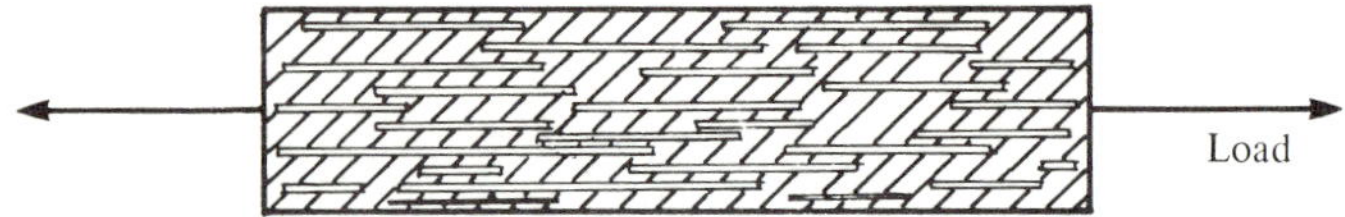

Fig. 9. Schematic representation of a composite in which the fibres are aligned but discontinuous. The applied load is transferred into the fibres by shear forces at the interface, relatively little load being transferred across the fibre ends. The fibre radius is r_0 and the mean centre to centre spacing is $2R$ (see eqn (12)).

Two-dimensional isotropy can be obtained in a composite by randomizing the orientations of fibres in the plane of the plate. The modulus is found by averaging E_θ over all values of θ

$$E_\theta = \frac{\int_0^{\pi/2} E_\theta \, d\theta}{\int_0^{\pi/2} d\theta} \tag{11}$$

The solution of eqn (11) for reinforced resins and several values of the modulus ratio, E_f/E_m, is shown in Fig. 7. Clearly, for composites manufactured from costly, high performance fibres like carbon, where the modulus ratio can be as high as 200, the effect of randomizing in this way on the rigidity of the composite would be disastrous. More realistic lamination procedures allow the design of composites with much improved torsional rigidity while retaining a substantial proportion of the longitudinal stiffness of the unidirectional composite, as shown in Fig. 8.

In composites containing discontinuous reinforcements, the short fibres can no longer be loaded directly through their ends and stress must be transferred into them from the matrix by shear forces at the fibre/matrix interface (Fig. 9). They may still carry a high proportion of the load on the composite, but there is a section at each end of each fibre where the fibre tensile stress is still building up to its equilibrium value and in which region its reinforcing efficiency is diminished. The variation of fibre tensile stress, σ, and interface shear stress, τ, along a fibre of aspect ratio l/d may be something like that shown in Fig. 10 (Kelly, 1973) and Kelly's model for the Young's modulus of an aligned, discontinuous fibre composite gives

$$E_{33} = E_m(1 - V_f) + E_f V_f \left[1 - \frac{\tanh(\beta l/2)}{\beta l/2}\right] \tag{12}$$

where
$$\beta = \left[\frac{2\pi(G_m/G_f)}{V_f \log_e(R/r_0)}\right]^{\frac{1}{2}}$$

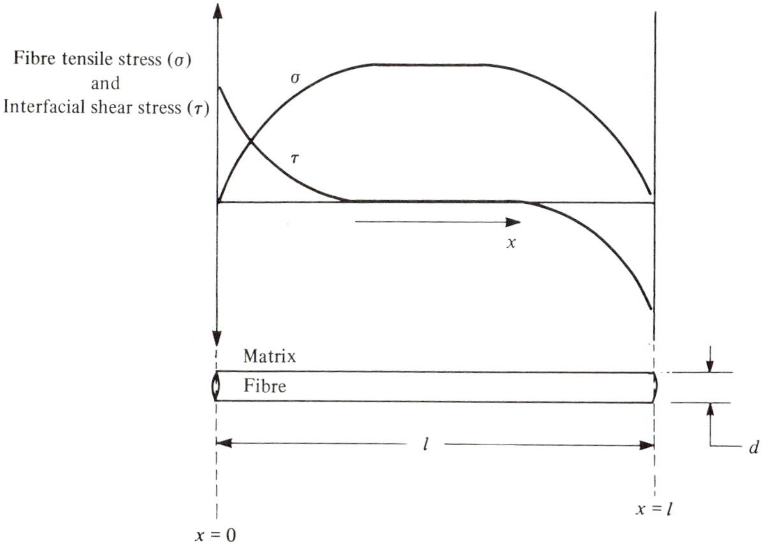

Fig. 10. Variation of fibre tensile stress, σ, and interface shear stress, τ, along a fibre of length, l, in a discontinuous fibre composite. After Kelly, 1973.

governs the rate of stress build up at the fibre ends. The fibre efficiency factor,

$$\left[1 - \frac{\tanh \beta l/2}{\beta l/2}\right],$$

is small if the fibres are long and eqn (12) then becomes the rule of mixtures once more. There are other, more rigorous, treatments of this problem, but the results usually differ from eqn (12) only in the final value of β.

The elastic behaviour of pliant composites is somewhat more difficult to characterize than that of rigid composites. When a typical uncoated woven fabric is stretched the effective modulus will include apparent strain contributions from several mechanisms, not present in rigid materials, other than the true elastic (or viscoelastic) extension of the fibres themselves. These may be briefly identified as follows:

(i) Yarn shear or rotation, if the principal stresses are not aligned with the warp and weft of the fabric.

(ii) Crimp interchange. For example, if a highly crimped weft yarn is directly loaded it will not begin to extend elastically until it has lost the crimp which will be, in some measure, transferred to the originally straight warp yarns.

(iii) Yarn flattening, leading to a reduction in effective yarn diameter.

(iv) Yarn compaction, resulting in an increase in packing density.
(v) Fibre straightening and/or rotation, within the yarn structure.

The first two of these occur at low stresses and depend on the principal stress ratios (not on their magnitudes), whereas the other mechanisms become active only at high stresses. In normal fabric behaviour true elastic extension of the individual filaments probably never occurs, and since this is the only mechanism considered in conventional netting analyses like that of Cox (1952) such analyses are not useful for predicting the behaviour of fabrics or pliant composites. Shanahan, Lloyd & Hearle (1978) have shown that in some respects fabrics may be treated as sheets of an elastic continuum and have applied the basic framework of laminate theory. For membrane strains, for example, the tensile modulus, E_θ, from eqn (10) gives a reasonable approximation to measured properties of some types of fabric, but coupling between membrane modes of deformation, bending and twisting, and large non-linear elastic strains, cannot be dealt with by the continuum model. The application of an extensible or rubbery coating to a woven fabric is normally considered, for design purposes, not to change the cloth characteristics. For small-strain extensional behaviour in the warp or weft directions this may be an acceptable assumption, but even a non-rigid matrix will significantly raise the in-plane shear resistance of a coated fabric. The shear stiffness may also be affected by a stress system which increases the inter-yarn frictional forces. The wall of internally pressurized tube, for example, appears to become stiffer as the internal pressure is increased and rotation of the yarns is inhibited by the increased inter-yarn friction (Topping, 1961). The generalized two-dimensional form of Hooke's Law for a textile or coated fabric (Alley & Faison, 1972) contains nine stiffness coefficients. Since the straining of such materials is not necessarily conservative (i.e. the values of the coefficients may be dependent on the state of strain in the material and its loading history) all nine coefficients are independent, necessary, and multi-valued. Furthermore, in a coated fabric some of the coefficients are likely to be time-dependent. In a non-linear anisotropic membrane, then, it is not surprising that the setting up of practicable constitutive relations should be an exceptionally unwieldy process and of little practical use for engineering design purposes.

Tensile strength

The simplest approach to estimating the strength of a composite is to consider the stress–strain curves of a fibre and matrix material and, by

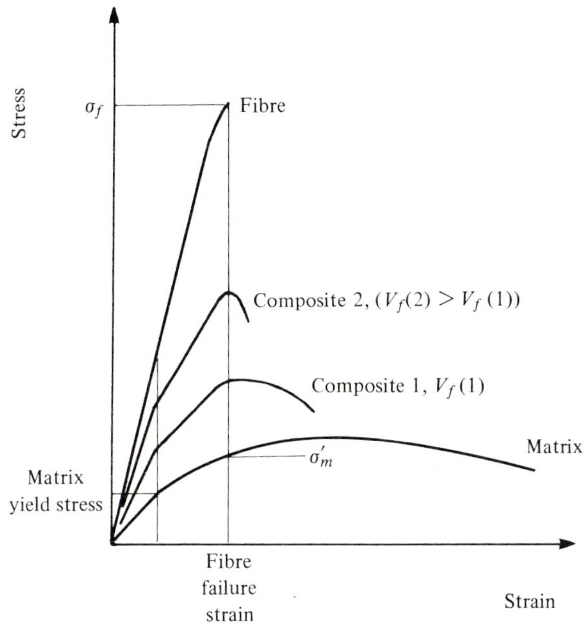

Fig. 11. Schematic stress/strain curves for fibre, matrix, and two composites with volume fractions $V_f(1)$ and $V_f(2)$.

interpolation, to derive the curve for a composite containing a given volume fraction of aligned, continuous, strong, rigid fibres. This is illustrated in Fig. 11, for the case where the low strength matrix is plastically deformable but the fibres fail after little or no non-elastic deformation. If the fibres and matrix are well bonded and deform together the load on the composite is shared between them in proportion to their cross-sectional areas. Any stress on the composite is therefore distributed as

$$\sigma = \sigma_f V_f + \sigma_m(1 - V_f)$$

where $\sigma_f = E_f\varepsilon$, $\sigma_m = E_m\varepsilon$, and ε is the strain level in the composite. This will hold until failure of one of the phases occurs. In many man-made composites it is the fibres which fail first, and at that point load is transferred back to the weaker matrix. This will be unable to sustain the whole of the composite load if there is a reasonably high fibre volume fraction and rapid failure of the composite ensues. If the stress carried by the matrix at the fibre failure strain is σ'_m, and we now let σ_f represent the mean tensile strength of the fibres, the tensile strength of such a composite is simply

$$\sigma_c = \sigma_f V_f + \sigma'_m(1 - V_f) \quad . \ . \ . \quad V_f > V_{\min}. \tag{13}$$

$V_{\min}$ is usually quite low and is determined by the difference between the matrix yield and tensile strengths. This rule-of-mixtures gives a reasonable approximation if the Poisson's ratios of the components are roughly equal, but it is an isostrain model which assumes that the fibre strength has a clearly defined and unique value. This assumption is invalid for brittle fibres like glass, boron and carbon, the strengths of which are statistically distributed. It is necessary, therefore, to consider what value of σ_f is appropriate for use in eqn (13) by referring to statistical theories of failure (Rosen, 1964; Zweben & Rosen, 1970). The simplest approach based on the Weibull distribution (Weibull, 1951) gives the cumulative distribution function for the probability of failure, $g(\sigma)$, as:

$$g(\sigma) = 1 - \exp\left\{-\varrho\left(\frac{\sigma - \sigma_u}{\sigma_0}\right)\right\}^m \tag{14}$$

where σ_u is the stress for zero probability of failure, σ_0 is a normalizing factor, ϱ is related to the volume of the sample, and m is determined by the scatter of the strength data or the flaw size distribution. If the strengths of fibres in a bundle are distributed according to such a relation the weaker fibres in the bundle will fail at loads lower than the mean fibre strength. As they break, the loads they carry will be transferred to the unbroken ones until, at some critical point, the failure of one more fibre results in an increment of load transfer that the remaining body of unbroken fibres can no longer support. This point will then define a bundle strength which must be lower than the mean fibre strength and a simple treatment (Corten, 1967) shows this bundle strength to be

$$\sigma_B = \sigma_0[m(l/d)]^{-1/m} \exp(-1/m) \tag{15}$$

where l/d is the aspect ratio of the fibres. The value of σ_B is strongly dependent on the flaw distribution parameter (which is related to the coefficient of variation of test results) as well as on the fibre length/diameter ratio (Fig. 12). For glass fibres, for which values of m between 5 and 15 are common, the bundle strength may thus be as low as 70% of the mean fibre strength. It is not satisfactory, however, to use σ_B in eqn (13) to obtain a composite strength such that, neglecting the small matrix contribution,

$$\sigma_c \approx \sigma_B V_f \approx \sigma_f V_f$$

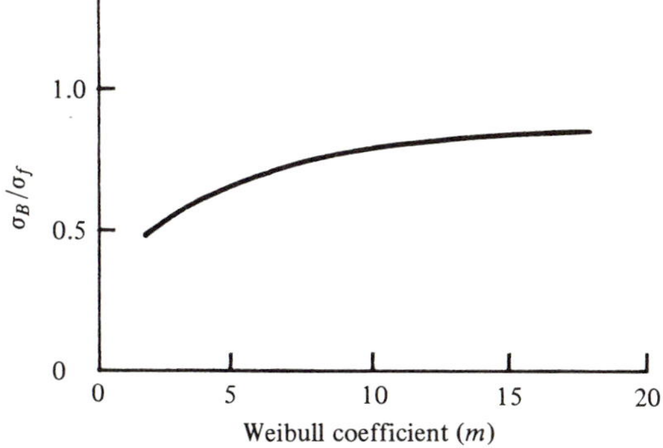

Fig. 12. Influence of scatter in strength measurements, as defined by the Weibull parameter, m, on the bundle strength, σ_B. σ_f is the average strength of fibres in the bundle. After Corten, 1967.

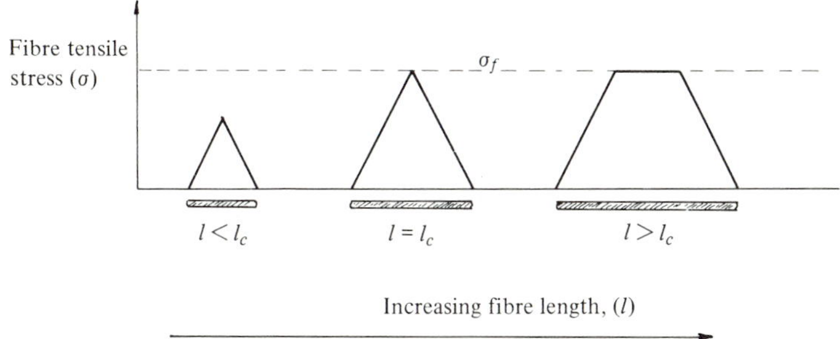

Fig. 13. Simplified schematic illustration of the variation of the tensile stress in a short fibre as a function of fibre length. σ_f is the fibre failure stress.

because the behaviour of a fibre bundle is completely altered when bonded by a resin matrix. In a rigid composite, a fibre which breaks at low load does not cease to contribute to the overall load-bearing ability. The stress distribution near the break is perturbed because the broken ends of the fibre carry less than the mean fibre load: but within a distance $l_c/2$ (known as the ineffective length) from each end the broken halves are again supporting their full share of the load. There will clearly be some critical length, l_c, however, which is too short to permit this, and fibres shorter than this critical length cannot be broken down by stress transfer in the composite, as shown schematically in Fig. 13.

Near a fibre break the matrix and neighbouring fibres must support the extra load shed by the broken fibre. In most practical composites this extra load does not cause rupture of adjacent fibres, leading to

Table 2. *Strengths of some unidirectionally reinforced plastics*

Composite	Volume fraction (V_f)	Strength (GPa)
Glass/epoxy	0.73	1.7
Carbon (type 1)/polyester	0.40	0.7
Carbon (type 2)/epoxy	0.60	1.4
Boron/epoxy	0.67	1.4
Kevlar 49/epoxy	0.60	1.3

catastrophic failure of the composite at that cross section, because the weakest places in those adjacent fibres are also statistically distributed. Such brittle failures occur only if the fibres are very close or touching and have a very narrow distribution of strengths. In general, fibre breaks accumulate at first throughout the whole sample, resulting in a cumulative weakening of the composite, and final failure occurs when the number of breaks in any given cross-section has effectively lowered the useful fibre volume fraction in that section below that needed to support the current applied load. The actual composite strength is thus bounded $V_f\sigma_B \lesssim \sigma_c \lesssim V_f\sigma_f$ on this simple model. As a consequence of the statistical flaw distribution the broken sections of a fibre become successively stronger each time the fibre ruptures (there is a dependence of strength on test length in many types of strong fibre, such as carbon). And since there may also be a large Poisson's ratio mismatch between matrix and fibre, composite strengths greater than the rule-of-mixture bound (eqn (13)) are sometimes observed. Typical strengths of some unidirectionally reinforced, high quality laminates are given in Table 2.

The strengths of unidirectional composites are highly anisotropic. Perpendicular to the fibres they are very weak and failure is controlled by the rupture (or plastic flow) of the matrix or by fibre/matrix debonding. In low V_f composites the constraint imposed by the relatively undeformable fibres causes triaxial strain concentrations in the matrix which may lead to brittle failure at low strains even in otherwise ductile matrix materials. In high V_f composites, however, transverse failure occurs simply by the linking up of fibre/matrix debonds (Fig. 14). Transverse failure of a similar kind also occurs at strains well below the overall composite tensile failure strain in the transverse plies of cross-plied composites, resulting in some loss of rigidity and the possibility of general failure through splitting generated by the transverse cracks. If the strengths of the composite parallel with the fibres, σ_c, perpendicular to the fibres, σ_t, and in shear parallel with the fibres, τ, are all known,

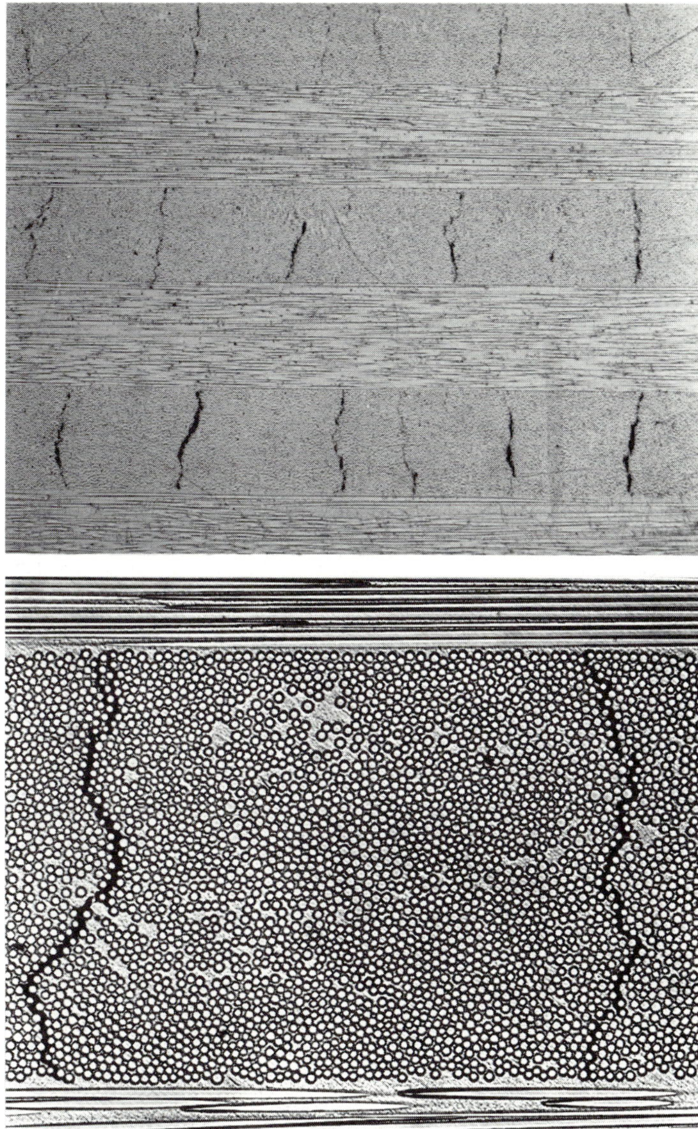

Fig. 14. Optical micrographs of transverse cracking in a grp laminate by joining up successive fibre/matrix debonds at interfaces normal to the applied stress. The mean ply thickness is about 0.6 mm. From Harris, Guild & Brown, 1979.

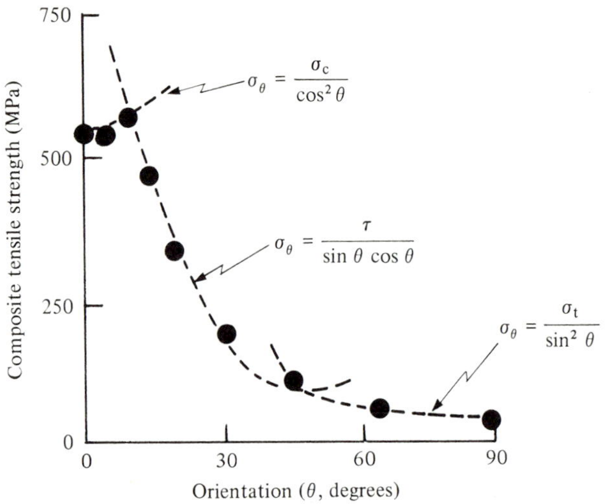

Fig. 15. Comparison of measured composite strengths (filled circles), as a function of the angle, θ, between the stress axis and the fibre direction, with theoretical predictions of the maximum stress theory. Carbon fibre/epoxy composites; $V_f = 0.57$. After Dimmock & Abrahams, 1968.

there are several ways of predicting the strength as a function of orientation. The simplest is the maximum stress theory of Stowell & Liu (1961), obtained simply by resolving forces, which defines three modes of failure, characterized by the three equations (16, *below*)

$$\begin{aligned} &\sigma_\theta = \sigma_c \sec^2\theta, \text{ for failure controlled by tensile fracture of fibres} \\ &\sigma_\theta = 2\tau \operatorname{cosec} 2\theta, \text{ for failure controlled by shear parallel with fibres} \\ &\sigma_\theta = \sigma_t \operatorname{cosec}^2\theta, \text{ for tensile failure normal to the fibres} \end{aligned} \tag{16}$$

Typical experimental results for a carbon fibre/epoxy resin composite are compared with predictions of the maximum stress theory in Fig. 15, and equally good agreement has been found for a wide range of other materials, including balsa wood (Soden & McLeish, 1976). Tsai (1968) has shown that in some grp materials a maximum distortional energy theory based on von Mises' criterion fits the data more closely, the variation of σ_θ with orientation being

$$\frac{1}{\sigma_\theta^2} = \frac{\cos^4\theta}{\sigma_c^2} + \frac{\sin^4\theta}{\sigma_t^2} + \cos^2\theta \sin^2\theta \left[\frac{1}{\tau^2} - \frac{1}{\sigma_c^2}\right] \tag{17}$$

The two models approach each other most closely when θ tends to 0° or 90° and differ most at $\theta = 45°$. The critical angle at which the fibre tensile fracture mode gives way to the shear mode is given by

$$\theta_{crit} = \tan^{-1}(\tau/\sigma_c)$$

and at orientations greater than θ_{crit} the composite strength falls rapidly.

There have been several attempts to develop failure criteria for complex stress systems and these are generally also similar to von Mises' criterion. A common modification of the plane stress distortional energy criterion frequently used for anisotropic materials is that of Azzi & Tsai (1965):

$$(\sigma_3/\sigma_c)^2 + (\sigma_1/\sigma_t)^2 + (\tau_{13}/\tau)^2 - \left(\frac{\sigma_1\sigma_3}{\sigma_c\sigma_t}\right) = 1 \qquad (18)$$

for the plane stress system (σ_3, σ_1, τ_{13}), the suffix 3 again indicating the fibre direction. Similar models have been used to predict the strength of more complex laminates with a reasonable degree of success. For example, Eckold, Leadbetter, Soden & Griggs (1978) have developed a laminate theory for the prediction of failure envelopes for filament wound structures under biaxial loads. They predict failure, using a maximum stress criterion from a ply-by-ply analysis, which allows for progressive failure of the laminate, shear non-linearity, and for different properties in compression and tension. The composite is assumed to be made up of a series of homogeneous, orthotropic plies, as is usual in laminate theory calculations. Good agreement was obtained between theoretical predictions and experimental results, but the authors point out that the arithmetic involved is complex. The same authors (Soden, Leadbetter, Griggs & Eckold, 1978) have also successfully applied an alternative treatment, in which the composite is assumed to be homogeneous and the failure criterion is applied to the material as a whole, to predict failure envelopes for pipes of $\pm 35°$ and $\pm 55°$ winding angle under a variety of residual stresses. This second method is easier than the first, since it does not require any knowledge of the stress distributions within the composite, but it lacks the generality of the first method since practical data must be obtained for each winding pattern.

A somewhat cruder but nonetheless useful approach is that based on the use of the rule of mixtures with the incorporation of an efficiency factor, η, which can be calculated from a netting type of analysis (Cox, 1952), thus

$$\sigma_c = \eta\sigma_f V_f + \sigma'_m(1 - V_f). \qquad (19)$$

Typical values of η calculated by Krenchel (1964) for various arrays of fibres are shown in Fig. 16.

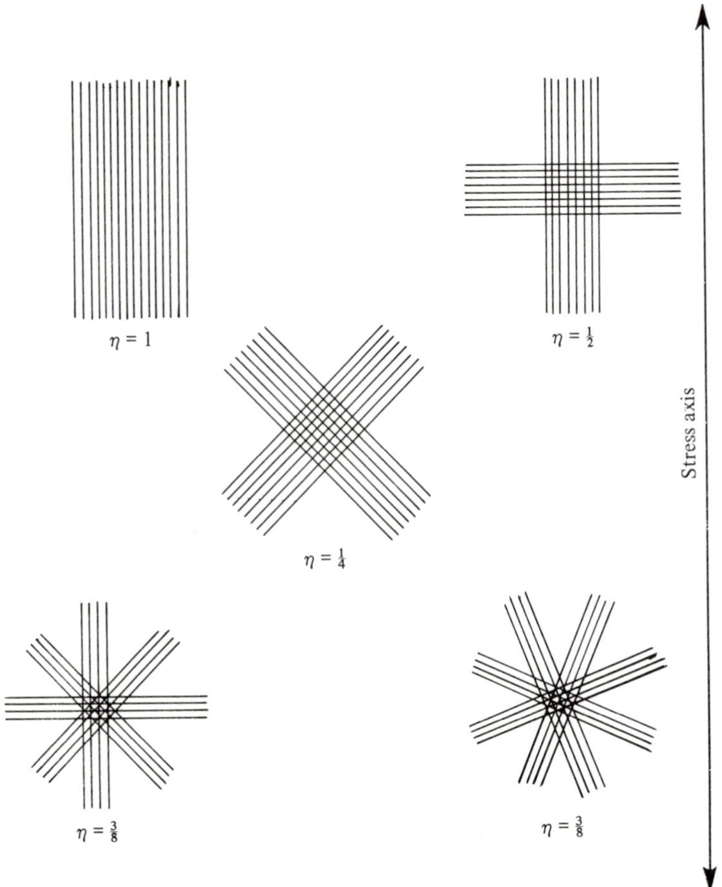

Fig. 16. Composite strength efficiency factors calculated by Krenchel (1964) for various schematic arrays of fibres.

When the fibres are discontinuous it is necessary to take account of the ineffective length. Short fibres with length, l, longer than l_c, the critical length, may be stressed to breaking point, but when they do break the *average* stress in the fibres, $\bar{\sigma}_f$, depends on the form of the stress distribution at the fibre ends for which several models have been proposed. For the crude linear form shown in Fig. 13 the mean fibre stress at failure is

$$\bar{\sigma}_f = (1 - l_c/2l)\sigma_f$$

and the composite tensile strength is therefore

$$\sigma_c = \sigma_f V_f(1 - l_c/2l) + \sigma'_m(1 - V_f) \quad . \ . \ . \ l > l_c \qquad (20)$$

In contrast if the fibres are shorter than l_c – an unusual, because uneconomic, situation – the fibres cannot be broken and their strengthening potential cannot be realized. Since the critical aspect ratio, l_c/d, is simply equal to $\sigma_f/2\tau_i$, where τ_i is the shear strength of the interface or the matrix, whichever is the weaker, the maximum fibre stress is only $2\tau_i l/d$, and the *mean* stress is only half this. The composite strength is then

$$\sigma_c = \left(\frac{\tau_i l}{d}\right) V_f + \sigma'_m(1 - V_f) \ldots l < l_c \qquad (21)$$

Although the strength of a composite is thus strongly dependent on the fibre length, some 95% of the available strength of continuous fibres can nevertheless be obtained with discontinuous fibres provided l/l_c is greater than about ten (Kelly & Davies, 1965).

The most complex composite will, typically, be an injection moulding of some thermoplastic containing short fibres whose orientation is randomly distributed and the lengths of which are also widely distributed. Models for the strength of such materials (Bader & Bowyer, 1973) must sum the contributions from the fibres over the whole range of lengths and must include an orientation factor, typically 0.375 for a planar random array and 0.167 for a true three-dimensional arrangement.

Compression and shear strength

Studies of the compression strength of fibre composites have been impeded by the fact that when testing small, free blocks failure modes occur which are not characteristic of the fibre behaviour, and the compression strength is frequently much lower than the tensile strength of the composite. The compression strength of wood, similarly, is only about a third of its tensile strength. This is not at all unexpected, since fibres work satisfactorily only in tension. In unidirectional composites the characteristic failure mode is by a kinking mechanism, apparently initiated by shear of the matrix (Fig. 17) similar to that which occurs in wood, in drawn polymer fibres, and also in single crystals of some metals. By contrast, bone may also deform in compression by a kinking mechanism but is usually much stronger in compression than in tension.

Current theoretical treatments of the compression strength of composites do not satisfactorily fit available experimental results. The model of Rosen (1965) most commonly used, which was developed for laminar

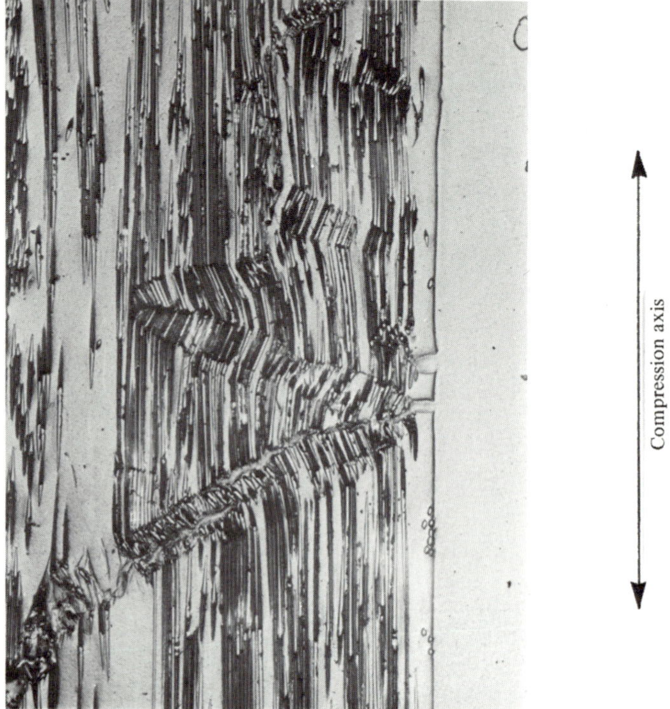

Fig. 17. Failure in compression of a glass-fibre reinforced polyester resin by the development of instability due to the formation of kinks by shear. The diameter of this sample (width of picture) is 6 mm.

composites, predicts a compression strength, governed by shear instability of the weaker phase and in-plane elastic buckling of the reinforcement, equal to $G_m/(1 - V_f)$, which consistently overestimates experimental values. Modifications to this theory use a post-yield tangent modulus, $d\sigma/d\varepsilon$, instead of G_m (Rosen, 1965); the use of an arbitrary factor to allow for out-of-plane buckling (Lager & June, 1969); and a reduced value of G_m resulting from a supposed dependence of the matrix shear modulus on stress (Foye, 1966; Hayashi & Koyama, 1971). Compression strength has often been shown to obey a mixture rule based, curiously, on the *tensile* strength of the fibres (de Ferran & Harris, 1970) and some recent work by Piggott & Harris (1979) shows that it also depends strongly on matrix yield strength. Their results lend no support to current theories of either elastic or plastic failure.

If the ends of a short compression test piece are not constrained, failure at low stress levels will be initiated by longitudinal splitting at the ends in contact with the test platens. Similarly, while a plate of multi-ply laminate may be resistant to low-strain kinking failure in the plane of the plate because of the constraints imposed by fibres acting in the various directions, such a plate would be susceptible to lateral delamination and the compression strength would again not reflect the true tensile capability of the composite. It is interesting to note that compression strength is higher in composites reinforced with larger diameter fibres, and it is perhaps only the superior compression strength of boron-fibre composites, with fibres at least five times the diameter of carbon and glass fibres, that enables them still to be regarded as competitors to cfrp and grp for aeronautical structures.

Since the matrix and interface are inevitably sources of weakness in fibre composites any stress system which imposes substantial shearing forces on them will result in premature failure of the composite. Interlaminar shear failure at low load levels is therefore another characteristic of some reinforced materials that must be allowed for when designing with composites. It can result in catastrophic splitting along the neutral plane of a beam loaded in three-point bending if the span-to-depth ratio of the beam is not greater than $\sigma_c/2\tau_I$, where σ_c is the composite tensile strength and τ_I is its interlaminar shear strength (ILSS). The ILSS may be substantially increased by improving the fibre/matrix bond, which clearly indicates that it is the interface rather than the matrix itself, in reinforced plastics, that is the controlling factor. Conversely poor adhesion, or the presence of voids resulting from poor manufacturing conditions, or the diffusion of water from the environment into the interface region, will result in serious deterioration of the composite shear strength. By contrast with man-made composites the shear strengths of natural composites like wood and bone are much higher because of the more intimate association of the phases.

Toughness and impact resistance

The cracking process in ideally brittle solids like ceramics is uncomplicated, amounting to little more than the breaking of interatomic or intermolecular bonds. Measurements of the work of fracture, or fracture energy, γ_f, show that the process must be somewhat more complex than this, for instead of a value of 0.5 J m^{-2}, which we would calculate for bond-breaking, γ_f is usually about 5 J m^{-2}. In metals and plastics the path of a crack is still easily defined, but crack tip shear deformation mechanisms, plastic or viscoelastic, raise the work of fracture by many orders of magnitude over the simple bond-breaking value. In heterogeneous composite materials crack propagation is usually a much more complicated phenomenon because of the presence of many well-defined interfaces. There are two levels of interfacial discontinuity. At the microscopic level there is the discontinuity between fibres and matrix which is a consequence of their disparate chemical and physical characteristics: and at the macroscopic level there are discontinuities like the interlaminar regions in a laminate. In attempting to assess the work of fracture of a fibre composite, then, we must consider a number of separate contributions to the overall value of γ_f, some of which relate to the properties of the components and some of which relate to mechanisms of cracking peculiar to this class of materials.

In composites like grp and cfrp the fibres have works of fracture of only about $10\,\mathrm{J\,m^{-2}}$ while typical matrices like unmodified polyester and epoxy resins, which are not brittle by ceramics standards, but nevertheless classed as brittle materials, have γ_f values of the order of only $100\,\mathrm{J\,m^{-2}}$. Good quality glass/epoxy and carbon/epoxy laminates, however, may have works of fracture as high as $10^5\,\mathrm{J\,m^{-2}}$, clearly indicating the importance of the mechanistic contributions to γ_f. Conversely, if metals like aluminium or tough thermoplastics like Nylon, with fracture energies of 10^4 to $10^5\,\mathrm{J\,m^{-2}}$, are reinforced with high modulus, high strength, low failure-strain fibres their toughness may be seriously impaired because their capacity to sustain large amounts of plastic flow is reduced. Reinforcing filaments like steel, or the Terylene fibres used in coated fabrics for air-supported structures, or the Kevlar 49 fibres used to reinforce brittle resins, may be classed as tough because they show extensive non-elastic deformation after yield. They are tolerant of defects and surface damage, unlike glass and carbon, so that it is unnecessary to resort to statistical treatments of bundle failure, and a substantial part of their high fracture energy can be transferred into a composite. Other highly extensible fibrous reinforcing elements that have been conceived by Morley (1971) contribute to the toughness of a composite by the unravelling of a helical spring-like component after failure of an outer sheath. This has a certain similarity to the manner in which the 'filament wound' layers of a soft wood tracheid unravel following tensile buckling of the cell during the fracture of wood (Jeronimidis, 1976).

Toughness arising from fibre–matrix interaction

A crack propagating in the matrix phase is effectively halted by a fibre (Fig. 18)

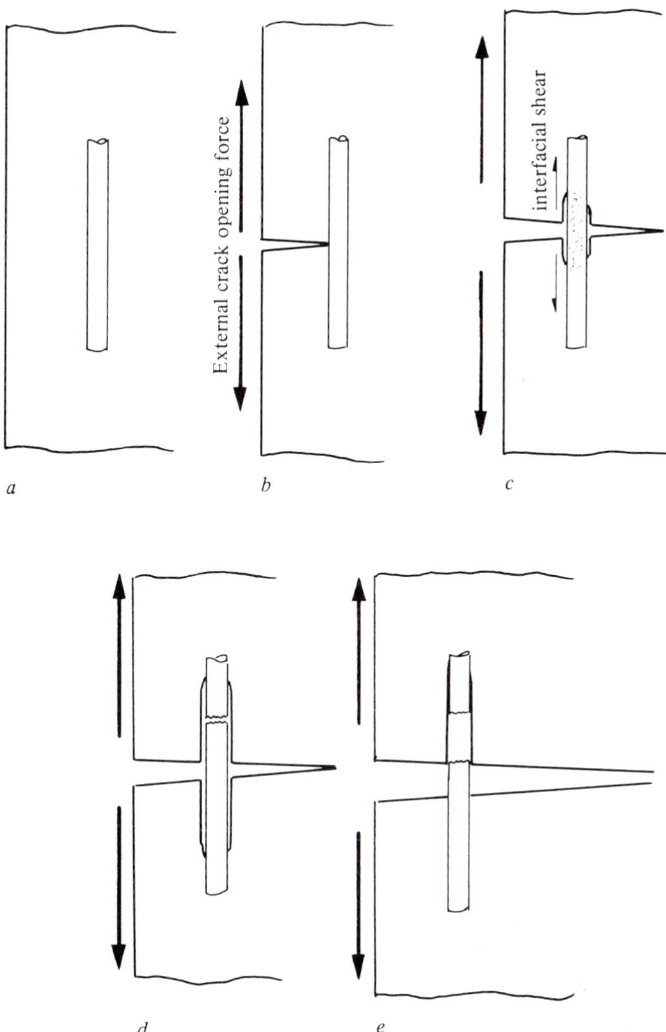

Fig. 18. Schematic illustration of the process of extension of a crack in a matrix containing a fibre. *a*, fibre gripped by shrinkage stresses; *b*, crack halted: opening displacement prevented by fibre; *c*, debonding: crack opening displacement increases and crack runs on. Strain energy stored in fibre; *d*, extensive debonding: fibre breaks in debonded region and cracks extend further; *e*, broken fibre still gripped: broken end pulled out, work being done against friction.

(a) because the higher stiffness of the fibre inhibits further opening of the matrix crack, and
(b) because the strength of the fibre is too high to permit of its being broken by the level of stress currently concentrated at the tip of the matrix crack.

The following discussion is based upon the treatment of Harris, Morley & Phillips (1975).

Debonding

As the load on the composite is increased, matrix and fibre at the crack tip attempt to deform differentially and a relatively large local stress begins to build up in the fibre. This stress causes local Poisson contraction which, exacerbated by the lateral concentration of tensile stress (normal to the interface) ahead of the crack tip (Cook & Gordon, 1964), may initiate fibre/resin decohesion, or debonding (Fig. 18*c*). The interfacial shear stress resulting from fibre/matrix modulus mismatch will then cause extension of the debond along the fibre in both directions away from the crack plane. This will permit further opening of the matrix crack beyond the fibre, and the process will be repeated at the next fibre. An upper limit to the energy of debonding (Outwater & Murphy, 1970) is given by the total elastic energy that will subsequently be stored in the fibre at its breaking load, $\sigma_f^2/2E_f$ per fibre per unit volume, or with N fibres bridging the crack,

$$W_{\text{Debond}} = \frac{N\pi d^2 \sigma_f^2 y}{8E_f}$$

for a composite with fibres of diameter d, breaking stress σ_f, and modulus E_f, y being the mean debonded length ($y/2$ on each side of the crack). Debonding can, in some circumstances, lead to the large-scale deviation of the crack tip parallel with the fibres, resulting in an effective blunting of the crack; cracking may then proceed on some other plane, remote from the original crack plane, with a resultant increase in the complexity of the fracture face and increase in composite toughness.

Frictional work following debonding

After debonding, the fibre and matrix move relative to each other as crack opening continues and work must be done against frictional resistance during the process. The magnitude of this work is hard to assess because we do not know with any certainty what level of frictional force is operating. One estimate, assuming that the interfacial frictional force, τ, acts over a distance equal to the fibre failure extension, suggests that this contribution

$$W_{\text{Friction}} = \frac{N\tau\pi dy^2\varepsilon_f}{2}$$

where ε_f is the fibre failure strain, contributes substantially to the toughness of glass/resin composites.

Fibre Pull-out

After debonding a continuous fibre is loaded to failure over a 'gauge length' equal to the debonded length and it may break at any point within this region (Fig. 18*d*), presumably with a strength statistically characteristic of that gauge length. The broken ends will retract and regain their original diameter, and they will be regripped by the resin. In order to permit further opening of the crack and, ultimately, to separate the two parts of the sample, these broken ends must be pulled out of the matrix (Fig. 18*e*). Further frictional work is needed to accomplish this, and the resulting fracture surface will often have the brushy appearance characteristic of many composites (Fig. 19). Crude estimates of the pull-out work give

$$W_{\text{pull-out}} = \frac{N\tau\pi dl_c^2}{12}$$

where the distance over which the fibre end is pulled out is given approximately (and probably incorrectly) by $l_c/4$. If the fibres are at an angle to the crack face, brittle ones will fail prematurely without much pull-out (Harris *et al.*, 1975) but plastically deformable ones will undergo extra work of shearing as they pull out, contributing extra toughness (Helfet & Harris, 1972).

It should be noted that composite toughness should always be increased by raising the fibre volume fraction, increasing the fibre diameter, or using stronger fibres. Contrary to expectation however, improving the fibre/matrix bond will usually reduce the toughness because it inhibits debonding and therefore reduces pull-out. The behaviour of many types of composite has been reasonably well explained in terms of some summation of the contributions from these mechanisms but it is not yet possible to design a composite to have a given toughness. Clearly ease of cracking (and therefore toughness) will be strongly dependent on orientation. The polar diagram in Fig. 20 shows, for example, that for grp and cfrp the fracture energy parallel

Fig. 19. Characteristic brushy fracture surface of a carbon-fibre composite showing fibre pull-outs. Original magnification ×30.

with the fibres is two orders of magnitude less than that normal to the fibres. It is common experience that the same is true of wood.

Toughness arising from laminate effects

Simple crack/fibre interactions rarely occur alone in practical composites made by laminating together sheets of pre-impregnated fibres. The variation of axes of anisotropy from lamina to lamina results in coupling shear stresses in the plane of the plate when the composite is loaded, and since the inter-laminar planes are always planes of weakness it follows that inter-laminar shearing stresses may easily become so large as to disrupt or delaminate the composite well before the overall tensile load on the material becomes high enough to cause fibre fracture. Cracks are also caused to deviate along these planes of weakness. Because of the increased surface area associated with this mode of cracking the resultant fracture mode may be a comparatively high

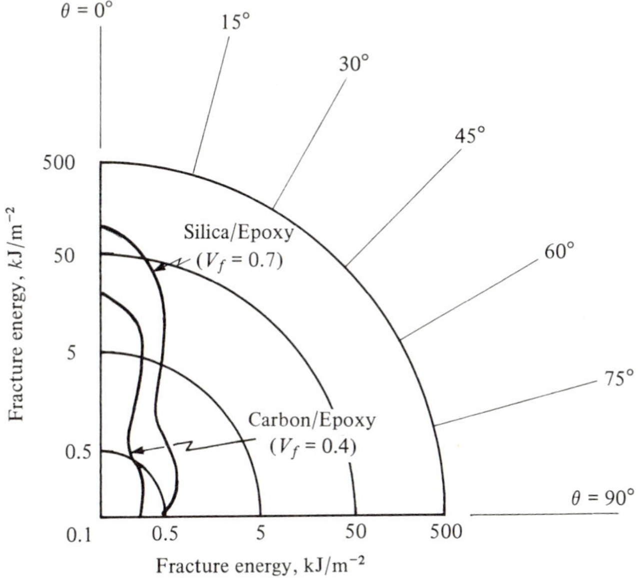

Fig. 20. Polar diagram showing the orientation-dependence of the fracture energy of unidirectional fibre composites. For $\theta = 0°$, the crack propagates perpendicular to the fibres. These results were obtained by Charpy impact tests on sharply notched samples. From Ellis & Harris, 1973.

energy mode and indeed experiments to encourage delamination have resulted in composites of higher toughness (Favre, 1977). It has been shown, for example, that in certain composites fracture accompanied by a single delamination can raise γ_f by a factor of three over the normal value for straightforward transverse crack propagation (Aveston, 1971). This kind of effect can also be seen in the fracture of wood which, because of its intermittent rate of growth, also exhibits a 'laminated' structure on the macroscopic scale. Cracking often seems to occur preferentially at late-wood/early-wood interfaces and fracture faces show evidence of roughness due to crack deviation in this way rather than to what was commonly thought to be tracheid pull-out (Fig. 21). It has also been found that the fracture-energy of woods such as Parana Pine, which does not show a clearly distinguished macroscopic ring-structure, is lower than that of temperate soft wood like Scots Pine in which the annual rings are more marked.

Attempts have also been made to analyse the fracture behaviour of compact bone in terms of composite failure mechanisms because of the apparently similar behaviour to grp. In any small volume of bone the collagen fibrils are roughly parallel and are surrounded by and filled

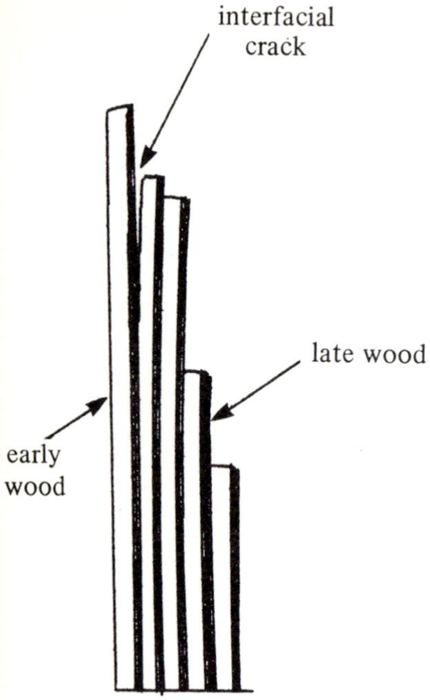

interfacial crack

late wood

early wood

a

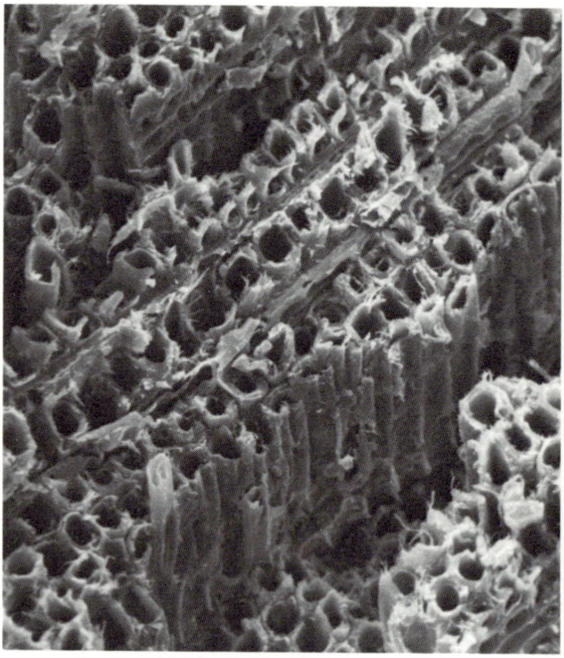

b × 70

c × 1200

with the hydroxyapatite needles (or platelets) which are said to be orientated with their long axes parallel to the local direction of the collagen fibrils. A variety of composite micro-structures occurs in woven and lamellar bone. A sample of bone, translucent prior to deformation, gradually becomes opaque throughout the tensile region during testing in a manner, identical with the behaviour of grp, that suggests the accumulation of damage by many, widely-distributed micro-failure events (Currey & Brear, 1974). In grp these events are fibre/resin debonds and fibre failures, and it has been suggested that in bone the corresponding mechanism is delamination of the Haversian systems (Cooke, Zeidman & Scheitele, 1973). At slow rates of crack growth high fracture energies are measured and the resulting rough surfaces have indicated that the high γ_f values are due to debonding at the weak osteon/matrix interfaces with subsequent pull-out of osteons (Pope & Murphy, 1974). It is also clear that in lamellar structures interlamellar weakness results in splitting of the kind found in grp laminates and the accompanying Cook–Gordon crack-tip blunting. The ability of the non-mineral components of bone to undergo substantial inelastic deformation before failure must also contribute to the large fracture energy of the material, and it is by no means certain what the relative contributions from these mechanisms are.

The fracture energies of wood and bone are both approximately 10 kJ m^{-2}. This is low by comparison with those of the best continuous fibre grp and cfrp, about 100 kJ m^{-2}, but much better than those of ordinary glassy plastics like PMMA (about 1 kJ m^{-2}) and chopped glass fibre polyester moulding compounds (up to 5 kJ m^{-2}).

Conclusion

This discussion has been limited to considerations of elastic behaviour, strength and toughness of structural composite materials, but this should by no means be thought to imply that time-dependent processes like creep and dynamic fatigue are not equally important. A great deal of experimental work has been done on synthetic composites in order to test theoretical predictions and, where appropriate, to set up model systems to determine specific micro-mechanical parameters required by the theories. Model work has also given useful insights into the

---

Fig. 21. *a*, schematic illustration of a typical interfacial fracture in Scots Pine; *b*, stepped fracture in Scots Pine late wood; *c*, helical fracture in Scots Pine late wood. Illustrations from Ansell & Harris, 1979.

mechanisms of composite behaviour on the micro-structural scale. This work has been reasonably successful. Attempts to apply composite theories to biological materials have so far been less successful, and perhaps this is not surprising as it is far less easy to set up sensible model experiments. As Wainwright *et al.* (1976) have shown, however, an important start has been made, and the prognostications are excellent.

References

ALLEY, V. L. & FAISON, R. W. (1972). Experimental investigation of strains in fabric under biaxial and shear forces. *Aircraft,* **9,** 55–60, 211–16.

ANSELL, M. P. & HARRIS, B. (1979). The relationship between toughness and fracture surface topography in wood and composites, *Proc. Third International Conference on Mechanical Behaviour of Materials (ICM3), Cambridge, August 1979, Vol. 3, pp. 309–18.* Oxford: Pergamon.

AZZI, V. D. & TSAI, S. W. (1965). Anisotropic strength of composites. *Experimental Mechanics,* **5,** 283–8.

AVESTON, J. (1971). Strength and toughness in fibre-reinforced ceramics, *Proceedings of the NPL Conference on the Properties of Fibre Composites*, pp. 63–73. London: IPC Press.

BADER, M. & BOWYER, W. H. (1973). An improved method of production for high strength fibre reinforced thermoplastics. *Composites,* **4,** 150–6.

BLACK, M. M., MELCHER, D. H., MELVILLE, H. A. H. & ALDERMAN, B. (1975). Developments in cervical dilation including the use of vibratory techniques. *Clinics in Obstetrics and Gynaecology,* **2,** 173–201.

COOK, J. & GORDON, J. E. (1964). A mechanism for the control of crack propagation in all-brittle systems. *Proceedings of the Royal Society of London, Series A,* **282,** 508–20.

COOKE, F. W., ZEIDMAN, H. & SCHEITELE, S. J. (1973). The fracture mechanics of bone – another look at composite modelling. *Journal of Biomedical Materials Research Symposium,* **4,** pp. 383–99. New York: Wiley.

CORTEN, H. T. (1967). Micromechanics and fracture behaviour of composites. In *Modern Composite Materials*, ed. L. J. Broutman & R. H. Krock, pp. 27–105. New York. Addison-Wesley.

COX, H. L. (1952). Elasticity and strength of paper and other fibrous materials. *British Journal of Applied Physics,* **3,** 72–9.

CURREY, J. D. & BREAR, K. (1974). Tensile yield in bone. *Calcified Tissue Research,* **15,** 173–9.

DIMMOCK, J. & ABRAHAMS, M. (1968). Prediction of composite properties from fibre and matrix properties. *Composites,* **1,** 87–93.

ECKOLD, G. C., LEADBETTER, D., SODEN, P. D. & GRIGGS, P. R. (1978). Lamination theory in the prediction of failure envelopes for filament wound materials subject to biaxial loading. *Composites,* **9,** 243–6.

ELLIS, C. D. & HARRIS, B. (1973). The effect of specimen and testing variables on the fracture of some fibre reinforced epoxy resins. *Journal of Composite Materials,* **7,** 76–88.

FAVRE, J.-P. (1977). Improving the fracture energy of carbon fibre reinforced plastics by delamination promoters. *Journal of Materials Science,* **12,** 43–50.

DE FERRAN, E. M. & HARRIS, B. (1970). Compression strength of polyester resin reinforced with steel wires. *Journal of Composite Materials,* **4,** 62–72.

FOYE, R. L. (1966). *Compression Strength of Unidirectional Composites*, paper to the AIAA Structural Composites Group. Columbus, Ohio: North American Aviation Inc.

GOATHAM, J. I. (1970). Materials problems in the design of compressor blades with fibrous composites. *Proceedings of the Royal Society of London, Series A,* **319,** 45–57.

HARRIS, B., GUILD, F. J. & BROWN, C. R. (1979). Accumulation of damage in GRP laminates, *Journal of Physics D: Applied Physics*, **12,** 1385–407.

HARRIS, B., MORLEY, J. & PHILLIPS, D. C. (1975). Fracture mechanisms in glass reinforced plastics. *Journal of Materials Science,* **10,** 2050–2061.

HAYASHI, T. & KOYAMA, K. (1971). Theory and experiments of compressive strength of unidirectionally fibre reinforced composite materials. *Proceedings of an International Conference on Mechanical Behaviour of Materials, Kyoto, 1971, Society of Materials Science, (Japan),* **5,** 104–12.

HELFET, J. & HARRIS, B. (1972). Fracture toughness of composites reinforced with discontinuous fibres. *Journal of Materials Science,* **7,** 494–8.

HILL, R. (1964). Theory of mechanical properties of fibre strengthened materials. *Journal of Mechanics and Physics of Solids,* **12,** 199–218.

JERONIMIDIS, G. (1976). The fracture of wood in relation to its structure. *Leiden Botanical Series,* **3,** 253–65.

JONES, R. M. (1975). *Mechanics of Composite Materials.* Washington D.C.: Scripta Book Co.

KASTELIC, J. & BAER, E. (1980). Relationships between structural hierarchies and mechanical properties of tendon collagen. In *The Mechanical Properties of Biological Materials* (34th Symposium of the Society for Experimental Biology) ed. J. F. V. Vincent & J. D. Currey, pp. 395–435. Cambridge University Press.

KATZ, J. L. (1980). The structure and biomechanics of bone. In *The Mechanical Properties of Biological Materials* (34th Symposium of the Society for Experimental Biology) ed. J. F. V. Vincent & J. D. Currey, pp. 137–168. Cambridge University Press.

KELLY, A. (1973). *Strong Solids* 2nd Edn. Oxford University Press.

KELLY, A. & DAVIES, G. J. (1965). The principles of the fibre reinforcement of metals, *Metallurgical Review,* **10,** 1–77.

KRENCHEL, H. (1964). *Fibre Reinforcement.* Copenhagen: Akodemisk Forlag.

LAGER, J. R. & JUNE, R. R. (1969). Compressive strength of boron–epoxy composites. *Journal of Composite Materials,* **3,** 28–56.

MORLEY, J. G. (1971). Duplex fibre reinforced composites. *Proceedings of the NPL Conference on the Properties of Fibre Composites*, pp. 33–5. London: IPC Press.

NIELSEN, L. E. & CHEN, P. E. (1968). Young's modulus of composites filled with randomly oriented fibres. *Journal of Materials,* **3,** 352–8.

OUTWATER, J. O. & MURPHY, M. C. (1970). Fracture energy of unidirectional laminates. *Modern Plastics,* **47,** 160–9.

PHILLIPS, D. C. & HARRIS, B. (1977). Strength, toughness and fatigue properties of composites. In *Polymer Engineering Composites,* ed. M. O. W. Richardson, pp. 45–154. Applied Science Publishers.

PIGGOTT, M. R. & HARRIS, B. (1981). Compression strength of fibre reinforced polyester resins, *Journal of Materials Science*, in press.

POPE, M. H. & MURPHY, M. C. (1974). Fracture energy of bone in a shear mode. *Medical and Biological Engineering,* **1,** 763–7.

ROSEN, B. W. (1964). Tensile failure of fibrous composites. *A.I.A.A. Journal,* **2,** 1985–91.

ROSEN, B. W. (1965), Mechanics of composite strengthening. In *Fibre Composite Materials,* pp. 37–75. American Society for Metals.

ROYAL SOCIETY, LONDON (1980). Proceedings of a Discussion Meeting on *Newer fibres and their composites, May 1978*, ed. W. Watt, B. Harris & A. Ham.

SHANAHAN, W. J., LLOYD, D. W. & HEARLE, J. W. S. (1978). Characterising elastic behaviour of textile fabrics in complex deformations. *Textile Research Journal,* **48,** 495–505.

SODEN, P. D. & MCLEISH, R. D. (1976). Variable affecting the strength of balsa wood.

Journal of Strain Analysis, **11,** 225–34.

Soden, P. D., Leadbetter, D., Griggs, P. R. & Eckold, G. C. (1978). The strength of a filament wound composite under biaxial loading. *Composites,* **9,** 247–50.

Stowell, E. Z. & Liu, T. S. (1961). On the mechanical behaviour of fibre reinforced crystalline materials. *Journal of Mechanics and Physics of Solids,* **9,** 242–60.

Swanson, S. A. V. (1980). Articular cartilage. In *The Mechanical Properties of Biological Materials* (34th Symposium of the Society for Experimental Biology) ed. J. F. V. Vincent & J. D. Currey, pp. 377–95. Cambridge University Press.

Topping, A. D. (1961). An introduction to biaxial stress problems in fabric structures. *Aerospace Engineering,* **20,** 18, 19, 53–8.

Tsai, S. W. (1964). Structural behaviour of composite materials, NASA report CR–71, July 1964. Washington: NASA.

Tsai, S. W. (1968). Strength theories of filamentary structures. In *Fundamental Aspects of Fibre Reinforced Plastics Composites*, ed. R. T. Schwartz & H. S. Schwarts, pp. 3–11. New York: Interscience.

Tsai, S. W., Adams, D. F. & Doner, D. R. (1966). Analysis of composite structures, NASA report CR–620, Nov. 1966. Washington: NASA.

Wainwright, S. A., Biggs, W. D., Currey, J. D. & Gosline, J. M. (1976). *Mechanical Design in Organisms*, London: Edward Arnold.

Weibull, W. (1951). A statistical distribution function of wide applicability. *Journal of Applied Mechanics,* **18,** 293–7.

Zweben, C. & Rosen, B. W. (1970). A statistical theory of material strength with application to composite materials. *Journal of Mechanics and Physics of Solids,* **18,** 189–206.

MECHANICAL PROPERTIES OF MOLLUSC SHELL

J. D. CURREY

Department of Biology, University of York, York YO1 5DD, UK

Introduction

It is a commonplace that the mechanical properties of a structure are determined both by the mechanical properties of its constituent materials and also by the way this material is arranged in space: the *build* of the structure. This commonplace, though true of skeletons of all animal groups, is exemplified particularly clearly by the shells of molluscs. The reason for this is that there are various types of molluscan shell material, all made of roughly the same chemical constituents, with characteristically different mechanical properties. These types of material may be made into shells that are in some cases very different, and in some cases very similar in their external form. In this review I shall briefly describe the various types of material and their mechanical properties, as far as they are known. Then I shall attempt to explain some of these properties in relation to their microstructure, and I shall try to show how, in some cases, we can begin to understand why some types of material are found in some shells and others in others.

Structural types of molluscan shell material

A typical mollusc shell has an outer organic layer, the periostracum, which is mainly protein. This acts as a basis on which the calcified part of the shell is laid down. By far the greater part of the molluscan shell is calcified. The mineral is calcium carbonate; it is usually in the form of aragonite, rather less commonly of calcite and occasionally it is amorphous. There is always an organic component, mainly proteinaceous, which varies, in different structural types, from about 0.1% to 5% by weight. The composition of the organic material is discussed by Wilbur & Simkiss (1968) and Grégoire (1972). The amino acids with small side-chains, glycine and alanine, predominate, accounting for about half the residues.

Molluscan shell is arranged in one of a number of rather well characterised types. Some of these types may intergrade to some extent, but in general a glance at a scanning electron-microscope image allows one to classify a shell type. Although each of the structures is found in more than one of the six molluscan classes, we shall discuss only those found in the three major classes: the cephalopods, the gastropods, and the bivalves.

The structural types of gastropod (snail) shells are discussed fully by Bøggild (1930) and those of bivalve (clam) shells by Taylor, Kennedy & Hall (1969). Here follows a very simplified description.

Prisms (Fig. 1*a*). Polygonal columnar crystals, of calcite or aragonite, run normally to the surface of the shell. The crystals are quite large, 10–200 μm across, and they may be several millimetres long. The organic sheet surrounding each crystal may be thick, up to 5 μm or so.

Nacre (mother of pearl) (Fig. 1*b*). The crystallites are flat tablets of aragonite, arranged in sheets. The sheets are more or less parallel to the surface of the shell. The tablets, and therefore the sheets, are very thin, about 0.3–0.5 μm. Between each pair of sheets is a thin organic layer, which also extends up between the tablets. The tablets in neighbouring sheets are staggered in relation to each other so there is no direct pathway through the sheets.

Crossed-lamellar (Fig. 1*c*). This is a lamellate arrangement rather like plywood in which each lamella, about 20 μm thick, is built of long aragonite needles all orientated the same way. In adjacent lamellae the predominant direction alters by about 90°. The protein matrix is very tenuous.

Foliated (Fig. 1*d*). This is made of long lath-like calcite crystals arranged in overlapping layers, with some change in orientation between layers.

Homogeneous (Fig. 1*e*). This is made of very small granules of aragonite, 0.5–3 μm in diameter, with no particular orientation. It is really a very fine-scale rubble. The organic matrix is extremely tenuous.

There are some other structural types which will not be discussed, since they are relatively less widespread, and we know nothing of their mechanical properties. The various types of structure shown in Fig. 1 are not dotted around at random. In the bivalves, for instance, there are three characteristic arrangements: prisms on the outside with nacre on the inside, foliated with a little crossed-lamellar, and crossed lamellar on its own or with some homogeneous material (Taylor, Kennedy & Hall, 1973). There are other arrangements, but these three between them make up the great bulk of bivalve shells. The structural type employed is

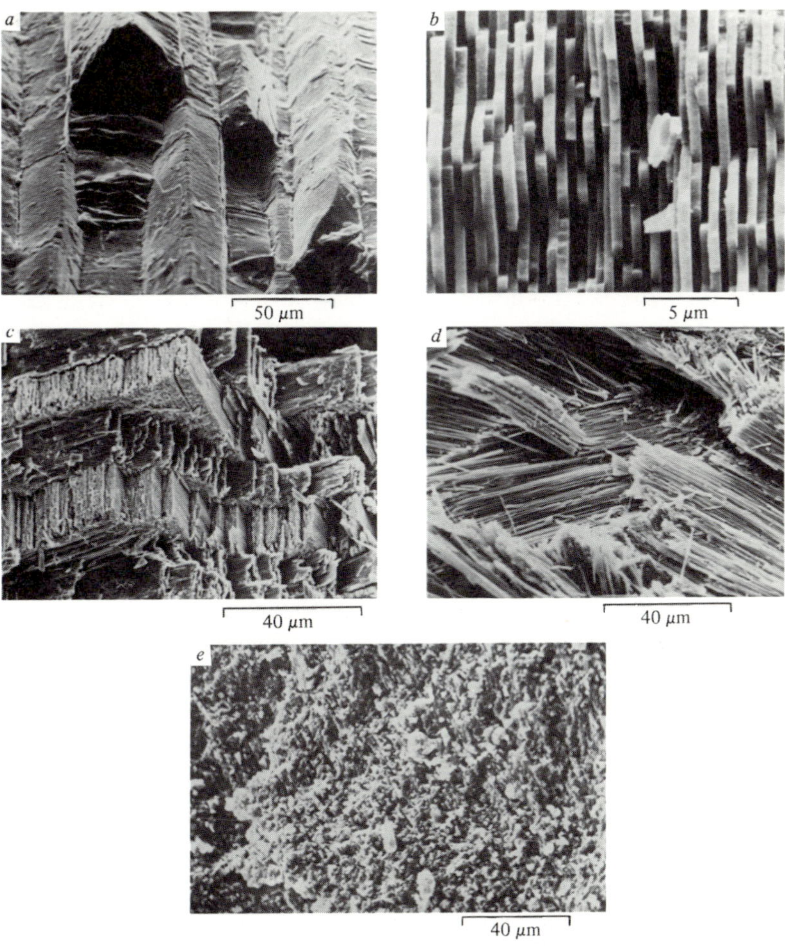

Fig. 1. Scanning electron micrographs, at various magnifications, of fracture surfaces of various structural types. *a*, prisms; *b*, nacre; *c*, crossed-lamellar; *d*, foliated; *e*, homogeneous.

usually constant within a family, and sometimes within a superfamily. It is not possible to make many generalisations concerning the types of shells in which different structural types are found, though very thin shells tend to be made of prisms plus nacre or to be foliated, and very thick ones to be of crossed-lamellar type. However, as we shall see later, it is possible to make some generalisations concerning the life habits of the animals using particular types.

These structural types are, of course, related in that they are all the product of crystallisation of calcium carbonate from a solution containing some organic material. Taylor (1973) has suggested ways in which they may also be evolutionarily related. Prisms with nacre seem to be the earliest type, and Taylor shows how the others can have evolved from it by changes in the conditions of crystallisation such as speed and the amount of supersaturation of the liquid. Much more work needs to be done on this most interesting topic.

Mechanical properties of molluscan shell materials

Molluscan shell presents some problems in mechanical testing.

(*a*). Most obvious is that testing is confined to fairly large shells. Even a shell that is several centimetres long may be useless for testing because the radius of curvature of the shell is everywhere so small. This difficulty is compounded when one tries to obtain information about anisotropic behaviour, because then one may be testing in a direction in which even a large shell can produce only very small specimens.

(*b*). Obtaining a uniform specimen can be awkward, and at times it is problematical whether a uniform sample is appropriate. This is seen well in crossed-lamellar structure, which has a fine-grained layering which alters in direction every 20 μm or so. It would seem sensible to test specimens containing many such layers; indeed it would be difficult not to. However, as explained below, many crossed-lamellar shells have also a much larger-scale alteration of direction of the lamellae. It is doubtful to what extent one should include only one, or several of these layers in a test specimen.

(*c*). Many specimens are difficult to machine into a reasonably standard shape, without inducing stress concentrations, because they are so weak. This is particularly true of specimens of homogeneous structure, and of the chalky deposits found in part of the shell of oysters.

(*d*). Finally, many of the specimens so far tested come from shells that have, at some time or another, been allowed to dry out; they are usually museum specimens. Although they can be wetted before being tested, it is possible that cracks may have developed in them while they were dry, or that their protein component may have been altered in a way affecting the mechanical properties of the material. Currey (1979) showed that, for some types of material, drying is unlikely to have a large effect on the strength of specimens. Nevertheless, this point does remain a worry.

Table 1. *Mechanical properties of molluscan shell materials. Each value is the mean for the species*

Class*	Species	Type†	Tensile Strength (MNm^{-2})	Compressive Strength (MNm^{-2})	Bending Strength (MNm^{-2})	Modulus of Elasticity (GNm^{-2})
G	*Turbo marmoratus*	N	116	353	267	54
G	*Trochus niloticus*	N	85	320	220	64
C	*Nautilus pompilius*	N	78	401	193	47
B	*Hyria ligatus*	N	79	382	211	44
B	*Anodonta cygnaea*	N	35	322	117	44
B	*Pinctada margaritifera*	N	87	419	208	34
B	*Atrina vexillum*	N	86	304	173	58
B	*Modiolus modiolus*	N	56	416	199	31
G	*Conus miles*	XL	46	278	63	50
G	*C. leopardus*	XL	31	297	130	56
G	*C. prometheus*	XL	38	271	134	58
G	*C. striatus*	XL	40	336	108	–
G	*C. virgo*	XL	52	323	165	–
G	*C. litteratus*	XL	49	301	80	–
G	*Cypraea tigris*	XL	25	208	150	41
G	*Strombus gigas*	XL	35	198	78	41
G	*S. costatus*	XL	60	280	58	49
G	*Lambis lambis*	XL	41	217	66	39
B	*Hippopus hippopus*	XL	9	229	35	50
B	*Mercenaria mercenaria*	XL	22	336	95	66
B	*Egeria* sp.	XL	43	163	106	77
B	*Ensis siliqua*	XL	–	196	85	55
B	*Chama lazarus*	CXL	–	222	36	82
B	*Pinna muricata*	P	62	210	–	12
B	*Atrina vexillum*	P	60	295	139	39
B	*Ostrea edulis*	F	–	82	93	34
B	*Pecten maximus*	F	42	133	110	30
B	*Saccostrea cucullata*	F	40	74	44	29
G	*Patella mexicana*	F	33	208	171	60
B	*Arctica islandica*	H	30	248	60	60

* G, Gastropoda; B, Bivalvia; C, Cephalopoda.
† N, Nacre; XL, crossed-lamellar; CXL, Complex crossed-lamellar; P, prisms; F, foliated; H, homogeneous.

These points must be borne in mind when examining the results illustrated in Table 1 and Fig. 2. In particular, some materials, particularly homogeneous ones, are grossly under-represented because it proved impossible to machine respectable specimens from them.

Some generalisations can be made about these results. There is considerable variation both within species and within structural types.

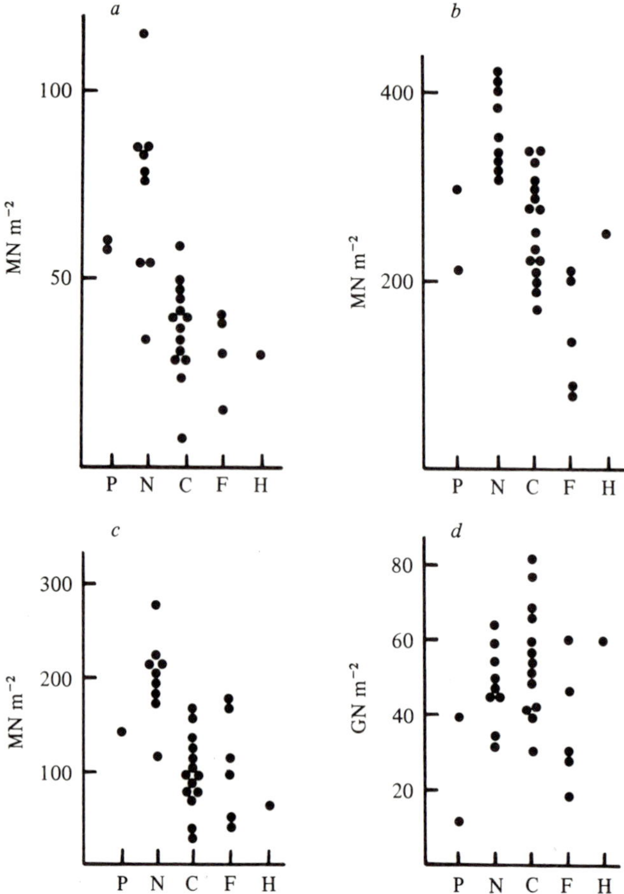

Fig. 2. Mechanical properties of different structural types. Each circle represents the mean value of all test specimens from a particular species. P, Prisms; N, Nacre; C, Crossed-lamellar; F, Foliated; H, Homogeneous. *a*, Tension, Tensile strength; *b*, Compression, Compressive strength; *c*, Bending, Bending strength; *d*, Stiffness, Young's modulus of elasticity.

Nacre is in general considerably stronger in tension, compression and bending than the other types, though the difference is less well marked in compression. In tension and bending there is little to choose between the other structural types, but in compression foliated shell is weaker than the others.

The values for modulus of elasticity are rather unsatisfactory, the variation within species being rather high. However, there is some indication that crossed-lamellar structure is rather stiffer than the others, and the prisms of *Pinna muricata* are definitely very compliant.

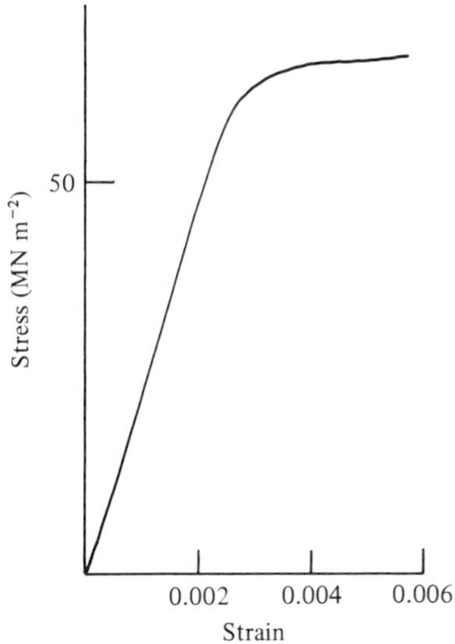

Fig. 3. Stress–strain curve of nacre in tension.

Perhaps the most interesting point about the mechanical behaviour of the various types is that the strongest, nacre, is not the type most commonly found in the molluscs and, indeed, seems to be more primitive than crossed-lamellar in that groups with nacre seem to have changed less from the earlier molluscs than have those with cross-lamellar (Taylor *et al.*, 1973; Taylor, 1973). Another point of great interest is that nacre, although being composed almost entirely of calcium carbonate, is rather strong, stronger than naturally occurring ceramics. For example, jade, bending strength about 70 MNm^{-2} (Rawcliffe & Fruhauf, 1977), flint 180 MNm^{-2}, slate 100 MNm^{-2}, chalcedony 165 MNm^{-2} (Iler, 1963), granite 150 MNm^{-2} (Gillam, 1969). Limestone never seems to reach high values. It is instructive to see how the strength of mollusc shell is achieved, and to compare the fracture process in nacre with that in crossed-lamellar structure.

Tensile and bending fracture in Nacre

Most of the work discussed here is from Currey (1977). A load–deformation curve of nacre loaded in tension usually shows an elastic region followed by a plastic region (Fig. 3). Typically there is a rather sharp yield point, occurring at a strain of about 0.002. The plastic portion of the curve (which has been shown to be plastic in that there is a residual strain if the load is removed) indicates a plastic strain of about

0.01. However, it is almost certain that locally, over gauge lengths of 1 mm or so, the plastic strain would be greater than this.

Nacre loaded in tension shows optical effects associated with the onset of plasticity. The material becomes opaque or reflecting, depending on whether it is seen by transmitted or reflected light, in the highly strained regions. Such effects remain, for the most part, when the load is removed (Fig. 4). Unfortunately, we have not been able to determine what is happening at the submicroscopic level, but it is very probable that the change in optical properties is caused by the appearance of tiny cracks or interfaces in the material.

The fracture surface is rough at the submicroscopic level (Fig. 1*b*). A most important question is: does the crack travel through the calcium carbonate blocks, or between them? We were able to investigate this by looking at a part of a specimen at the root of a severe stress concentration, which was the site of crack initiation, before and after loading. The crack travelled through 65 sheets of the nacre, and in all but seven of the sheets it went between the blocks, not through them. In order to do this the crack had to deviate markedly from its general direction of travel. It seems clear, therefore, that it is failure of the protein matrix, presumably mostly in shear, that allows the crack to travel.

There are several structural features of nacre that make it strong:

(*a*). The thickness of the sheets. The sheets are of the order of 0.5 μm thick in the usual direction of crack travel. Taking the tensile stress at fracture as 100 MNm^{-2}, we can calculate the critical crack length as $c = 2\gamma E/(\pi\sigma^2)$. Unfortunately the surface energy, γ, of aragonite seems not to be known. Gilman (1960) gives the surface energy of *calcite*, which is probably not too different, as 0.23 Jm^{-2}. Hearmon (1946) reports the modulus of elasticity of aragonite to be 100 GNm^{-2}. The critical crack length, c, is therefore about 1.5 μm. This is a very rough calculation, but it shows that the sheets are not thick enough to contain dangerous flaws.

(*b*). The uniformity of thickness of the plates. A low average thickness of the sheets would be no good if there were an occasional very

---

Fig. 4. Optical effects in nacre. A prismatic specimen is loaded in three-point bending. The pictures are stills from a cine film. The load is increasing from the top (unloaded) to the fourth. In the fifth the specimen has broken and is flying apart. The diffuse patches of light seen in the top picture are irrelevant reflections. The optical effects associated with plastic deformation run across the specimen and, as seen in the bottom, remain after fracture.

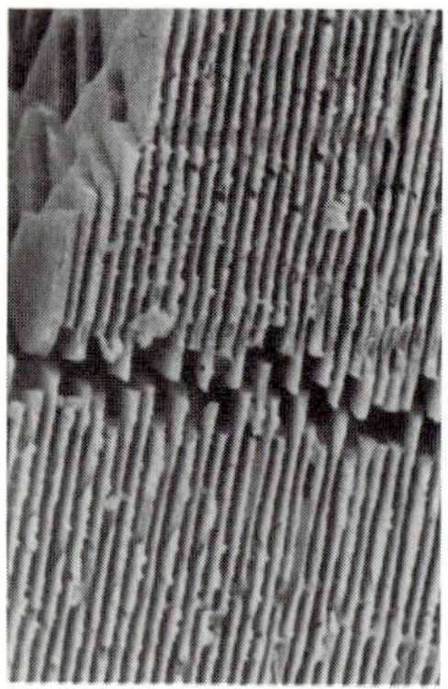

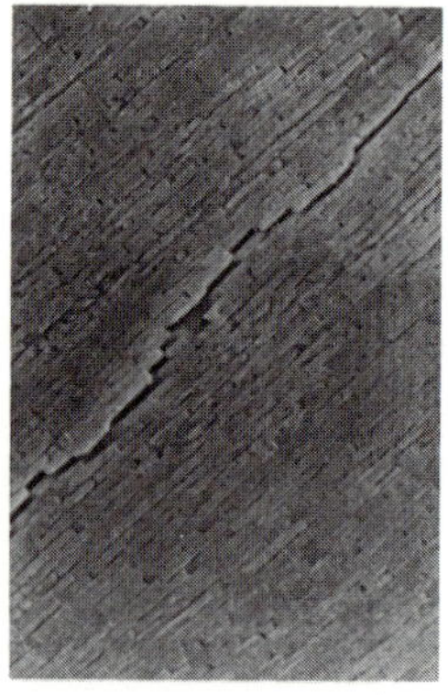

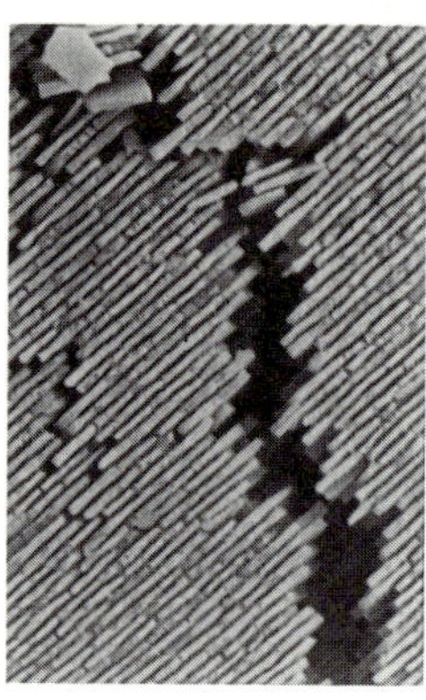

Fig. 5. Polished, etched, specimens of nacre which have been fractured so that the general direction of crack travel is different. Note how the crack is much more direct when it runs predominantly between the plates than when it has to run across them.

thick sheet, because dangerous cracks could start in such a sheet. However, in most nacres I have examined the coefficient of variation (standard deviation/mean) of thickness is between 0.15 and 0.2, implying that 95% of all sheets are within 30% of the mean values, and there seem never to be any really thick sheets.

(*c*). The staggered arrangement of the blocks. The crack travelling through nacre infrequently goes through the calcium carbonate itself. Because the blocks in neighbouring layers are not laid with their edges in line, a crack must travel through a series of nearly right-angled bends to get through the material. Each time the crack comes to the broad face of a tablet it will be blunted. This will make it difficult for the crack to continue, both because the crack growth will be energetically unfavourable, and because the stress concentration at the tip will become small (Cook & Gordon, 1964; Kendall, 1975). The shape of the resulting crack is shown in Fig. 5. Using a scanning electron microscope we can examine the relation between the angle a crack makes with the plane in which the sheets lie and the actual length of the crack, including all the deviations. Table 2 shows this for *Pinctada margaritifera*, the pearl oyster.

The total length of the path, and hence the work that must be done to disrupt the protein, increases markedly with angle. In fact the work of fracture of nacre is markedly anisotropic, being about 1600 Jm^{-2} when

Table 2. *Relation between angle of crack travel and the tortuosity of the crack path in nacre, for* Pinctada margaritifera, *the pearl oyster*

Angle between the direct fracture path and the plane of the sheets	$\frac{\text{Length of actual path}}{\text{Length of direct path}}$
2°	1.10
14°	1.25
61°	2.36
90°	2.57

the crack travels across the sheets, and 150 Jm^{-2} when it travels between them. At least part of this difference must be attributable to the greater amount of matrix that must be sheared when the crack travels across the sheets.

The crack-stopping abilities of nacre are seen clearly if a specimen is loaded in bending with the prisms, which are usually found outside the nacre in a complete shell, left in place on the tension side. The load–deformation curve is linear until, at a fairly low load, there is a sudden drop in load of 10–15%; as the load is increased the curve rises and then drops sharply again, in a series of saw-teeth. If the specimen is removed when the load drops, it is found that the drop is associated with a crack that has travelled right through the prismatic layer, presumably very rapidly, and which has then been brought to a halt in the nacre within 50–100 μm, having travelled through the prisms possibly for several millimetres. Nacre loaded on its own in bending shows a load–deformation curve rather like bone, with considerable plastic deformation after an initial linear region. Of course the amount of plasticity is not as great as in bone.

Fracture in crossed-lamellar structure

Shells made from nacre tend to be rather thin-walled in relation to their overall size. The sheets of the nacre conform in their general orientation to the shape of the shell. Because most cracks run through the thickness of the shell, they must break the nacre in the most difficult direction. This is not necessarily the case with shells made of crossed-lamellar structure. In the tropical cone shells of the genus *Conus*, for instance, the shell usually has three layers (Fig. 6). In all layers the lamellae run

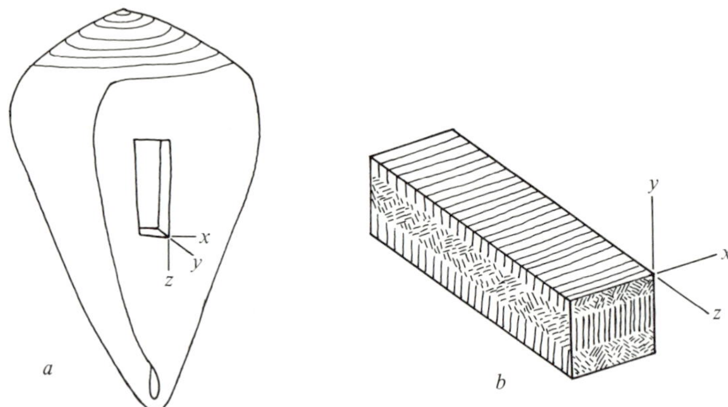

Fig. 6. *a*. Diagram of a *Conus* shell. A specimen has been cut out. *b*. Diagram of the specimen cut from the whole shell. The lamellae of the inner and outer layers lie in the *XY* planes; those of the middle layer in the *ZY* planes.

parallel to the thickness of the shell wall (the *Y* direction). In the inner and outer of the three layers, however, the lamellae run towards the lip, and so lie in to the *XY* plane. In the middle layer, the lamellae run parallel to the lip, and are, therefore, in the *ZY* plane. This arrangement of the layers obviously has mechanical implications. The mechanical properties of cone shells were investigated by Currey & Kohn (1976), and much of the following account is based on their work. The results given in Table 1 for *Conus* refer only to specimens orientated with their longitudinal axes in the *Z* direction, because this was usually the only orientation which would yield large enough specimens. Furthermore, the specimens were machined to include a large proportion of the thickness of the shell wall, without regard for the orientation of the layers in the specimens. In particular, a bending specimen will have the greatest tensile stresses across the *XY* planes. If the layers on the outside are parallel to these planes, it is likely that they will split rather easily.

We found that this was indeed so. The load–deformation curve in bending has a characteristic shape (Fig. 7). The first part of the curve is saw-toothed. We found that each of the little dips, of which there were five on average, was associated with the appearance of a new crack that ran from the tensile surface of the specimen into the middle layer, where it stopped. (This is closely analogous, of course, to the situation when cracks travelling through prisms are brought to a halt after a few tens of microns of travel in nacre.) With each new crack the specimen became more compliant. Finally, the load–deformation curve had a fairly

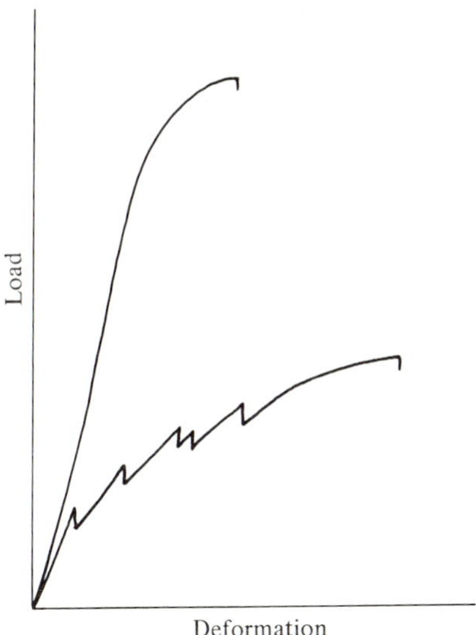

Fig. 7. Load–deformation curve for crossed-lamellar structure. Lower curve: tensile surface has lamellae in the *XY* planes (the 'easy' directions). Upper curve: tensile surface has lamellae in the *YZ* planes (the 'difficult' directions).

smooth region before catastrophic failure occurred. This smooth region was associated with the slight extension of one of the cracks into the middle layer. This crack would suddenly extend, and the specimen fail.

The implication of these observations is that crossed-lamellar structure is weak if loaded so that the tensile stresses act *across* the planes in which the layers lie. We checked that this was the case in two experiments. In one we prepared matched pairs of bending specimens, of similar size. One member of each pair had the inner layer intact except that it was machined flat. The other member of the pair had the inner layer completely ground off. In the latter specimens the greatest tensile stresses would be across the *XY* planes, but the part of the specimen subjected to the stresses would have layers orientated in the *ZY* planes. The mean bending strength of the standard specimens was 54 MNm^{-2}, but in those in which the middle layer was subjected to the greatest stress the bending strength was 195 MNm^{-2}. The latter value is greater than the mean value for any species with crossed-lamellar structure that we tested in the ordinary way. It could, perhaps, be argued that the middle layer was for some reason inherently stronger than the inner layer, and that anisotropy is not important. However, the following experiment shows that this is not the case. *Conus striatus* has a sufficiently large shell for it to be possible to test specimens prepared with their long axes in the *X* direction, as well as specimens orientated in the *Z* direction. The mean value of bending strength for the latter was 108 MNm^{-2}, for those cut in the *X* direction the mean value was 207 MNm^{-2}.

The load–deformation curves for these strong specimens are quite unlike those of the weak ones. The curve is initially straight, with no saw-teeth. Towards the greatest load the curve bends over somewhat, but there is no clear yield point (Fig. 7). Crossed-lamellar specimens, however, never showed the quite long, almost horizontal, region of the load–deformation curve that was often seen in nacre in bending.

These experiments show that in bending crossed-lamellar structure is rather anisotropic and that, in some orientations, it has a bending strength about the same as that of nacre. (However, nacre is not anisotropic in this way, it is equally strong whatever the orientation of the specimen in the *XZ* plane.) The reason for this anisotropy is fairly obvious from a consideration of the structure and an examination of fracture surfaces. Tension in one orientation is like pulling apart sheets of plywood. Once a crack has started to run, there is little to prevent it going further. There are no crack-blunting mechanisms and no devices for making the crack alter its direction. The situation is analogous to that seen where sheets of nacre were pulled apart in work of fracture experiments. The fracture surface in this direction is rather smooth, with only small steps and discontinuities on it (Fig. 8). When the direction of the layers changes, however, things are more complex. In general, for any short length on the crack front there will be a direction, at about 45° from the direction in which it was travelling previously, in which it will be relatively easy to continue, because this will involve pulling laths apart, but not breaking across them. However, the local crack cannot travel more than a few micrometres in this direction, because neighbouring lengths of the front will be travelling off in a different direction. In fact the two directions will be at about right angles to each other (Fig. 9). The crack will be brought to a halt, or will have to break through laths in order to continue. This will be energetically much more expensive, and explains why cracks travel easily through one layer of a crossed-lamellar specimen, but are brought to a halt by the next layer. Only a considerable increase in the applied load allows the crack to travel again.

The tensile and compressive strengths of crossed-lamellar structure show anisotropy, though only in a few species was it possible to prepare specimens that had all laths orientated in one direction. Baldly, in tensile specimens, the more nearly the lamellae were in line with the

Fig. 8. The fracture surface of a specimen from *Conus*. On the right the crack is travelling between the lamellae. The surface is fairly smooth. On the left, the direction of the layers has changed, making it more difficult for the crack to travel.

tensile forces the greater the strength. This effect was marked. In *Conus litteratus* it was possible to prepare almost uniform specimens, and the tensile strength fell from about 120 MNm^{-2} when the lamellae were not more than 10° from the line of action of the force, to very low values, 10 MNm^{-2} or less, when the angle was greater than 60°. The value of 120 MNm^{-2} is more than the strength of many nacre specimens. In *Volutocorona nobilis* the corresponding values were about 95 MNm^{-2} at 5° to 20 MNm^{-2} at 50°.

In compression the crossed-lamellar specimens tend to be *stronger* when the lamellae make a large angle with the line of the compressive forces. The effect is not so marked, however, as it is in tension.

Fracture in other types of material

The strength of materials other than nacre and crossed-lamellar has hardly been investigated in relation to microstructure. In particular, it would be good to have a better understanding of the fracture behaviour of prismatic type because it is such a common constituent of shells, and because it has a relatively high content of organic material. It is

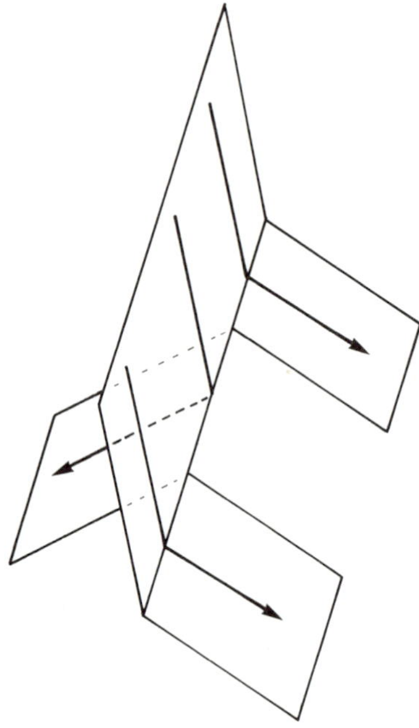

Fig. 9. Diagram showing how a crack travelling in the easy direction tends to break up when it comes to the difficult direction because of the mutually perpendicular arrangement of the laths.

apparent from scanning electron micrographs of fracture surfaces that the structure always breaks between the prisms, through the organic matrix.

Homogeneous structure is notable for being particularly weak in tension, so that for only one species, *Arctica islandica*, could we prepare tensile specimens. It is not surprising that it is so weak, because it has very little organic material, and the mineral seems to consist entirely of small more-or-less isodiametric blocks, so that a crack travelling through it has to deviate little from the direct path, so there are no effective crack stoppers.

Modulus of elasticity

The values of modulus of elasticity are more or less what might be expected from a material composed almost entirely of calcium carbonate with a very small amount of organic matrix. The moduli of aragonite and calcite, the principal mineral components are 100 GNm^{-2} and 140 GNm^{-2} respectively (Hearmon, 1946; Bhimasenachar, 1945). Most of the molluscan materials have values between 30 and 70 GNm^{-2}. This spread of values shown in Table 1 is, unfortunately, probably more a function of the inadequacy of the testing techniques than of real differences between the materials.

The only type to have really low values for modulus is prismatic shell. This accords well with its relatively high organic content, and the way the structure is organised; much of the relatively thick organic interprismatic material is loaded in series with the mineral component when the shell is loaded in life and it will, therefore, have a disproportionate effect on the compliance of the whole material. This, as will be seen later, has important adaptive consequences.

The functions of molluscan shell materials

So far we have considered the mechanical properties of the different materials, and the possible structural reasons for these differences, without discussing function. Perhaps the best way into the problem of function is to ask the question: why do all molluscs not have the same structural material?

It is important to be clear that the possession of different structural types is not an historical accident. There is a strong body of evidence showing that the primitive molluscan pattern of shell was of a layer of nacre inside a layer of prisms. Taylor (1973) has shown that, in bivalves at least, shells with homogeneous structure, with crossed lamellar structure, and possibly with foliated structure, have evolved independently several times. This phenomenon of the independent evolution of different structures, under the influence of natural selection, is also seen in the gastropods.

It is perhaps not immediately helpful to state that the shells of molluscs are, like every other part of the animals' bodies, adapted to maximise the animals' genetic contribution to the next generation. Yet to state this does emphasise that no particular property of the shell need be paramount; every character must be weighed in the balance of selection. To give a particularly clear example: in land pulmonate gastropods, particularly in regions where calcium is difficult to obtain, it is often advantageous to have no shell at all. Hence the evolution of the slug-like habit. The characteristic of the shell most important here is its absence.

Nevertheless, it is reasonable to suppose that the mollusc shell is adapted to provide protection for the body. The problem is that it is not at all clear how the shell should best perform this function. Let us hypothesise a generalised bivalve mollusc and consider the ways in which the shell can best protect it. If it is attacked by a crab or a fish, then its resistance to static loading will be important. If it is to resist blows from the bill of an oystercatcher (*Haematopus ostralegus*), or from rocks tossed about by waves, then resistance to dynamic loading will be important. This is also true if it is carried up in the air by a gull and then dropped, though in this case it would also be an advantage to be very heavy, so the gull could not lift it. If it is attacked by a whelk boring a hole into it, then it will be good to be resistant to abrasion and to chemical dissolution, both these modes of attack being used by whelk. Moreover, in attack by drilling predators it is also advantageous to be thick-shelled so the predator, whose depth of drilling is limited by its anatomy, cannot reach right through to the animal. If the bivalve is to be attacked by a starfish, which will attempt to prise apart the valves a crack in order to extrude its stomach into the gap, it will be advantageous for the shell to be stiff.

All these mechanical properties of the shell will be a function both of the properties of the shell material and of the build of the shell: its size and shape. Obviously, the thicker and larger the shell the stronger and stiffer it will be; which is usually desirable. However, most molluscs do not have extremely large shells, and some shells are quite small in comparison with the size of the body. The reason is obvious: size has costs associated with it. The cost may be the metabolic energy required to produce the material, which energy could otherwise be used for producing reproductive material. It may be the time taken to achieve the mature size; it may be that size itself hinders locomotion, whether because of the weight associated with the mass, or because of the bulk.

If different shell material types have different mechanical properties, which we have seen is so in molluscs, then the amount of material needed to produce shells with particular properties will be different, yet it may well also be true that the cost of different materials, either in metabolic energy, or in time, will be different. There must, therefore, for a mollusc (or any other animal with a skeleton) be performed a complex set of optimisations. These optimisations are in fact performed by the action of natural selection over many generations. Since we have a remarkably imperfect knowledge of the selective pressures acting on any molluscs it will be difficult to explain why different materials are found in various molluscs.

Consider first, however, two cases in which particular characteristics of the shell are rather obviously important and then we can consider the problem more generally. There are some bivalve molluscs for which it is advantageous for the shell to be flexible. This is notably so in *Solemya*. Members of this genus live in fine sand and, uniquely for bivalves, are rarely in contact with the water above the sand. Their locomotion involves alterations of the volume of the shell with its enclosed space in order to flush water through the sand and make it into a slurry (Beedham & Owen, 1965). This flexibility is produced by having a very thick, pleated, periostracum which overhangs the free margin of the shell and also by having a prismatic shell containing a remarkably large amount of organic material (Taylor *et al.*, 1969; Yonge, 1953). The interprismatic walls are particularly thick and the organic component of this part of the shell is about 4.5% by weight, which is high. Unfortunately we have no data on the modulus of elasticity of *Solemya* prismatic shell material, but it is clearly very low.

The low modulus of the prismatic shell material of *Pinna* (12 $\mathrm{GNm^{-2}}$) is also adaptive. This animal lives in mud with a significant portion of its fan-shaped shell exposed to predatory fish. If the animal is attacked the body tissues are withdrawn below the level of the surface of the sea bed and the adductor muscles press the two valves together. The valves are so flexible that they become apposed on their inner surfaces. If the fish breaks the shell it merely gets a mouthful of prisms and it does not gain access to the inside of the shell (Yonge, 1953). After the disgruntled fish has gone, the bivalve can build new shell remarkably quickly, a band a centimetre wide being produced in a few hours.

In many situations it may be possible to trade off having a relatively weak material by having more of it. This is particularly so if the mass of the shell is not important. When this is the case it is very difficult for us, as observers, to analyse what is being selected for. However, another example of a shell in which the optimisation problems are more than usually obvious is that of *Nautilus*, which uses its shell as a buoyancy tank. The animal, which is rather like an octopus, lives in the last chamber of a multichambered shell. The other chambers are filled with gas at about atmospheric pressure. *Nautilus* shells can stand pressure equivalent to a depth of 600 m, at which depth the pressure is 60 atmospheres (Denton & Gilpin-Brown, 1966). The design of the shell is beautifully adapted to withstanding high external pressures, but here we are concerned only with the shell material itself. The shell is thin-walled and it is possible to apply the theory of thin-walled vessels. The shell will fail, at depth, either by fracture or buckling; in what follows it happens not to matter which. For thin-walled pressure vessels the fracture or buckling criteria have a term t/D, where t is the thickness and D is the diameter of the shell. (What this implies is that a small vessel can be thinner walled yet withstand the same pressure.) If the value of mass/volume is such as to produce neutral buoyancy, then if the density

Table 3. *Numbers of families of Bivalvia with shells of particular structural types having particular habits. Modified from Taylor & Layman, 1972*

	Prisms and Nacre	Foliated	Crossed-lamellar	Homo-geneous
Free-living epifaunal	0	2	0	0
Byssate	5	3	5	0
Cemented	3	3	1	0
Boring	1	0	5	1
Burrowing	14	0	29	10

of the material is constant, which is effectively true for molluscan shell materials, reduction in strength/mass or stiffness/mass will *ipso facto* reduce the depth to which the animal can dive. For suppose the shell, enclosing some volume of diameter D, were made of a material half as strong as some other. Then to bear the same pressure the thickness would have to be increased by a factor of two, but would be enclosing the same volume as before, so the whole system would be negatively buoyant and so sink. The value of t/D must remain constant, so reduction in strength or stiffness merely reduces the pressure that can be borne. It is not surprising that the shell of the nautilus is almost entirely made of nacre, the strongest of the materials.

Unfortunately, for analytical purposes, most mollusc shells have less obviously constraining functions. It is possible, however, to gain some insight into the reasons for the presence of particular structure in a particular group by looking at the relationship between the presence of the structure and the mode of life. This has been analysed for the families of bivalve molluscs by Taylor & Layman (1972). Table 3 is derived from their findings.

Nacre with prisms, with which it is nearly always associated, seems to be distributed more or less according to the overall frequencies. Foliated structure, however, is never found in burrowing bivalves, whereas homogeneous structure is never found in families that live above or on the surface. Crossed-lamellar structure is perhaps found more often than would be expected in burrowing families. It seems, therefore that the weak homogeneous structure is found only where the animal will be safe from predatory attack. The fact that foliated structure, which is not particularly strong, is always found in the open might seem somewhat of a puzzle, however. It is found frequently in bivalves with the oyster-like habit and it is becoming clear that such

Table 4. *Percentage loss of weight of standard-sized specimens treated in particular ways*

Abrasion*		Acid		EDTA		Protease	
H	4.3	P	0.7	P	13.9	P	0.0
XL	4.7	XL	0.7	N	22.0	XL	0.2
P	5.4	N	0.7	XL	23.3	N	0.3
N	5.8	H	0.9	H	25.4	H	0.5
F	9.8	F	1.1	F	31.5	F	0.6
O	17.2	O	10.4	O	94.2	O	3.8

*Symbols as in Table 1; O, oyster material.

animals have often evolved a shell mainly adapted for purposes other than being strong. Many mollusc shells fail, not because they fracture but because a predator, usually a gastropod, manages to bore through them and reach the soft tissues through the hole. The boring is carried out by a mixture of mechanical rasping and chemical attack (Carriker, 1978; Carriker & Williams, 1978). Gabriel (unpublished) has investigated the weight loss of standard-sized blocks of various materials to abrasion (in a gem polisher) and to various chemical agents. The different materials can be arranged in a league table (Table 4). (The material, so far not discussed, from the oyster *Ostrea edulis* has thin layers of foliated material separated by large, soft, chalky deposits.)

Abrasion resistance is greatest in a homogeneous structure which, because it is buried, is less likely than materials on the surface to be attacked by boring organisms, but which is also continually surrounded by abrasive sand. Prisms, which are typically found as a layer on the outside of shells consisting mainly of nacre, is particularly good at resisting chemical attacks of the types that may be used by borers.

What is particularly striking is the hopeless performance of the oyster material, because oysters are very frequently attacked by boring animals. They are attractive objects for the borers because they are cemented in one spot, and have no means of escape. It is at first sight surprising that they have a shell which is so vulnerable. However, there are two things that make this state of affairs less surprising. Oysters live in dense colonies, and once settled cannot move. There is therefore intense selection on the spat to grow quickly before they are smothered. The chalky shell of oysters is very unorganised, and is presumably capable of being laid down very quickly. It may be, therefore, that the mechanical properties of the shell are of small importance. There is another advantage in growing quickly. Boring organisms, such as snails, are limited in the depth of the hole they can make. The thickness of the shell of the oyster will protect it from many of its potential predators even though the material of this shell is extremely vulnerable to boring. This strategy of the oysters: overcoming the apparent disadvantage of a bad material by making much of it, of making an effective *construction* by making it large, is not, of course, confined to them. Although nacre is the strongest molluscan shell material it is clear that many molluscs have such thick shells that, although they are not made of nacre, they can survive virtually any attack. The shells of many adult cone shells, or tridancnas, for instance can rarely be broken. Although by no means always true, it is clear that often molluscs that reach some particular size are thereafter more-or-less immune from predatory attack (Hancock, 1970; Edwards & Huebner, 1977). However, this option of substituting quantity for quality is open only to animals that are not very active in locomotion. Wherever strength/mass ratios become important, high quality materials must be used. This is shown in *Nautilus*.

This brief survey of mollusc shells has, I hope, brought out two points. One is the remarkable *strength* that can be obtained from a rather unpromising basic material, the other is the remarkable *range* of shell mechanical properties that natural selection has imposed upon different molluscs.

References

Beedham, G. E. & Owen, G. (1965). The mantle and shell of *Solemya parkinsoni* (Protobranchia: Bivalvia). *Proceedings of the Zoological Society, London*, **145,** 405–30.

Bhimasenachar, J. (1945). Elastic constants of calcite and sodium nitrate. *Proceedings of the Indian Academy of Sciences*, **22A,** 199–208.

Bøggild, O. B. (1930). The shell structure of mollusks. *Kongelige Danske Videnskabernes Selskabs Skrifter, Copenhagen*, **2,** 232–325.

Carriker, M. R. (1978). Ultrastructural analysis of dissolution of shell of the bivalve *Mytilus edulis* by the accessory boring organ of the gastropod *Urosalpinx cinerea. Marine Biology*, **48,** 105–34.

Carriker, M. R. & Williams, L. G. (1978). Chemical mechanism of shell penetration by *Urosalpinx*: an hypothesis. *Malacologia*, **17,** 142–56.

Cook, J. & Gordon, J. E. (1964). A mechanism for the control of cracks in brittle systems. *Proceedings of the Royal Society of London, Series A*, **282,** 508–20.

Currey, J. D. & Kohn, A. J. (1976). Fracture in the crossed-lamellar structure of *Conus* shells. *Journal of Materials Science*, **11,** 1615–23.

Currey, J. D. (1977). Mechanical properties of mother of pearl in tension. *Proceedings of the Royal Society of London, Series B*, **196,** 443–63.

Currey, J. D. (1979). The effect of drying on the mechanical strength of mollusc shells. *Journal of Zoology, London*, **188,** 301–8.

Denton, E. J. & Gilpin-Brown, J. B. (1966). On the buoyancy of the pearly nautilus. *Journal of the Marine Biological Association, UK*, **46,** 723–69.

EDWARDS, D. C. & HUEBNER, J. D. (1977). Feeding and growth rates of *Polinices duplicatus* preying on *Mya arenaria* at Barnstable Harbor, Massachusetts. *Ecology*, **58,** 1218–36.

GILLAM, E. (1969). *Materials under stress*. London: Newnes-Butterworths.

GILMAN, J. J. (1960). Direct measurements of the surface energy of crystals. *Journal of Applied Physics*, **31,** 2208–18.

GRÉGOIRE, C. (1972). Structure of the molluscan shell. In *Chemical Zoology Vol. VII. Mollusca*, ed. M. Florkin & B. T. Scheer, pp. 45–102. New York: Academic Press.

HANCOCK, D. A. (1970). The role of predators and parasites in a fishery for the mollusc *Cardium edule* L. In *Proceedings of the Advanced Study Institute on Dynamics of Numbers in Populations*, ed. P. J. den Boer & G. R. Gradwell, pp. 419–43. Wageningen: Centre for Agricultural Publishing and Documentation.

HEARMON, R. F. S. (1946). The elastic constants of anisotropic materials. *Reviews of Modern Physics*, **18,** 409–40.

ILER, R. K. (1963). Strength and structure of flint. *Nature, London*, **199,** 1278–9.

KENDALL, K. (1975). Transition between cohesive and interfacial failure in a laminate. *Proceedings of the Royal Society of London, Series A*, **344,** 287–302.

RAWCLIFFE, D. J. & FRUHAUF, V. (1977). The fracture of jade. *Journal of Materials Science*, **12,** 35–42.

TAYLOR, J. D. (1973). The structural evolution of the bivalve shell. *Palaeontology*, **15,** 73–87.

TAYLOR, J. D., KENNEDY, W. J. & HALL, A. (1969). The shell structure and mineralogy of the bivalvia. Introduction. Nuculacea-Trigonacea. *Bulletin of the British Museum (Natural History): Zoology, Supplement*, **3,** 1–125.

TAYLOR, J. D., KENNEDY, W. J. & HALL, A. (1973). The shell structure and mineralogy of the bivalvia. II Lucinacea-Clavagellacea Conclusions. *Bulletin of the British Museum (Natural History): Zoology*, **22,** 256–94.

TAYLOR, J. D. & LAYMAN, M. (1972). The mechanical properties of bivalve (Mollusca) shell structures. *Palaeontology*, **15,** 73–87.

WILBUR, K. M. & SIMKISS, K. (1968). Calcified Shells. In *Comprehensive Biochemistry*, Vol. 26A, ed. M. Florkin & E. H. Stotz, pp. 229–95, Amsterdam: Elsevier.

YONGE, C. M. (1953). Form and habit in *Pinna carnea* Gmelin. *Philosophical Transactions of the Royal Society, Series B*, 335–74.

SOME MECHANICAL AND PHYSICAL PROPERTIES OF TEETH

N. E. WATERS

Royal Dental Hospital School of Dental Surgery, Leicester Square, London WC2 7LJ, UK

Introduction

Interest in the mechanical properties of tooth material has a long history; many hundreds of papers have been written on some aspects, for example hardness, whilst others, because of the experimental difficulties, have received little attention. Knowledge of the mechanical behaviour of these materials is, of course, essential from the point of view of the repair of teeth in that ideally the basic properties, moduli, strength, hardness, as well as the coefficient of expansion, thermal conductivity, refractive index, colour, etc. of the restorative material must all be as close as possible to those of the tissues they are replacing.

In the last few years academic interest has also been shown in these materials for the light which may be thrown on the way synthetic composite materials may be improved.

Structure of teeth

The complete human adult dentition consists of 32 teeth; 16 maxillary teeth (i.e. in the upper jaw) and 16 mandibular teeth in the lower jaw. Teeth are normally divided into groups by their function. Each complete arch has, from the mid-line and working towards the back of the mouth, 4 incisors (the cutting teeth), 2 canines (the tearing teeth), 2 premolars (for grasping) and 6 molars (for grinding or masticating the food).

The principal morphological features of a developed erupted tooth are shown in Fig. 1 which depicts a mid-plane section through an anterior tooth. The main part of any tooth is a bone-like substance known as dentine which is covered, over that part which protrudes into the mouth, by a highly mineralized layer called enamel. The remainder of the tooth is embedded in the alveolar bone being covered by a layer of cementum and attached to the bone by the periodontal membrane, a

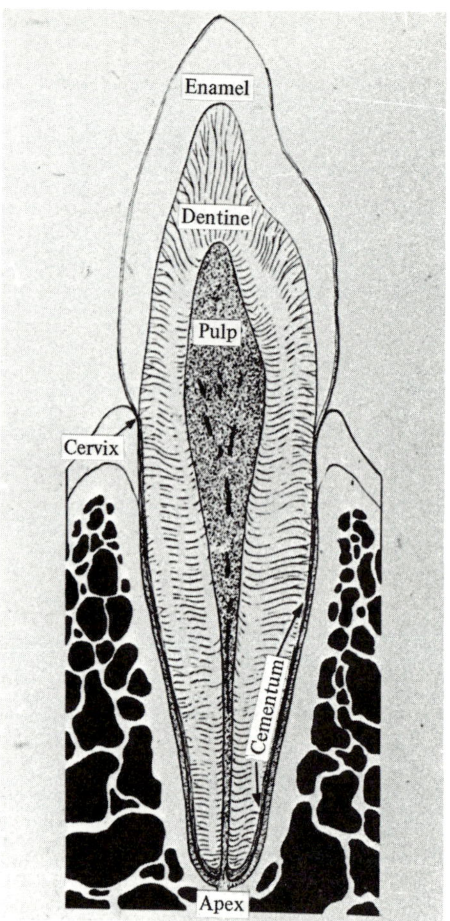

Fig. 1. Mid-plane section through an anterior tooth in its socket showing the principal morphological features. Diagrammatic.

layer of fibrous connective tissue. Embedded in the dentine and within the crown is the pulp, a filament of soft tissue containing cells and also nerves and blood vessels which enter the tooth through an opening at the apex of the root. Also shown diagrammatically in the figure are the dentinal tubules, tube-like spaces which radiate outwards from the pulp and extend to the dentine–enamel or the dentine–cementum interface.

The approximate chemical composition of dentine and enamel in terms of the basic phases present is given in Table 1, together with the comparable data for bone. It may be observed that bone and dentine are the nearest in composition, although dentine has slightly higher mineral

Table 1. *Composition of Hard Tissues by Volume (%). Weight (%) in brackets*

	Bone	Dentine	Enamel
Mineral	41	48	92
(density, 3000 $kg\,m^{-3}$)	(64)	(69)	(97)
Organic	48	29	2
(density, 1400 $kg\,m^{-3}$)	(31)	(20)	(1)
Water	11	23	6
(density, 1000 $kg\,m^{-3}$)	(5)	(11)	(2)

and water contents and hence a lower organic content. Enamel differs in having a very high mineral and low organic content. The layer of cementum surrounding the root varies in thickness from 20–50 μm at the cervix to 150–200 μm at the apex. Roughly half of the cementum is inorganic and half is composed of organic material and water. Because of the experimental difficulties involved, its mechanical properties, like those of the periodontal membrane, are largely unknown. X-ray diffraction studies have shown unequivocally an apatite structure for the mineral phase, chiefly in the form of hydroxyapatite, $Ca_{10}(PO_4)_6(OH)_2$, although the actual Ca/P ratio is generally lower than the value required by this empirical formula, 1.67, and varies from 1.3 to 2.0. According to the available evidence from X-ray diffraction the mineral in both dentine and bone is in the form of hexagonal shaped crystals approximately 3 nm in diameter and 64 nm long. The crystallites in enamel are thought to have the same general shape but by comparison are much larger being approximately 25 nm thick, 40–120 nm wide and 160–1000 nm in length, although the possibility that they are ribbon-like structures and are considerably longer cannot be ruled out. The crystallographic axis lies along the crystals.

The dentinal tubules contain tissue and cellular processes connected to the odontoblast cells lining the surface of the pulp. As may be seen in Fig. 2, a transverse section, the tubules have lateral branches (canaliculi) which are 1 μm or less in diameter. The tubules also show interbranching. Initially at the pulpal surface there are approximately 75 000 per mm^2, each 4 μm in diameter. Their number per mm^2, as well as their size, decreases progressively until at the outer surface there are about 20 000 per mm^2 each approximately 1 μm across. Within each tubule and surrounding the odontoblast process is an annular layer, highly calcified, referred to as peritubular dentine (Fig. 3). As a tubule is followed to the outer surface, the diameter of the odontoblast process becomes narrower and the peritubular zone becomes gradually wider.

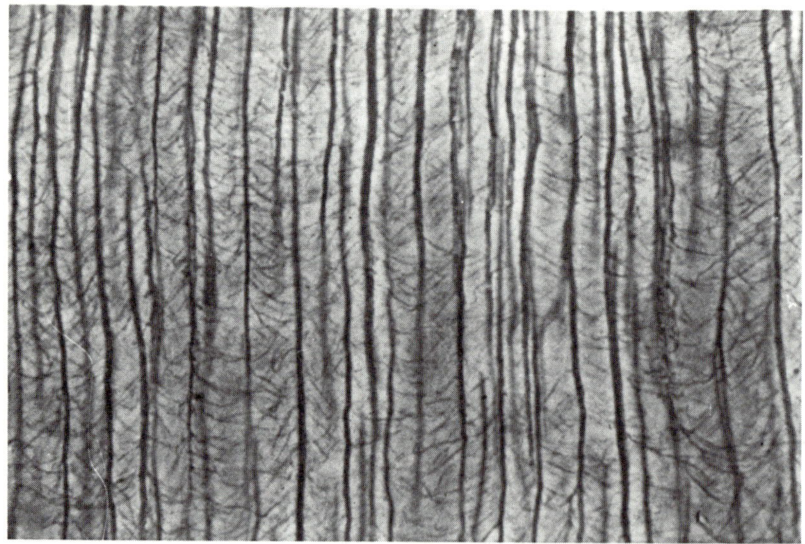

Fig. 2. Transverse section through dentine showing tubules and interbranching (×800).

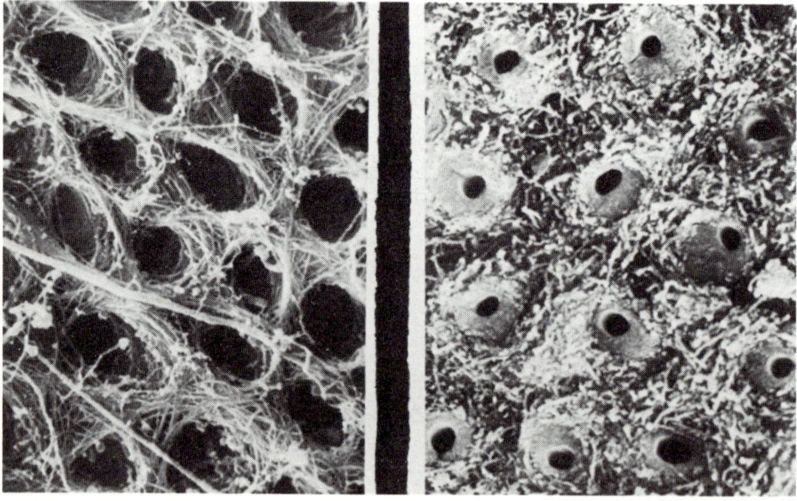

Fig. 3. Section perpendicular to the dentinal tubules. Right-hand side: peritubular dentine within tubules. Left-hand side: collageneous fibrils observable in predentine surface (×600). Reproduced from Scott, Simmelink & Nygaard (1974) with kind permission of the authors and the publisher.

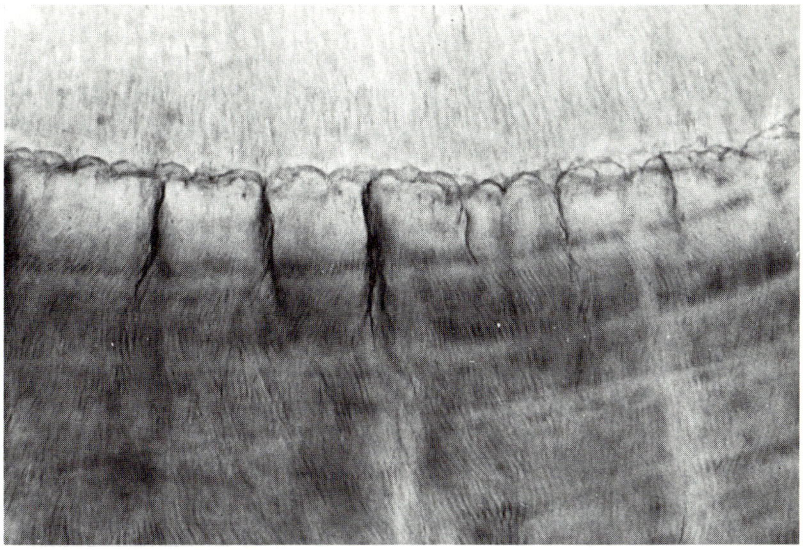

Fig. 4. Transverse section under the optical microscope, showing dentino-enamel junction and below this the wavy nature of the course of the prisms (×250).

The organic substance between the tubules within which most of the mineral is dispersed consists of collageneous fibrils approximately 0.3 μm thick and a ground substance of proteoglycans. The arrangement of these fibrils may be seen on the left-hand side of Fig. 3, where the odontoblasts have been removed from a predentine surface. In the main the fibrils appear to be arranged in planes perpendicular to the tubules. X-ray diffraction studies indicate that the mineral crystals in dentine are in general randomly orientated. Certain less highly mineralized regions between normally calcified globules may occur in crown dentine which are referred to as 'interglobular dentine'.

The main features observed under an optical microscope (see Fig. 4) in the bulk of the enamel are the well-defined rod- or prism-like structures, approximately 4 μm in diameter, which follow an undulating course from the dentino-enamel junction to the enamel surface, the mean diameter increasing slightly as they approach the outer surface. In cross-section the prisms usually appear horse-shoe shaped (Fig. 5) but neither the shape nor the pattern in which they are disposed may be termed regular. Over the last 10–15 years it has become accepted that a key-hole shaped structure is the most typical, as shown diagrammatically in Fig. 6. If a section of enamel is viewed in reflected light a series of bands is seen, the *Hunter–Schreger* bands (see Fig. 7), which are believed to be a result of the orientations of successive layers of prisms (Fig. 8). The disposition of the hydroxyapatite crystallites within a prism

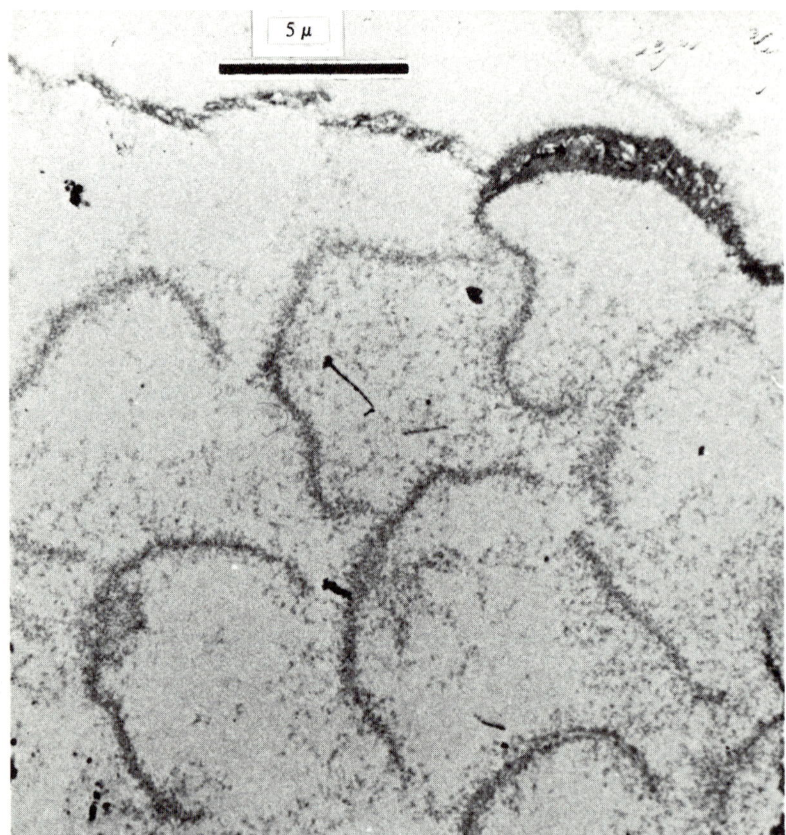

Fig. 5. Section of human enamel at right-angles to prism axis. Darker areas, produced by light etching, demark prism boundaries.

is believed to be as shown in Fig. 9. In the central region in the 'head' of the key-hole the crystallites are orientated with their long axis parallel to the prism direction. Moving away from this region the orientation of the crystallites gradually changes and is approximately 40° to the prism axis in the 'tail' region. Because of the tilting of the crystallites near a boundary it seems plausible that this region will contain more organic material than the centre. The prism boundaries also define a surface across which there is a marked change in direction of the crystallites. Prisms may not be uniform in cross-section; some fractographs appear to show regular undulations at intervals of about 4 μm in the cross-sectional plane. This periodic variation in the width, which is thought to be due to rest phases which occurred during the formation of the enamel, also gives rise to the dark bands or cross-striations in sections cut in the plane of the prisms, which are seen with the light microscope.

Fig. 6. Key-hole model for the interlocking of enamel prisms.

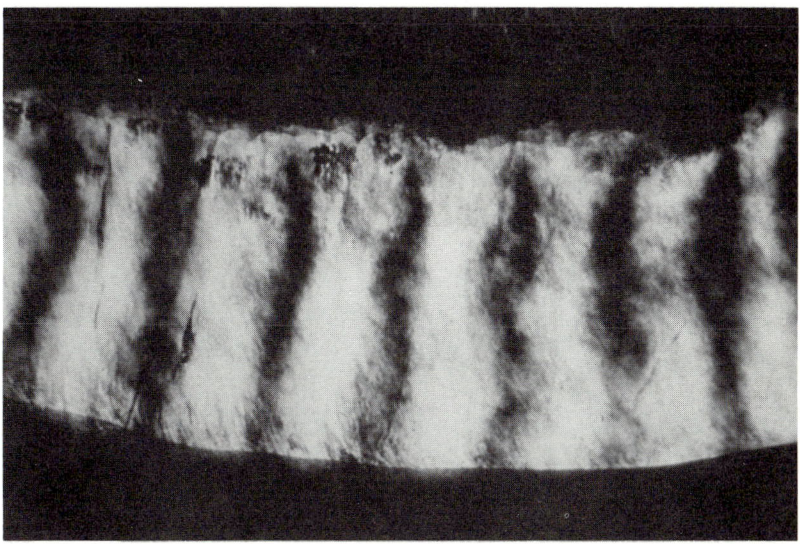

Fig. 7. Enamel section viewed in reflected light showing the Hunter-Schreger bands (× 80).

Various structural features and imperfections can be seen in enamel. These include enamel tufts and spindles to be observed at the dentino-enamel junction and enamel lamellae which extend through the enamel and appear to be cracks filled with organic material. More regular and periodic structural features which occur in both enamel and dentine are the incremental growth lines of Retzius and von Ebner respectively. Fig. 10 is a microphotograph of a section showing the brown striae of Retzius. Enamel tufts and a lamella may also be seen.

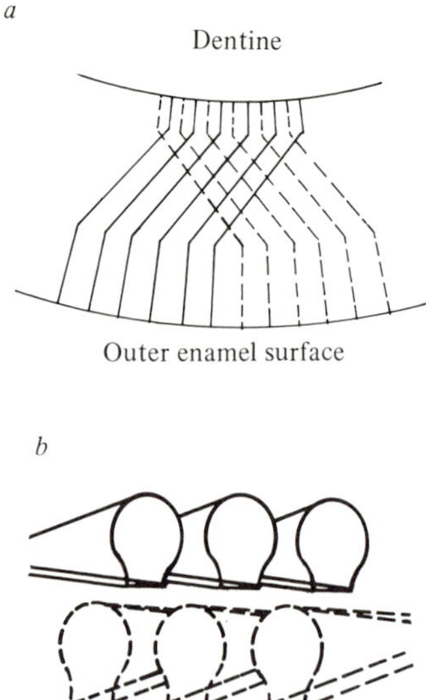

Fig. 8. Diagrammatic representation of the orientation of successive layers of enamel prisms believed to be the cause of the fringe system seen in reflected light (see Fig. 7). *a*, section at right-angles to tooth axis; the full lines show the course of prisms in this plane. The course of prisms beneath this plane is shown by the dashed lines. *b*, 'end-on' view of prisms.

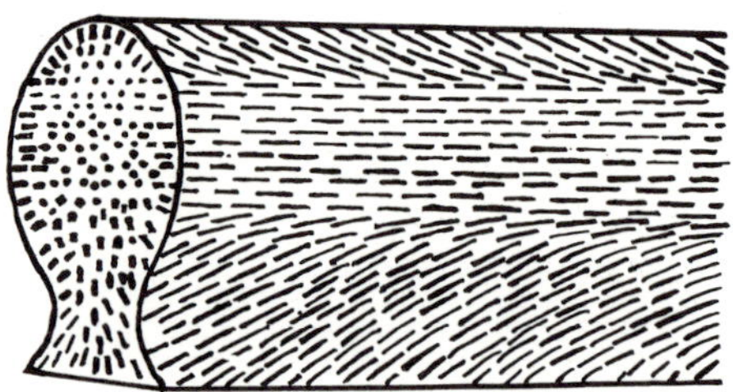

Fig. 9. Orientation of c-axis of the crystallites within an enamel prism.

In general, particularly after exposure to the oral fluids, the outer enamel surface is more highly mineralized and contains less organic material. The amount of organic material, however, slowly increases as the dentino-enamel junction is approached. The organic material in enamel is made up of soluble and insoluble proteins in roughly equal amounts, peptides and citric acid. Most of the insoluble protein material is thought to reside in the lamellae and in tufts in the inner part of enamel.

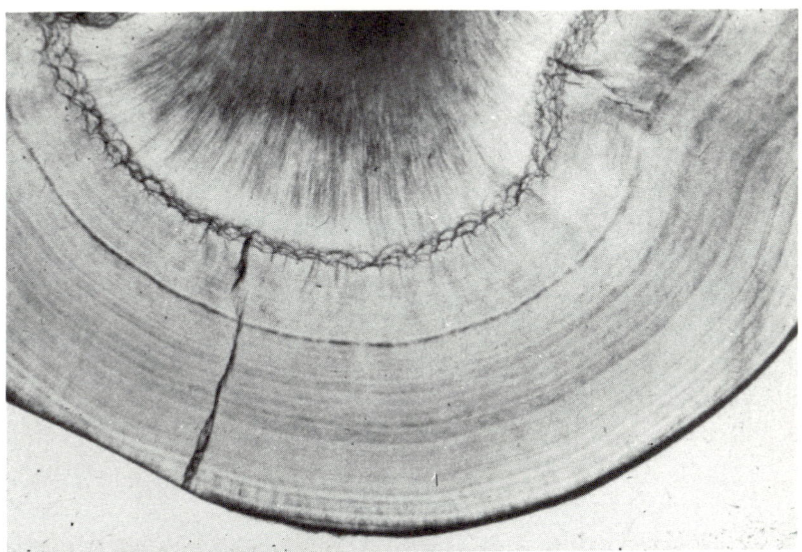

Fig. 10. Transverse section through tooth showing the incremental growth lines (brown striae of Retzius) within the enamel (×40).

Table 2. *Conditions under which teeth operate*

Rate of chewing	60–80 cycles min^{-1}
Forces: maximum biting force overall teeth	640 N
maximum biting force on a single tooth	265 N
normal forces on a single tooth	3–18 N
Contact time:	~0.2 s
time maximum pressure sustained (20% of one masticatory cycle)	0.07 s
Total contact time: normal	10 min
(over 24 h) Bruxist	30 min–3 h
Mean sliding distance	1.0 mm
Contact area (1st molar region)	15 mm^2
Maximum stress	20 MNm^{-2}

Conditions under which teeth operate

A summary of the relevant aspects of the biomechanics of mastication and tooth movement is given in Table 2, which is mainly based on an excellent review by Harrison & Lewis (1975). These data, although approximate and subject to considerable variation, serve to remind us of the stringent mechanical conditions to which teeth are subject, with maximum stresses under normal conditions applied to molar teeth of 20 MNm^{-2} some 3000 times per day for a life-time without, as far as we know, fatigue failures occurring and with only moderate attrition or wear, usually more apparent on anterior teeth. Where involuntary grinding of the teeth occurs, usually at night (a condition known as Bruxism) the total contact time is considerably increased and greater attrition is observed.

Naturally there are occasions when pieces of stone, metal or bone may be involuntarily masticated. Fracture of sound whole natural teeth however would seem to be extremely rare. Gross attrition, attributable to diet and abnormal amounts of grit in the food, is apparent in the dentition of primitive man.

Problems associated with testing

The problems associated with the mechanical testing of tooth structure and material are considerable. With the greater awareness of the public of the need for oral hygiene suitable teeth are not always easy to obtain. There is then the difficulty of assessing whether the tooth is both sound, i.e. has not been damaged during extraction and is free from caries, and is normal, i.e. has not suffered trauma during its developmental stages.

Additionally there are the problems of cutting extremely small specimens from the tooth without damaging them either mechanically or thermally and of keeping the material during the storage, preparative and testing stages in as near as possible its normal condition. Storage is a particular difficulty. After extraction a tooth requires careful cleaning to remove any tissue (including the pulp) which may putrefy. Fungal growth may occur, particularly in solutions containing phosphate, unless thymol is added, and algae may also form unless the tooth is kept in the dark. Storage in distilled water, tap water or 'physiological' salt solutions is common. During prolonged storage, dissolution of mineral is possible and a change in water content may occur if the storage solution sets up an unfavourable osmotic gradient. Storage in formalin will cause dehydration and also cross-linking of proteinaceous material, and commercial formalin generally is acidic due to the formation of formic acid, which will result in dissolution of the mineral component.

The mechanical properties of dentine and enamel

All the stress–strain measurements on enamel and dentine reported in the literature, whether in tension, compression, bending or shear,

Table 3. *Compressive properties of dentine*

Author(s)	No. of specimens	Proportional limit MNm^{-2}	Compressive strength MNm^{-2}	Young's modulus GNm^{-2}
Black (1895)	70	–	256 ± 30.3	–
Peyton *et al.* (1952)	10	161 ± 23.7	250 ± 22.6	11.5 ± 1.2
Stanford *et al.* (1958)	8	173 ± 15.5	348 ± 24.5	11.4 ± 1.7
Craig & Peyton (1958)	10	167 ± 20	297 ± 24.8	16.6 ± 18.5
Stanford *et al.* (1960)	65	crown 143.3 root 105.9	crown 282.6 root 233.5	crown 12.5 root 7.75

appear to have been made at room temperature and under conditions (not always given in detail) which may be termed quasi-static. Initial, essentially linear, regions have been observed by all workers using low strain rates for dentine from which a modulus and proportional limit could be deduced. It should be noted, however, that various time-dependent effects have been observed (see below, p. 125) and that the linear region of the true stress–strain curve in flexure and shear is reported to be small. Initial linear stress–strain behaviour has also been observed with enamel. Divergences from linearity have been reported when intact crowns of teeth have been subject to compression (Newman & Di Salvo, 1957; Haines, 1968).

Compressive properties of dentine

The values reported in the literature are presented in Table 3. The earliest systematic measurements on dentine were reported by Black (1895) who, ignoring any possible location or directional effects, obtained a mean value for the compressive strength of 256 MNm^{-2} with a coefficient of variation of 11.8%. As this is within the accepted scatter for this type of test it may tentatively be concluded that neither location nor orientation with respect to the tubules is a major variable. Black's value for the compressive strength is close to the average obtained by other workers. Peyton, Mahler & Hershenov (1952) identified the teeth used for examination and also attempted to look for anisotropy by recording tubule direction within their cylindrical specimens, but the small number of specimens and other variables in the testing procedure obscured any possible trend.

Stanford, Paffenberger, Kampula & Sweeney (1958), on the basis of a limited number of tests, originally concluded that dentine exhibited anisotropy with regard to tubule orientation, but in a more extensive study (Stanford, Weigel, Paffenberger & Sweeney, 1960), this conclusion was not substantiated. These workers were, however, able to conclude that the compressive strength and also the modulus and proportional limit did not appear to be affected by the type of tooth examined and that the properties of dentine from deciduous teeth were similar to those of the dentine from permanent teeth. On average these authors also found that root dentine gave lower values than coronal dentine for all three compressive properties.

Craig & Peyton (1958) obtained values for the proportional limit and compressive strength in reasonable agreement with other workers, but their value for the modulus is, for some unaccountable reason, considerably higher.

Compressive properties of enamel

Most workers have commented upon the extreme difficulties encountered in either preparing or testing enamel and only two groups of workers have examined the compressive properties (Table 4). Stanford *et al.* (1960) found it necessary to reject more than 50% of the prepared specimens after visual examination for chipping or cracks at a thirtyfold magnification. Craig, Peyton & Johnson (1961) indeed reported that it was impossible to produce cylindrical specimens with the general prism orientation perpendicular to the axis and found it necessary to reject other specimens showing imperfections after microscopic examination. The difficulty of producing plane parallel ends to the cylindrical specimens such that the stress is applied uniformly was noted by Black (1895) who experimented with the use of gold foil at the ends of the specimen to minimize stress concentrations due to these inaccuracies. Strangely no other workers have commented on this problem with enamel although it is mentioned by Craig & Peyton (1958) in their compression tests on dentine. In view of the obvious difficulties it is perhaps not surprising that the total number of tests carried out is relatively small, and that these show a considerable variation. Stanford *et al.* (1958, 1960) found that the compressive properties as a whole were similar for cuspal enamel from either canines or molar teeth. The values obtained for the other locations examined, *viz.* side and occlusal enamel and also enamel from the incisal edge, were generally much lower and showed greater variation. Stanford *et al.* suggested that decalcification in the mouth may account for the lower figures. The considerable

Table 4. *Compressive properties of enamel*

Author(s)	Type of tooth	No. of specimens	Prism orientation w.r.t. axis of specimen	Proportional limit MNm^{-2}	Compressive strength MNm^{-2}	Young's modulus GNm^{-2}
Stanford *et al.* (1958)	not recorded	4 (cusp)	variable	236 ± 21	277 ± 12.4	47.5
		4 (side)	across	145 ± 28	194 ± 15.2	30.3
		6 (occlusal)	along	116 ± 28	134 ± 27.9	8.96
Stanford *et al.* (1960)	canine	7 (cusp)	variable	194 ± 18.6	288.2 ± 47.6	47.5 ± 5.5
		6 (side)	across	183.4 ± 12.4	253 ± 35.2	33 ± 2.1
		4 (incisal edge)	variable	91 ± 10.3	220 ± 13.1	20 ± 6.2
Stanford *et al.* (1960)	molar	19 (cusp)	variable	224.1 ± 25.5	261 ± 41.4	46.2 ± 4.8
		11 (side)	across	186.2 ± 17.2	250 ± 29.6	32.4 ± 4.1
		16 (side)	along	70.3 ± 22	94.5 ± 32.4	9.65 ± 3.45
		6 (occlusal)	along	98.6 ± 25.5	126.9 ± 30.3	11.0 ± 2.1
Craig *et al.* (1961)	molar	7 (cusp)	along	353 ± 77.6	384 ± 85.3	84.1 ± 6.2
		10 (side)	along	336 ± 61	372 ± 56	77.9 ± 4.8

differences between the proportional limit and the compressive strength in some cases, which could be due to porosity, would seem to support this possibility. As has been noted, Craig *et al.* (1961) carried out tests only with the general prism direction parallel to the axis. The values obtained for all the compression properties of molar cusp and side enamel were similar and significantly higher in each case than the values for these properties obtained for molar enamel by Stanford *et al.* (1960). The puzzling feature in this disparity is the self-consistency of the data from these two laboratories, which seems to rule out the possibility that either damage to the specimens or non-alignment in testing was a factor responsible. As the methods of test and also the specimen size and shape were similar, which of these two sets of figures is the more typical will probably remain unresolved until other data become available. It is perhaps worth adding that Stanford *et al.* (1960) found little difference between enamel taken from teeth from an area with endemic fluoride and those which had developed in a non-fluoride area, so that a systematic difference in the nature of the material used also appears to be precluded.

Finally, although from Stanford *et al.*'s experiments it would appear that the three compressive properties in the direction of the prisms are lower than that at right-angles, the disparity with the results of Craig *et al.* makes the difference somewhat uncertain. It should also be noted that from the sonic velocity data of Lees & Rollins (1972) bovine enamel shows considerable anisotropy.

The tensile behaviour of enamel and dentine

The tensile strength of dentine (see Table 5) has been evaluated by direct tension using cylindrical dumb-bell shaped specimens and by the diametral compression method. In most cases, as far as can be assessed, the tensile strength was measured at right-angles to the tubules. The direct tension method of Bowen & Rodriguez (1962) gave a mean value of 52 MNm^{-2} which is in reasonable agreement with the mean value of 41 MNm^{-2} obtained by Lehman (1967). The diametral test in which a thin cylinder of material is loaded perpendicular to its axis has been used by Hannah (1970). The mean value was comparable to the values obtained in direct tension. Overall, a figure of about 45 MNm^{-2} (approximately one-sixth of the compression strength) appears to be typical, but the scatter reported for all studies and the difficulties of specimen preparation suggest that the true figure may be higher. The values for Young's modulus in tension derived by Bowen & Rodriguez

Table 5. *Tensile properties of dentine and enamel*

Author(s)	Material	Method	No. of specimens	Tensile strength MNm^{-2}	Young's modulus GNm^{-2}
Bowen & Rodriguez (1962)	dentine	tension	9	51.7 ± 10.3	19.3 ± 5.4
Lehman (1967)	dentine	tension	82	41 (21.3 → 53.2)[a]	–
			12	–	11.0 ± 5.5
Hannah (1970)	dentine	diametral compression	not given	30–65[a]	–
Bowen & Rodriguez (1962)	enamel	tension	9	10.3 ± 2.6	–
Hannah (1970)	enamel	diametral compression	not given	33–35[a]	–

[a] Range

Table 6. *Flexural properties of dentine and enamel*

Author(s)	Material	Method	No. of specimens	Flexural strength MNm^{-2}	Young's modulus GNm^{-2}
Tyldesley (1959)	dentine	4 point bending	52	268 ± 12.4[a]	12.3
Renson (1970)	dentine	3 point bending		245	–
		cantilever		–	12.0
Tyldesley (1959)	enamel	4 point bending	not given	75.8 ± 12.4	131 ± 16

[a] Range.

(1962) and Lehman (1967) span the range of values for this property found in compression.

Measurements that have been made of the flexural strength (see Table 6) (the maximum tensile strength in bending), however, all give mean values closely approaching the compression strength. Although it is not uncommon for the tensile strength in bending to be considerably higher than that in direct tension the reasons for the discrepancy in the present case are unclear. One factor which may be of importance is that whereas Lehman reported a linear stress–strain curve to fracture, Tyldesley (1959), who used the method suggested by Nadia (1950) to derive stress–strain curves from his data from tests in bending, found very low proportional limits (mean value 66 MNm^{-2}) in comparison with the fracture stress (mean value 268 MNm^{-2}) with the main part of the curve showing evidence of plastic flow. The different strain system imposed seems therefore to cause the material to exhibit different structural behaviour.

It is of interest to note that the results for three- and four-point bending are in close agreement. As the strain is uniform over a much longer section of the specimen in four-point loading, this suggests first that the flaws from which failure is initiated are structural rather than due to surface imperfections produced during specimen preparation and second, that the flaw distribution is reasonably uniform. The results for Young's modulus determined by bending tests (Tyldesley, 1959; Renson, 1970) are in good agreement with the majority of individual measurements made in both tension and compression and also with an independent measurement of the modulus by Renson & Braden (1971) who obtained a mean value of 12.0 GNm^{-2} from the analysis of the results obtained with an indentation test.

Tyldesley's results (Tyldesley, 1959) for enamel using a four-point bending test are somewhat puzzling. Specimens approximately 11×10 mm long composed of both enamel and dentine and cut from the labial plates of upper incisors were used. The values for both the flexural strength and the Young's modulus are very high; the latter figure indeed being comparable to or slightly above the accepted figure for pure hydroxyapatite, namely, 114 GNm^{-2}. Although Tyldesley admits of difficulty in obtaining regular shaped specimens and the method involves both tensile and compressive strains no satisfactory explanation of these high values has been found.

The tensile strength of enamel has been measured by Hannah (1970) using diametral compression. The value obtained, 30–35 MNm^{-2}, is only slightly lower than that obtained on dentine either by the same method or by direct tension.

The shear strength of enamel and dentine

The available data in the literature on the shear properties of enamel and dentine are given in Table 7. The punch technique for shear strength is not an absolute method because of the possibility that the specimen, which must be in sheet form, may bend and also because of friction between the punch and the die. The most systematic work reported is that of Smith & Cooper (1971) who found larger values with smaller diameter punches. The shear strength also varied unpredictably with the thickness of the sample. Nevertheless, the punch shear strength of dentine was always higher than that of enamel.

The shear modulus of dentine and derived values of Poisson's ratio

The only quasi-static method of measuring the shear modulus recorded in the literature is that of Renson & Braden (1975). By means of dead loads and a pulley system known torques were applied to rectangular prisms of dentine and the rotations measured using a lamp and scale. A mean shear modulus (for nine results) of 6.19 GNm^{-2} was recorded, it being noted that the reversible linear portion of the deformation curve was limited to small torques indicating that this material has a very low elastic limit in shear.

Assuming dentine to behave as a homogeneous ideally elastic and isotropic solid, values for Poisson's ratio (the ratio of the lateral to longitudinal strain) were calculated for each individual specimen. Some of the values of Poisson's ratio reported appear to have been

Table 7. *Shear properties of dentine and enamel*

Author(s)	Material	Method	No. of specimens	Shear modulus (GNm^{-2})	Proportional limit (MNm^{-2})	Punch shear strength (MNm^{-2})
Renson & Braden (1975)	dentine	torsion	9	6.19 ± 0.98	60	
Cooper & Smith (1968)	dentine	punch[a]				
		100 μm dia				132
		200 μm dia				101
Roydhouse (1970)	dentine	punch 1.1 mm				69–147
Cooper & Smith (1968)	enamel	punch[a]				
		100 μm dia				93
		200 μm dia				64

[a] Variation with specimen thickness observed.

Table 8. *Moduli from specific acoustic impedance* (GNm^{-2})

Author(s)	Material	Young's modulus	Shear modulus	Bulk modulus	Poisson's ratio
Gilmore, Pollack & Katz (1969)	bovine dentine (longitudinal)	21	8	18	0.31
Reich, Brendan & Porter (1967)	human enamel	76.5	29.3	65.3	0.3
Lees (1968)	bovine enamel	73.0	29.7	45.0	0.23
Gilmore *et al.* (1969)	bovine enamel (longitudinal)	74.0	30.0	46.0	0.23

miscalculated; assuming the original data to be correct the mean value of 0.12 originally obtained should be 0.23, with a standard deviation of 0.19. This large standard deviation is due to the fact that any errors in Young's modulus and the shear modulus are amplified by the calculation. Any anisotropy present in dentine would naturally invalidate the method of calculating Poisson's ratio from the other elastic constants. What is also problematical is whether specimens of dentine with open-ended tubules will behave in the same way as dentine *in vivo*. Haines (1968) has obtained a low value (which appears to be a single result) of 0.014 by direct measurement. His value, however, may be subject to error because the dentine block used was surfaced by enamel on three sides which would have restricted dimensional changes.

Specific acoustic impendance measurements of the moduli

Measurement of the specific acoustic impendance of dental tissues to ultrasonic waves can be used to determine all the principal elastic moduli for an elastic continuum, or in certain cases the elastic constants of an anisotropic material. This method appears to give similar mean values for both bovine and human dentine and enamel (see Table 8). For enamel the Young's modulus values are comparable with the highest values obtained in compression tests, and give a Poisson's ratio of between 0.23 and 0.3. The higher values obtained in general for the moduli of dentine at the ultrasonic frequencies used (30–35 MHz) are not too unexpected in view of the known increase in stiffness of molecular segments of polymeric materials with frequency (Dorrington,

Table 9. *Estimated values for the elastic constants of calcified tissues based on experimental sonic velocities* (GNm^{-2})

	C_{11}	C_{12}	C_{13}	C_{33}	C_{44}	C_{66}	Density
Fluorapatite[a]	143.4	44.5	57.5	180.5	41.5	49.5	
Hydroxyapatite[b]	137	42.5	54.9	172	39.6		3.17
Enamel[c]	115	42.4	30	125	22.8		2.9
Dentine[c]	37	16.6	8.7	39	5.7		2.2
Bone[b]	31	14.7	11.3	33	6.2		1.96

[a] Data of Yoon & Newnham (1969).
[b] Computed for a pseudo-crystal with hexagonal symmetry by Katz & Ukraincik (1971).
[c] Computed by Lees & Rollins (1972).

1980). Dentine, with the higher organic fraction, would be expected to show a greater frequency dependence than enamel.

The value obtained for Poisson's ratio of bovine dentine by this method is in fair agreement with the mean value of 0.23 calculated from quasi-static tests (Renson & Braden, 1975: *vide supra*) although at high frequencies it might be expected that Poisson's ratio would be smaller.

This technique is the only method really capable of investigating the variation in the elastic constants as a function of location and orientation. Lees & Rollins (1972) have investigated the use of critical-angle reflection techniques specifically for this purpose and their results appear to show that anisotropy at 45 MHz is more marked in enamel than in dentine, a result which is in accordance with the known more ordered structure of the crystallites in enamel, but which is still in doubt from quasi-static measurements (*vide supra*). This anisotropy was established for bovine incisors by systematically removing material layer by layer from the outer surface. In this way the variation of ultrasonic velocity (both longitudinal and shear) with depth and orientation was followed. This particular method is, according to the authors, capable of further development and if it could be used in conjunction with localized determinations of the density and composition it would be an extremely powerful tool for elucidating the relationship between structure and properties.

Estimated values from sonic velocity measurements of the elastic constants of fluoroapatite (Yoon & Newnham, 1969) hydroxyapatite (Katz & Ukraincik, 1971) and for enamel, dentine and bone (Lees & Rollins, 1972) are presented in Table 9.

Hardness of dental tissues

As hardness measurements are simple and rapid to carry out, require minimal preparation of the specimen and can be conducted at specific areas in the tissue it is understandable that this type of test has been popular.

According to Braden (1976) there have been over 50 papers on the subject since 1912. Not all the so-called hardness measurements of course give unambiguous information about a specific physical property. These include various sclerometers and pendulum hardness testers (Braden, 1976). More recent tests have utilized conventional Brinell, Vickers and Knoop hardness testers, the last being most widely used presumably on the basis that the incidence of shattering for brittle materials like enamel should be reduced by the stress pattern imposed by this indenter. No reference to fracture occurring on enamel with the Vickers test can, however, be found in the literature.

Few of the measurements reported for conventional as opposed to microhardness tests are likely to yield data completely independent of the specimen thickness. As has been shown for elastic indentations the stress will not be negligible at the specimen surface remote from the indenter unless the ratio of the thickness of the specimen to the actual radius of contact is large (greater than eight for a spherical indenter) (Waters, 1965). Similar considerations will clearly apply to plastic yielding. A further problem which applies to microhardness measurements (as indeed to 'strength' tests in general) is the extent to which the surface is affected by sectioning, grinding and polishing. Finally, since both dentine and enamel contain water, the conditions under which the material has been stored, machined and tested may influence the water content and hence the results obtained (see also p. 108).

Using spherical, cylindrical and cone-shaped indenters Renson & Braden (1971) were able to show that for sufficiently low loads the indentation was elastic and fulfilled the predictions of classical elasticity theory which, of course, assumes a homogeneous isotropic elastic continuum. With higher loading the deformation was shown to be in accord with the general theory for the plastic yielding of metals due to sub-surface shear, a conclusion supported by the low elastic limit in shear found by these authors (Renson & Braden, 1975). Hardness measurements are by no means standardized and there is a number of possible variables. The variation of Vickers hardness with load has been examined in two studies (Ryge, Foley & Fairhurst, 1961; Hegdahl & Hagebö, 1972) using the procedure adopted by Meyer (1908) in which exponent n in the relationship $L = ad^n$ is determined where L is load, d the diagonal of the indentation and a and n are constants. There are discrepancies between the two studies, more particularly for dentine, which may reflect the different times for which the loads were applied and its viscoelastic nature. Both studies agree, however, that n for both dentine and enamel is close to the value of 2.0 which indicates that measured hardness is independent of the load used.

The tentative conclusion to be drawn from the large number of studies made on dentine is that there is little difference in hardness with age, sex, location within a tooth or type of tooth, provided that the material is sound and that measurements are within the bulk material, remote from either the pulp chamber or the dentino-enamel junction, and not in regions containing interglobular dentine or secondary dentine.

A careful and systematic series of measurements carried out on dehydrated sections cut from the roots of teeth by von Komiya & Kröncke (1968) indicated a slight increase in hardness of the bulk dentine as measurements were taken further from the pulp. This could reflect the increasing amount of peritubular dentine in this direction.

The conclusion regarding the effects of the possible variables on the hardness of dentine seems to apply equally well to sound enamel from erupted teeth, although there is evidence that the hardness of the surface can be affected by exposure to the oral environment (von der Fehr, 1967). A recent study by Davidson, Hoekstra & Arends (1974) in which Knoop measurements were taken in three different directions with respect to the orientation of the enamel prisms showed that sound enamel could be regarded as a mechanically isotropic material for the test provided the resulting impression covered several enamel prisms.

Results from specific studies which appear to be typical for the two tissues are given in Table 10. These figures show that enamel is some five times harder than dentine.

Fracture

It is only very recently that a direct measurement of the fracture properties of human enamel and dentine has been attempted. This study by Rasmussen, Patchin, Scott & Hener (1976) has considerably advanced our knowledge of the relationship between the microstructure and mechanical properties of these materials. The measurement selected by these authors, namely the work to fracture, W_f, is clearly the most suitable for anisotropic brittle materials available only in small quantities, and the technique followed that used in earlier controlled

Table 10. *Typical Knoop hardness values for dentine and enamel ($kg\,mm^{-2}$)*

Authors	Material	No. of specimens	Knoop hardness number
Craig & Peyton (1958)	crown dentine	4	74 ± 6
Caldwell, Muntz, Gilmore & Pigman (1957)	anterior tooth enamel	15	365 ± 35
Caldwell *et al.* (1957)	posterior tooth enamel	9	393 ± 50

Table 11. *Comparison of the properties of a sintered and compacted hydroxyapatite and enamel*

	HAP–60K–1200C[a]	Enamel
Mineral loading (v/v%)	98	92
Density ($g\,cm^{-3}$)	3.1	2.96
Knoop hardness ($kg\,mm^{-2}$)	450	380
Young's modulus (GNm^{-2})	121	78
Compressive strength (MNm^{-2})	376	380
Tensile strength (MNm^{-2})	–	35

[a] Data of Rootare, Powers & Craig (1978).

fracture studies on various materials including bone and ceramics (Tattersall & Tappin, 1966). The specimen (see Fig. 11) is subjected to 3-point bending until fracture with a constant deformation-rate testing machine. The total work to fracture is then determined from the load versus extension curve and divided by the projected normal area of the two fracture surfaces. The grooving of the specimen to give a central V-shaped section ensures that the crack initiates in the tip of the section (at the apex) and that as it advances the length of the crack front increases so that catastrophic growth of the crack is limited. Thereafter

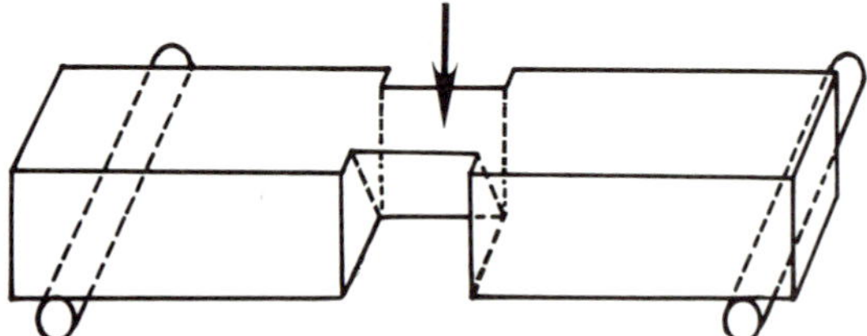

Fig. 11. Specimen for controlled fracture studies.

crack growth occurs under controlled conditions. The teeth used were sound premolars extracted for orthodontic reasons from 12–14 yr old children. Specimens of both enamel and dentine were cut such that the crack growth occurred essentially parallel or perpendicular to either the enamel rods or the dentinal tubules respectively.

W_f for dentine was found to be of the same magnitude as the high strain rate W_f value of 200 Jm^{-2} found for bone by Piekarski (1970), being 550 Jm^{-2} for fracture parallel to the tubular direction and 270 Jm^{-2} for fracture at right angles to the tubules. These values of W_f for dentine are some ten times larger than those of many ceramic materials, but some hundred times lower than typical W_f values for ductile materials such as copper (50 kJm^{-2}) or brass (30 kJm^{-2}). Rasmussen *et al.* (1976) concluded that although considerable non-recoverable energy absorbing processes were occurring, these were small in comparison with those occurring during the fracture of ductile materials and that in consequence dentine must be considered a brittle material. Support for this contention, they suggested, was provided by fractographs obtained by scanning electron microscopy which showed no evidence of blunting of crack tips or of macroscopic displacement prior to initiation of the crack. Examination of the fractographs of dentine provided evidence that the strength of the bond between the apatite crystals in peritubular dentine is poor. Fractographs perpendicular to the tubules showed limited evidence of pullout, and cleavage of the peritubular dentine was observed in parallel fractographs. In general the impression which emerged from study of the fractographs was that the bonding was such that the peritubular dentine acted as a cylindrical void rather than as an integral unit or fibre with respect to parallel fracture.

The difference between the W_fs for the two modes of fracture seems to be reasonably accounted for in terms of the orientation of the collageneous network between the tubules which, as has already been noted, appears to form in planes approximately perpendicular to the

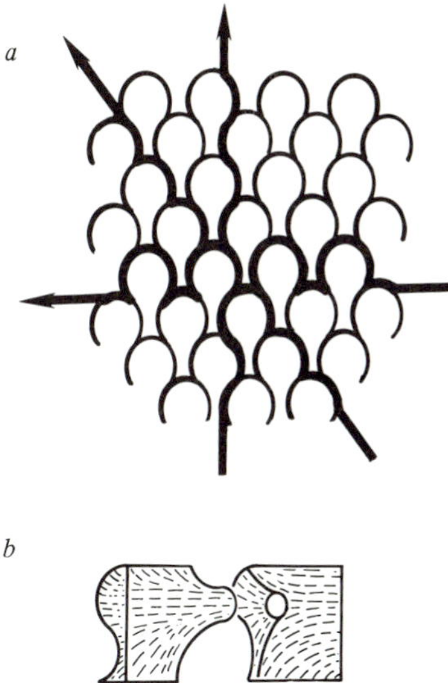

Fig. 12. *a*, fracture modes between prisms; *b*, cone-type fracture across prisms. After Rasmussen *et al.*, 1976.

tubules. Thus the energy required for a crack to grow perpendicular to these networks would be expected to be greater than for crack growth between the networks. The results of Rasmussen *et al.* (1976) for the fracture properties of enamel were equally interesting. For fracture parallel and perpendicular to the prism orientation, mean W_fs of 13 and 200 J m^{-2} were obtained. In the parallel direction the W_f is thus similar to brittle materials like alumina (40 J m^{-2}) beryllia (20 J m^{-2}) and magnesia (10 J m^{-2}). The gross anisotropy of enamel in this respect and the susceptibility to fracture parallel to the prism direction have long been appreciated by the dental profession and operative procedures followed for the preparation of cavities in carious teeth are designed to obviate this weakness in enamel (Pickard, 1961).

Fractographs of parallel specimens clearly indicated that fracture occurred around the enamel rods rather than through them. Fractographs of the perpendicular specimens were extremely rough and showed evidence that the prisms were acting as strong integral units or fibres. The enamel rods did not fracture perpendicularly to their long axis but exhibited cone-shaped fractures, the apex of the cones pointing towards the dentino-enamel junction. Rasmussen *et al.* (1976) explain this phenomenon, very plausibly, in terms of the known microstructure. The fracture surfaces observed parallel to the enamel rods were consistent with one or other of the three possible fracture modes between prisms depicted in two dimensions for the ideal key-hole shaped structure in Fig. 12a. As has already been noted there is an abrupt change in orientation of the crystallites on either side of a prism boundary and it is also very likely that there is more organic material in this region. The crystallite reinforcement of the structure as a whole is therefore weakest here and fracture is likely to occur principally, if not entirely, through the organic material. The cone type fracture observed across the prisms (see Fig. 12b) again emphasizes the relative weakness of the organic material; that the crack surfaces should follow the direction in which the long axes of the crystallites are orientated until the stress distribution at the crack tip is such that fracture perpendicular to the crystallites in the central core of each prism is more favourable on energy considerations. This concept is, of course, based on the reasonable assumption that cleavage down the long axes of the crystallites does not take place, and that crystal orientation within the prisms conforms to the model proposed by Helmcke (1967).

A comparison of the fracture toughness of enamel and dentine has also been carried out by Rasmussen *et al.* (1976). Using typical data for Young's modulus and W_f the ratio of the critical fracture stresses of the two materials when possessing the same geometry and crack depth may be deduced on the assumption that the Griffith equation is applicable. On this basis the critical fracture stress for dentine is between 1.4 and 4 times higher than that of enamel. If it is also taken into account that surface defects or cracks are more likely to occur in enamel the general clinical observation that enamel is the weaker material is substantiated.

At high crack propagation rates in dentine a different picture emerges. From an analysis of the direction taken by fractures within 33 incisor teeth broken by an impact, Andreasen (1972) concluded that the direction of fracture was unaffected by any structural orientation within the dentine (see Fig. 13). A similar conclusion was deduced by Renson, Boyde & Jones (1974) from a scanning electron microscope study of fracture surfaces produced when specimens subject to bending and shear failed catastrophically. Clearly the difference between the rates of crack propagation for these two sets of results must have some bearing on the totally different modes of fracture observed in that stress–relaxation mechanisms operating at the crack tip at low rates of propagation may well be virtually absent during catastrophic rupture.

The considerable fracture anisotropy of enamel reported by Rasmussen *et al.* (1976) was also observed in Andreasen's analysis of incisors fractured by impact. The fracture paths on both entering and leaving the teeth were always parallel to the general prism direction (Fig. 13).

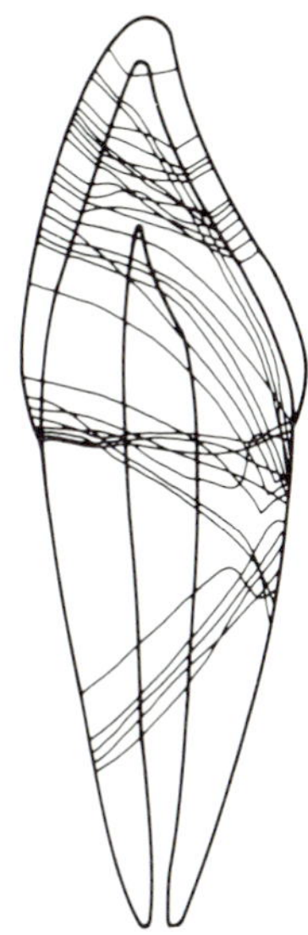

Fig. 13. Fracture paths through 33 incisor teeth broken by an impact. Reproduced from Andreasen (1972) with the kind permission of the author.

Viscoelastic properties

In view of the relatively high proportion of organic material and water that analysis shows to be present in dentine and the known viscoelastic behaviour of collageneous tissue (Viidik, 1968), it would be surprising if time-dependent mechanical behaviour were not exhibited by this material under certain stress conditions. In a study of the compression properties of dentine Craig & Peyton (1958) reported that on loading up to a stress below the proportional limit and unloading again a hysteresis loop was obtained. On examining the effect of load recycling on two specimens it was found that the modulus calculated from both the loading and unloading curves of the second cycle were equal and some 14% higher than that from the loading curve of the first cycle. Some slight creep was also observed in a specimen held under moderate static loading for 20 min and delayed elastic effects were discernible after unloading. Both effects were considerably enhanced when the stress applied was greater than the proportional limit. To date the only reasonable systematic study of the viscoelastic properties of human dentine reported in the general literature is that of Duncanson & Korostoff (1975). These authors examined the stress–relaxation behaviour in compression of dentine machined from the roots of maxillary incisors and canines and mandibular canines and premolars. These particular teeth were selected because their favourable cylindro-conical root formation enabled reasonably large thick-walled cylindrical specimens to be obtained with radial orientation of the dentinal tubules. Testing was carried out in a conditioning chamber at 37°C and 100% relative humidity. A specific constant compressive strain was applied to a cylindrical specimen loaded axially and the decay in stress noted as a function of time. The load decrease was, in all cases examined, exponential in character, and the normalized relaxation modulus $E_r(t)/E_r(0)$ when plotted against $\log t$ gave linear plots over up to four decades. After 6 h no observable change in load occurred. From this limiting stress the relaxed modulus, $E_r(\infty)$, was evaluated. Following the standard procedure for slowly varying $E_r(t)$ a first approximation to the logarithmic distribution of relaxation times $H_1(\tau)$ was calculated from the expression:

$$H_1(\tau) = -\frac{1}{2.303}\left[\frac{dE_r(t)}{d\log_{10}t}\right]_{t=\tau}$$

where the term in brackets may be found from the slopes of the above plots. The calculated $H_1(\tau)$ values ranged from 192–656 MNm^{-2} for the 25 specimens tested, with a mean of 380 ± 136. Attempts to relate the value of H_1 determined for a given specimen with age of patient, sex, % organic phase, site of tooth, reason for extraction and with the value of the relaxed modulus $E_r(\infty)$ were unsuccessful. However, the mean value obtained for the relaxed modulus $E_r(\infty)$ of 12.0 GNm^{-2} is in very good agreement with the majority of values for Young's modulus obtained in compression, tension and flexure by other workers.

An attempt by the authors to compare a theoretical retardation-time distribution $L_1(\tau)$ deduced on the basis of the applicability of linear viscoelastic theory from the experimentally determined relaxation-time distribution $H_1(\tau)$, with the $L_1(\tau)$ distributions obtained from creep measurements, appeared to be reasonable although because of the range of results for $H_1(\tau)$ experiment and $L_1(\tau)$ experiment some doubt exists as to whether these two viscoelastic functions are directly interconvertible (see Ferry, 1972). In a further paper which utilized some of the same data Korostoff, Pollack & Duncanson (1975) suggest that their results show that at the strain used, 0.6%, which from the results of other workers is known to be approximately 50% of the known ultimate compressive strain, permanent changes could occur within the dentine, the most probable being the collapse of the dentinal tubules. The worrying feature of this work is the large range in the values of $H_1(\tau)$

observed. However, in order to explain this behaviour more data are clearly required.

The mechanical behaviour of whole teeth

The structural behaviour of whole teeth *in vivo* when subject to loads is a difficult and complex subject, both experimentally and theoretically. The vitality of teeth, their size and involvement with one another and the difficulties of gaining access to them create enormous problems from the experimental point of view. Theoretically the generally irregular geometrical form, the number of different structures involved, and the uncertainty of some of the physical data, particularly for the periodontal membrane, are limiting factors.

In general this interest in the load–deformation characteristics of teeth stems from the orthodontists' need first to be able to predict and control the translational and rotational movements a tooth will be subject to under a prescribed force system and second to be able to assess whether the stress generated by the loaded tooth in the surrounding bone is likely to be greater than the capillary blood pressure and hence likely to interfere with normal resorption processes by restricting the flow of blood to the area.

Certainly the most elegant and informative of these analyses is that of Synge (1933). The three basic assumptions of the original theory are that the periodontal membrane is elastically (a) homogeneous, (b) isotropic and (c) incompressible. It is also implicitly assumed that both the tooth and the bony socket have infinite rigidity. The assumption that the membrane is incompressible appears to be all important for it leads to the result that the displacement of a tooth due to the application of a force to the crown is very small and varies as the cube of the membrane thickness. This is in contra-distinction to the displacement which would be obtained theoretically if the membrane were either a compressible elastic layer or, alternatively, if it behaved as a system of elastic cords, since in either case analysis shows that the displacement for a given force would be directly proportional to the thickness of the membrane. On the further assumption that the membrane had the same modulus of rigidity as natural rubber, Synge (1933) calculated that an axial load of 1 N would cause a displacement of 4 nm and a transverse load of 1 N applied to the incisal edge, a displacement of 260 nm which appears to justify his reference to the tightness of teeth in the title of his paper to the Royal Society (see also Waters, 1975).

Whereas the application of classical methods of stress analysis to teeth

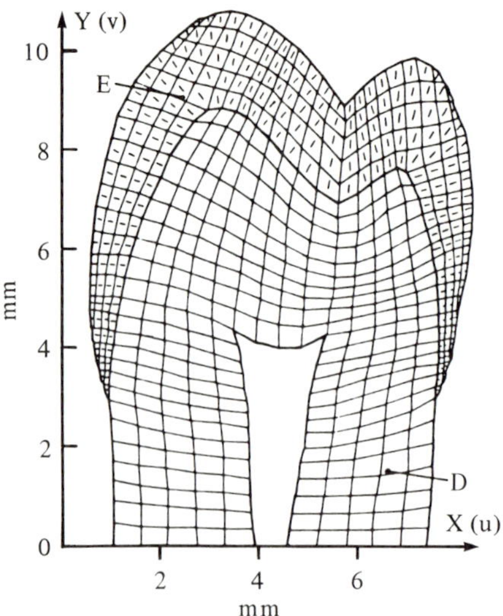

Fig. 14. Finite element mesh of model bucco-lingual section of crown of second mandibular premolar used by Yettram *et al.* (1976). D, dentine; E, enamel. The directions of the enamel prisms are shown by lines within the elements.

is necessarily limited in scope, the finite element technique can be used to advantage for irregular structures, such as teeth, which are composed of materials with differing properties. An interesting example of this technique is that of Yettram, Wright & Pickard (1976) who determined the stress distribution in a two-dimensional representation of a second premolar under masticatory forces; Fig. 14 shows the mesh used for the analysis. Fig. 15 shows the principal stress distribution for two point occlusal loading under conditions in which the enamel is assumed to be orthotropic. A similar pattern of behaviour was found when the enamel was assumed to be isotropic.

Clearly the enamel cap, by virtue of its higher modulus, is able to distribute the load to the root uniformly such that the dentine core remains relatively lightly stressed and the pulp is protected from undue stress concentrations. The way in which the masticatory forces flow around the enamel cap is shown diagrammatically in Fig. 16.

Prediction of mechanical behaviour by model systems

No account of the mechanical properties of tooth material would be complete without some reference to the careful and systematic analysis of the applicability of the various theoretical models of composite behaviour to the hard tissues, bone, enamel and dentine by Katz (1971).

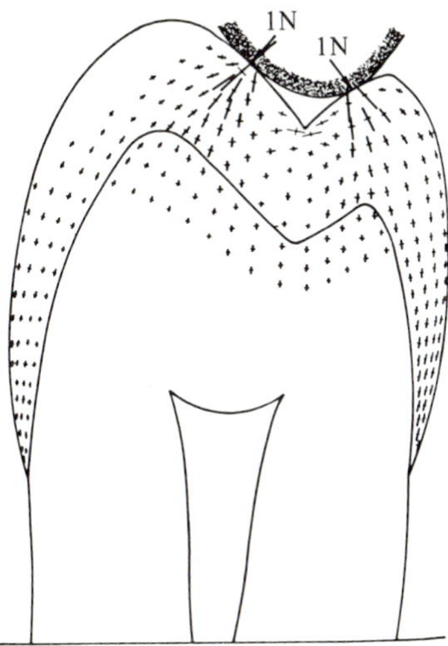

Fig. 15. Principal stress distribution for two point occlusal loading. Young's moduli of 70 and 23 GNm^{-2} along and perpendicular to the prisms were assumed. The magnitudes of the maximum and minimum principal stresses are represented by the lengths of the lines and the directions in which they act by their orientation. Compressive stresses are shown by thick lines and tensile stresses by thin lines. Only values above 16 kNm^{-2} are plotted. From Yettram *et al.* (1976).

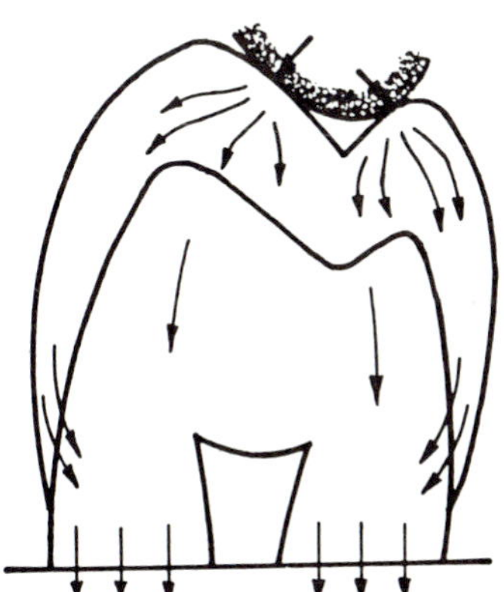

Fig. 16. Flow of masticatory forces through tooth structure. From Yettram *et al.* (1976). Figs. 14, 15 and 16 are reproduced with the kind permission of the authors and the publisher.

In brief, although more reliable data for the elastic constants of the various forms of organic matrix material and of the mineral fraction of hard tissues are required, it is concluded that as far as hard tissues in general are concerned the variation of modulus with volume concentration of the mineral component is not well represented by the Hashin spherical particle model (Hashin, 1962) and at best the Voigt, Reuss and Hashin–Shtrikman models (Hashin & Shtrikman, 1963) are capable only of giving the upper and lower bounds for the elastic modulus of the hard tissue concerned. It is true that the Hashin–Shtrikman upper and lower bounds are closer than the upper and lower bounds predicted respectively by the Voigt and Reuss models but the range between the bounds is so wide that application of this model provides very little useful information for hard tissues in general. However, considering the critical dependence of the modulus predicted by the Hashin-Shtrikman or Reuss models above 90% mineral volume concentration and the possible variation in the mineral loading for enamel, it would appear that, of all the hard tissues, enamel is the tissue most closely fitted by these latter models. The possible role played by the presence of free water within enamel is considered below.

The composite nature of dental hard tissues

Although the structure of the organic material in enamel is unknown, its chemical composition has been carefully analysed. It is known, for example, that certain phosphoproteins are present and that these have a very high affinity for hydroxyapatite (HAP). Certain other protein groups are also present which could interact electrostatically with the crystal surface and Fig. 17 shows the various possible combinations as proposed by Rölla (1977). Although these bonds are weaker than covalent bonds the crystals of HAP are so small that the interfacial area between the crystals and the matrix is about 10 $m^2 cm^{-3}$; the number of such bonds per unit volume is therefore large and the cohesive strength of the bulk material high. The smallness of the crystals and the fact that they are so tightly packed within the enamel prisms ensures that there are large numbers of interfaces or planes of potential cleavage preventing a crack propagating across the prisms by a simple Griffith process (Cook & Gordon, 1964) and increasing the energy required for fracture.

Crack propagation in the direction of the prisms requires less energy because there are fewer interfaces but the relatively small size of the prisms, their disposition, undulations in diameter, interweaving and interlocking, are all factors which will cause roughening of the crack tip and hence an increase in fracture energy.

Although attention has been drawn by Katz and his co-workers (Grenoble, Katz, Dunn, Gilmore & LingaMurty, 1972) to the importance of pores in influencing the elastic properties of hard tissues, the role played by the water in enamel, as far as the mechanical properties

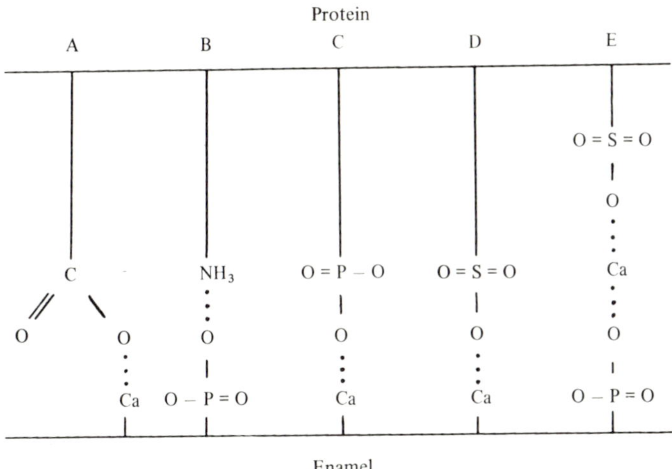

Fig. 17. Chemical groups (A–E) on macromolecules which are known to interact with hydroxyapatite. After Rölla (1977).

of enamel are concerned, seems to have been neglected. Studies of the pore structure by Dibdin (1969) and Moreno & Zahradnik (1973) indicate an extensive network of pores with a fairly broad distribution of pore radius between 1 and 10 nm but with the majority ranging between 0.9 and 2.5 nm.

The exact water content of enamel and the ratio of loosely to firmly bound water is still uncertain, but the 'free' or loosely bound water will obviously be present in these fine capillaries and the more strongly bound to the surfaces of the hydroxyapatite crystals and the organic material. That portion associated with the organic material will presumably help to characterize the modulus of the organic matrix. The presence of loosely bound water will ensure that the material as a whole has no internal stress which, although it can prove advantageous and is, of course, often deliberately built in, is nevertheless often a potential source of weakness. Fox (personal communication) has drawn attention to the fact that if enamel is stressed water will be forced out of the pores and that this constitutes an energy absorbing mechanism. Fox suggests that energy involved in this process is enhanced because the electrokinetic effects associated with the electrical double layer present on the pore walls (Klein & Amberson, 1929; Waters, 1968, 1971) will in effect increase the viscosity of the liquid. Although it is too early to say how this idea, if substantiated, will modify our ideas of the behaviour of enamel to stress, the principles involved could be important in extend-

Table 11. *Comparison of the properties of a sintered and compacted hydroxyapatite and enamel*

	HAP–60K–1200C[a]	Enamel
Mineral loading (v/v%)	98	92
Density ($g\,cm^{-3}$)	3.1	2.96
Knoop hardness ($kg\,mm^{-2}$)	450	380
Young's modulus GNm^{-2}	121	78
Compressive strength (MNm^{-2})	376	380
Tensile strength (MNm^{-2})	–	35

[a] Data of Rootare, Powers & Craig (1978).

ing the range of mechanical behaviour exhibited by synthetic composite materials.

The effect of incorporating, as nature does, roughly 2% by volume of a suitable organic material and approximately 6% of water with the mineral can be seen by comparing the properties of enamel with those for a highly compacted and sintered HAP (Rootare, Powers & Craig, 1978) (Table 11). It may be seen that whereas the hardness or Young's modulus of enamel are some 25% and 30% lower respectively, the compressive strength is as high. Unfortunately the fracture properties of this compact HAP have not been measured but Rootare *et al.* (1978) specifically mention its brittleness in comparison with enamel when discussing the results of a scratch test. However, a crude comparison of the modulus of resilience (i.e., energy to fracture per unit volume) calculated assuming failure to occur close to the proportional limit, gives 0.58 and 0.93 MJm^{-3} for the HAP composite and enamel, respectively; an improvement of about 50%. Nature has thus managed to retain comparable load bearing ability whilst making sacrifices in the hardness and the stiffness in order to improve the toughness. Moreover, nature's composite was formed at 37°C and atmospheric pressure whereas this man-made equivalent from (nearly) the same mineral involves compaction at 4 GNm^{-2} and sintering at 1200°C.

Viewed in terms of their mechanical properties alone teeth are remarkable structures, and with care will withstand the rigours of mastication for many years. It has to be borne in mind, however, that they are not solely designed to withstand the stresses of masticating and cutting food and distributing these stresses as uniformly as possible to the bony support but are so shaped that together, because of their disposition and the angular inclinations of their occlusal surfaces, they are capable of interdigitating and thus producing an extremely efficient masticatory system.

The author wishes to express his appreciation of the help given to him in producing this review by various colleagues, in particular Mrs M. Bradley, for her assistance with the literature survey, and Drs Moira Meredith Smith and A. F. Hayward of the Oral Anatomy Department for providing microphotographs and for the time they spent discussing various aspects of the structure of dental tissues. Finally, the author wishes to acknowledge his debt to Braden (1976) which was an invaluable source of information.

References

Andreasen, J. O. (1972). *Traumatic Injuries of the Teeth*. Copenhagen: Munksgaard.

Black, G. V. (1895). An investigation into the physical characters of the human teeth in relation to their diseases and to practical dental operations, together with the physical characters of filling materials. *Dental Cosmos*, **37**, 353–421, 469–84, 553–71, 637–61, 737–57.

Bowen, R. L. & Rodriguez, M. S. (1962). Tensile strength and modulus of elasticity of tooth structure and several restorative materials. *Journal of the American Dental Association*, **64**, 378–87.

Braden, M. (1976). Biophysics of the tooth. In *Frontiers of Oral Physiology*, ed. Y. Kawamura, Vol. 2, 1–37. Basle: S. Karger.

Caldwell, R. C., Muntz, M. L., Gilmore, R. W. & Pigman, W. (1957). Microhardness studies of intact surface enamel. *Journal of dental Research*, **36**, 732–8.

Cook, J. & Gordon, J. E. (1964). A mechanism for the control of crack propagation in all-brittle systems. *Proceedings of the Royal Society of London, Series A*, **282**, 508–20.

Cooper, W. E. G. & Smith, D. C. (1968). Determination of the shear strength of enamel and dentine. *Journal of dental Research*, **47**, (Abstract) 997.

Craig, R. G. & Peyton, F. A. (1958). Elastic and mechanical properties of human dentin. *Journal of dental Research*, **37**, 710–18.

Craig, R. G., Peyton, F. A. & Johnson, D. W. (1961). Compressive properties of enamel dental cements and gold. *Journal of dental Research*, **40**, 936–40.

Davidson, C. L., Hoekstra, I. S. & Arends, J. (1974). Microhardness of sound decalcified and etched tooth enamel related to the Ca content. *Caries Research*, **8**, 135–44.

Dibdin, G. H. (1969). The internal surface and pore structure of enamel. *Journal of dental Research*, **48**, 5, 771–6 (Part 1).

Dorrington, K. L. (1980). The theory of viscoelasticity in biomaterials. In *The Mechanical Properties of Biological Materials* (34th Symposium of the Society for Experimental Biology) ed. J. F. V. Vincent & J. D. Currey, pp. 289–314. Cambridge University Press.

Duncanson, M. G. & Korostoff, E. (1975) Compressive visco-elastic properties of human dentine: 1. Stress-relaxation behaviour. *Journal of dental Research*, **54**, 1207–12.

VON DER FEHR, F. R. (1967). The ^{32}P uptake and the hardness of unabraded, abraded and exposed human enamel surfaces. *Archives of oral Biology,* **12,** 623–38.

FERRY, J. D. (1972). *Viscoelastic Properties of Polymers*, 2nd edn. New York: John Wiley.

GILMORE, R. S., POLLACK, R. P. & KATZ, J. L. (1969). Elastic properties of bovine dentine and enamel. *Archives of oral Biology*, **15,** 787–96.

GRENOBLE, D. E., KATZ, J. L., DUNN, K. L., GILMORE, R. S. & LINGAMURTY, K. (1972). The elastic properties of hard tissues and apatites. *Journal of Biomedical Materials Research*, **6,** 221–33.

HAINES, D. J. (1968). Physical properties of human tooth enamel and enamel sheath material under load. *Journal of Biomechanics*, **1,** 117–25.

HANNAH, C. McD. (1970). *The tensile properties of human enamel and dentine*. I.A.D.R., Abstr. No. 113.

HARRISON, A. & LEWIS, T. T. (1975). The development of an abrasion testing machine for dental materials. *Journal of Biomedical Materials Research*, **9,** 341–53.

HASHIN, Z. (1962). The elastic moduli of heterogeneous materials. *Journal of Applied Mechanics*, **84,** 143–50.

HASHIN, Z. & SHTRIKMAN, S. (1963). A variational approach to the theory of elastic behaviour of multiphase materials. *Journal of the Mechanics and Physics of Solids*, 11, 127–40.

HEGDAHL, T. & HAGEBÖ, T. (1972). The load dependence in micro indentation hardness testing of enamel and dentin. *Scandinavian Journal of dental Research*, **80,** 449–52.

HELMCKE, J. G. (1967). Ultrastructure of enamel. In *Structure and Chemical Organization of Teeth*, ed. A. E. W. Miles, Vol. 2, pp. 135–63. New York: Academic Press, Inc.

KATZ, J. L. (1971). Hard tissue as a composite material. I. Bounds on the elastic behaviour. *Journal of Biomechanics*, **4,** 455–73.

KATZ, J. L. & UKRAINCIK, K. (1971) On the anisotropic elastic properties of hydroxyapatite. *Journal of Biomechanics*, **4,** 221–7.

KLEIN, H. & AMBERSON, W. R. (1929). A physico-chemical study of the structure of dental enamel. *Journal of dental Research*, **9,** 667–88.

VON KOMIYA, T. & KRÖNCKE, A. (1968). Mikrohärtemessungen in Wurzeldentin von Zähnen mit gesunder und gangränöser pulpa. *Dtsch. zahnärztel. Z.*, **23,** 975–80.

KOROSTOFF, E., POLLACK, S. R. & DUNCANSON, M. G. (1975). Visco-elastic properties of human dentin. *Journal of Biomedical Materials Research*, **9,** 661–74.

LEES, S. (1968). Specific impedance of enamel and dentine. *Archives of oral Biology*, **13,** 1491–500.

LEES, S. & ROLLINS, F. R. (1972). Anisotropy in hard dental tissues. *Journal of Biomechanics*, **5,** 557–66.

LEHMAN, M. L. (1967). Tensile strength of human dentine. *Journal of dental Research*, **46,** 197–201.

MEYER, E. (1908). Untersuchungen über Härteprüfung und Härte. *Z. Ver. dent. Ing.*, **52,** 645–835.

MORENO, E. C. & ZAHRADNIK, R. T. (1973). The pore structure of human dental enamel. *Archives of oral Biology*, **18,** 1063–8.

NADIA, A. (1950). *Theory of flow and fracture in solids*. 2nd edn. Vol. 1, London: McGraw Hill.

NEWMAN, H. H. & DISALVO, N. A. (1957). Compression of teeth under the load of chewing. *Journal of dental Research*, **36,** 2, 286–90.

PEYTON, F. A., MAHLER, D. B. & HERSHANOV, B. (1952). Physical properties of dentine. *Journal of dental Research*, **31,** 366–70.

PICKARD, H. M. (1961). *A manual of operative dentistry*. London: Oxford University Press.

PIEKARSKI, K. (1970). Fracture of bone. *Journal of applied Physiology*, **41,** 215–23.

RASMUSSEN, S. T., PATCHIN, R. E., SCOTT, D. B. & HENER, A. H. (1976). Fracture properties of human enamel and dentin. *Journal of dental Research*, **55,** 154–64.

Reich, F. R., Brenden, B. B. & Porter, N. S. (1967). *Ultrasonic imaging of teeth*. Report, Batelle Memorial Institute, Pacific Northwest Laboratory, Richland, Washington.

Renson, C. E. (1970). An experimental study of the physical properties of human dentine. PhD Thesis. University of London.

Renson, C. E., Boyde, A. & Jones, S. J. (1974). SEM of human dentine specimens fractured in bend and torsion tests. *Archives of oral Biology*, **19**, 447–57.

Renson, C. E. & Braden, M. (1971). The experimental deformation of human dentine by indentors. *Archives of oral Biology*, **16**, 563–72.

Renson, C. E. & Braden, M. (1975). Experimental determination of the rigidity modulus, Poisson's ratio and elastic limit in shear of human dentine. *Archives of oral Biology*, **20**, 43–7.

Rölla, G. (1977). Formation of dental integuments – some basic considerations. *Swedish dental Journal*, **1**, 241–51.

Rootare, H. M., Powers, J. M. & Craig, R. G. (1978). Sintered hydroxyapatite ceramic for wear studies. *Journal of dental Research*, **57**, 777–83.

Roydhouse, R. H. (1970). Punch shear tests for dental purposes. *Journal of dental Research*, **49**, 131–6.

Ryge, G., Foley, D. E. & Fairhurst, C. W. (1961). Micro-indentation hardness. *Journal of dental Research*, **40**, 1116–26.

Scott, D. B., Simmelink, J. W. & Nygaard, V. (1974). Structural aspects of dental caries. *Journal of dental Research*, **53**, 165–78.

Smith, D. C. & Cooper, W. E. G. (1971). The determination of shear strength. A method using a micro-punch apparatus. *British dental Journal*, **130**, 333–7.

Stanford, J. W., Paffenberger, G. C., Kampula, J. W. & Sweeney, A. B. (1958). Determination of some compressive properties of human enamel and dentine. *Journal of the American dental Association*, **57**, 487–95.

Stanford, J. W., Weigel, K. V., Paffenberger, G. C. & Sweeney, W. T. (1960). Compressive properties of hard tooth tissues and some restorative materials. *Journal of the American dental Association*, **60**, 746–51.

Synge, J. L. (1933). The tightness of teeth, considered as a problem concerning the equilibrium of a thin elastic membrane. *Philosophical Transactions of the Royal Society of London, Series A*, **231**, 435–70.

Tattersall, H. G. & Tappin, G. (1966). The work of fracture and its measurement in metals, ceramics and other materials. *Journal of Materials Science*, **1**, 296–301.

Tyldesley, W. R. (1959). Mechanical properties of human dental enamel and dentine. *British dental Journal*, **106**, 269–78.

Viidik, A. (1968). *Function and structure of collageneous tissue*. Goteborg: Elanders Boktryckeri Aktiebolag.

Waters, N. E. (1965). The indentation of thin rubber sheets by spherical indentors. *British Journal of applied Physics*, **16**, 557–63.

Waters, N. E. (1968). Electrochemical properties of human dental enamel. *Nature, London*. **219**, 62–3.

Waters, N. E. (1971). The selectivity of human dental enamel to ionic transport. *Archives of oral Biology*, **16**, 305–22.

Waters, N. E. (1975). *Scientific aspects of dental materials*, ed. J. A. von Fraunhofer. London: Butterworths.

Yettram, A. L., Wright, K. W. J. & Pickard, H. M. (1976). Finite element stress analysis of the crowns of normal and restored teeth. *Journal of dental Research*, **6**, 1004–11.

Yoon, H. S. & Newnham, R. E. (1969). Elastic properties of fluorapatite. *American Mineralogist*, **54**, 1193–7.

THE STRUCTURE AND BIOMECHANICS OF BONE

J. LAWRENCE KATZ

The Center for Biomedical Engineering, Rensselaer Polytechnic Institute, Troy, New York, USA

Introduction

Bone is a dynamically adaptable tissue whose form is continually undergoing subtle remodeling in order to conform to its functions. This 'form–function' relationship is the key to understanding the material properties and behavior of bone. Conversely, a complete description of the material properties and behavior of bone, especially when correlated with structure, can provide critical information necessary to help unravel the mechanisms which dictate its formation and growth.

Human bones may be described as belonging to one of the five general categories: long bones such as the femur (and other leg bones) and humerus (and other arm bones); short bones such as found in the wrist and ankle; flat bones such as the skull; irregular bones; sesamoid bones. However, in order to understand the properties and behavior of bone it is necessary to consider other levels of structure in addition to the macroscopic size and shape. Indeed, it is important to examine bone structure on many levels since all these levels interact in providing bone with its characteristic properties. A description found quite reasonable in illustrating the hierarchical aspects of the organization of bone, especially the long bones of the body, includes four levels of structure (Fig. 1).

On the molecular level, bone is unique within the body in that it contains a mineral-like component in addition to the organic components. The major organic component of bone, as of all the connective tissues of the body, is collagen. Additional organic components (protein polysaccharides, lipids, etc.), while only present in small amounts, play important roles in stabilizing the structural organization. This organic structure by itself would behave as a compliant material with good energy-absorbing characteristics (toughness) and other properties generally characteristic of polymers. The major inorganic phase of bone is hydroxyapatite present in the form of small crystallites about

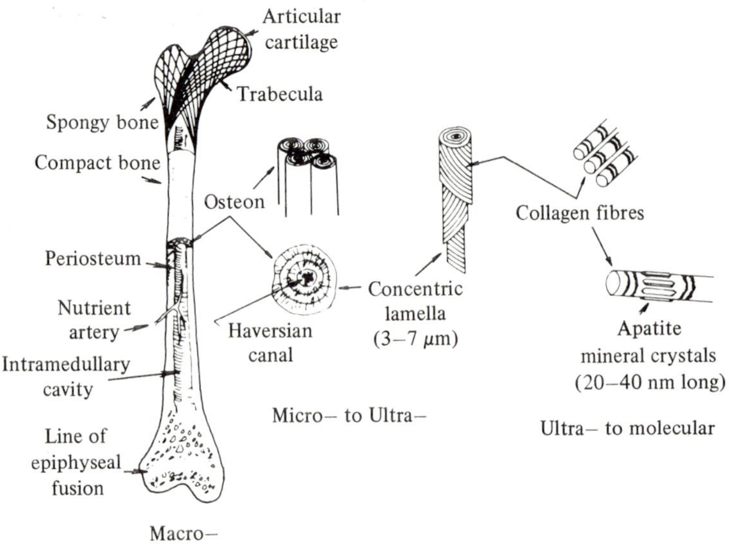

Fig. 1. Hierarchical levels of structural organization in a human long bone (femur). (From Park, 1979. Courtesy of Plenum Press and Dr. J. B. Park.)

5 nm × 20 nm × 40 nm (Glimcher & Krane, 1968). This material by itself is quite rigid and stiff compared to collagen; indeed its stiffness is approximately two-thirds that of steel. However, it is also quite brittle with poor impact resistance, i.e. it fractures quite readily. The tiny mineral crystallites are in intimate juxtaposition known as epitaxy in a highly organized geometrical arrangement with the collagen fibrils.

It is this admixture of organic and inorganic constituents which then comprises the second level of complex hierarchical structure, one which is heterogeneous and anisotropic (different properties in different directions). This superposition of two quite diverse materials with significantly different properties results in what is known as a composite material, i.e. a material whose properties differ significantly from those of the original components. Thus, bone represents an important materials concept which evolved naturally over millions of years, long before being rediscovered by engineers in modern times.

In mature cortical bone the collagen-hydroxyapatite composite is formed into densely packed concentric lamellar structures, the Haversian system or osteon (Fig. 2A). This is in contrast with another form of lamellar bone, called plexiform, which is often found in immature mammals such as young steers or pigs (Fig. 2B). The osteon is essentially a cylindrical structure whose long axis courses somewhat

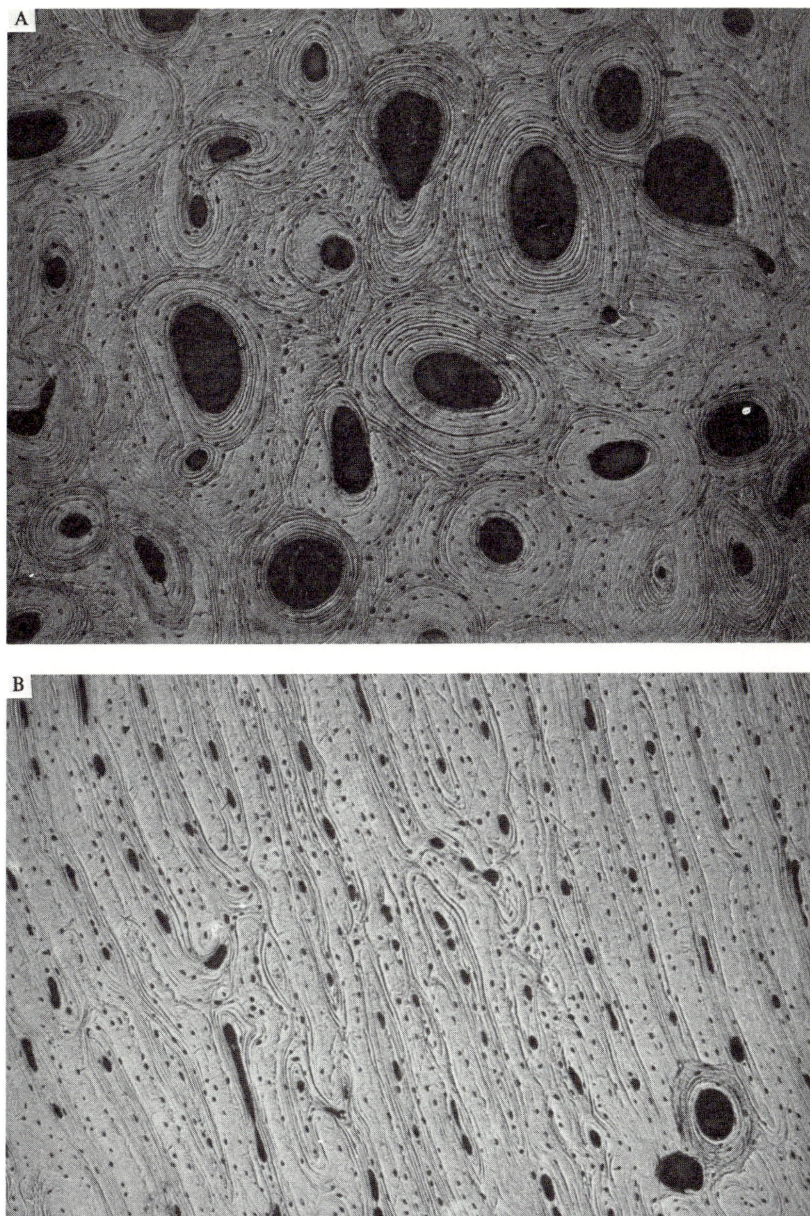

Fig. 2. Optical micrographs of transverse cross-sections showing the microstructure of human and bovine femora. Originals 75×. (A). Human Haversian bone; (B). Bovine plexiform bone. (From Yoon & Katz, 1976*a*. Courtesy of *Journal of Biomechanics*.)

irregularly along the long axis of the bone; it interfaces either with other osteons or with non-concentric lamellar groups, the interstitial lamellae. These highly organized lamellar groupings represent still another level of structural organization, denoted here as microstructural. The osteon represents a hierarchical composite material system as well. The concentric lamellar structures strongly bonded to one another and to the interstitial lamellae through the ground substance (cement line) may be viewed to form a composite material (a fiber reinforced composite); again, in this instance, a naturally formed system which predates engineering concepts by millions of years. Recent experiments appear to confirm that it is this microstructural organization which is in part responsible for some of the observations made regarding both the elastic and viscoelastic (or time-dependent deformational) behavior of macroscopic bone.

The final level of structure is of course the macroscopic. On this level, the size and shape of the whole bone are considered.

While each type of bone is of interest biomechanically, the long bones of the lower body, femur, tibia and fibula, are of significant interest because of their role as the structural supports of the body, i.e. they are weight-bearing. Thus, these long bones would appear to be particularly suited for studies aimed at elucidating 'form–function' relationships. Such studies entail several aspects: the first is the detailed investigation of the structure of the particular bone on both the ultra- and microstructural levels of organization; the second is a set of experiments aimed at delineating the anisotropic elastic and viscoelastic properties of specimens of compact bone. Finally, theoretical models can be applied to relate structure to function as well as to describe what mechanisms are involved in dissipating the mechanical energy applied to the bone.

Ultra- and micro-structure of single osteons and lamellar bone

Historical review

The structure of human osteon has been of considerable interest to many investigators for over a century (Howship, 1817; Todd & Bowman, 1845; Kolliker, 1873; Cohen & Harris, 1958; Cooper, Milgram & Robinson, 1966). The use of light microscopy easily established several quite indisputable structural features, i.e. the presence of the central Haversian canal, with its somewhat circumferentially disposed lamellae, as well as the presence of lacunae and canaliculae, but provided very little further information with any degree of certainty.

Attempts to determine the spatial orientations of collagen fibers in these osteons have also been many (Ranvier, 1875; von Ebner, 1887; Gebhardt, 1906; Ziegler, 1906; Weidenreich, 1923*a, b*; Ruth, 1947; Rutishauser, Huber, Kellenberger, Majno & Rouiller, 1950; Kellenberger & Rouiller, 1950; Smith 1960; Ascenzi & Bonucci, 1968). These primarily employed light microscopy, which suffers severely due to very small depth of focus. Even the advent of transmission electron microscopy (TEM) did not considerably increase our knowledge of this aspect of ultrastructural studies. Both replication techniques as well as direct thin sectioning techniques have been employed in preparing specimens for these investigations.

It was not until the successful application of scanning electron microscopy (SEM) by Boyde (1968, 1972), Boyde & Hobdell (1968, 1969*a, b*), Hobdell & Boyde (1969), Swedlow, Harper & Katz (1972) as well as the more recent work by Frasca, Harper & Katz (1976, 1977*a*, 1978*a, b*) that the complex relationship of collagen fibers could be appreciated; evidence of interconnecting fiber bundles and of regions of fibers, limited in extent, containing similarly oriented fibers (domains), has led investigators to reconsider some of the concepts concerning collagen fibers and the lamellar nature of bone. However, the determination of collagen fiber orientations necessitates a study of interlamellar surfaces, which are not readily accessible as are such natural surfaces as Haversian and Volkman canal linings and lacunar walls.

The recent development of sample preparation techniques has permitted SEM studies of interlamellar surfaces previously inaccessible by fracture or by standard cross-sectional cutting preparations (Frasca, Rao, Harper & Katz, 1976; Frasca *et al.*, 1978*a*). These studies have led to the determination of collagen fiber orientations in osteons distinguished by their appearance in polarized light and have yielded information which confirms and supplements the latest views on the lamellar nature of the osteonic bone specifically and lamellar bone in general (Frasca, Harper & Katz, 1976, 1977*a*, 1978*b*).

Collagen fiber orientations

Scanning electron microscopy studies

By compressing an isolated decalcified osteon between glass slides it is possible to cause many lamellae to detach themselves from each other and thus expose previously hidden lamellar surfaces (Fig. 3). The SEM study of osteons which when viewed with crossed polaroids appear characteristically bright, dark or somewhat intermediately so (Ascenzi & Bonucci, 1968) reveals correlations of fiber orientations and patterns

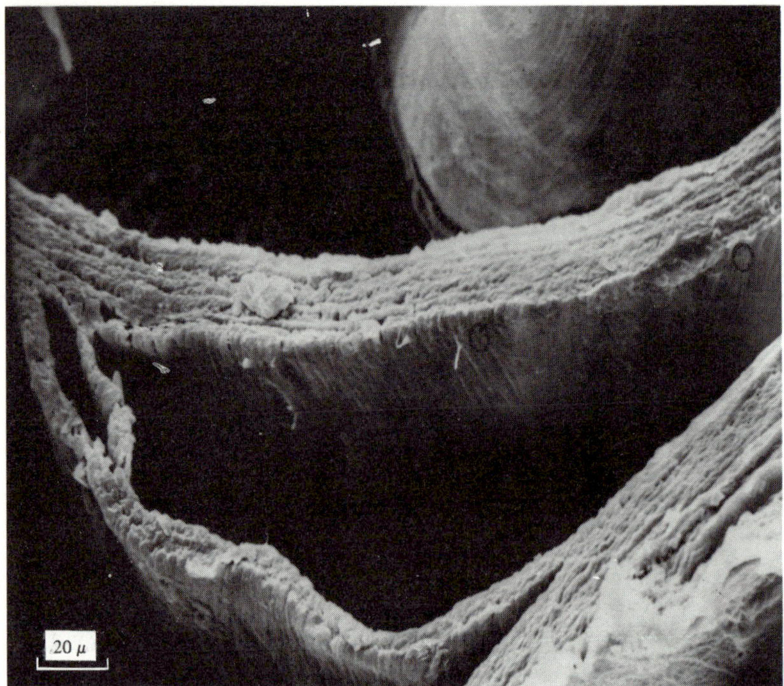

Fig. 3. Scanning electron micrograph of a specimen prepared by EDTA decalcification followed by mechanical manipulation. Portion of a single osteon showing separation of the lamellae and vertical fibers in some areas.

with their appearance in polarized light (Frasca *et al.*, 1977*a*). 'Dark' osteons have revealed a great predominance of longitudinal and nearly longitudinal fibers. 'Bright' osteons on the other hand have revealed the additional presence of circumferential fibers, which in many instances were found to coexist with longitudinal fibers. For so-called 'intermediate' osteons, longitudinal and steeply oblique fibers were found to dominate and occasionally oblique fibers, circumferential fibers, and coexisting fibers of various orientations were observed (Fig. 4).

Periosteal, endosteal and trabecular bone were similarly studied; in all three cases the predominance of fibers oriented in the direction of the bone axis or along the axis of the trabecula was observed (Frasca, Harper & Katz, 1976, 1977*a*, 1978*a, b*).

X-ray microdiffraction studies

X-ray diffraction studies of single osteons provide results statistical in nature and thus serve to complement the SEM observations of only a fraction of the lamellae constituting an osteon. Wide-angle diffrac-

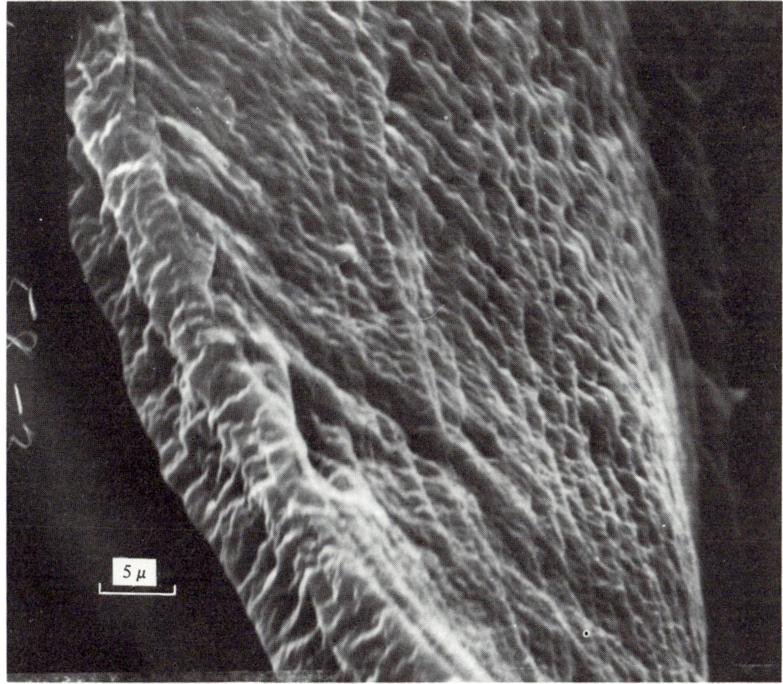

Fig. 4. Scanning electron micrograph of a specimen prepared by EDTA decalcification followed by mechanical manipulation. Two lamellae showing coexisting vertical and transverse fibers. (From Frasca *et al.*, 1978. Courtesy of *Journal of Dental Materials Research.*)

tion patterns due to the mineral phase are much easier to obtain than low-angle patterns for the collagenous phase, since apatitic crystallites are oriented epitaxially with their crystallographic axis (C-axis) along the direction of collagen fibrils and thus along collagen fiber orientations.

Wide angle patterns of 'dark' osteons taken with the incident x-ray beam perpendicular to the osteon axes, reveal a much greater statistical predominance of orientation of the C-axis of the apatitic crystallites along the osteon axis than is shown for 'bright' osteons, in agreement with SEM observations. 'Intermediate' osteons generally reveal a pattern intermediate between those obtained for 'dark' and 'bright' osteons. Similar diffraction patterns reveal a high degree of statistical crystallite orientations similar to that obtained for 'dark' osteons, adding confidence to the SEM observations (Frasca *et al.*, 1978*b*).

Contrary to previous views, no osteons were found which contained only circumferential fibers nor were lamellae observed which alternated regularly in their possession of longitudinal and circumferential fibers.

Structure of human lamellar bone

Interlamellar region

The SEM studies show the presence of collagen-rich lamellae 5–7 μm wide separated from each other by a much narrower region 1–2 μm wide. This interlamellar region is traversed by a low concentration of collagen fibers and fibrils and appears also to contain ground substance of unidentified biochemical nature (Swedlow *et al.*, 1972; Swedlow, Frasca, Harper & Katz, 1975).

The collagen fibers are continuous across lamellae; they extend across the tops of several joined lamellae and broken fibers originate from the surfaces of separated lamellae. This evidence thus establishes three-dimensional continuity of collagen fibers in lamellar bone first proposed by Rouiller (1956). The relatively low concentration of interconnecting fibers results in the separation of lamellae when samples are subjected to stresses during sample preparation.

The lamella

The SEM studies also reveal the lamella as consisting of several layers of fibers between which are sandwiched extremely thin layers of ground substance. These layers are apparently one fiber thick and are connected to each other across the ground substance by means of interconnecting fibers and fibrils, probably collagenous. Each fiber sheet is generally observed to contain fibers of one orientation or pattern, but in rare instances fiber orientations change even on what appears to be the same continuous surface (fiber sheet) (Swedlow *et al.*, 1972; Frasca *et al.*, 1976, 1978*a*, *b*) (Fig. 5).

Mechanical properties of single osteons

Historical review

Comprehensive studies on the mechanical properties of portions of single osteons using static experiments were first undertaken by Ascenzi & Bonucci (1964, 1965, 1967, 1968, 1971, 1972, 1976*a*, *b*) and Ascenzi, Bonucci & Simkin (1973). Among other things, they investigated the tensile properties of portions of longitudinally sectioned osteons, the compressive properties of portions of osteons cut out from thin cross-sections subjected to stresses directed both along the osteon axis and perpendicular to it, and the shear properties of portions of single osteons as they were slowly punched out of thin cross-sections; the

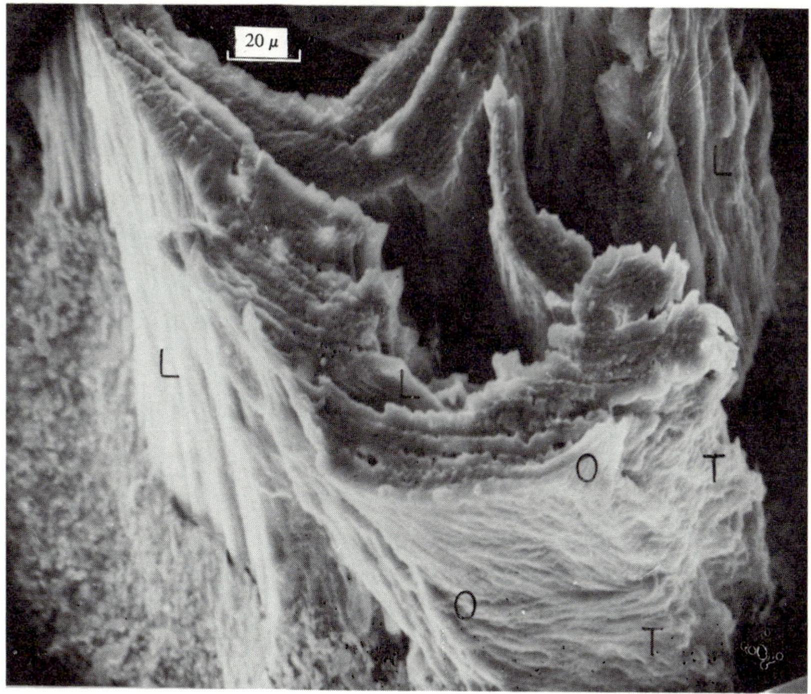

Fig. 5. Scanning electron micrograph of a specimen (courtesy of *Acta Anato.*) prepared by EDTA decalcification followed by mechanical manipulation. Portion of a single osteon showing several fiber orientations coexisting in what appears to be a single lamella. (From Frasca *et al.*, 1977*a*.)

maximum sample lengths were of the order of half a millimeter. The osteon samples studied were monitored for their degree of calcification and for their appearance in polarized light in an effort to relate structure with their mechanical properties. Unfortunately, the fact that these samples were not complete osteons limits the usefulness of these data in modeling calculations.

The techniques to isolate single osteons of considerable lengths mentioned above permitted cyclic dynamic studies in both the linear and torsional modes, not previously accessible with very short osteon samples. Samples were isolated by propagating fractures as closely as possible around the cement lines. For osteons longer than 2 mm, isolation of osteons at their natural boundaries becomes more difficult with increasing length, but even for osteons up to several centimeters long, most of the osteon was included in the sample and little additional bone was present (Frasca, 1974) (Fig. 6). However, these recent studies

Fig. 6. Osteons of various lengths isolated from cross-sections of varying thicknesses obtained from human tibial bone.

lack the sample control on calcification level and appearance in polarized light exercised in the previously mentioned studies.

Single osteons – shear storage and Young's moduli

Isolation of single osteons with aspect ratios (ratio of length to diameter) near to ten is a relatively difficult and time-consuming process. Consequently, few single osteon samples were tested; six in torsion and 12 in cyclic tension–compression. Of these some were tested only either wet or dry; only some of them were tested up to the onset of plasticity. These osteon samples were dissected from a tibial specimen taken from a normal 12-yr old male; the bone was frozen while fresh and stored until dissection.

Examination by light microscopy of the fracture site of a single osteon broken by stress overload and not by fatigue shows a somewhat spiral fracture inclined at about 30° with the vertical (Fig. 7) (Frasca, Harper & Katz, 1977*b*). It is interesting to note that torsional fractures of whole dog tibiae and femurs studied by Sammarco, Burstein, Davis & Frankel (1971) display angles of fracture of 40°. An analogy can be drawn, to a certain degree, between the structure of the osteon and that of the whole long bone and apparently the torsional fracture behavior of these two structures is quite similar.

Torsional studies of dry and wet bone microsamples containing

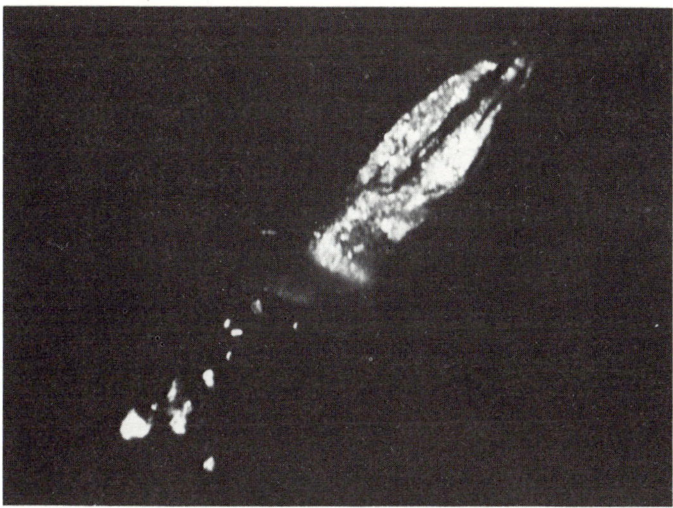

Fig. 7. Portion of a single human osteon which failed due to a stress overload in cyclic torsion. The fracture surface is inclined at about 30° with the vertical exposing the Haversian canal. Original × 75. (From Frasca *et al.*, 1977*b*. Courtesy of *Fracture* (1977) **3**.)

between one and a dozen osteons have provided results which are dependent upon the number of osteons in the microsample. These studies have provided some insight on the mechanical consequences of cement lines (Frasca, 1974).

(a). The shear modulus is highest for single osteons and decreases with increasing number of osteons in the microsample, reaching its lowest value in microsamples containing a half dozen or more osteons (Fig. 8).

(b). Dry single osteons exhibit a linear region in the stress–strain relationship up to strains of $\lesssim 10^{-2}$. Also, the greater the number of osteons in the microsample, the lower the strain value at which linearity ends; for samples containing about 10 osteons this strain value is as low as 10^{-4}. For wet samples this trend is almost totally destroyed for all number of osteons in the specimen tested and non-linearity sets in at the lower strain value of $\sim 10^{-4}$.

(c). A decrease in the ratio of the wet to dry values of the storage moduli and an increase for the loss moduli with increasing number of osteons is observed. These trends indicate that as the number of osteons in the sample increases, water becomes more effective in reducing the elastic modulus and in increasing the viscoelastic nature of the bone sample.

For osteonic bone, size and/or number and hydration effects seem to

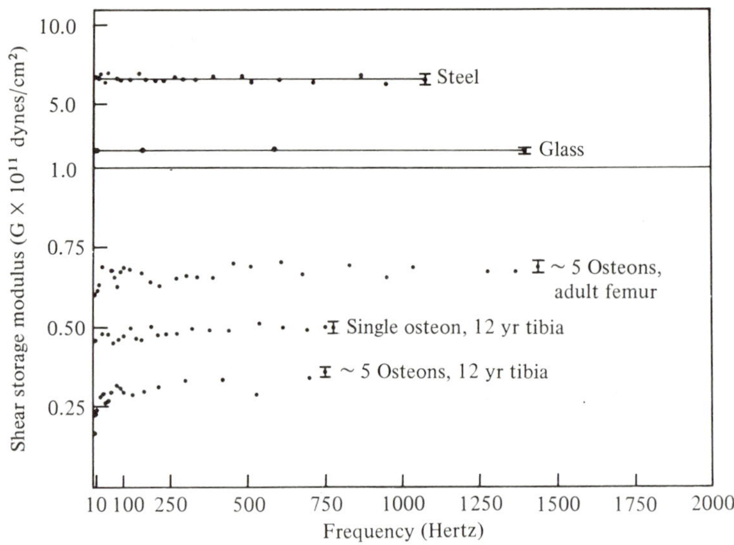

Fig. 8. Frequency dependence of the shear storage modulus for a single osteon and groups of osteons compared with a thin glass wire and a steel wire.

exist for onset of non-linear and plastic regions, shear storage modulus and internal energy loss. It is suspected that the reduction in elastic modulus as well as the reduction in strain values for onset of non-linearity and plasticity with increasing number of osteons is due to the general weakness of the osteon boundaries (cement line) relative to the osteon itself. Ground substance which acts as the cementing material is thought to be present in these boundaries and this along with the existing discontinuities in structure results in an area of relative weakness. Piekarski (1970) has shown that cracks tend to propagate preferentially along the osteon boundaries rather than through the osteons themselves during slow fracture.

These studies indicate a definite frequency-dependent shear storage modulus over the frequency range of 1–10^3 Hz. No such frequency dependence was noticed for either glass or steel, purely elastic materials, whereas decreases of ~10% for a single osteon sample and of ~10%–35% for three microsamples containing several osteons were noticed. These data point to the possibility of a frequency variation in modulus with number of osteons in the microsample (Fig. 8).

Such observations, although sparse, point to the same conclusions derived from the strain dependence studies, i.e., that (in the torsional mode at least) cement lines contribute significantly to the viscoelastic properties of bone (Lakes & Saha, 1979).

Mechanical properties of compact bone

Historical review

Studies of the mechanical properties of bone also go back over 150 years (Bevan, 1826; Wertheim, 1847; Rauber, 1876; Koch, 1917). In modern times the number of papers published is so large that no attempt will be made here to cover the field. However, much of this literature has been cited in reviews by Evans (1967, 1973), Yamada (1970) and Katz & Mow (1973). While many of these studies are concerned with the strength of bone, of particular interest herein are investigations of the anisotropic elastic and viscoelastic properties of specimens of fresh compact bone from the long bones of the lower body.

The earliest studies of the anisotropy in the elasticity of fresh compact bone were performed by Hulsen (1896); he found the average bending modulus of transverse specimens of tibial bone was 74% of the average longitudinal modulus. Yokoo (1952) observed, in compression tests on short cylinders of human compact bone, that the Young's modulus transverse to the bone axis was 56% of that for specimens taken along the axis; Dempster & Liddicoat (1952) studied samples in compression with orientations along both the radial and tangential directions; the average elastic moduli were 43% and 48% respectively of the stiffness for longitudinal specimens. Subsequent experiments by other investigators have confirmed these observations for both bovine (Burstein, Reilly & Frankel, 1973) and human cortical bone (Bargren, Bassett & Gjelsvik, 1974).

Effects of microstructure

Many studies attempted to relate such observations to microscopic structure. Walmsley & Smith (1957) related differences in Young's modulus in equine bone of different histological types. In a later study, they reported a relationship between Young's modulus and the degree of vascularity in going from the periosteal to endosteal surfaces for several species (Smith & Walmsley, 1959). Evans & Bang (1966, 1967) in their studies of lower extremity human bone suggested that increase in size or proportion of osteons (or osteon fragments), cement lines, and especially Haversian and Volkmann's canals reduces the stiffness of bone. Reilly, Burstein & Frankel (1974) reported that remodeled bone was more compliant (had a lower Young's modulus of elasticity) with an average modulus approximately 83% that of the mean value obtained from studies on primary compact bone. Currey (1975) reported similar observations where the elastic modulus of completely Haversian bone averaged only 69% of the value for essentially laminar specimens.

Effects of collagen fiber orientation

Olivo (1937) found higher stiffness in tension associated with specimens in which the collagen fibers in the osteons were presumably vertical or steeply spiralling (longitudinal) as compared to fibers which were circular or obliquely oriented (transverse) within the osteons. Of course the extensive studies by Ascenzi & Bonnuci cited earlier have had a great influence on 'form–function' studies at the microstructural level. However, since this work was on portions of single osteons rather than on standard specimens, no further discussion of these studies will be included here.

Effects of degree of mineralization

There have been many studies of the dependence of the elastic properties of the bone on the degree of mineralization. With one exception (Vincentelli & Evans, 1971; Evans & Vincentelli, 1974) the studies show an increase in elastic modulus with the degree of mineralization in humans and various other species (Vose & Kubala, 1959; Vose, 1962; Mather, 1968; Currey 1969*a*, *b*; Bonfield & Clark, 1973). This dependence will be discussed again later in the theoretical analysis of the behavior of bone as a composite material.

Viscoelastic properties of cortical bone

Strain rate dependence

It is only in recent years that serious consideration has been given to examining the time-dependent properties, i.e. viscoelastic properties of bone. McElhaney (1966) studied wet bovine bone in compressional constant strain–rate experiments at six different strain rates between 10^{-3} and $1.5 \times 10^{3}\,s^{-1}$. Since the leading portion of each of the stress–strain curves was linear, elastic moduli could be obtained; a threefold increase in the modulus was observed with increasing strain rate. Lakes & Katz (1974) transformed these data into the complex modulus representation in order to compare McElhaney's results with those of the dynamic experiments.

The work by Bird, Becker, Healer & Messer (1968) showed that the anisotropy in elastic modulus is not altered by strain rate, i.e. the increasing order of Young's modulus for specimens of radial, circumferential, and longitudinal orientation did not change with an increase in the rate of deformation. Wood (1971) observed an increase of 100% in elastic modulus over a strain rate range of 5×10^{-3} to $1.5 \times 10^{-2}\,s^{-1}$ for compact cranial bones. Crowninshield & Pope (1973) observed corresponding results for bovine tibial bone as did Bonfield & Clark (1973) for rabbit tibia.

More recently, Bonfield & Datta (1974) studied wet bovine tibial bone in compression at room temperature. At a strain of $3 \times 10^{-4}\,s^{-1}$, their results differed from those of McElhaney. In their analysis, Bonfield & Datta postulate that their results imply nonlinear behavior for bone. Tennyson, Evert & Niranjan (1972) utilized a split Hopkinson bar apparatus to study the compressive properties of wet bovine femoral bone at strain rates between 10 and $4 \times 10^{2}\,s^{-1}$. Lewis & Goldsmith (1975) also used this technique on bovine tibiae in compression. They derived an elastic modulus and viscosity coefficient from rheological models. Their results for elastic modulus compared favorably with those of Tennyson *et al.* (1972) while the discrepancy in viscous component was attributed to differences in respective rheological models. Currey (1975) obtained a strain-rate-dependent equation for the Young's modulus of wet bovine femoral bone in tension based on his experiments examining the relation of properties of bone to the extent of Haversian remodeling.

Wright & Hayes (1976) observed a more than doubling in elastic modulus with increasing strain rate on wet bovine femoral specimens in tension. They proposed a linear relationship between elastic modulus and log (strain) similar to the analysis of Crowninshield &Pope (1973) in tension, and McElhaney (1966) in compression. Wright & Hayes also observed the inverse relationship of elastic modulus to the proportion of secondary Haversian bone tissue reported by Currey (1975).

In addition to these strain-rate studies focussing on specimens of compact bone, there have been several investigations of the strain-rate sensitivity of whole bone. Burstein & Frankel (1968) and Sammarco, Burstein, Davis & Frankel (1971) studied both human and canine cortical bone over four decades in strain rate; no elastic moduli were reported. Panjabi, White & Southwick (1973) found only 5% increase in stiffness with increase in strain rate over their range in whole rabbit femora and ulnae.

Stress–relaxation

Lugassy & Korostoff (1969) performed stress–relaxation experiments on wet bovine femoral bone and whole dentin in compression. Lakes &

Katz (1974) transformed their data into the dynamic modulus representation again for comparison with other viscoelastic experiments. One important aspect of these results was the confirmation that bone is a non-linear viscoelastic material.

Creep

Apparently Wertheim (1847) was the first to report observations of viscoelastic behavior in bone, although he did not provide any data. The first data from creep tests on human femoral bone were reported by Yokoo (1952), Ko (1953) and Tsuda (1957). Craig & Peyton (1958) studied creep in dentin, while Smith & Walmsley (1959) did the same for bone. Sedlin (1965) studied fresh compact bone and developed rheological models to describe his data.

Currey (1964) studied the long-term creep behavior of wet bovine tibial and metacarpal bone in cantilever bending experiments. Creep deflection at a given time was found to be proportional to load. Although dry bone displayed less creep than wet bone, irreversible effects due to drying were not large.

Dynamic experiments

In addition to the type of viscoelastic studies described above, there are also dynamic experiments. These are especially important because an understanding of the physiological behavior of bone in motion requires analysis of the response of bone to dynamic loading conditions. Such experiments require the application of a sinusoidal strain to the specimen. Since the resulting stress is out of phase with the strain the dynamic response of the stress in terms of strain is then a complex mathematical function as is also the modulus of elasticity, E^* (Dorrington, 1980).

Smith & Keiper (1965) studied the dynamic mechanical behavior of wet specimens of human cortical bone over a frequency range of 5×10^2 to 3.5×10^3 Hz. They obtained both the real part (storage modulus) and imaginary part (loss modulus) of the complex modulus of elasticity. Thompson (1971) obtained the complex modulus using dynamic techniques on canine radial bone specimens. Bargren *et al.* (1974) studied the dynamic moduli of wet femoral bone at three frequencies 0.66, 5.2 and 7.5 Hz. Laird & Kingsbury (1973) studied the dynamic moduli of wet bovine humeral bone over the range 1–16 Hz. Black & Korostoff (1973) studied human tibial bone and observed some anomalous behavior near 2×10^2 Hz.

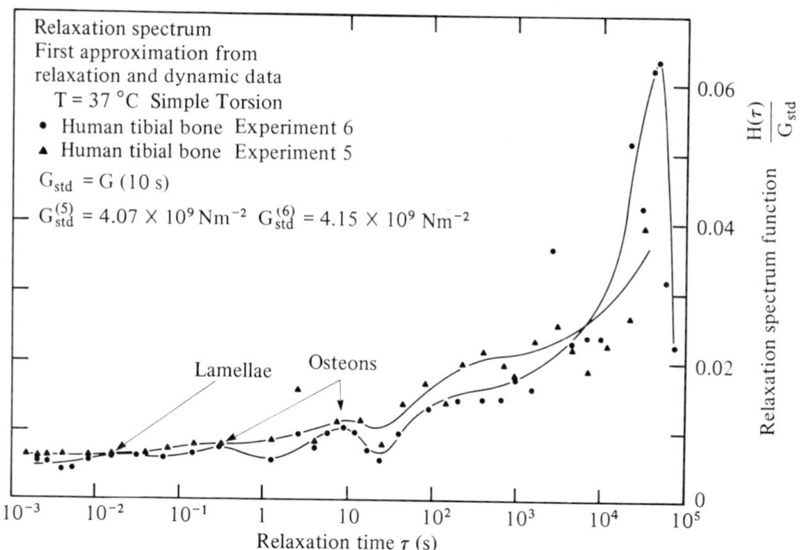

Fig. 9. The relaxation spectra for two specimens of wet human tibial bone in simple torsion. (From Lake & Katz, 1979*a*. Courtesy of *Journal of Biomechanics*.)

Shear

Interestingly enough, the earliest attempt to study the viscoelastic properties of bone in shear go back over 100 years to Rauber (1876). However, he did not note the time-dependent behavior probably due to the inadequacies of his instrumentation. Ninety years later Bonfield & Li (1967) measured shear moduli in the range of 5.6 to 6.1 GNm^{-2} for filaments of bovine tibial bone in torsion about the long bone axis. Thompson (1971) used a resonant torsional vibration technique to measure the shear properties of dry canine radii.

Recently, Lakes, Katz & Sternstein (1975, 1979) performed dynamic and stress-relaxation measurements in uniaxial torsion and biaxial torsion-tension on specimens of wet human and bovine compact bone at several temperatures. In addition to observing nonlinear behavior at long times, they observed that bone is a thermorheologically complex material. Data for two of the specimens of human bone have been used to establish a relaxation spectrum (Fig. 9) (Lakes & Katz, 1979*a*). Assuming that the bumps in the curve represent real physiological responses, Lakes & Katz (1979*a*) calculated the contributions to the loss tangent of cortical bone due to various dissipative effects as a function of time; these are plotted on Fig. 10. It is seen that several of the effects do give rise to peaks at times and with amplitudes which correspond to

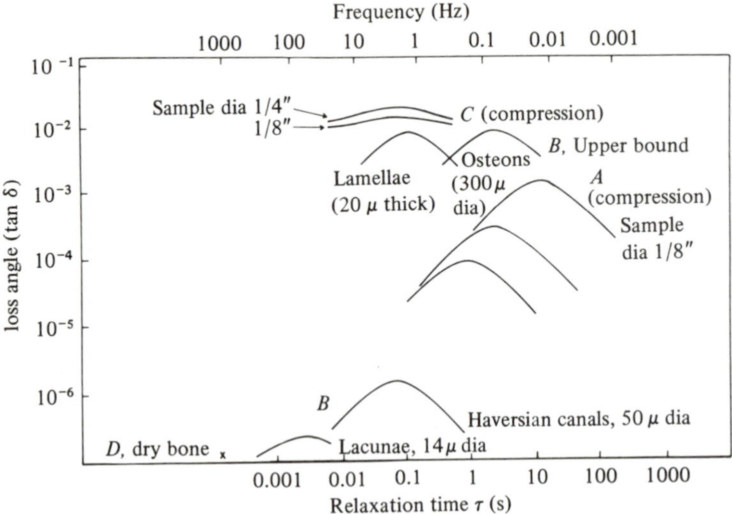

Fig. 10. The contribution of several relaxation mechanisms to the loss tangent of human cortical bone. *A*, homogeneous thermoelastic effect; *B*, inhomogeneous thermoelastic effect; *C*, fluid flow effect; *D*, piezoelectric effect. (From Lakes & Katz, 1979*a*. Courtesy of *Journal of Biomechanics*.)

bumps in the relaxation spectra plotted on Fig. 9. In addition, Lakes & Katz (1979*b*) obtained a constitutive equation to describe the dynamic and relaxation data for the human tibial bone at small strains and at body temperature.

The large viscoelastic effects observed by Currey (1964) in bending and by Lakes *et al.* (1979) in torsion for the long time experiments supported the present author's thesis that cement-line motion was important in establishing the viscoelastic properties of bone. Earlier, in microscopic studies of human bone in cantilever bending, Tischendorff (1951, 1952, 1954) reported observing small motions of the lamellae. Unfortunately, the time scale associated with motion was not reported. Lakes & Saha (1979) have recently reported cement line displacements of over 6 μm following prolonged loading of wet compact bone at constant torsional stress.

Ultrasonic wave propagation in cortical bone

Extensional waves

While, in effect, ultrasonic measurements are dynamic in nature due to the high frequencies involved, they provide some additional insight into 'form-function' relationships in bone not readily obtainable with mechanical experiments.

Flioriani, Debevoise & Hyatt (1967) determined Young's modulus in whole guinea-pig femur using extensional ultrasonic waves. Abendschein & Hyatt (1970) also used extensional waves at 10^5 Hz to determine Young's modulus, E, in specimens of human femoral and tibial cortical bone. The relationship between E and the ultrasonic wave velocity v_l is given by $E = \varrho v_l^2$ where ϱ is the density of the specimen. Their results correlated well with the Young's moduli they measured in static mechanical experiments. Greenfield *et al.* (1975) used ultrasonic methods to determine the elastic modulus of the human proximal radius *in vivo*. Their measurements also correlated well with *in vitro* studies.

Recently, Bonfield & Grynpas (1977) used extensional wave propagation to study the angular-dependent elastic behavior of both wet and dry specimens of bovine cortical bone cut at various angles relative to the long bone axis. They then compared their experimental data with a calculation by Currey (1969*b*) based on a fiber reinforced composite model of Cox (1952). Since there was poor agreement between the two, Bonfield & Grynpas (1977) concluded '. . . an alternative model is required to account for the dependence of Young's modulus on orientation'. This point will be returned to later in the section on composite modeling.

Bulk waves

However, the real *forte* of the ultrasonic technique is its use in examining the anisotropic elastic properties of bone on the microstructural level of organization. In this case bulk wave propagation of longitudinal (dilational) and transverse or shear (equivoluminal) waves are employed. Here, the velocities are related to the elements of the elastic stiffness (actually a fourth rank tensor). The Young's, shear and bulk (so-called technical) moduli can be obtained as a function of the angle of orientation between the specimen and stress axes directly from the elastic stiffnesses (Boas & Mackenzie, 1950). Thus symmetry-related properties can be compared both with structural studies and the results of mechanical measurements.

Lang (1970) used 5 MHz longitudinal and 2.25 MHz transverse waves to measure the elastic constants of bovine phalanx and femur. Lang chose hexagonal symmetry for his measurements based on considerations of molecular and ultrastructural symmetry. However, such conditions do not necessarily establish the appropriate symmetry at the level of resolution measured at the ultrasonic frequencies he employed.

Ambardar & Ferris (1976) in a less detailed study on bovine and ovine femora and tibiae, also assumed hexagonal symmetry and measured only two of the five constants necessary to describe completely the elastic properties of their specimens. Again such *a priori* assignment of symmetry, while possibly being correct, is improper without a complete ultrasonic analysis.

This consideration led Yoon & Katz (1976*a*, *b*) to perform a series of experiments on specimens of Haversian bone from human femora. After a critical theoretical review of ultrasonic propagation in materials with hexagonal symmetry (equivalent to transverse isotropy in microtextural systems) (Yoon & Katz, 1976*a*), they used pulse transmission techniques at 5 MHz to measure longitudinal and transverse velocities in eight symmetry-specified directions (Yoon & Katz, 1976*b*). The redundancies observed in certain specified equivalent directions verified the use of hexagonal symmetry and confirmed the microstructural observations made by them and others (Swedlow *et al.*, 1972; Frasca, 1974). Yoon & Katz (1976*c*) also showed that for these very low strain experiments there was a negligible piezoelectric contribution to the stiffness. In addition, they studied the dispersion of both the longitudinal and transverse wave propagation in the eight symmetry-related directions used in the above measurements (Yoon & Katz, 1976*d*).

In addition to the studies on whole bone specimens Katz and his colleagues also used ultrasonic wave propagation to study bone mineral alone obtained from deorganified samples of bone as well as to study synthetic and natural apatites (Gilmore & Katz, 1968; Grenoble *et al.*, 1972). These studies were undertaken to obtain the elastic properties of the mineral phase of the bone in order to model bone as a composite material; moduli determined from these experiments have been used in the composite modeling discussed below.

Surface waves

In addition to the two types of wave which can propagate through an extended solid medium, elastic waves may also be propagated along the surface of a solid; one type of such wave is known as Rayleigh waves. Lees & Rollins (1972) used an ultrasonic goniometer with a right-angle reflector to measure the surface elastic properties of dentin and enamel at 15 MHz or greater; they compared their experimental and calculated values with those predicted for bone by Katz & Ukraincik (1971). Rollins (personal communication) used the same technique to measure the longitudinal and surface wave velocities in bovine femur along three directions; parallel, at 45% and perpendicular to the bone axis. The velocities decreased with increasing angle to the long bone axis, confirming the anisotropy measured in mechanical experiments.

Whiting (1977) and Garcia, McNeill & Cobbold (1978) have also used such critical angle techniques to study wave propagation in bovine tibial specimens; while Adler (1977) also reported on surface wave measurements in bovine bone.

Recently, Lees (1974 – personal communication) and Yoon & Katz (1975 – work in progress) have initiated surface wave studies on human cortical bone in order to analyze the surface structure of such bone relative to the inner Haversian-type structure. Again the desire to understand 'form–function' relationships provided the motivation for such studies.

Modeling of bone as a hierarchical composite material

Introduction

Currey (1964) in his original discussion of bone as a composite material laid the groundwork for many of the subsequent treatments. Bonfield & Li (1967) attempted to use the linear rule of mixtures to model the Young's modulus of bone in terms of the relative volumes (V_{min}, V_{coll}) and elastic moduli (E_{min}, E_{coll}) of the mineral and collagenous components, respectively, i.e. $E_{bone} = V_{min}E_{min} + V_{coll}E_{coll}$ where $V_{min} + V_{coll} = 1$. However, the lack of data concerning the elastic properties of the component phases of bone proscribed any meaningful calculations. Katz (1971*a*) using the ultrasonic data obtained by his group as discussed earlier was able to provide bounds to the elastic properties of the various calcified tissues; Piekarski (1973) independently arrived at the same analysis. Both of them used both the Voigt (so-called parallel or uniform strain) model (Voigt, 1910) and Reuss (so-called series or uniform stress) model (Reuss, 1929) to obtain these bounds. In addition, they used more sophisticated models which provided tighter bounds on the elastic properties (Katz, 1971*b*). However, none of these models provided any real insight into the behavior of bone as a composite. Even Currey (1969*b*) in his elegant attempt to introduce both the effects of orientation and structural considerations into a model of bone as a fiber-reinforced composite did not prove successful, as was alluded to earlier when describing the experiment of Bonfield and Grynpas.

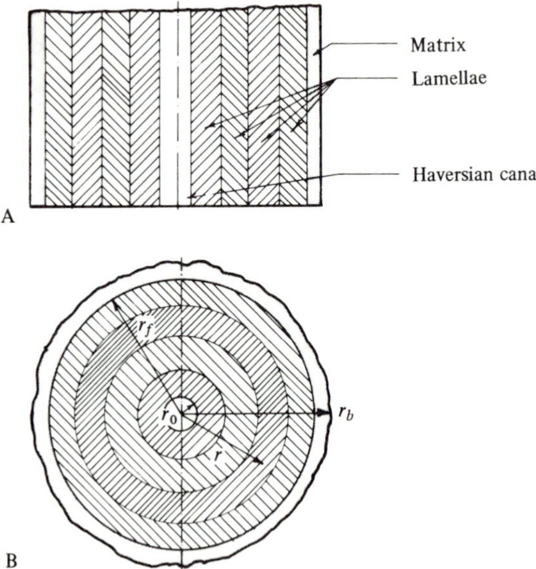

Fig. 11. A single osteon modeled as a hollow fiber for the Hashin–Rosen (1964) calculation. A, longitudinal section; B, transverse section. r_0 is the radius of the Haversian canal (hollow), r_f is the radius of the single osteon (fiber) and r_b is the radius of the osteon plus cement line (fiber and matrix). (From Katz, 1980. Courtesy of *Nature*.)

The present model

Katz (1976, 1980) and his colleagues (Katz & Ukraincik, 1971, 1972; Katz, Ukraincik & Swedlow, 1972; Katz & Thompson, 1977; Ukraincik & Katz, 1971) have suggested that it is insufficient to model compact bone as a superposition of mineral and collagen even if structure and orientation are taken into account. They proposed instead a hierarchical model based on two levels of structure. On the first level the single osteon (Haversian system) is assumed to be a coherent mechanical entity. It is modeled as a hollow right circular cylinder comprised of concentric lamellae surrounding the Haversian canal (Fig. 11). At the ultrastructural level the apatite crystallites are assumed to be infused throughout the collagen fibrils in some form of bonded relationship (the exact nature of which is not of concern here) with the crystallographic C axis of the apatite crystallites aligned epitaxially with the collagen fibrillar axes (Glimcher & Krane, 1968). On this basis the elastic modulus for such an osteon can be calculated as a function of the structural organization and relative orientations of the mineral–collagen assemblages. Such calculations are in good agreement with the values

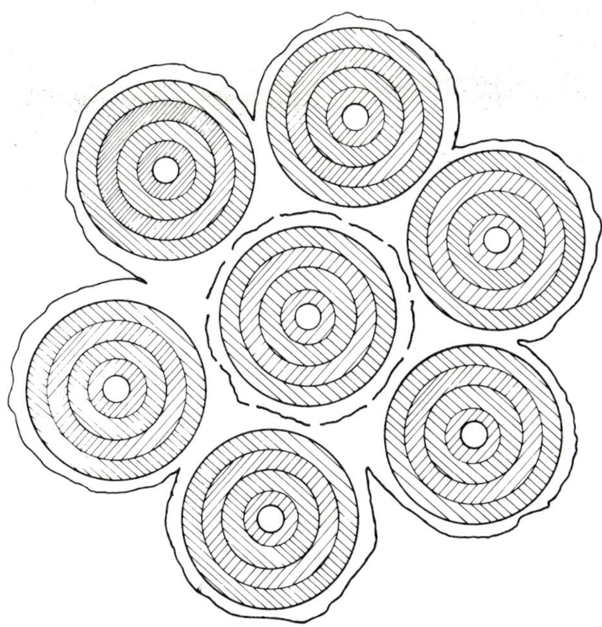

Fig. 12. A transverse cross-section of seven hollow fibers in a near-hexagonal array modeling a specimen of cortical Haversian bone. (The number of lamellae depicted for each osteon is for artistic purposes only and does not affect the calculation.) (From Katz, 1980. Courtesy of *Nature*.)

measured by Frasca (1974). These values are also comparable to the Young's moduli obtained for similar orientations using the model of Currey (1969*b*).

It is here that the second hierarchical level of modeling, the microstructural level, is introduced. On this level the osteons are assumed to be packed in a near-hexagonal arrangement (Fig. 12) as has been suggested as a result of the microscopic observations and ultrasonic wave propagation measurements described earlier. Here the osteon is treated as a hollow fiber immersed in a matrix (or binder) of ground substance, collagen and mineral in the cement line and adjacent interstitial lamellae.

Hashin & Rosen (1964) modeled the elastic behavior of hollow fiber-reinforced composite materials for both hexagonal and relaxed hexagonal symmetry of the fibers. This calculation, which generates sets of the five effective elastic constants necessary for hexagonal symmetry has been adapted by Katz and his colleagues in the studies referred to above. With parameters from the literature as well as those inferred from the results of trial calculations, a series of computer simulations

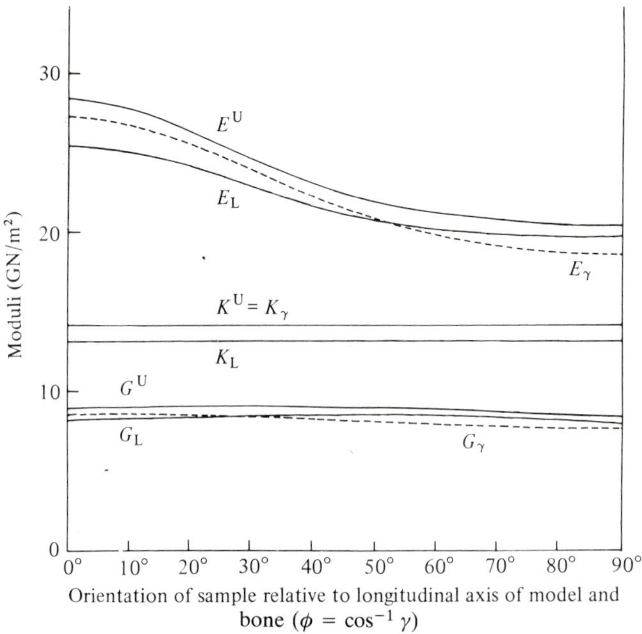

Fig. 13. The angular-dependent variation of the elastic moduli for human cortical bone with respect to the orientation of the long axis of the bone (ϕ) calculated from the ultrasonic data of Yoon & Katz (1976*a, b*) (Young's modulus, E_γ; bulk modulus, K_γ; and shear modulus, G_γ, are denoted by interrupted lines) compared with the theoretical curves representing the upper ^{U} and lower $_L$ bounds on the angular-dependent behavior of the elastic moduli using the Hashin-Rosen (1964) calculation (Young's moduli E^U, E_L; bulk moduli, K^U, K_L: and shear moduli, G^U, G_L respectively are denoted by full lines). Input parameters are: $E_{OST} = 40.0\,GNm^{-2}$; $E_{GS} = 10.0\,GNm^{-2}$; $\upsilon_{OST} = 0.30$; $\upsilon_{GS} = 0.25$; $V_0 = 0.10$ and $V_T = 0.70$. OST represents the single osteon; GS represents the matrix material; $V_0 = r_0^2/r_f^2$; $V_T = r_g^2/r_b^2$. (From Katz, 1980. Courtesy of *Nature*.)

generating upper and lower bounds for the elastic behavior of bone has been run using the Hashin–Rosen formalism.

The angular dependent elastic properties calculated from the five effective elastic constants obtained from the computer simulations were analyzed by comparison with the angular dependent behavior of the technical moduli obtained from the ultrasonic experiments of Yoon & Katz (1976a, b). The sets of parameters which provided the best fit to the ultrasonic properties were then used in subsequent calculations and analyses. Such a set of parameters and the curves derived therefrom are shown on Fig. 13. It will be seen that the results from ultrasonic measurements lie very close to the theoretical bounds over the entire range of angular orientation of specimens with respect to the long bone axis (Katz, 1980).

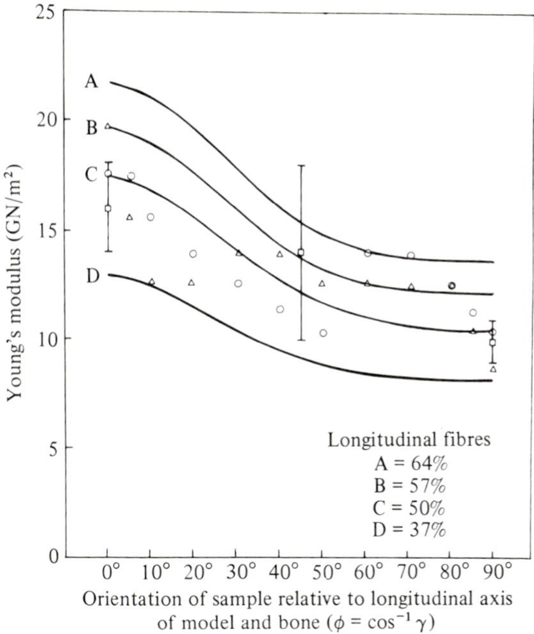

Fig. 14. The variation of Young's modulus for bovine femoral specimens with respect to the orientation of specimen axes cut at the angle ϕ to the long axis of the bone for wet (circles) and dry (triangles) conditions from the work of Bonfield & Grynpas (1977) compared with the angular-dependent behavior of Young's modulus (full line) computed using the Hashin–Rosen (1964) calculation. Each curve represents a different lamellar configuration of collagen fibers and hydroxyapatite crystallites within a single osteon. Curve A, 65% longitudinal fibers; curve B, 57% longitudinal fibers; curve C, 50% longitudinal fibers; and curve D, 37% longitudinal fibers. The remainder of the fibers are assumed to be horizontally arranged in each case. Other input parameters are: $E_{GS} = 5.5$ GNm^{-2}; $V_0 = 0.20$; and $V_T = 0.80$. The square symbols are readings taken from the study by Lugassy (1968). (From Katz, 1980. Courtesy of *Nature*.)

These parameters were then used to estimate appropriate input parameters for calculations to model the Bonfield & Grynpas (1977) data. In this case, several first-level calculations were first performed, e.g., the Young's modulus for osteons of several different collagen fiber–apatite crystallite orientations was computed. These were then used with the other parameters necessary to generate several sets of effective elastic constants by means of the Hashin–Rosen formalism. Curves A through D on Fig. 14 represent the angular-dependent behavior of Young's modulus for hexagonal microstructural assemblages of osteons of four different ultrastructural organizations, respectively. The circles plotted on the same figure represent data from Bonfield & Grynpas (1977) on wet specimens; the triangles represent their data on dry specimens.

It is clear that Bonfield & Grynpas' data are bounded by curves A to D. This is in contrast with the results of Currey's (1969*b*) model where the calculated values of E from 30° or so onwards are well below the experimental points. Even in the present modeling, one curve is insufficient to describe all the data of Bonfield & Grynpas. The reasons for this are many (Katz, 1981). One is due to the limits of the model; only one size of osteon can be handled in a given calculation and certain of the input parameters must be inferred since there are no experimental values for the mechanical properties of proteoglycans and the other organic components of bone. A second is that even specimens of bone cut from regions quite close together may differ sufficiently in micro- and/or ultra-structure to give rise to slightly different properties. Thus, it is not to be expected that all the experimental points should of necessity lie on a single curve. The additional points (square symbols) plotted on Fig. 14 are from the study by Lugassy (1968); good agreement with the simulated models is again observed.

At present the model is being extended to incorporate osteons of different sizes and structure as well as to include interstitial lamellae simultaneously. Further plans include use of the Correspondence Principle to extend the elastic calculation to model the viscoelastic properties of bone as a hierarchical composite.

Summary

An attempt has been made to present a uniform view of studies of bone which have as their *raison d'etre* an understanding of the 'form–function' relationship. This is the key to understanding the mechanisms controlling the growth, development and behavior of bone as a dynamic biological system in its response to the mechanical stresses imposed through both normal and pathological conditions.

Studies of structure at various levels of organization were presented both in themselves as well as where correlated with mechanical measurements; similarly, mechanical measurements alone were described as well as studies in which the mechanical properties were correlated with structural studies. These studies are presenting sufficient details for a picture to begin to form with enough resolution to provide a standard against which to test models which attempt to describe Wolff's Law mechanisms (form predicating function and vice versa) as well as models which attempt to understand what is responsible for the material properties of bone.

The hierarchical composite modeling which closes this paper is such an attempt at the latter problem. The modeling is rooted in both the structural studies and mechanical measurements thus closing the loop.

The background for this material was prepared during 1978 while the author was on sabbatical leave at Harvard Medical School and Children's Hospital, Boston, Mass. For the hospitality extended to him he wishes to thank Dr Melvin Glimcher, Harriet Peabody Professor of Orthopaedic Surgery. The author also wishes to thank the John Simon Guggenheim Foundation for appointing him a Fellow for 1978.

The original work from the author's laboratory quoted herein is the result of the dedication and devotion to research of many colleagues and former students too numerous to name; the author wishes to take this opportunity to thank them all again. Finally, such efforts require considerable financial support. The author would like to thank the National Institute of Dental Research, NIH, which for the past 16 years has provided this support thus making the efforts possible.

References

Abendschein, W. & Hyatt, G. W. (1970). Ultrasonics and selected physical properties of bone. *Clinical Orthopaedics*, **69**, 294–301.

Adler, L. (1977). Ultrasonic surface wave propagation in bone. In *Proceedings of the 30th Annual Conference on Engineering in Medicine and Biology*, p. 364. Chevy Chase, MD: Alliance for Engineering in Medicine and Biology.

Ambardar, A. & Ferris, C. D. (1976). A simple technique for measuring certain elastic moduli in bone. *Biomedical Sciences Instrumentation*, **12**, 23–7.

Ascenzi, A. & Bonucci, E. (1964). The ultimate strength of single osteons. *Acta Anatomica*, **58**, 160–3.

Ascenzi, A. & Bonucci, E. (1965). The measurements of the tensile strength of isolated osteons as an approach to the problem of intimate bone texture. *Calcified Tissue Research*, **31**, 325–35.

Ascenzi, A. & Bonucci, E. (1967). The tensile properties of single osteons. *Anatomical Record*, **158**, 375–86.

Ascenzi, A. & Bonucci, E. (1968). The compressive properties of single osteons. *Anatomical Record*, **161**, 377–92.

Ascenzi, A. & Bonucci, E. (1971). A micromechanic investigation on single osteons using a shearing strength test. *Israel Journal of Medical Sciences*, **7**, 471–2.

Ascenzi, A. & Bonucci, E. (1972). The shearing properties of single osteons. *Anatomical Record*, **172**, 499–510.

Ascenzi, A. & Bonucci, E. (1976*a*). Relationship between ultrastructure and pin test in osteons. *Clinical Orthopaedics*, **121**, 275–94.

Ascenzi, A. & Bonucci, E. (1976*b*). Mechanical similarities between alternate osteons and cross-ply laminates. *Journal of Biomechanics*, **9**, 65–71.

Ascenzi, A., Bonucci, E. & Simkin, A. (1973). An approach to the mechanical properties of single osteonic lamellae. *Journal of Biomechanics*, **6**, 277–35.

Bargren, J. H., Bassett, C. A. L. & Gjelsvik, A. (1974). Mechanical properties of hydrated cortical bone. *Journal of Biomechanics*, **7**, 239–45.

Bevan, B. (1826). On the strength of bone. *Philosophical Magazine*, **68**, 181–2.

Bird, F., Becker, H., Healer, J. & Messer, M. (1968). Experimental determination of the mechanical properties of bone. *Aerospace Medicine*, **39**, 44–8.

Black, J. & Korostoff, E. (1973). Dynamic mechanical properties of viable human cortical bone. *Journal of Biomechanics*, **6**, 435–8.

Boas, W. & MacKenzie, J. K. (1950). Anisotropy in metals. In *Progress in metal physics*, ed. B. Chalmers, pp. 90–120. New York: Interscience.

BONFIELD, W. & CLARK, E. A. (1973). Elastic deformation of compact bone. *Journal of Materials Science*, **8,** 1590–4.

BONFIELD, W. & DATTA, P. K. (1974). Young's modulus of compact bone. *Journal of Biomechanics*, **7,** 147–9.

BONFIELD, W. & GRYNPAS, M. D. (1977). Anisotropy of Young's modulus of bone. *Nature, London*, **270,** 453–4.

BONFIELD, W. & LI, C. H. (1967). Anisotropy of nonelastic flow in bone. *Journal of Applied Physics*, **38,** 2450–5.

BOYDE, A. (1968). SEM of collagen-free calcified connective tissues. *Beitr Elektronenmikroskop Direktabb Oberfl.*, **1,** 213–22.

BOYDE, A. (1972). SEM studies of bone. In *The biochemistry and physiology of bone*, ed. G. H. Bourne, vol. 1, pp. 259–310. New York: Academic Press.

BOYDE, A. & HOBDELL, M. H. (1968). SEM of bone. *Calcified Tissue Research*, **2,** 4–5.

BOYDE, A. & HOBDELL, M. H. (1969*a*). SEM of lamellar bone. *Zeitschrift für Zellforschung*, **93,** 213–31.

BOYDE, A. & HOBDELL, M. H. (1969*b*). SEM of primary membrane bone. *Zeitschrift für Zellforschung*, **99,** 98–108.

BURSTEIN, A. H. & FRANKEL, V. H. (1968). The viscoelastic properties of some biological materials. *Annals of the New York Academy of Sciences*, **146,** 158–65.

BURSTEIN, A. H., REILLY, D. T. & FRANKEL, V. H. (1973). Failure characteristics of bone tissue. In *Perspectives in Biomedical Engineering*, ed. R. M. Kenedi, pp. 131–4. Baltimore: University Park Press.

COHEN, J. & HARRIS, W. (1958). The three-dimensional anatomy of Haversian systems. *Journal of Bone and Joint Surgery*, **40A,** 419.

COOPER, R. R., MILGRAM, J. W. & ROBINSON, R. A. (1966). Morphology of the osteon: an electron microscope study. *Journal of Bone and Joint Surgery*, **48,** 1239–71.

COX, H. L. (1952). The elasticity and strength of paper and other fibrous materials. *British Journal of Applied Physics*, **3,** 72–9.

CRAIG, R. G. & PEYTON, F. A. (1958). Elastic and mechanical properties of human dentin. *Journal of dental Research*, **37,** 710–18.

CROWNINSHIELD, R. D. & POPE, M. H. (1973). The response of compact bone in tension at various strain rates. *Annals of Biomedical Engineering*, **2,** 217–25.

CURREY, J. D. (1964). Three analogies to explain the mechanical properties of bone. *Biorheology*, **2,** 1–10.

CURREY, J. D. (1969*a*). The mechanical consequences of variation in the mineral content of bone. *Journal of Biomechanics*, **2,** 1–11.

CURREY, J. D. (1969*b*). The relationship between the stiffness and the mineral content of bone. *Journal of Biomechanics*, **2,** 477–80.

CURREY, J. D. (1975). The effects of strain rate, reconstruction, and mineral content on some mechanical properties of compact bone. *Journal of Biomechanics*, **8,** 81–6.

DEMPSTER, W. T. & LIDDICOAT, R. T. (1952). Compact bone as a nonisotropic material. *American Journal of Anatomy*, **91,** 331–62.

DORRINGTON, K. (1980). The theory of viscoelasticity in biological materials. In *The Mechanical Properties of Biological Materials* (34th Symposium of the Society for Experimental Biology) ed. J. F. V. Vincent & J. D. Currey, pp. 289–314. Cambridge University Press.

EVANS, F. G. (1967). Bibliography of the physical properties of the skeletal system. *Art. Limbs*, **11,** 48.

EVANS, F. G. (1973). *Mechanical properties of bone*. Springfield: Thomas.

EVANS, F. G. & BANG, S. (1966). Physical and histological differences between human fibular and femoral compact bone. In *Studies on the anatomy and function of bones and joints*, ed. F. G. Evans, pp. 142–55. Berlin: Springer-Verlag.

EVANS, F. G. & BANG, S. (1967). Differences and relationships between the physical properties and microscopic structure of human femoral, tibial, and fibular cortical bone. *American Journal of Anatomy*, **120,** 79–88.

Evans, F. G. & Vincentelli, R. (1974). Relations of the compressive properties of human cortical bone to histological structure and calcification. *Journal of Biomechanics*, **7,** 1–10.

Flioriani, L. P., Debevoise, N. T. & Hyatt, G. W. (1967). Mechanical properties of healing bone by the use of ultrasound. *Surgical Forum*, **18,** 468–70.

Frasca, P. (1974). Structure and mechanical properties of human single osteons. Ph.D. Dissertation. Rensselaer Polytechnic Institute, Troy, N.Y.

Frasca, P., Harper, R. A. & Katz, J. L. (1976). Isolation of single osteons and osteon lamellae. *Acta Anatomica*, **95,** 122–9.

Frasca, P., Rao, H., Harper, R. A. & Katz, J. L. (1976). New method of bone preparation for collagen fiber orientation studies by means of SEM. *Journal of Dental Research*, **55,** 372–5.

Frasca, P., Harper, R. A. & Katz, J. L. (1977*a*). Collagen fiber orientation in human secondary osteons. *Acta Anatomica*, **98,** 1–13.

Frasca, P., Harper, R. A. & Katz, J. L. (1977*b*). Mechanical failure on the microstructural level in Haversian bone. *Fracture*, **3,** 1167–72.

Frasca, P., Harper, R. A. & Katz, J. L. (1978*a*). A new technique for studying collagen fibers and ground substance in bone with scanning electron microscopy. *Microscopica Acta*, **3,** 211–14.

Frasca, P., Harper, R. A., & Katz, J. L. (1978*b*). Mineral and collage fiber orientation in human secondary osteons. *Journal of dental Research*, **57,** 526–33.

Garcia, B. J., McNeill, K. G. & Cobbold, R. S. C. (1978). Propagation of ultrasound in bone: longitudinal and shear wave reflection and transmission coefficients. *Program and Abstracts, 3rd Intern. Symp. on Ultrasonic Imaging and Tissue Characterization*, pp. 47–51. Gaithersburg, Maryland: National Bureau of Standards.

Gebhardt, W. (1906). Uber funktionell Wichtige Anordnungsweisen der grosseren und feineren Bauelemente des Wirbeltierknochens. *Archiv für Entwicklungsmechanik*, **20,** 187–322.

Gilmore, R. S. & Katz, J. L. (1968). Elastic properties of apatites. *Proc. Int. Symp. Structural Properties of Hydroxyapatite and Related Compounds*. Washington, D.C.: National Bureau of Standards.

Glimcher, M. J. & Krane, S. M. (1968). The organization and structure of bone and the mechanism of calcification. In *Treatise on collagen*, ed. G. N. Ramachandran & B. S. Gould, Vol. 2B, pp. 68–251. New York: Academic Press.

Greenfield, M. A., Craven, J. D., Wisko, D. S., Huddleston, A. L., Friedman, R. & Stern, R. (1975). The modulus of elasticity of human cortical bone: an in vivo measurement and its clinical implications. *Radiology*, **115,** 163–6.

Grenoble, D. E., Katz, J. L., Dunn, K. L., Gilmore, R. S. & Murty, K. L. (1972). The elastic properties of hard tissues and apatites. *Journal of Biomedical Materials Research*, **6,** 221–33.

Hashin, Z. & Rosen, B. W. (1964). The elastic moduli of fiber reinforced materials. *Journal of Applied Mechanics*, **31,** 223–32.

Hobdell, M. H. & Boyde, A. (1969). Microradiography and SEM of bone sections. *Zeitschrift für Zellforschung*, **94,** 487–94.

Hulsen, K. K. (1896). Specific gravity, resilience and strength of bone. *Bulletin of Biological Laboratories*, **1,** 7–35.

Katz, J. L. (1971*a*). Hard tissue as a composite material I. Bounds on the elastic behavior. *Journal of Biomechanics*, **4,** 455–73.

Katz, J. L. (1971*b*). Elastic properties of calcified tissues. *Israel Journal of Medical Science*, **7,** 439–41.

Katz, J.L. (1976). Hierarchical modeling of compact Haversian bone as a fiber reinforced material. In *Advances in Bioengineering*, ed. R. E. Mates & C. R. Smith, pp. 18–19. New York: American Society of Mechanical Engineers.

Katz, J. L. (1980). Anisotropy of Young's modulus of bone. *Nature, London*, **283,** 106–7.

Katz, J. L. (1981). Hierarchical modeling of compact Haversian bone as a fiber reinforced material. In press.

Katz, J. L. & Mow, V. C. (1973). Mechanical and structural criteria for orthopaedic implants. *Biomat. Med. Dev. Art. Org.*, **1,** 575–634.

Katz, J. L. & Thompson, W. A. (1977). A composite microstructural model of the elastic behavior of cortical bone. *23rd Annual Meeting Orthopaedic Research Society*, Las Vegas.

Katz, J. L. & Ukraincik, K. (1971). On the anisotropic elastic properties of hydroxyapatite. *Journal of Biomechanics*, **4,** 221–7.

Katz, J. L. & Ukraincik, K. (1972). A fiber-reinforced model for compact Haversian bone. *16th Annual Meeting of Biophysical Society FPM-C15*, Toronto.

Katz, J. L., Ukraincik, K. & Swedlow, D. B. (1972). Morphological basis for modeling the elastic behavior of compact Haversian bone. *4th International Biophysical Congress International Union Pure Applied Biophysics E19B1/7*. Moscow.

Kellenberger, E. & Rouiller, C. (1950). Die Knochemstruktur, untersucht mit dem Elektronenmikroskon. *Schweizerische Zeitschrift für Allgemeine Pathol. Bakteriol*, **13,** 783.

Ko, R. (1953). The tension test upon the compact substance of the long bones of human extremities. *Journal Kyoto Prefecture Medical University,* **53,** 505–25.

Koch, J. C. (1917). The laws of bone architecture. *American Journal of Anatomy,* **21,** 177–298.

Laird, G. W. & Kingsbury, H. B. (1973). Complex viscoelastic moduli of bovine bone. *Journal of Biomechanics,* **6,** 59–67.

Lakes, R. S. & Katz, J. L. (1974). Interrelationships among viscoelastic functions for anisotropic solids: application to calcified tissues and related systems. *Journal of Biomechanics,* **7,** 259–70.

Lakes, R. S. & Katz, J. L. (1979*a*). Viscoelastic properties and behavior of cortical bone: Part II: Relaxation mechanisms. *Journal of Biomechanics,* **12,** 679–87.

Lakes, R. S. & Katz, J. L. (1979*b*). Viscoelastic properties and behavior of cortical bone: Part III: A non-linear constitutive equation. *Journal of Biomechanics,* **12,** 689–98.

Lakes, R. S., Katz, J. L. & Sternstein, S. S. (1975). Torsional and biaxial dynamic and stress relaxation properties of bovine and human cortical bone. ASME Biomechanics Symposium. AMD Vol. 10, ed. R. Skalak & R. M. Nevern, pp. 133–4. New York: American Society of Mechanical Engineers.

Lakes, R. S., Katz, J. L. & Sternstein, S. S. (1979). Viscoelastic properties and behavior of cortical bone: Part I: Torsional and biaxial studies. *Journal of Biomechanics,* **12,** 657–78.

Lakes, R. S. & Saha, S. (1979). Cement line motion in bone. *Science,* **204,** 501–3.

Lang, S. B. (1970). Ultrasonic method for measuring the elastic coefficients of bone and results in fresh and dried bovine bones. *IEEE Transactions Biomechanical Engineering,* **17,** 101–5.

Lees, S. & Rollins, Jr., F. R. (1972). Anisotropy in hard dental tissues. *Journal of Biomechanics,* **5,** 557–66.

Lewis, J. L. & Goldsmith, W. (1975). The dynamic fracture and prefracture response of compact bone by split Hopkinson bar methods. *Journal of Biomechanics,* **8,** 27–40.

Lugassy, A. A. (1968). Mechanical and viscoelastic properties of bone and dentin in compression. Ph.D. Dissertation, Metallurgy and Materials Science, University of Pennsylvania.

Lugassy, A. A. & Korostoff, E. (1969). Viscoelastic behavior of bovine femoral cortical bone and sperm whale dentin. In *Research in dental and medical materials*, ed. E. Korostoff, pp. 1–17. New York: Plenum Press.

McElhaney, J. H. (1966). Dynamic response of bone and muscle tissue. *Journal of Applied Physiology*, **21**, 1231–6.

Mather, B. S. (1968). The effect of variation in specific gravity and ash content on the mechanical properties of human compact bone. *Journal of Biomechanics*, **1**, 207–10.

Olivo, O. M. (1937). Rispondenza della funzione meccanica varia degli osteoni con la loro diversa minuta architettura. *Boll. Soc. Ital. Biol. Sper.*, **12**, 400–1.

Panjabi, M. M., White, A. A. & Southwick, W. O. (1973). Mechanical properties of bone as a function of rate of deformation. *Journal of Bone and Joint Surgery*, **55**, 322–30.

Park, J. B. (1979). Biomaterials: An Introduction. New York: Plenum Press.

Piekarski, K. (1970). Fracture of bone. *Journal of Applied Physics*, **41**, 215–33.

Piekarski, K. (1973). Analysis of bone as a composite material. *International Journal of Engineering Science*, **11**, 557–65.

Ranvier, L. (1875). *Traite technique d'histologie*. Paris: Savy.

Rauber, A. A. (1876). *Elasticität und festigkeit der Knochen*. Leipzig: Engelmann.

Reilly, D. T., Burstein, A. H. & Frankel, V. H. (1974). The elastic modulus for bone. *Journal of Biomechanics*, **7**, 271–5.

Reuss, A. (1929). Berechnung der fliessgrenze von Mischkristallen auf Grund der plastizitatsbedingung für Einkristalle. *Z. Ungew. Math. Mech.*, **9**, 49–58.

Rouiller, C. (1956). Collagen fibers of connective tissue. In *The Biochemistry and Physiology of Bone*, ed. G. H. Bourne, Chapter 5, pp. 107–47. New York: Academic Press.

Ruth, E. B. (1947). Bone studies: I. Fibrillar structures of adult human bone. *American Journal of Anatomy*, **80**, 35–53.

Rutishauser, E., Huber, L., Kellenberger, E., Majno, G. & Rouiller, G. (1950). Etude de la structure de l'os au microscope electronique. *Archives des Sciences*, **3**, 175–81.

Sammarco, G. J., Burstein, A. H., Davis, W. L. & Frankel, W. H. (1971). The biomechanics of torsional fractures: the effect of loading on ultimate properties. *Journal of Biomechanics*, **4**, 113–17.

Sedlin, E. D. (1965). A rheological model for cortical bone. *Acta Orthop. Scand. Suppl.*, **83**.

Smith, J. W. (1960). The arrangement of collagen fibers in human secondary osteons. *Journal of Bone and Joint Surgery*, **42**, 588–605.

Smith, J. W. & Walmsley, R. (1959). Factors affecting the elasticity of bone. *Journal of Anatomy*, **93**, 503–23.

Smith, R. W. & Keiper, D. A. (1965). Dynamic measurement of viscoelastic properties of bone. *American Journal of Med. Elec.*, **4**, 156–60.

Swedlow, D. B., Frasca, P., Harper, R. A. & Katz, J. L. (1975). Scanning and transmission electron microscopy of calcified tissues. *Biomat. Me. Dev. Art. Org.*, **3**, 121–53.

Swedlow, D. B., Harper, R. A. & Katz, J. L. (1972). Evaluation of a new preparative technique for bone examination in the SEM. *SEM*, **72**, 335–42.

Tennyson, R. C., Evert, R. & Niranjan, V. (1972). Dynamic viscoelastic response of bone. *Experimental Mechanics*, **12**, 502–7.

Thompson, G. (1971). Experimental studies of lateral torsional vibration of intact dog radii. Ph.D. Thesis, Biomedical Engineering, Stanford University.

Tischendorff, F. (1951). Das verhalten der Haverschen system bei belastung. *Roux' Archives*, **145**, 318–32.

Tischendorff, F. (1952). Quantitative Beobachtung über das Verhalten de Haverschen Lamellen bei Belastung. *Roux' Archives*, **146**, 1–20.

Tischendorff, F. (1954). Die mechanische Reaktion der Haverschen system under ihrer Lamellen auf experimentelle Belastung. *Roux' Archives*, **146**, 661–704.

TSUDA, K. (1957). Studies on the bending test and impulsive bending test on human compact bones. *Journal Kyoto Prefecture Medical University*, **61**, 1001–25.

UKRAINCIK, K. & KATZ, J. L. (1971). Elastic properties of bone as a fiber-reinforced composite material. *49th Annual Meeting International Association Dental Research*, pp. 71. Chicago.

VINCENTELLI, R. & EVANS, F. G. (1971). Relations among mechanical properties, collagen fibers, and calcification in adult human cortical bone. *Journal of Biomechanics*, **4**, 193–201.

VOIGT, W. (1910). *Lehrbuch der Kristallphysik*. Berlin: Teubner.

VON EBNER, V. (1887). Sind die Fibrillen des Knochengewebes verkalt oder nicht. *Arch. Mikroscop. Anat.*, **29**, 213–36.

VOSE, G. P. (1962). The relation of microscopic mineralization to intrinsic bone strength. *Anatomical Record*, **144**, 31–6.

VOSE, G. P. & KUBALA, A. L. (1959). Bone strength – its relationship to x-ray determined ash content. *Human Biology*, **31**, 261–70.

WALMSLEY, R. & SMITH, J. W. (1957). Variation in bone structure and the value of Young's modulus. *Journal of Anatomy*, **91**, 603.

WEIDENREICH, F. (1923*a*). Knochstundien: Uber Aufbau und entwicklung des Knochens und den charakter des Knochengewebes. *A. Anat. Entwicklungsgesch.*, **69**, 382–466.

WEIDENREICH, F. (1923*b*). Knochenstudien: Uber sehnenverknocherungen und faktoren der Knochenbildung. *Z. Anat. Entwicklungsgesch.*, **59**, 558–97.

WERTHEIM, M. G. (1847). Memoire sur l'elasticite et la cohesion des principaux tissus de corps humain. *Annales de Chimie et de Physique*, **21**, 385–414.

WHITING, J. F. (1977). Ultrasonic critical angle reflection goniometer for *in vivo* bone mineral analysis. In *Ultrasound in Medicine*, ed. D. White & R. E. Brown, pp. 1629–43. New York: Plenum Press.

WOOD, J. L. (1971). Dynamic response of human cranial bone. *Journal of Biomechanics*, **4**, 1–12.

WRIGHT, T. M. & HAYES, W. C. (1976). Tensile testing of bone over a wide range of strain rates: Effects of strain rate, microstructure and density. *Medical and Biological Engineering*, **14**, 671–9.

YAMADA, H. (1970). *Strength of biological materials*, ed. F. G. Evans, pp. 19–75. Baltimore: Williams & Wilkins.

YOKOO, S. (1952). The compression test upon the diaphysis and the compact substance of the long bone of human extremities. *Journal Kyoto Prefecture Medical University*, **51**, 291–313.

YOON, H. S. & KATZ, J. L. (1976*a*). Ultrasonic wave propagation in human cortical bone: I. Theoretical considerations for hexagonal symmetry. *Journal of Biomechanics*, **9**, 407–12.

YOON, H. S. & KATZ, J. L. (1976*b*). Ultrasonic wave propagation in human cortical bone: II. Measurements of elastic properties and microhardness. *Journal of Biomechanics*, **9**, 459–64.

YOON, H. S. & KATZ, J. L. (1976*c*). Ultrasonic wave propagation in human cortical bone: III. Piezoelectric contribution. *Journal of Biomechanics*, **9**, 537–40.

YOON, H. S. & KATZ, J. L. (1976*d*). Dispersion of the ultrasonic velocities in human cortical bone. *Proc. Ultrasonics Symp. 76Ch1120-5SU*, pp. 48–50.

ZIEGLER, D. (1906). Studien über die feinere Structure des Rohrenknochens und desses polarization. *Deutsche Zeitschrift für Chirurgie*, **85**, 248–63.

WOOD, ONE OF NATURE'S CHALLENGING COMPOSITES

G. JERONIMIDIS

Department of Engineering,
University of Reading, Reading, UK

Introduction

When we think of the mechanical properties of biological materials we often pay more attention to animals than to plants. The requirements for stiffness and strength seem perhaps more obvious in organisms that run, jump, swim or fly and we seldom consider that plants in general and tall plants in particular must also resist considerable mechanical forces in order to survive. The way in which such 'static' structures are designed may appear self-evident; in fact, it is only comparatively recently that we are beginning to understand how such things work.

The relationship between the basic structural material, cellulose, and the end-product, wood, is very complex and in order to discuss the mechanical properties of wood we must remember that we are dealing with a natural system optimized over a very long period of time. Indeed, it is difficult to make a clear distinction between material and structure; this is often arbitrary but it seems to be particularly so in the case of biological tissues.

Plants are not mobile and this constraint will be reflected in the way in which mechanical properties and design interact to produce structures capable of performing their function with a considerable degree of safety, a measure of which is, perhaps, the lifespan of trees, hundreds if not thousands of years.

Wood versus tree or material versus structure

If a tree were considered simply as a cantilever made of a standard material called 'wood' we would run the risk of falling into a circular argument. Wood is what it is because it is made by trees. Professor Wainwright makes the plea that 'someone must find out what wood is doing to trees' (Wainwright, 1980). Perhaps the real question is: what are trees doing to wood?

It is true that the great proportion of load-bearing tissue in higher plants is dead wood, that is a lignified skeleton of hollow tubular cells, but it is important to remember that this material has been produced by a living organism capable of detecting changes in load patterns and hence of taking appropriate action by altering the properties of the material or by design. Conditions of growth have a profound effect on the mechanical behaviour of wood; reaction wood, for example, is generally a nuisance for the technological use of timber but in trees it is the logical answer to a particular set of circumstances.

Structure and mechanical properties

Wood has a cellular composite structure in which it is possible to identify four levels of organisation: molecular, fibrillar (or ultrastructural), cellular (or microscopic) and macroscopic. This paper will be concerned mainly with the relationships between the mechanical properties of individual cells and wood as a whole and in the way in which the microfibrillar nature of the plant cell walls affects the behaviour of wood in relation to trees.

The major chemical constituent of plants is cellulose, a high molecular weight polysaccharide which is directly responsible for stiffness and strength. The cellulose molecules are arranged in microfibrils (units about 10–20 nm in diameter) in which crystalline and amorphous regions coexist. There are also varying quantities of low molecular weight sugars, hemicelluloses and lignin which can be considered as a binding matrix. During growth, the cellulose produced by the cells is deposited forming different walls at different stages, each being characterized by a distinct microfibrillar pattern. Fig. 1 shows a simplified and generalized model of a wood tracheid. A more comprehensive account of cellular ultrastructure can be found in Mark (1967) and Preston (1974). Some aspects are of particular interest for the purpose of this paper.

The predominance of the S2 wall in terms of relative cross-sectional area (typically 80% of the total cell wall area) and its low microfibrillar angle mean that it is the major load bearing component.

The pattern of microfibrils in S2 is a single helix, generally right-handed; this may seem a curious arrangement for an element which must resist tensile and compressive forces directed along the cell axis (cross-plies are favoured by engineers). It is however a characteristic geometry which has been discussed by Cave (1968) from the point of view of elastic behaviour and by Gordon & Jeronimidis (1974) in relation to the fracture properties of wood.

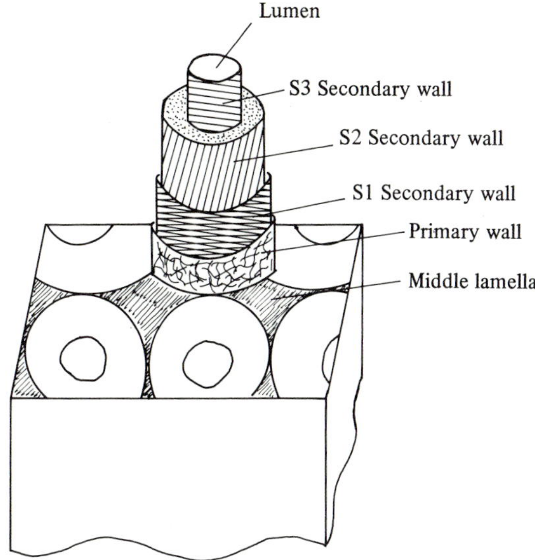

Fig. 1. Simplified model of a wood tracheid.

There is a correlation between the microfibrillar angle in S2 and the cell length: the longer the cell the steeper the angle (Preston, 1934).

Nearly 90% of the cells are aligned in one direction forming a honeycomb structure with very anisotropic mechanical properties. The alternation of spring and summer growth in species from temperate climates produces well-known ring patterns which introduce a further element of complexity.

The stiffness and strength of cellulose itself are considerable, the theoretical values for Young's modulus and tensile strength being of the order of 250 GPa (Gillis, 1969) and 25 GPa (Mark, 1967) respectively. It is interesting to compare the experimental values obtained for different plant materials with the predicted limits. This is shown in Table 1 where the reported values are the average of many measurements with a fair degree of variability. It can be seen that the degradation of stiffness and strength with respect to cellulose is much greater in wood cells than in other fibres. This is due, to a large extent, to higher microfibrillar angles. Although this may appear a rather inefficient way of utilising the potential of cellulose, it provides a defence mechanism against the propagation of cracks and catastrophic failure. Owing to their size and the loads experienced in practice, this is more likely to be an important consideration in trees than in other plants. An indication of this is given by the values of E/σ_b in the third column of Table 1. In ideal solids, this

Table 1. *Comparison of stiffness and strength for various cellulose-based tissues. Where appropriate, the mean microfibrillar angle (m.m.a.) has been reported in brackets*

	Young's modulus E(GPa)	Tensile strength σ_b (GPa)	E/σ_b
Cellulose (theoretical)	250	25	10
Bast fibres (m.m.a. ≃5°)	75	0.9	83
Wood cells (m.m.a. ≃25°)	25	0.5	50

ratio is of the order of 10 (Cottrell, 1964) but in practice it is never achieved because of the presence of defects and stress concentrations. However, the smaller E/σ_b, the better the material will be in resisting crack propagation. From this point of view it appears that, although stronger in the absolute, fibres with steep microfibrillar angles may well be less tough.

It is possible to follow to a certain extent the loss of stiffness and strength of wood with respect to cellulose taking into account the various structural levels and their interaction. This is illustrated in Fig. 2. More detailed calculations have been carried out by Mark (1967), Cave (1969) and Page *et al.* (1977).

Mechanical anisotropy of wood

The mechanical anisotropy of wood is due to a great extent to the fact that trees must resist mainly tensile and compressive bending loads in one direction. Because of this, the mechanical properties of wood measured along the grain are different from those in the transverse directions. The longitudinal Young's modulus is between 20 and 60 times greater than the radial or tangential moduli and, similarly, the tensile strength parallel to the grain is about 50 times the value across the grain. This is a reflection of the microfibrillar orientation in S2 and of the poor mechanical properties of the middle lamella (Fig. 1).

More important, both biologically and technologically perhaps, is the difference in behaviour in tension and compression along the grain. The compressive strength of wood is about one third (≃30 MPa) of its tensile strength and in this respect wood is no better than artificial fibrous composites. This difference is inherent in the structure of the cells themselves. A hollow tube can fail in compression by Euler buckling

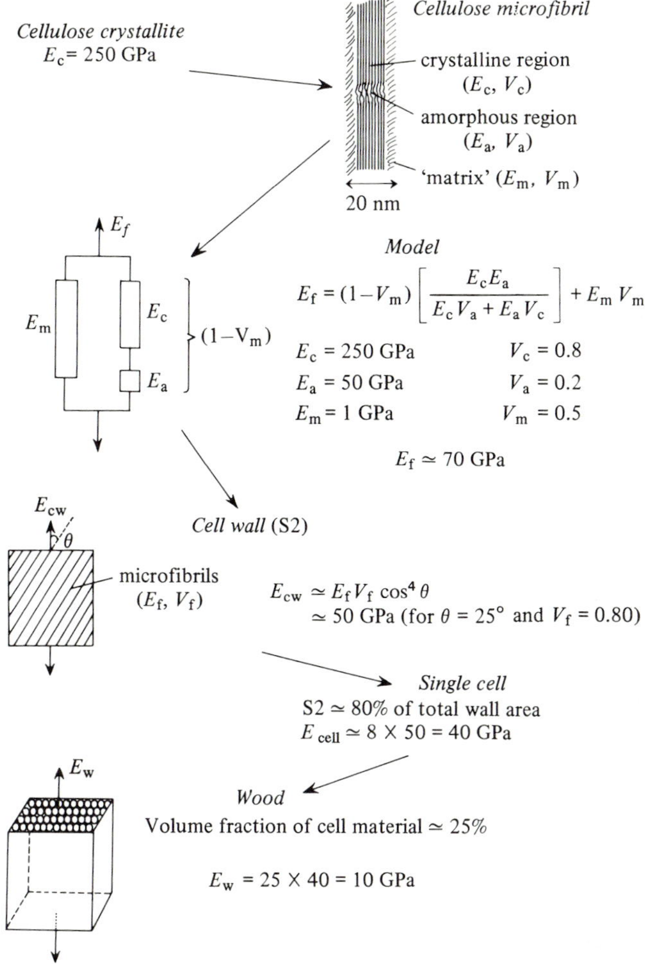

Fig. 2. The degradation of Young's modulus from cellulose to wood.

over its entire length or by local buckling in the wall but the support given by neighbouring cells, restricting lateral movement will, in general, prevent the first mechanism from occurring in wood. It has been shown that failure in compression is initiated by localized damage in the S2 wall (Dinwoodie, 1968) followed at higher loads by a macroscopic 'compression crease'. Local buckling in a tube wall occurs when the compressive stress is of the order of Et/d, where E is the Young's modulus of the material, t the wall thickness and d the tube diameter. In order to achieve a higher strength for a given diameter it is therefore necessary to increase E, t or both. As far as wood cells are concerned, an increase in thickness is probably metabolically expensive and an increase in stiffness means either more cellulose (also expensive) or a decrease of the angle between microfibrils and cell axis in the S2 layer. This, however, will have the effect of promoting buckling at the microfibrillar level in spite of the partial support given by the S1 and primary walls. Except in special circumstances such as reaction wood, the solution which seems to be present in trees is that of prestressing.

Prestressing in trees

In a tree subject to bending forces, I have calculated that the average tensile and compressive stresses are of the order of 20 MPa and whereas this value is well below the tensile strength of wood, it is dangerously close to its compressive strength. A further element of risk is introduced by the fact that a compression crease, even if not critical by itself, can act as an initial crack when the load is reversed and hence lead to fracture in tension. An economical way of compensating for the lack of sufficient compressive strength and providing a greater safety factor is to make more efficient use of the excess of tensile strength which is available. This is achieved by prestressing in tension the critical areas, i.e. the periphery of the trunk where the bending stresses are higher. The tension near the bark will have to be balanced by compressive forces in the centre of the tree. Although these forces are occasionally sufficient to induce failure in the core (a condition known as 'brittleheart') there is little danger for the living plant because the bending stresses are negligible near the neutral axis.

Comparatively little work has been done on growth stresses (Boyd, 1950; Dinwoodie, 1966) but their origin has been associated with longitudinal shrinkage of the cells following the deposition of the S2 layer. The measurements which have been reported suggest that the tensile stresses can be in excess of 10 MPa with corresponding compressive stresses in the centre of the order of 18 MPa. The point of zero stress is generally at a distance of one third of the radius from the periphery.

The combined effect of bending and prestressing is illustrated in Fig. 3 where it can be seen that the net result is a reduction of the maximum compressive stress of the order of 50% at the expense of a similar increase in maximum tensile stress which is, however, still below the critical level. It is interesting to note that there is a correspondence between the variation of microfibrillar angle in S2 from centre to periphery (Cowdrey & Preston, 1966) and the stress distribution of

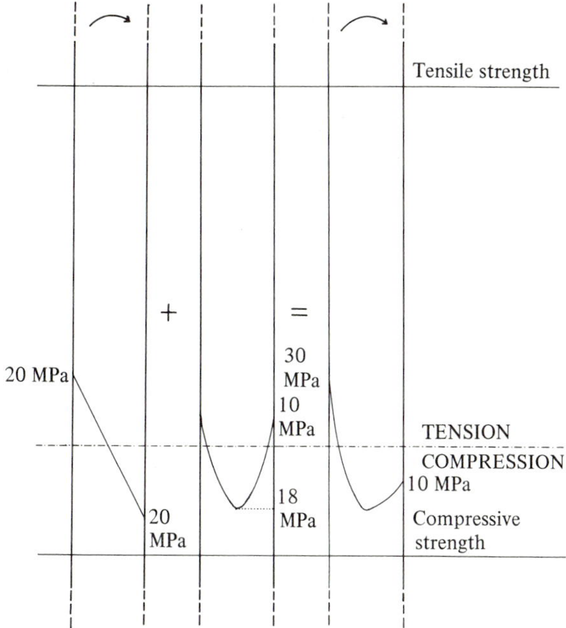

Fig. 3. The effect of prestressing on the stress distribution in bending of a tree trunk.

Fig. 3, the steeper helical angles being associated with the areas in tension and the flatter with those in compression.

Fracture of wood

In common with unidirectional fibrous composites, the fracture properties of wood are strongly dependent on the direction of crack propagation. Because of symmetry, there are in theory six different systems of crack growth (Schniewind & Centeno, 1973), but for practical purposes they can be grouped into two main types, fracture along and across the grain. The resistance to crack propagation can be measured by the work of fracture or fracture energy γ_f, a quantity which is related to the stress σ in the material, the Young's modulus E and the critical crack length l_c by a failure criterion developed originally by Griffith (1920):

$$l_c = 2E\gamma_f/\pi\sigma^2 \tag{1}$$

When wood is split longitudinally it behaves in a brittle manner and the values which have been measured for γ_f are of the order of 0.1–0.2 KJ m^{-2} (Atack *et al.*, 1961; Debaise, Porter & Pentoney, 1966), typical of polymeric materials such as 'Perspex' ('Plexiglass'). This comparatively low value of the work of fracture means that even for moderate stresses the critical crack lengths corresponding to unstable fracture are likely to be of a few millimetres only (splitting during drying, etc.). The fracture across the grain, on the other hand, requires energies of about 10 KJ m^{-2} (Tattersall & Tappin, 1966; Schniewind & Pozniak, 1971; Jeronimidis, 1976; Ansell & Harris, 1979), a value which is high in absolute terms and exceptionally high if one considers the low density of wood. Using eqn (1) it is possible to calculate that, for the average bending stress in a tree, l_c is of the order of 0.20–0.40 m. When prestressing is taken into account, increasing σ by a factor of 2, l_c is reduced by a factor of four but this still leaves a comfortable degree of safety.

The importance of a high work of fracture of wood cannot be overemphasised because modest reduction in γ_f can result in critical crack lengths which would be unacceptable and very dangerous for the living tree.

The reason for the large difference in fracture energy between the two modes of fracture is to be sought in the energy absorbing mechanisms which can operate during crack propagation. Splitting along the grain involves mainly the middle lamella and the interface between the primary and secondary walls (Fig. 1) and the crack path does not affect in general the S2 layer itself. The only sources of work of fracture are the breaking of chemical bonds across the crack plane (which can contribute only a few J m^{-2}) and small amounts of plastic deformation of the polymeric components. In that respect there is a similarity between wood and artificial composites.

However, in crack propagation normal to the fibre direction there is a difference between these materials and wood. In composites, the work of fracture is due mainly to energy dissipated as friction during pull-out (Kelly, 1973) but this kind of mechanism does not seem possible in wood, at least if one considers the cells as the reinforcing elements (Jeronimidis, 1979). Also, even if simple pull-out did take place, the calculated fracture energy would be one order of magnitude lower than the experimental values.

It is possible to account for the observed results in terms of a model which was proposed a few years ago (Jeronimidis, 1976) and which is based on the tensile properties of wood cells and other fibres. It has been shown that the stress–strain behaviour of single tracheids (Page, El-Hosseiny & Winkler, 1971; Hardacker & Brezinski, 1973) and of cell bundles (Spark, Darnbrough & Preston, 1958) is strongly dependent on the angle that the microfibrils make with the cell axis in the S2 wall. At low microfibrillar angles ($\leqslant 10°$), the load–extension curves are quasi-linear with breaking strains of 1% or 2%. In this case little or no irreversible energy absorption is possible before fracture. As the angle increases, the stress–strain curves show increasing amounts of 'plastic' deformation and a more or less defined instability point, similar to the yield point in ductile metals. The limit of elastic behaviour is controlled by cell-wall thickness and initial microfibrillar angle. For angles of more than 45°, ultimate strains of 25% or more have been recorded. This type of behaviour has been called 'tension buckling' (Pagano, Halpin & Whitney, 1968) and is due to the asymmetric filament winding of the S2 layer. Considerable amounts of energy can be absorbed irreversibly during the post-buckling deformation (the elastic limit being of the order of 2%) and, from the results which have been published, it is possible to estimate the energies involved and convert them into work of fracture. For breaking strains of 20%, which may be typical of wood cells with microfibrillar angles of 25°–30°, the theoretical value for γ_f is in excess of $100\,\mathrm{KJ\,m^{-2}}$, one order of magnitude higher than the experimental measurements. However, the calculations have been based on the assumption that all the cells on a given cross-section deform in this manner, reaching the same maximum failure strain. The results obtained represent therefore an upper limit.

Work carried out on models simulating the S2 cell walls (Gordon & Jeronimidis, 1980) has shown that when the helical angle of the fibres is 15°–20°, there is an optimum compromise between energy absorption and reduction in axial stiffness. Tension buckling is initiated by micro-cracks in the S2 layer, parallel to the direction of the microfibrils. Figs. 4, 5 and 6 illustrate some aspects of the fracture morphology of wood. Recent work by Ansell & Harris (1979) on the acoustic emission associated with flaw growth may help to clarify further the relationship between work of fracture and cell deformation.

The model which has been proposed predicts that the fracture energy of wood for crack propagation across the grain should be virtually independent of temperature because the mechanisms involved are mainly controlled by geometry. This has been confirmed experimentally (Jeronimidis, 1976; Bennett, 1977; B. Harris, personal communication) and is an advantage for the species growing at latitudes where extreme temperatures are common. It is worth remembering that many materials which are tough at ambient conditions because they can deform plastically show dramatic reductions in their work of fracture at low temperatures, becoming very brittle.

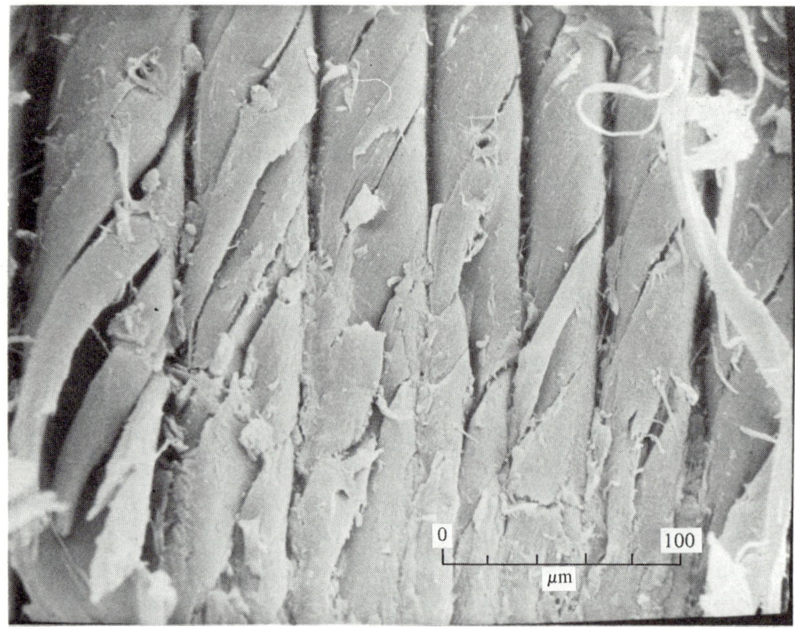

Fig. 4. Tensile failure in Sitka spruce (*Picea sitchensis*). Folding inward of the S2 wall as a result of cracks parallel to the microfibrillar direction.

Reaction wood

One of the advantages of using composite materials for structural applications is that it is possible, in theory, to tailor their mechanical properties in order to meet specific needs for strength and stiffness. This is done by changing the orientation of the fibres so that they can carry the loads more efficiently.

Living organisms have the great advantage that their load bearing materials are very often fabricated and deposited under stress, a condition which produces a blueprint of the design requirements. Reaction wood is an example of how this principle works.

When trees are displaced from the vertical position, they become subject to a system of forces acting in one particular direction, superimposed on the more random loading due to the wind. They respond by producing specialized tissues paticularly suited to resist tensile or compressive stresses, recovering in the limit a symmetrical posture (Scurfield, 1973).

In hardwoods (angiosperms), 'tension wood' appears on the side of the tree in tension and its morphology differs from what we may

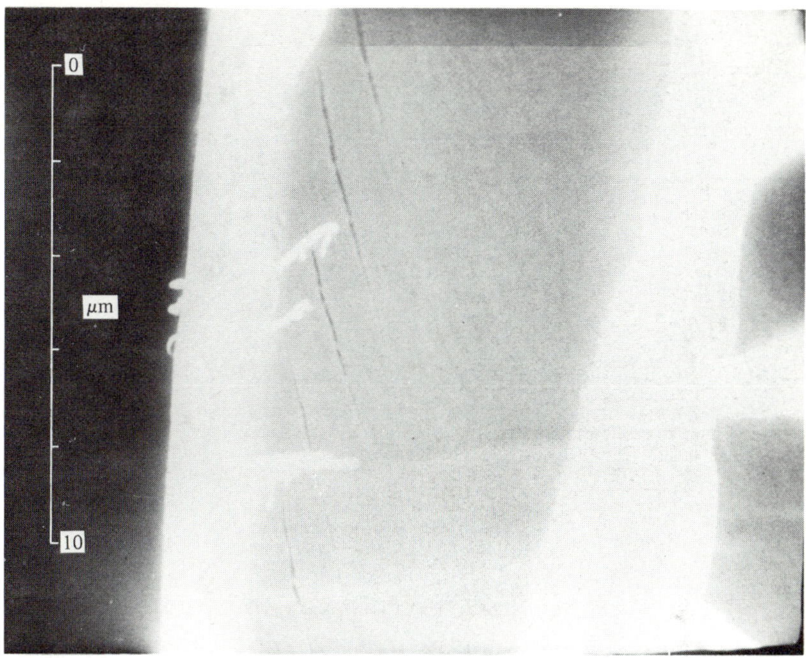

Fig. 5. Spiralling fractures in Sitka spruce (*Picea sitchensis*).

consider normal wood. The basic features shown in Fig. 1 are still present but there is now an innermost layer (G layer) which fills the lumen cavity, in which the cellulose microfibrils run almost parallel to the cell axis. Softwoods (gymnosperms) produce instead 'compression wood' on the side in compression of a leaning tree. In this case also, there are major differences in cell-wall architecture. As one might expect, the tracheids are much thicker-walled than normal and the microfibrillar angle in S2 is very flat ($\geqslant 50°$), presumably to prevent local buckling. Often there are radial cracks starting from the lumen.

Both varieties of reaction wood can be considered as extreme designs where the balance of properties of a normal wood cell has been sacrificed due to special circumstances. Compression wood, for example, is more brittle (Koehler, 1933) and is often a cause of premature failure in timber if undetected. Less is known about the mechanical properties of tension wood but it is likely that the increased axial stiffness will result in lower toughness.

Fig. 6. Tension buckling in the cell wall of Sitka spruce (*Picea sitchensis*) fractured in tension.

Conclusions

As a composite material, wood has a combination of stiffness, strength, toughness and low density which is unique. Modern techniques of investigation such as polymer physics, composite theory and fracture mechanics have provided us with the means for understanding how this is achieved. Although a great deal is now known about the relationships between mechanical properties and morphology, the picture is by no means complete. Much remains to be done in many areas, in particular on the force detection mechanisms of plants and on the multifunctional aspects of design. In systems as sophisticated as biological structures, it is rare that one can isolate a part and correlate it with a single specific function.

References

Ansell, M. P. & Harris, B. (1979). The relationship between toughness and fracture surface topography in wood and composites. In *Proceedings of the Third International Conference on the Mechanical Properties of Materials*, ed. K. J. Miller & R. F. Smith, **3,** 309–18. Toronto: The Pergamon Press.

Atack, D., May, W. D., Morris, E. L. & Sproule, R. N. (1961). The energy of tensile and cleavage fracture of black spruce. *Tappi,* **4,** 555–67.

Bennett, C. P. (1977). Wooden pipelines. *Chart. Mech. Eng.,* June 1977, 70–3.

Boyd, J. D. (1950). Tree growth stresses. I. Growth stress evaluation. *Australian Journal of Scientific Research*, 294–309.

Cave, I. D. (1968). The anisotropic elasticity of the plant cell wall. *Wood Science and Technology,* **2,** 268–78.

Cave, I. D. (1969). The longitudinal Young's modulus of *Pinus radiata. Wood Science and Technology,* **3,** 40–8.

Cottrell, A. H. (1964). *The mechanical properties of matter.* New York: John Wiley & Sons.

Cowdrey, D. R. & Preston, R. D. (1966). Elasticity and microfibrillar angle in the wood of Sitka spruce. *Proceedings of the Royal Society of London, Series B,* **166,** 245–72.

Debaise, G. R., Porter, A. W. & Pentoney, R. E. (1966). Morphology and mechanics of wood fracture. *Materials Research and Standards,* **6,** 493–9.

Dinwoodie, J. M. (1966). Growth stresses in timber. A review of literature. *Forestry,* **39,** 162–70.

Dinwoodie, J. M. (1968). Failure in tember. 1. Microscopic changes in cell-wall structure associated with compression failure. *Journal of the Institute of Wood Science* **21,** 37–53.

Gillis, P. P. (1969). Effect of hydrogen bonds on the axial stiffness of crystalline cellulose. *Journal of Polymer Science, A* **2,** 783–94.

Gordon, J. E. & Jeronimidis, G. (1974). The work of fracture of natural cellulose. *Nature, London,* **252,** 116.

Gordon, J. E. & Jeronimidis, G. (1980). Composites with high work of fracture. *Philosophical Transactions of the Royal Society of London, Series A,* (in press).

Griffith, A. A. (1920). The phenomenon of rupture and flow in solids. *Philosophical Transactions of the Royal Society of London, Series A,* **221,** 163–98.

Hardacker, K. W. & Brezinski, J. P. (1973). The individual fiber properties of commercial pulps. *Tappi,* **56,** 154–7.

Jeronimidis, G. (1976). The fracture of wood in relation to its structure. In *Wood structure in biological and technological research*, ed. P. Baas, A. J. Bolton & D. M. Catling, Leiden botanical series No. 3, 253–65. Leiden: The University Press.

Jeronimidis, G. (1979). Morphological aspects of wood fracture. In *Proceedings of the Third International Conference on the Mechanical Properties of Materials*, ed. K. J. Miller & R. F. Smith, vol. 3, pp. 329–40.

Kelly, A. (1973). *Strong solids*, 2nd edn. Oxford: The Clarendon Press.

Koehler, A. (1933). Causes of brashness in wood. *Technical Bulletin, Forest Products Laboratory*, 342, Madison, USA.

Mark, R. E. (1967). *Cell wall mechanics of wood tracheids*. New Haven: Yale University Press.

Pagano, N. J., Halpin, J. C. & Whitney, J. M. (1968). Tension buckling of anisotropic cylinders. *Journal of Composite Materials*, **2,** 154–00.

Page, D. H., El-Hosseiny, F. & Winkler, K. (1971). Behaviour of single wood fibres under axial tensile strain. *Nature, London,* **229,** 252–3.

Page, D. H., El-Hosseiny, F., Winkler, K. & Lancaster, A. P. S. (1977). Elastic modulus of single wood pulp fibers. *Tappi,* **60,** 114–17.

Preston, R. D. (1934). The organisation of the cell wall in the conifer tracheid. *Philosophical Transactions of the Royal Society of London, Series B,* **224,** 131–74.

Preston, R. D. (1974). *The physical biology of plant cell walls*. London: Chapman & Hall.

Schniewind, A. P. & Pozniak, R. A. (1971). On the fracture toughness of Douglas fir. *Engineering Fracture Mechanics,* **2,** 223–33.

Schniewind, A. P. & Centeno, J. P. (1973). Fracture toughness and duration of loading factor. I. Six principal systems of crack propagation and the duration factor for cracks propagating parallel to the grain. *Wood and Fiber,* **5,** 152–9.

Scurfield, G. (1973). Reaction wood, its structure and function. *Science*, **179,** 657–9.

Spark, L. C., Darnbrough, G. & Preston, R. D. (1958). Structure and properties of vegetable fibres. II. A micro-extensometer for the automatic recording of load-extension curves of single fibrous cells. *Journal of the Textile Institute, transactions*, **49,** 309–16.

Tattersall, H. G. & Tappin, G. (1966). The work of fracture and its measurement in metals, ceramics and other materials. *Journals of Materials Science,* **1,** 296–301.

Wainwright, S. A. (1980). Adaptive materials: a view from the organism. In *The Mechanical Properties of Biological Materials* (34th Symposium of the Society for Experimental Biology) ed. J. F. V. Vincent & J. D. Currey, pp. 437–453. Cambridge University Press.

INSECT CUTICLE: A PARADIGM FOR NATURAL COMPOSITES

J. F. V. VINCENT

Biomechanics Group, Department of Zoology,
The University, Reading RG6 2AJ

Cuticle as a composite

Insect cuticle is a classic composite of chitin and protein plus a few other components (mostly lipids and some salts) which contribute only a few percent to the total dry matter. The chitin is present as microfibrils typically 2.8 nm in diameter containing 18–20 chitin chains running antiparallel to each other (i.e. alpha chitin) (Neville, 1975). The amount of chitin present in the cuticle of an arthropod can vary from about 4% dry weight in the tick, *Boophilus* (Hackman, 1974) to about 60% dry weight in some pliant cuticles. In general cuticles which are pliant contain more chitin (50–60% dry weight) than cuticles which are sclerotised and stiff (30–40% dry weight). The orientation of the chitin may vary from being totally preferred through alternating preferred layers arranged in a plywood fashion to alternating preferred and helicoidal layers to all helicoidal (Fig. 1; see also Barth, 1973). For simple mechanical considerations, chitin is a very stiff fibre, akin to cellulose but more tightly hydrogen-bonded (Minke & Blackwell, 1978) and so more stable so that, for example, it can withstand boiling in 5% NaOH and is soluble only in strong inorganic acids. This chitin is embedded in a matrix of protein. There are many different fractions to this protein (sometimes called *arthropodin*) and it is a moot point whether one should refer to the entire matrix as a protein composed of many different sub-units or as composed of many different proteins. It seems very likely that there will be some interactions between the protein components, but whether these will turn out to be sufficiently specific for each component to be considered a specific sub-unit of a single large protein remains to be seen. Gel electrophoresis of suitable samples can separate very many of these proteins: up to 30 have been isolated from pharate locust femur (Andersen, personal communication) and there is some evidence that these are all very similar in their amino-acid composition (Andersen, 1977). Various people have sug-

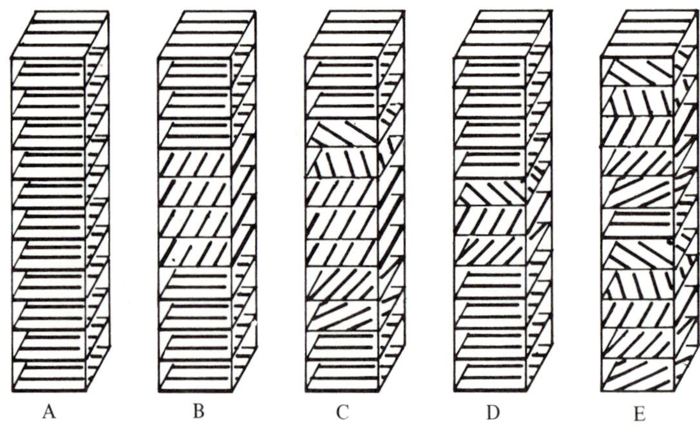

Fig. 1. Some possible combinations of chitin fibre orientations in insect cuticle; each column represents a vertical section through the cuticle with the orientation of chitin shown in a series of representative planes. A, preferred orientation (e.g. *Schistocerca* and *Locusta* apodemes); B, plywood orientation (e.g. adult *Oryctes rhinoceros* pronotum); C, D, possible orientations, probably found only in intersegmental membranes; E, typical 'helicoidal' orientation with rotation of the direction of chitin fibres, often found alternating with type A (e.g. in the locust type A is deposited during the day, type E at night). After Barth (1973).

gested that this could be achieved either by limited cross-linking of one or two basic units producing a range of molecular weights (e.g. Mills, Greenslade, Fox & Nielsen, 1967) or by gene duplication (Hackman, 1976). Thus the heterogeneity of arthropodin may be only apparent in that the number of different amino-acid sequences occurring in any one cuticle may be quite few. This view is not universally held and there are several reports that soluble and insoluble fractions of the proteins, irrespective of the solvents used, have different amino-acid compositions (Andersen, 1971; Fristrom, Hill & Watt, 1978). There seems to be no general agreement on this.

The conformation of these proteins within the cuticle is unknown. The major problem is that the chitin is highly crystalline and has an amino group on it. This makes infra-red (IR) or X-ray analysis well nigh impossible on most machines, though Fourier transform IR seems promising. Protein from aqueous extracts of maggot (*Calliphora vomitoria*) cuticle, which is unsclerotised, will take up beta-sheet conformation on drying as shown by X-ray (Fraenkel & Rudall, 1947) and IR (Hillerton & Vincent, 1979; Hackman & Goldberg, 1979) analysis. The same protein in solution shows no secondary structure in 0.1 M phosphate buffer, 15% alpha helix and 25% beta sheet when an equal volume of ethanol is added and precipitates at 60% ethanol concentra-

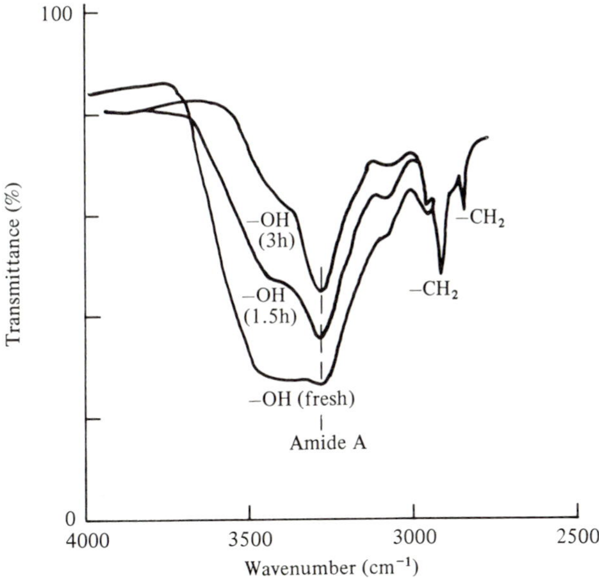

Fig. 2. Infra-red spectra of locust hind wing showing steady reduction in water content. The specimen was left in the beam of the spectrophotometer and spectra recorded at the times shown. Loss of water is indicated by the progressive reduction of the —OH peak. The amide A peak is dropping simply because the —OH peak is so broad and is added to it. The —CH_2 peaks are unaffected.

tion (Hillerton & Vincent, 1979; Hackman & Goldberg, 1979), has very similar results. Water-soluble proteins from another unsclerotised cuticle, that of larval *Rhodnius*, a blood-sucking bug, give similar results for conformation in solution (measured by circular dichroism) but no sign of beta sheet in the IR (Hillerton & Vincent, 1979). The conformation of these proteins in the cuticle is unknown. Weis-Fogh (1961) has shown that the conformation of resilin in pre-alar arm is random coil (see Gosline, 1980) but this is a very specialised example. Some clues as to the conformation of cuticle proteins can be obtained from consideration of their overall hydrophobicity (p. 199).

Thus highly crystalline chitin combines with protein of unknown conformation, after the fashion of a classical fibre-in-matrix composite (Harris, 1980). There is very little good information on the mechanical properties of insect cuticle (Table 1). Those familiar with the subject will be aware of the work of Hepburn, who has published many papers on the subject. Unfortunately it was not until late 1976 that he tested cuticle in a controlled environment (Hepburn & Chandler, 1976). His previous samples were drying out during testing and so becoming stiffer. This led him to report such phenomena as work-hardening, which is

Table 1. *Studies on the mechanical properties of insect cuticle, the results of which are reliable*

Authors	Cuticle source	Type[a]	Test solution	Specimen shape	Test performed	Relevance
Weis-Fogh (1961)	*Aeshna*	Alar tendon resilin	Saline	Cylinder	Stress–strain	Flight
Jensen & Weis-Fogh (1962)	*Schistocerca*	Pre-alar arm resilin	Saline	Various	Stress–strain	Flight
Wood (1974)	*Locusta* adult	i.s.m.	Saline	Rectangle	Stress–strain Stress–relaxation	Oviposition Development
Reynolds (1975*a*, *b*)	*Rhodnius* larva	abd. seg. IV	Liquid paraffin	Loop	Stress–relaxation Stress–strain	Feeding
Vincent (1975*a*)	*Locusta* adult female	i.s.m.	Saline	Rectangle	Stress–strain	Oviposition
Vincent (1975*b*)	*Rhodnius* adult	Pleural pleat	Saline	Rectangle	Stress–strain	Feeding
Hepburn & Chandler (1976)	*Locusta* adult	i.s.m.	Saline	Rectangle	Stress–strain	Material properties
	Pachynoda adult	i.s.m.	Saline	Rectangle	Stress–strain	Material properties

	Bombyx larva	Tergite	Saline	Waisted	Stress–strain	Material properties
	Limulus	Leg a.m.	Saline	Waisted	Stress–strain	Material properties
	Scylla	Leg a.m.	Saline	Waisted	Stress–strain	Material properties
Vincent (1976)	*Locusta* adult female	i.s.m.	Saline	Rectangle	Stress–relaxation	Oviposition
Tychsen & Vincent (1976)	*Locusta* adult female	i.s.m.	Saline	Rectangle	Stress–strain	Development
Ker (1977)	*Schistocerca*	Apodemes	Liquid paraffin saturated with water	Rectangular	Creep Stress–strain	Material properties Locomotion
Thompson & Hepburn (1978)	*Apis* pharate adult	abd. terg. III	Saline	Rectangular	Stress–strain	Development
Hillerton (1979*a*)	*Rhodnius* larva	abd. terg. IV & V	Liquid paraffin	Waisted	Creep	Feeding Development
Vincent & Hillerton (1979)	*Calliphora* larvae	Entire	Saline or air	Rectangular	Stress–strain	Material properties

[a] i.s.m., intersegmental membrane; a.m., arthrodial membrane; abd., abdomen; seg., segment; terg., tergite.

obviously an artifact of his experimental method. Only two of his papers so far take account of water content (Hepburn & Chandler, 1976; Thompson & Hepburn, 1978) and I have used some of these data. Jensen & Weis-Fogh (1962) tested, amongst other samples of cuticle, the wing of the locust, *Schistocerca*. But a locust wing has quite a significant amount of water in it when taken from the insect and it dries out relatively quickly in the beam of an IR spectrophotometer (Fig. 2). However, a fair proportion of the water is probably not in the cuticle but in the cells and haemolymph between the wing laminae. Fraenkel & Rudall (1947) reported that locust wing cuticle contains 53% water but did not mention whether they cleaned off the cells and washed out the haemolymph before making their measurements. Testing the cuticle in a bath of saline does not entirely overcome this problem, as noted by Hepburn & Chandler (1976): and maggot cuticle dissected in water contains about 70% water (Fraenkel & Rudall, 1940) but about 55% water if it is dissected in oil (Dennell, 1946). The effect of these different levels of hydration on the mechanical properties of cuticle has not been investigated. A third testing medium has been used, namely liquid paraffin (Reynolds, 1975*a*), preferably saturated with water or saline (Ker, 1977) which will stabilise the water content.

As to the shape of the specimen, much as been said of the necessity of using waisted or dumb-bell shaped specimens. I consider this to be a counsel of perfection at this state of the art, more particularly because it does not make much difference in a pliant specimen, provided that the specimen is long enough for end effects due to clamping to be neglected and that fracture loads (which are mostly 'non-biological') are not approached. Ker (1977; 1980) used parallel-sided specimens of stiff cuticle and found that he needed to make a small end correction (Ker, 1980). The strip is probably better than the loop (used by Reynolds, 1975*a, b*) since experience with loop-shaped specimens in Reading (testing pig aorta) shows that strain can be over-estimated by 10 to 15% (Jeronimidis, in Hillerton, 1979*a*). All this reduces the number of useful and reliable studies on insect cuticle to those shown in Table 1, producing the values shown in Table 2.

Two major attempts to analyse the composite mechanical behaviour of insect cuticle have been made. Hepburn and associates tested films of chitin made by removing protein from cuticle and by regenerating chitin as films from a chitin xanthate dispersion (Joffe & Hepburn, 1973; Hepburn & Joffe, 1976) and used values obtained for the stiffness of these preparations in calculations of the factors involved in the control of stiffness of insect cuticles. Not surprisingly they have not got far in this analysis which is based on an approach too indirect to warrant further comment. Happily, Ker (1977) has analysed cuticle as a composite using composite theory (Harris, 1980) and has derived some sensible answers. Taking as 17% the volume fraction of chitin in his test piece, the tibial extensor apodeme from the hind leg of the locust *Schistocerca*, which has the chitin orientated preferentially along its length, Ker's results give matrix stiffness, E_m, = 0.12 GNm^{-2} and fibre stiffness, E_f, = 86 GNm^{-2} (ranging between 70 and 90 GNm^{-2}). This figure for the stiffness of chitin fibres is about the same as for wet cellulose (Wainwright, Biggs, Currey & Gosline, 1976) and so is very reasonable. As a cross-check Ker (1977) increased the modulus of the matrix by about 25% by cross-linking the matrix in glutaraldehyde. This enabled him to find, independently, the length of the chitin microfibrils which he estimated as 0.36 μm. This represents a degree of polymerisation of about 700 sugar residues which agrees very well with biochemical data by Strout, Lipke & Geoghegan (1976) on the degree of polymerisation of chitin from maggot cuticle measured by gel filtration. Ker (1977) also measured the stiffness of locust tibial cuticle which contains 29% chitin and obtained a comparable value for E_f. He calculated the fracture stress of the chitin fibrils as about 4GNm^{-2} and the fracture strain as about 4.2%. These values for chitin presumably hold true for other insect cuticles since there is no evidence that chitin is different in other cuticles although its molecular weight may well vary. So for stiff cuticles one can say that there is little doubt that the material is acting as a classic fibrous composite and that the chitin fibrils are the main load bearers, the load being transferred *via* the matrix whose shear stiffness is therefore also important. This is in direct contradiction to the interpretation put forward, with no mathematical analysis, by Hepburn & Chandler (1976) that in stiff cuticles the matrix is stiffer than the chitin. This view is totally untenable in the light of Ker's analysis.

For pliant cuticles the situation is somewhat different. Table 3 shows values of E_m from three comparable systems which have the chitin orientated preferentially. The white puparial cuticle is from the earliest stage of puparium formation of the fly, *Calliphora vomitoria*. In this state the matrix contains water so that the volume fraction of the chitin is considerably less than its dry weight fraction, which is about 37%. The chitin, being crystalline and tightly H-bonded with no water of crystallisation, probably binds insignificant amounts of water; in addition it is closely associated with protein and so not exposed to free water. Quite by chance the stiffness of locust intersegmental membrane (i.s.m.) parallel to the chitin (Hepburn & Chandler, 1976) is almost identical to

Table 2. *Values for some mechanical properties of insect cuticle*

Insect		Stiffness (Nm^{-2})		Chitin orientation[a]	Water content	
Locusta						
Immature female adult	(i.s.m.)	10^6	Wood (1974)	Perp	65%	Tychsen & Vincent (1976)
		3.2×10^6	Hepburn & Chandler (1976)	par		
		7.42×10^6	Hepburn & Chandler (1976)	Par		
Mature female adult		10^3	Vincent (1975*a*)	Perp	75%	Tychsen & Vincent (1976)
Mature male adult		2.04×10^6	Wood (1974)	Perp	–	
Schistocerca						
Pretarsal flexor	(apodeme)	1.3×10^{10}	Ker (1977)	Par	–	
Tibial extensor	(apodeme)	1.1×10^{10}	Ker (1977)	Par	–	
		1.5×10^8	Ker (1977)	Perp		
Tibia (post-ecdysial)		1.7×10^{10}	Ker (1977)	Par		
Pre-alar arm (resilin)		2×10^6	Jensen & Weis-Fogh (1962)	P		
Rhodnius						
Larva	abdominal cuticle unplasticised	2.43×10^8	Hillerton (1979*a*)	P	25.8%	Reynolds (1975*a*, *b*)
		6.2×10^7	Reynolds (1975*a*, *b*)	P		
	plasticised (5HT)[b]	9.45×10^6	Reynolds (1975*a*, *b*)	P		
	(pH5)	2.46×10^6	Reynolds (1975*a*, *b*)	P	31.3%	Reynolds (1975*a*, *b*)
	[illegible]	[illegible]	[illegible]	P		

Adult	pleural pleat	6×10^6	Vincent (1975*b*)	P	–	
Apis						
Pharate adult	abdominal sclerite	$10^8 - 10^9$	Thompson & Hepburn (1978)	P	–	
Calliphora						
Larva	wet	$\sim 10^7$	Vincent & Hillerton (1979)	P	–	
	dry	$\sim 10^9$	Vincent & Hillerton (1979)	P		
White puparium	wet	7.34×10^7	Vincent & Hillerton (1979)	Par		
	dry	2.2×10^9	Vincent & Hillerton (1979)	Par	70%	Fraenkel & Rudall (1940)
Tanned puparium	wet	2.45×10^8	Vincent & Hillerton (1979)	Par	12%	Fraenkel & Rudall (1940)
	dry	3.05×10^9	Vincent & Hillerton (1979)	Par		
Aeshna	resilin	2.2×10^6	Weis-Fogh (1961)	No chitin	62%	Weis-Fogh (1961) (original figure quoted was 'corrected for swelling' due to water!)
Bombyx						
Larva		2.2×10^7	Hepburn & Chandler (1976)	P	–	
Scylla	leg a.m.	2.4×10^8	Hepburn & Chandler (1976)	P	–	
Pachynoda	i.s.m.	5.05×10^8	Hepburn & Chandler (1976)	P	–	
Limulus	leg a.m.	3.8×10^7	Hepburn & Chandler (1976)	P	–	

[a] Par, strain parallel to chitin fibres; Perp, strain perpendicular to chitin fibres; P, chitin fibres in other arrangements (see Fig. 2) to give an effective plywood-like effect.
[b] 5HT, 5-hydroxytryptamine.

Table 3. *Modulus of the cuticular matrix for some different cuticles*

Source	E_m (Nm^{-2})	Reference
Calliphora		
White puparial cuticle, wet	$1.302\ (\pm 0.193) \times 10^6$	
White puparial cuticle, dry	$9.04\ (\pm 2.73) \times 10^7$	
Locusta		
Male intersegmental membrane	1.8×10^6	Wood (1974)
	2.7×10^6	Hepburn & Chandler (1976)
Schistocerca		
Tibia	1.8×10^9	Ker (1977)
Apodeme	1.2×10^8	Ker (1977)

that of *Calliphora* (Vincent & Hillerton, 1979). The stiffness of locust i.s.m. perpendicular to the chitin gives estimates for the stiffness of the matrix which are a little higher than those obtained from *Calliphora* white puparial cuticle using a torsional pendulum but the difference is probably not very important in this context. Using standard composite theory (Harris, 1980) it is now possible to calculate the transfer length of the chitin fibres in these pliant cuticles. It turns out to be about 11 μm which is about fifteen times the length of chitin microfibrils determined from similar material by Strout *et al.* (1976). So it can safely be said that the chitin fibres in these cuticles are not carrying much of the load since the matrix has too low a shear stiffness to transmit the load from one fibre to the next. The chitin is acting more as a high aspect ratio filler. This generalisation could probably be made for most, if not all, pliant cuticles.

Sclerotisation

One of the most dramatic features of the life of any arthropod is ecdysis. The cuticle undergoes rapid and extreme changes during this period, being initially pliant and then, after being stretched, being stiffened (at least, in most insects). This process of stiffening has been called sclerotisation or tanning or hardening and darkening. The latter term is somewhat confusing as 'hardness' is a different quantity from stiffness (see Vincent & Hillerton, 1979). Pryor (1940) maintained that when cuticle is sclerotised the proteins are cross-linked by phenols and that this process stiffens the matrix. This idea has since been elaborated upon by many people and there is an impressive body of evidence for the introduction of phenols of some sort into the cuticle at this time. Beyond that, the evidence is vague. No-one has isolated a phenol attached to a protein, no-one has isolated two proteins attached to each other by any means, in fact no-one has ever demonstrated the presence of new covalent cross-links in cuticle after sclerotisation has occurred. It is probably true to say that although there is much more detailed knowledge of biochemical events, the essential understanding of the processes occurring at sclerotisation is no more advanced after forty years' work. This is probably because biochemists are trying to understand an event whose main significance is mechanical but for which there is precious little mechanical information.

A crude beginning has been made towards understanding the mechanical processes occurring during sclerotisation (Vincent & Hillerton, 1979; Vincent, unpublished observations). The initial problem is the separation of effects due to the chitin from effects due to the matrix. This is necessary because according to all available evidence only the protein matrix is changed during sclerotisation. So in order to derive comparable results from different samples and types of cuticle it is necessary to extract a value of E_m from mechanical data using composite theory. The torsional pendulum is extremely useful for such tests since it imposes such small strains on the specimen. This results in very small stresses being generated, which in turn means that it is simple to make a sufficiently stiff machine, one which will not deflect by more than a fraction. In addition the same test piece can be used for several tests since the deformation is well within the elastic limit of the specimen. Thus one of the major problems of experiments with biomaterials, reproducibility of results, is largely circumvented. In Table 3 results are presented for the E_m of wet and dry untanned white puparial cuticle from *Calliphora*. The dry samples were the same as the wet, the cuticle was mounted in the pendulum, tested, allowed to dry and tested again. The E_m for dry white puparial cuticle is not significantly different from the value for the matrix stiffness of tendon reported by Ker (1977). More important, Table 2 shows that there is no difference in stiffness (E_c) between dry tanned puparial cuticle and dry *un*tanned puparial cuticle. So tanning, if it occurs, can have its effects at least mimicked by drying. But if *Calliphora* larval cuticle is tanned in catechol not only does the protein become less soluble (Fraenkel & Rudall, 1947) but the cuticle becomes thinner (Fig. 3) which is very probably due to loss of water from the matrix. But just so long as there is water in the matrix the cuticle does not behave in a Hookean manner as does stiff, sclerotised cuticle (Fig. 4 and Vincent & Hillerton, 1979). All this

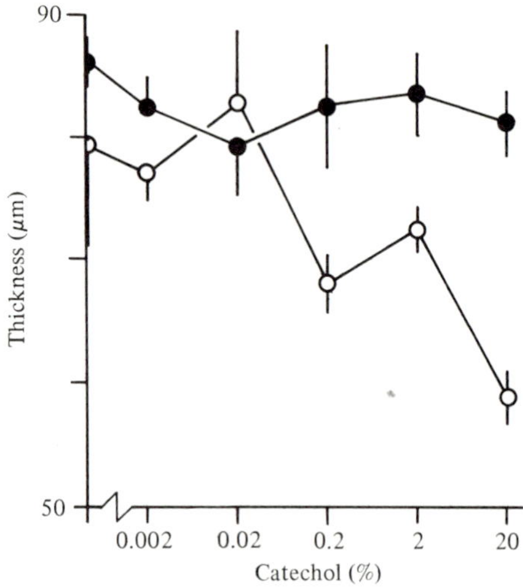

Fig. 3. Reduction in thickness of fully hydrated larval *Calliphora* cuticles as a result of being tanned in catechol for 1.5 h. Each point is the mean ± S.E. of 7 cuticles. Filled circles, before tanning; open circles, the same cuticles after tanning. Cuticle thickness was measured with a micrometer screw gauge. From Vincent & Hillerton (1979).

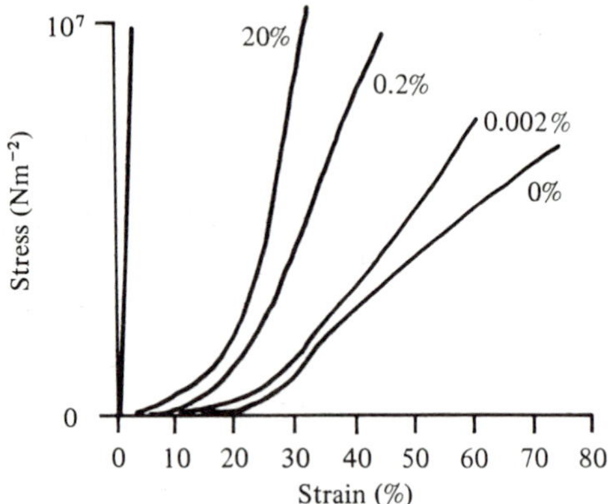

Fig. 4. Stress–strain curves of wet larval cuticles tanned in various concentrations of catechol (as indicated) for 1.5 h. The straight line represents the Hookean response of the same material dry. Adapted from Vincent & Hillerton (1979).

suggests that the phenols are not performing the function which Pryor (1940) envisaged and that the water content of the cuticle should be taken much more into consideration.

It is possible to examine Pryor's theory more closely from the mechanical point of view. What difference does the addition of phenolic cross-links make to the stiffness of the matrix in the absence of a change in water content? Since such a process seems not to occur in nature it is necessary to resort to calculation. Resilin (see also Gosline, 1980) is a type of rubbery cuticle whose proteins can be shown to be interacting only at the sites of covalent cross-linking. Hydrated resilin has a E_m of about 2 MN^{-2} and contains about 5% dry weight of cross-linker. Jensen & Weis-Fogh (1962) soaked some resilin in a saturated solution of quinone and found that the cuticle became deeply and permanently brown but that its rubbery nature was unchanged. The swelling properties changed (by which presumably is meant that the resilin swelled less, since the change was attributed to quinones having blocked the amino groups which would presumably reduce the amount of water bound into the resilin) and the stiffness doubled. Unfortunately Jensen & Weis-Fogh (1962) did not report how much quinone was absorbed. If the modulus doubled and the quinone were truly cross-linking, one would expect that about 3 to 5% dry weight of quinone had been absorbed, assuming rubber elasticity theory (Gosline, 1980). But the associated reduction in swelling would increase the volume fraction of both protein and chitin which would further increase the stiffness, so it is impossible to say how much quinone would have been absorbed. There is another possible approach. When it is sclerotised the puparial cuticle of *Calliphora* increases in dry weight by a massive 25% (Hillerton & Vincent, 1979). If resilin were to have this amount of cross-linker added than it would increase its stiffness five- or sixfold, assuming rubber elasticity theory holds. Of course not only is this approaching a high degree of cross-linking but resilin is by no means a flexible molecule (Gosline, 1980) so it is unlikely that the Gaussian model would hold. If a more detailed calculation is made taking these facts into account it is found that the increase in stiffness is of the order of 11.2 giving a value for E_m of 22.4 MN^{-2}. An increase of five to tenfold on this figure is needed to account for the stiffnesses actually observed (Table 3). This also assumes that the proteins are actually being cross-linked by the phenols. There is no good evidence for this and it is doubtful whether there are sufficient cross-linking sites (Hillerton & Vincent, 1979). The most widely used of the experiments purporting to show that covalent cross-links exist in insect cuticle is the insolubility of the protein in anything less severe than boiling M-NaOH (Hackman & Goldberg, 1958). But a formamide extract of *Calliphora* larval cuticle, dialysed against ammonium acetate and freeze-dried, can be as insoluble as sclerotin without the introduction of covalent cross-links (Vincent & Hillerton, 1979). Moreover, how does the chitin resist such treatment if it is 'only' hydrogen-bonded? The plain truth is that such experiments

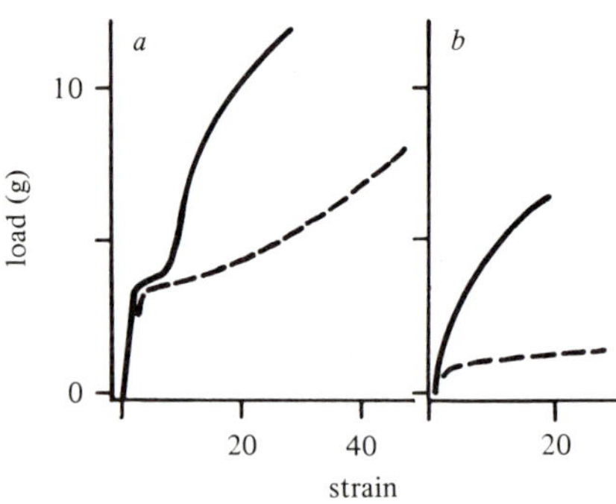

Fig. 5. Force–strain curves of single strands of tanned (solid lines) and untanned (broken lines) silks from *Antheraea pernyi* (*a*) and *Samia cynthia* (*b*). After Brunet & Coles (1974).

on the 'extractability' of proteins rely not only on the strength of the bonds but also on the number of the bonds and their availability to the solvent for the amount of protein which can be taken into solution by any one solvent: a classic example of one experiment with two or three variables. In fact resilin is the *only* cuticle type which has been shown, unequivocally, to be cross-linked covalently (Gosline, 1980). The importance of non-covalent cross-links has been remarked by Lipke & Strout (1972), Lipke & Geoghegan (1976) and Hackman & Goldberg (1977, 1978), but in the absence of mechanical tests none of them has been able to go as far as Vincent & Hillerton (1979) and suggest that there is evidence that covalent cross-links, even if they occur, are irrelevant to the development of stiffness in maggot cuticle. In fact this has also been shown, unwittingly, by Brunet & Coles (1974) who were investigating tanned silks. The initial stiffness of both untanned silk and silk which has subsequently been tanned is the same (Fig. 5). The yield behaviour is different. Brunet & Coles (1974) attribute this to cross-linking, but it could just as well be due to the filler effect (Mullins, 1980) of a polymerised phenol. The important point is that in this dehydrated system the addition of phenolic compounds makes no difference to the initial stiffness: the untanned *dry* material is as stiff as the tanned material. Perhaps the tanning is water-proofing the silk.

There is some information on the correlation of water content with mechanical properties of cuticle (Table 2) but in only one study (Zdarek & Fraenkel, 1972) has the change in mechanical properties been directly correlated with the change in water content (Fig. 6). These authors also observe that when tanning of the puparium of *Sarcophaga* is inhibited by the injection of carbidopa, an inhibitor of dopa-decarboxylase and thus an inhibitor of normal tanning, the water content of the cuticle

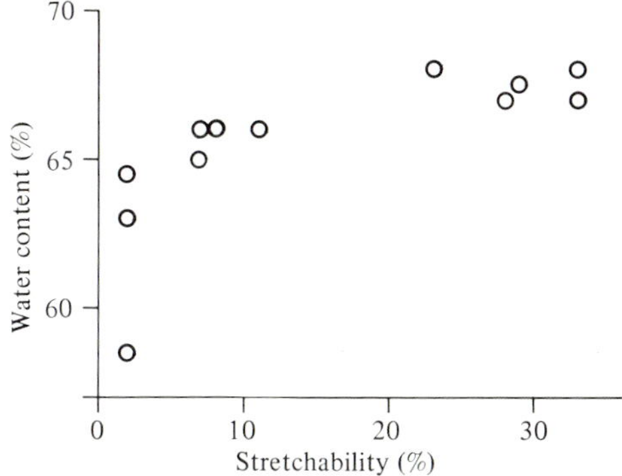

Fig. 6. The correlation between water content of white puparial cuticle of *Sarcophaga bullata* and its extensibility. After Zdarek & Fraenkel (1972).

remains high and constant. As a corollary to this, Reynolds (1976) showed that the change in plasticity which occurs in adult blowfly cuticle and which allows the insect to expand is not associated with tanning as Cottrell (1962) suggested. The change in plasticity is presumably effected by increased water content. Thus the one anomaly disappears and there seems to be no instance where stiffening of the cuticle is not accompanied by loss of water from the protein matrix; and whenever cuticle is sclerotised by phenols water is lost from the cuticle at the same time. Combine these observations with experiments on model systems by Pryor (1940) who showed that gelatin tanned in quinone becomes stiffer and dehydrated; and by Jensen & Weis-Fogh (1962) tanning resilin in quinine; and by Vincent & Hillerton (1979) who showed that tanning *Calliphora* larval cuticle in catechol causes it to become thinner (Fig. 4), presumably due to loss of water, and the possibility emerges that the process of sclerotisation is not necessarily one of covalent cross-linking but also, or perhaps mainly, of changing the water content, and hence the degree of permanence and density of secondary interactions, of the matrix. Fraenkel & Rudall (1940) encapsulated this idea by saying 'The water content (of puparial cuticle being sclerotised) decreases from about 70% in the larval cuticle to about 40% in the 36 hr old puparium. This dehydration is considered as having a chemical basis. Later a further purely physical drying takes place, and in the air-dried puparium there is a water content of about 12%'. If one takes into account the observations of Dennell (1946), mentioned above, then the actual figures for water content may well be different but the general trend towards a reduction in water content during sclerotisation remains.

Table 4. *Average hydrophobicity* ($H\varphi$ ave) *of cuticular proteins from female* Schistocerca gregaria *based on amino-acid analyses of whole cuticle except where stated (J. E. Hillerton, personal communication, based on Andersen, 1971) and the matrix modulus of two typical cuticles and resilin*

Source	$H\varphi$ (average)			E (Nm^{-2})
Femur	1182	↑	↑	
Femur (formamide extract)	1170			
Femur (insoluble residue)	1211			
Cornea	1165	Stiff cuticles	Relatively	1.8×10^9
Tibia	1157		insoluble	Ker (1977)
Thorax (ventral)	1152			
Thorax (lateral)	1123		↓	
Abdomen (sternite)	1087		↑	
Abdomen (tergite)	1049	↓		1.8×10^6
Abdomen (intersegmental membrane)	1025	↑		Wood (1974)
Neck membrane	994	Pliant cuticles	Relatively	
Coxal membrane	977	↓	soluble	
Resilin (Clypeo-labral spring)	725			
		Rubbery	Covalently	2×10^6
Resilin (pre-alar arm)	583	cuticle	cross-linked	Jensen & Weis-Fogh (1962)

Then the model for the sclerotisation of insect cuticle matrix, and possibly for other sclerotised systems, is that the protein matrix contains a small amount of loosely bound water. If water is lost from the matrix either by ordinary drying or by polymerising phenols within the water-binding areas (c.f. Pryor, 1940; Jensen & Weis-Fogh, 1962), then secondary interactions become much more favourable and the density of the bonding increases greatly. At the same time the volume fraction of the chitin increases greatly since the hydrated protein is reducing its volume as it loses water. The matrix also has an increasing shear modulus and can transmit forces to the chitin more effectively, which then can receive greater loads from the matrix and realise more completely its full potential as a reinforcement. The argument now turns from the presence and quantity of phenolic materials within the matrix to the presence, quantity and interactions of water with the proteins and with the phenols.

The amino-acid composition of the matrix can give some indications here: both Welinder (1975) and Andersen (1979) have shown that pliant cuticles tend to be relatively rich in polar and charged amino-acid

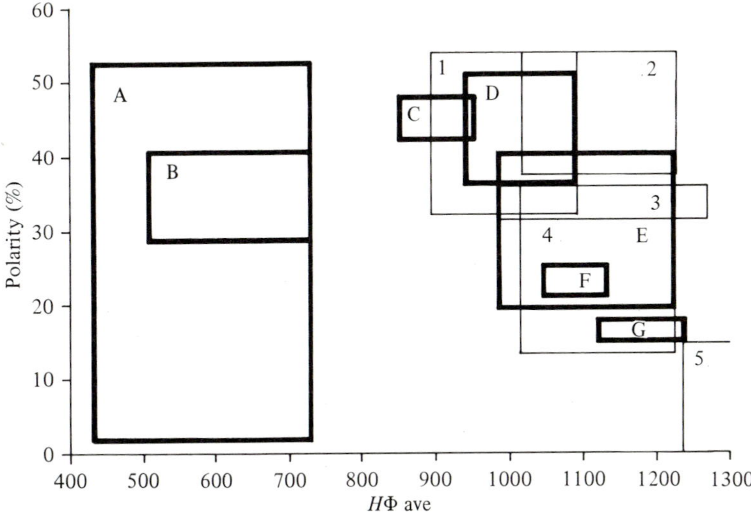

Fig. 7. Variation in hydrophobicity ($H\varphi$ ave) and polarity of cuticle proteins from insects (heavy boxes) and other sources (light boxes). Heavy boxes: A, insect silks; B, resilins; C, fly larval and puparial cuticles; D, termite cuticles; E, beetle cuticles; F, *Rhodnius* and *Boophilus* extensible cuticles; G, locust sclerite cuticles. Light boxes: 1, high-sulphate keratins; 2, globulins and albumens; 3, collagens; 4, high-tyrosine keratins; 5, elastins. Adapted from J. E. Hillerton (personal communication).

residues, whereas stiff cuticles and cuticles destined to be stiff tend to have more non-polar residues. It is possible to calculate an index of hydrophobicity ($H\varphi$): this has been done by Andersen (1979) using the method of Levitt (1976) and by J. E. Hillerton (personal communication) using the methods of Bigelow (1967) and of Capaldi & Vanderkooi (1972). Some of Hillerton's results (J. E. Hillerton, personal communication) are presented in Table 4 and Fig. 7. In general, stiffness and insolubility of cuticles tend to be associated with a higher hydrophobicity index and with a lower polarity index. Resilin is in a class of its own: its low hydrophobicity index suggests that it is, in fact, a soluble and hydratable protein. This may well be the reason why it is covalently cross-linked; in an aqueous environment it would not form internal hydrophobic interactions and so would not be stable. Indeed it is the only cuticle matrix material to be crossed-linked as soon as it is secreted and this could well be because otherwise it would simply not be sufficiently stable to remain in the matrix in the presence of solvating water. A more hydrophobic protein will tend towards a more globular conformation. There are two estimates of E_m, all there are at present, in Table 4. There are more estimates available of the stiffness of whole cuticle but these will be greatly affected by the chitin phase (its volume fraction, orientation) and its interactions with the matrix (shear modulus, transfer length to fibre – see Harris, 1980) and so may be expected not to bear so much relationship to the hydrophobicity index. Also, of course, the estimate of E_m for resilin is high precisely because it *is* cross-linked: in the un-cross-linked state it would probably simply be in solution. When $H\varphi$ and the polarity index are used as the axes of a graphical presentation (Fig. 7) a distinct pattern emerges which not only differentiates between pliant and stiff cuticles but in which cuticular proteins from insects of different orders are grouped separately. This strongly suggests that the mechanisms of stabilisation of the matrix may be different in different orders of insect, in emphasis of the different processes available if not in absolute biochemistry (i.e. they may all use phenols but the precise type of phenol used may vary). In marked contrast this graph has suggested that cuticular proteins often considered to be fundamentally different may well be more similar than suspected. Hillerton (1979*b*) extracted a hydrophobic protein from the cuticle of larval *Rhodnius* and he has succeeded in making similar preparations from pharate sclerite of *Locusta* and *Periplaneta* (J. E. Hillerton, personal communication) suggesting that the extensible cuticle of *Rhodnius* may be most closely akin to a pharate cuticle. In which case the expansion of the abdomen of *Rhodnius* may well be a paradigm for the expansion of insects at moulting (see also Glaser & Vincent, 1979).

Work in progress in my laboratory strongly suggests that even the larval cuticle of *Calliphora* can be considered to be rather hydrophobic. The amount of water which a material or surface will adsorb from atmospheres of different relative humidity varies in a typical way depending on the hydrophobicity of that surface (Zettlemoyer, Micale & Klier, 1975); larval and puparial cuticle both seem fairly hydrophobic under these conditions in that they will not adsorb water from an atmosphere unless the relative humidity is approaching 50% or even 80% if the cuticle were first dried. Differential scanning calorimetry (DSC) confirms this in that there is no evidence that water is bound into the matrix as it might be with hair or collagen (d'Arcy & Watt, 1970). However, sclerotised *Calliphora* puparial cuticle (taken about a day after pupariation) not only adsorbs water from atmospheres of *lower* relative humidity but also shows evidence of binding of water in DSC. At first sight this seems perverse, but it seems very likely that this represents water which is bound to the dixydroxy phenols. In turn, this suggests the possibility that the phenols are binding water more strongly than the protein and so encouraging further protein–protein interactions. Be that as it may, the evolutionary tendency is for cuticle proteins to become less hydrophobic in the 'higher' orders of insects so that the crude model put forward here will have to be refined to account for sclerotisation in different insects. Also, if sclerotisation does not necessarily involve such unique biochemical reactions as might be demanded by specific interactions with various proteins but involves a more general type of interaction which will cause the expulsion of water, then it seems quite possible that there will be several different ways of achieving the same effect.

The hydrophobicity of the matrix proteins will also affect the amount of water which the unsclerotised cuticle will bind. Andersen (personal communication) has shown that the femur cuticle of the locust not only has only 40% water at the start of ecdysis but that the water content goes down to only slightly less than 30% in the fully tanned material. However he is here dealing with endocuticle as well as exocuticle and so not entirely with sclerotised material. His experiments lead him to the conclusion that the solubility and swelling behaviour of sclerotised locust cuticle can best be explained by assuming the presence of covalent cross-links between protein molecules. He says that dehydration and polymerisation of phenolic material will have some role to play but maintains that they can hardly be the only factors of importance. The difficulty is that one cannot solve a problem in mechanics with biochemical techniques; the inadequacy of current theories of sclerotisation is ample witness to that. Dorrington (1980) may have the answer when he points out the effect of water on the glass transition of biomaterials. Sclerotisation could be regarded as the raising of the glass transition temperature of the matrix proteins by the removal of water. This suggests that careful manipulation of water content and temperature will enable the position of the glass transition temperature at different states of hydration to be estimated. If this behaviour is not affected by tanning then the role of the phenols will have to be reviewed. Clearly there is much to be done to decide how much of sclerotisation is phenolic cross-linking and how much is control of water content.

Cuticle as a layered structure

As far as the scleroprotein matrix is concerned sclerotisation is a fairly clear-cut concept and sclerotised article can be interpreted in terms of classic composite theory (Harris, 1980). But as far as the entire thickness of the cuticle is concerned the importance of sclerotisation has hardly been touched on in the literature. Most sclerotised insect cuticles are not sclerotised all the way through but are bi-layers with stiff, sclerotised, exocuticle on the outside and relatively pliant, more or less unsclerotised, endocuticle on the inside. Jensen & Weis-Fogh (1962) maintained that the stiffness of these two layers is the same, but they may well have allowed their samples to dry, which would certainly make the endocuticle as stiff as the exocuticle. Hepburn and his colleagues (see Hepburn & Joffe, 1976) in their numerous experiments noted that sclerotised cuticle fails at a higher stress and lower strain than endocuticle suggesting that the endocuticle is, indeed, more pliant. That this system can confer mechanical advantages is not obvious, but model systems of this type have been examined (Guild, Harris & Atkins, 1978) and it appears that if two layers of materials of different toughness but similar (though not necessarily the same) stiffness are stuck together, the resulting laminate will be tougher in tearing than one would expect from taking a simple mean value of the toughness values for the two materials separately. In addition the endocuticle tends to stop cracks travelling through the exocuticle, as is often observed in histological sections of cuticle, so the endocuticle is probably acting as a crack-stopper by blunting the leading edge of the crack. Finally Ker (1977) has shown that if the endocuticle is about half the stiffness of the exocuticle then the amount of energy per unit volume which can be stored by the laminate in bending, such that the exocuticle is in compression, is much greater (see also below, p. 206). This is because addition of material on the outside of the curve moves the neutral axis of the beam further towards the more pliant material (Fig. 8), thus throwing more of the stiffer material into compression. Since the stiffer material is quite safe in compression and the pliant material is much safer in tension than is the stiffer material the laminate can be bent further and store more strain energy before it breaks. Ker (1977) has shown that this type of design is used in the scroll springs which power the jump of certain flea beetles, but the property is of general use in that cuticle constructed in this fashion can take higher bending loads before it breaks and so the structure is tougher than a single layer of sclerotised cuticle would be. So perhaps the final point about sclerotisation is that it must be very controllable, since too stiff and brittle a cuticle would reduce its performance as a tough outer covering. Perhaps this is why dehydration at sclerotisation is a chemical rather than a physical process as it might be if the water were removed by an osmotic process. The biochemical process may be more easily controlled though it may still, of course, be backed up by an osmotic mechanism. Which returns this review to the

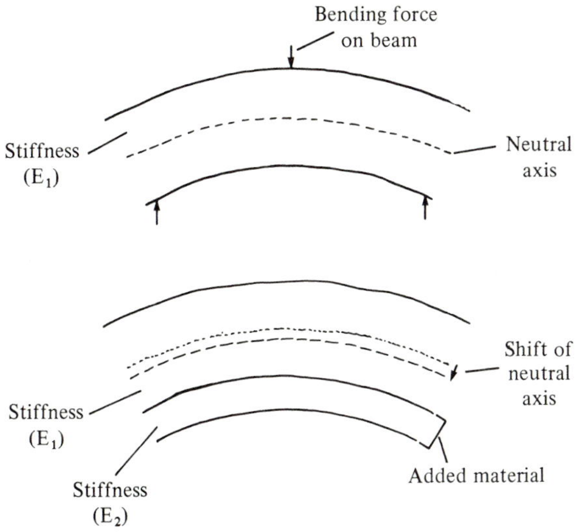

Fig. 8. Shift in the neutral axis of a beam when a layer of more compliant material is added (see text). After Ker (1977). (Energy is stored by *flattening* the beam, thus putting the added layer into tension.)

opening sentence: the few percent of salts found in insect cuticles has always been disregarded. But what if the salts were fine-tuning the water content of the cuticle?

Viscoelasticity

The viscoelastic properties of insect cuticle have hardly been studied at all. Jensen & Weis-Fogh (1962) reported that stiff cuticles show creep and also show internal losses when tested dynamically. Ker (1977) has tested *Schistocerca* tibial extensor apodeme in longitudinal creep, parallel to the chitin (Fig. 9) and shown that although the material *does* creep (the matrix proteins are not cross-linked), the original length is recovered virtually completely and immediately if the load is applied for only a short time (say a minute). The creep rate perpendicular to the fibres is about five times faster. High loads and/or long times under load in the longitudinal direction caused some permanent deformation and creep stiffening (somewhat analogous to the work hardening or cycle hardening noted by Hepburn & Chandler (1976) but noted by them to be due, after all, to dehydration; also related to stress-softening (Vincent, 1975*a*, 1976; Mullins, 1980) in the absence of dehydration) but Ker (1977) did not report systematic results for creep as he was more

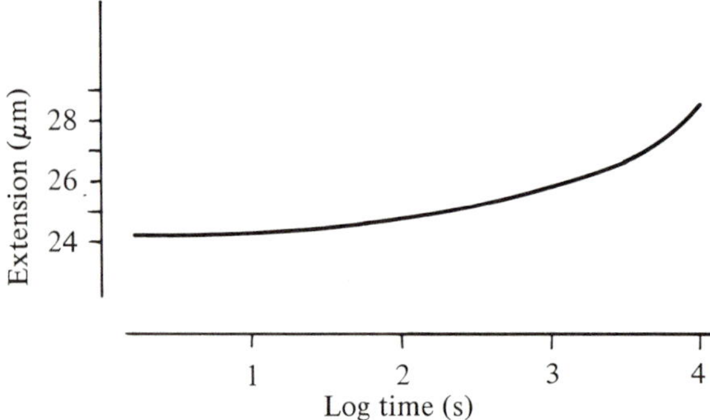

Fig. 9. Creep of a tibial extensor apodeme (post-ecdysial cuticle only). Load 100 g, length 2.85 mm. From Ker (1977).

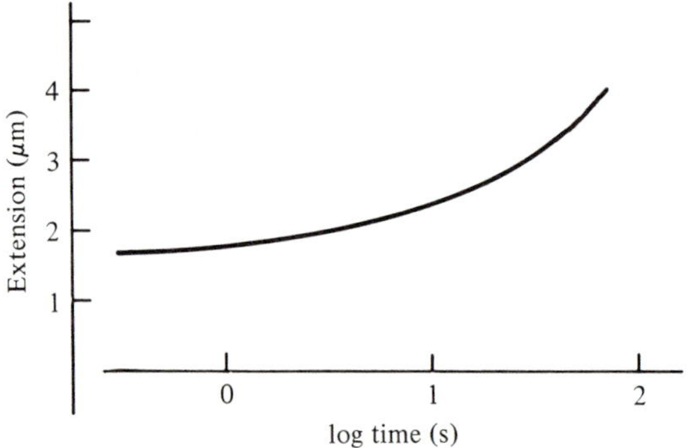

Fig. 10. Creep of tibial cuticle from a locust treated with diflubenzuron, thus inhibiting the deposition of chitin in the sample. Load 0.3 g, length 0.68 mm. From Ker (1977).

concerned with producing isochronal stress–strain curves (Dorrington, 1980). Ker (1977) also investigated the creep properties of *Schistocerca* tibial cuticle which had been treated with diflubenzuron to inhibit the deposition of chitin and found the rates to be much higher (Fig. 10).

Creep and stress–relaxation have also been studied in the larval cuticle of *Rhodnius* (Reynolds, 1975*a*; Hillerton, 1979*a*), in the intersegmental membranes of mature adult female *Locusta* (Wood, 1974; Vincent, 1976) and the cuticle of larval *Calliphora* (Vincent, unpublished). In all instances the curves are straight and fairly flat suggesting a broad spectrum of relaxation times. This can be accounted for simply by the concept of pliant cuticle as a filled rubber system (Vincent, 1975*a*, 1976; Mullins, 1980). In *Rhodnius* larval cuticle the rate of relaxation shifts by an order of magnitude or more in cuticle plasticised by the injection of 5-hydroxytryptamine into the larva (Reynolds, 1975*a*), an effect which is attributable to a change in water content of 25.8% to 31.3% (Reynolds, 1975*b*). This is entirely consistent with the interpretation that increased solvent reduces the glass transition temperature (Dorrington, 1980). The factors controlling the change in water content are still the subject of some contention (Hillerton, 1979*a*). *Rhodnius* cuticle has never been allowed to creep or relax for sufficiently long to show an equilibrium modulus; with extensible locust i.s.m. the equilibrium modulus is zero (Wood, 1974; Vincent, 1976), the reduction in modulus being governed by processes with relatively long relaxation times.

As is to be expected in a material with a broad spectrum of relaxation times, the response of the shear (elastic) modulus of *Locusta* female i.s.m. to temperature is fairly flat, increasing suddenly at about −18°C presumably due to water freezing in the matrix, and increasing more gradually from about 50°C (Wood, 1974) though Vincent & Prentice (1973) noted a slight temperature effect at 44°C. Apart from these experiments the effect of temperature on cuticle mechanics has not been investigated.

Energy storage and toughness

Cuticle is an important energy store and is necessary for small arthropods to jump (Bennet-Clark, 1975*a*) and fly (Jensen & Weis-Fogh, 1962). In order to store energy without loss and return it quickly the cuticle must have very low internal viscous losses, i.e., a low loss modulus or a low tan δ. Resilin fulfils this criterion (Gosline, 1980) but is not the only type of cuticle to store energy for use in locomotion. The apodeme of the locust hind leg forms a major part of the energy storage system for the locust jump (Bennet-Clark, 1975*b*, 1976; Ker, 1977) where, because it is so stiff (Table 2), it stores energy at relatively low strains – about 3% – but can store twice the amount of energy compared with resilin (Bennet-Clark, 1976) since it is 200 times as strong as resilin and works at 1% of the strain at which resilin works. A great problem with such energy storage is, however, that if the material is brittle it cannot dissipate as much strain energy as tough material can. So if a small crack or imperfection is initiated in the material the energy to continue the growth of that crack can be quickly fed to the crack tip with no loss and the material breaks with relative ease. Looked at in another way, the size of imperfection which such an energy-storing material can accommodate is much smaller than that which a piece of skeletal cuticle, which shows internal energy losses, can sustain with safety. Nevertheless the tendon in the hind leg of the locust is loaded to within 15% of its ultimate strength. At a rough estimate, using data from Bennet-Clark (1976) the size of crack which a piece of resilin could accommodate would probably be, at 0.25 mm, rather larger than that which a locust apodeme could accommodate (about 0.1 mm) for a similar amount of energy stored. These figures are not necessarily smaller than the size of the pieces of cuticle in question, so they are at risk from catastrophic fracture only in larger insects. This is not too bad if they are kept out of the way as, say, is the tendon down the middle of the hind leg; Bennet-Clark (1976) reports difficulty in loading the extensor tibiae apodeme to anything like its natural strength and attributes this to the likelihood of surface cracks initiated during the preparation of the specimen. A crude calculation of the critical crack length (Gordon, 1978) of the larval cuticle of *Rhodnius* using a figure of 1.4 kJm^{-2} for the tearing energy (Purslow, personal communication) gives a critical crack length at least twice that for resilin but the cuticle is probably never loaded to the same extent of its strength. In the extremely pliant mature female locust i.s.m., which has a lower tearing energy of about 0.4 kJm^{-2} it is quite common to find a healed area in the membrane of old individuals. This indicates that the membrane has split (up to half-way around the abdomen) without the insect having suffered unduly. However, the stress–strain curve of this material is markedly non-linear (Vincent, 1975*a*) almost approaching the surface-tension stress–strain behaviour mentioned by Gordon (1978).

It seems likely that the flea-beetle has overcome part of the energy storage problem by using essentially a composite beam to store energy for its jump (see above, p. 202). Ker (1977) calculated that if the stiffness of the untanned cuticle is about half that of the tanned cuticle and the untanned cuticle is put into tension as the beam bends (Fig. 8), then the beam can store $2\frac{1}{4}$ times as much energy per unit volume of cuticle as could tanned cuticle on its own. This improvement is due to the shift of the neutral axis within the tanned cuticle which effectively allows a greater distortion for a given volume of tanned cuticle. This is because the ultimate strain in tension of the untanned cuticle is 5.5%, that of the tanned cuticle only 3%, but 4.5% in compression.

Hardness

The hardness of insect cuticle has been commented upon briefly by Neville (1975) but no hardness measurements have been published, although Neville (1975) mentions a single report of scratch tests as used by geologists. Hillerton (1980) has performed tests on locust mandible and produced results which agree very well with some communicated by S. E. Reynolds (personal communication). Gardiner & Khan (1979) have shown that in the hard part of the locust mandible the diameter of the fibres is much greater (up to $0.5\,\mu$m) which apparently leads to greater compressive strength in composites (C. R. Chaplin, personal communication). Insects seem not to incorporate salts into cuticle to increase hardness in the way that other arthropods and molluscs do (Wainwright *et al.*, 1976) but there is evidence that the type of sclerotisation is different (J. E. Hillerton, personal communication) and that more melanin is incorporated.

In conclusion . . .

This review has not attempted to be fully balanced, it has been concerned with aspects of cuticle mechanics which seem ripe for further experiment. Biochemists have for too long pronounced on the mechanical properties of insect cuticle. It's time the *experiments* were done!

I thank Dr G. Jeronimidis and Dr J. E. Hillerton for much stimulating discussion and Professor S. O. Andersen for stimulating argument. Some of this work was carried out during tenure of grants from the S.R.C. and the A.R.C.

References

Andersen, S. O. (1971). Resilin. In *Comprehensive Biochemistry*, vol. 26C, ed. M. Florkin & E. H. Stotz, pp. 633–57. Amsterdam: Elsevier.

Andersen, S. O. (1977). Arthropod cuticles: their composition, properties and functions. *Symposium of the Zoological Society of London*, **39**, 7–32.

Andersen, S. O. (1979). Biochemistry of insect cuticle. *Annual Review of Entomology,* **24,** 29–61.

d'Arcy, R. L. & Watt, I. C. (1970). Analysis of sorption isotherms of non-homogeneous sorbents. *Transactions of the Faraday Society,* **66,** 1236–45.

Barth, F. G. (1973). Microfiber reinforcement of an arthropod cuticle. Laminated composite material in biology. *Zeitschrift für Zellforschung,* **144,** 409–33.

Bennet-Clark, H. C. (1975*a*). Energy storage in jumping animals. In *Perspectives in Experimental Biology, Volume 1, Zoology,* ed. P. Spencer Davies, pp. 467–79. Oxford: Pergamon.

Bennet-Clark, H. C. (1975*b*). The energetics of the jump of the locust, *Schistocerca gregaria. Journal of experimental Biology,* **63,** 53–83.

Bennet-Clark, H. C. (1976). Energy storage in jumping insects. In *The Insect Integument*, ed. H. R. Hepburn, pp. 421–43, Amsterdam: Elsevier.

Bigelow, C. C. (1967). On the average hydrophobicity of proteins and the relation between it and protein structure. *Journal of Theoretical Biology,* **16,** 187–211.

Brunet, P. C. J. & Coles, B. C. (1974). Tanned silks. *Proceedings of the Royal Society of London, Series B,* **187,** 133–70.

Capaldi, R. A. & Vanderkooi, G. (1972). The low polarity of many membrane proteins. *Proceedings of the National Academy of Science, USA,* **69,** 930–2.

Cottrell, C. B. (1962). The imaginal ecdysis of blowflies. Evidence for a change in the mechanical properties of the cuticle at expansion. *Journal of experimental Biology,* **39,** 449–58.

Dennell, R. (1946). A study of an insect cuticle: the larval cuticle of *Sarcophaga falculata* Pand. (Diptera). *Proceedings of the Royal Society of London, Series B,* **133,** 348–73.

Dorrington, K. L. (1980). The theory of viscoelasticity in biomaterials. In *The Mechanical Properties of Biological Materials* (34th Symposium of the Society for Experimental Biology) ed. J. F. V. Vincent & J. D. Currey, pp. 289–314. Cambridge University Press.

Fraenkel, G. & Rudall, K. M. (1940). A study of the physical and chemical properties of the insect cuticle. *Proceedings of the Royal Society of London, Series B,* **129,** 1–35.

Fraenkel, G. & Rudall, K. M. (1947). The structure of insect cuticles. *Proceedings of the Royal Society of London, Series B,* **134,** 111–43.

Fristrom, J. W., Hill, R. J. & Watt, F. (1978). The procuticle of *Drosophila*: heterogeneity of urea-soluble proteins. *Biochemistry,* **17,** 3917–24.

Gardiner, B. C. & Khan, M. F. (1979). A new form of insect cuticle. *Zoological Journal of the Linnean Society,* **66,** 91–4.

Glaser, A. E. & Vincent, J. F. V. (1979). The autonomous inflation of insect wings. *Journal of Insect Physiology,* **25,** 315–18.

Gordon, J. E. (1978). *Structures, or why things don't fall down.* Harmondsworth: Penguin.

Gosline, J. M. (1980). The elastic properties of rubber-like proteins and highly extensible tissues. In *The Mechanical Properties of Biological Materials* (34th Symposium of the Society for Experimental Biology) ed. J. F. V. Vincent & J. D. Currey, pp. 331–57. Cambridge University Press.

Guild, F. J., Harris, B. & Atkins, A. G. (1978). Cracking in layered composites. *Journal of Materials Science,* **13,** 2295–9.

Hackman, R. H. (1974). The soluble cuticular proteins from three arthropod species, *Scylla serrata, Boophilus microplus* and *Agrianome spinicollis. Comparative Biochemistry and Physiology,* **49,** 457–64.

Hackman, R. H. (1976). The interactions of cuticular proteins and some comments on their adaptation to function. In *The insect integument*, ed. H. R. Hepburn, pp. 107–20, Amsterdam: Elsevier.

Hackman, R. H. & Goldberg, M. (1958). Proteins of the larval cuticle of *Agrianome spinicollos* (Coleoptera). *Journal of Insect Physiology,* **2,** 221–31.

Hackman, R. H. & Goldberg, M. (1977). Molecular crosslinks in cuticles. *Insect Biochemistry,* **7,** 175–84.

Hackman, R. H. & Goldberg, M. (1978). The non-covalent binding of two insect cuticular proteins by a chitin. *Insect Biochemistry,* **8,** 353–7.

Hackman, R. H. & Goldberg, M. (1979). Some conformational studies of *Calliphora vicina* larval cuticular protein. *Insect Biochemistry,* **9,** 557–61.

Harris, B. (1980). Theory of composite materials. In *The Mechanical Properties of Biological Materials* (34th Symposium of the Society for Experimental Biology) ed. J. F. V. Vincent & J. D. Currey, pp. 37–74. Cambridge University Press.

Hepburn, H. R. & Chandler, H. D. (1976). Material properties of arthropod cuticles: the arthrodial membranes. *Journal of Comparative Physiology,* **109,** 177–98.

Hepburn, H. R. & Joffe, I. (1976). On the material properties of insect exoskeletons. In *The insect integument*, ed. H. R. Hepburn, pp. 207–35. Amsterdam: Elsevier.

Hillerton, J. E. (1979*a*). Changes in the mechanical properties of the extensible cuticle of *Rhodnius* through the fifth larval instar. *Journal of Insect Physiology*, **25,** 73–7.

HILLERTON, J. E. (1979*b*). The water-soluble proteins from the extensible cuticle of *Rhodnius* larvae. *Insect Biochemistry,* **9,** 143–148.

HILLERTON, J. E. (1980). The hardness of locust incisors. In *The Mechanical Properties of Biological Materials* (34th Symposium of the Society for Experimental Biology) ed. J. F. V. Vincent & J. D. Currey, pp. 483–84. Cambridge University Press.

HILLERTON, J. E. & VINCENT, J. F. V. (1979). The stabilisation of insect cuticles. *Journal of Insect Physiology,* **25,** 957–63.

JENSEN, M. & WEIS-FOGH, T. (1962). Biology and physics of locust flight. V Strength and elasticity of locust cuticle. *Philosophical Transactions of the Royal Society of London, Series B,* **245,** 137–69.

JOFFE, I. & HEPBURN, H. R. (1973). Observations on regenerated chitin films. *Journal of Materials Science,* **8,** 1751–4.

KER, R. F. (1977). Some structural and mechanical properties of locust and beetle cuticle. D. Phil. thesis, Oxford.

KER, R. F. (1980). Small-scale tensile tests. In *The Mechanical Properties of Biological Materials* (34th Symposium of the Society for Experimental Biology) ed. J. F. V. Vincent & J. D. Currey, pp. 487–89. Cambridge University Press.

LEVITT, M. (1976). A simplified representation of protein conformations for rapid simulation of protein folding. *Journal of molecular Biology,* **104,** 59–108.

LIPKE, H. & GEOGHEGAN, T. (1976). Enzymolysis of sclerotized cuticle from *Periplaneta americana* and *Sarcophaga bullata. Journal of Insect Physiology,* **17,** 415–25.

LIPKE, H. & STROUT, V. (1972). Peptidochitodextrins of sarcophagid cuticle. II Studies in the linkage region. *Israel Journal of Entomology,* **7,** 117–28.

MILLS, R. R., GREENSLADE, F. C., FOX, F. R. & NIELSEN, D. J. (1967). Purification of KCl-soluble proteins from the cuticle of *Acheta domesticus* (L.). *Comparative Biochemistry and Physiology,* **22,** 327–32.

MINKE, R. & BLACKWELL, J. (1978). The structure of α-chitin. *Journal of molecular Biology,* **120,** 167–81.

MULLINS, L. (1980). Theories of rubber-like elasticity and the behaviour of filled rubbers. In *The Mechanical Properties of Biological Materials* (34th Symposium of the Society for Experimental Biology) ed. J. F. V. Vincent & J. D. Currey, pp. 273–88. Cambridge University Press.

NEVILLE, A. C. (1975). *Biology of the arthropod cuticle.* Berlin: Springer-Verlag.

PRYOR, M. G. M. (1940). On the hardening of the ootheca of *Blatta orientalis. Proceedings of the Royal Society of London, Series B,* **128,** 378–93.

REYNOLDS, S. E. (1975*a*). The mechanical properties of the abdominal cuticle in *Rhodnius. Journal of experimental Biology,* **62,** 69–80.

REYNOLDS, S. E. (1975*b*). The mechanism of plasticization of the abdominal cuticle in *Rhodnius. Journal of experimental Biology,* **62,** 81–98.

REYNOLDS, S. E. (1976). Hormonal regulation of cuticle extensibility in newly emerged adult blowflies. *Journal of Insect Physiology,* **22,** 529–34.

STROUT, V., LIPKE, H. & GEOGHEGAN, T. (1976). Peptidochitodextrins of *Sarcophaga bullata*: molecular weight of chitin during pupariation. In *The Insect Integument* ed. H. R. Hepburn, pp. 43–61. Amsterdam: Elsevier.

THOMPSON, P. R. & HEPBURN, H. R. (1978). Changes in chemical and mechanical properties of honeybee (*Apis mellifera adansonii* L.) cuticle during development. *Journal of Comparative Physiology,* **126,** 257–262.

TYCHSEN, P. H. & VINCENT, J. F. V. (1976). Correlated changes in mechanical properties of the intersegmental membrane and bonding between the proteins in the female adult locust. *Journal of Insect Physiology,* **22,** 115–25.

VINCENT, J. F. V. (1975*a*). Locust oviposition: stress-softening of the extensible intersegmental membranes. *Proceedings of the Royal Society of London, Series B,* **188,** 189–201.

VINCENT, J. F. V. (1975*b*). The mechanism of extension of the pleural pleat in adult *Rhodnius prolixus*. *Journal of Entomology (A)*, **50**, 183–5.

VINCENT, J. F. V. (1976). Design for living – the elastic-sided locust. In *The Insect Integument*, ed. H. R. Hepburn, pp. 401–19. Amsterdam: Elsevier.

VINCENT, J. F. V. & HILLERTON, J. E. (1979). The tanning of insect cuticle – a critical review and a revised mechanism. *Journal of Insect Physiology*, **25**, 653–8.

VINCENT, J. F. V. & PRENTICE, J. H. (1973). Rheological properties of the extensible intersegmental membrane of the adult female locust. *Journal of Materials Science*, **8**, 624–30.

WAINWRIGHT, S. A., BIGGS, W. D., CURREY, J. D. & GOSLINE, J. M. (1976). *Mechanical design in organisms*. London: Edward Arnold.

WEIS-FOGH, T. (1961). Molecular interpretation of the elasticity of resilin, a rubber-like protein. *Journal of molecular Biology*, **3**, 648–67.

WELINDER, B. S. (1975). The crustacean cuticle. III Composition of the individual layers in *Cancer pagurus* cuticle. *Comparative Biochemistry and Physiology*, **52 (A)**, 659–63.

WOOD, S. D. E. (1974). The mechanical properties of the intersegmental membrane of the migratory locust. M.Sc. Thesis, Reading University.

ZDAREK, J. & FRAENKEL, G. (1972). The mechanism of puparium formation in flies. *Journal of experimental Zoology*, **179**, 315–24.

ZETTLEMOYER, A. C., MICALE, F. J. & KLIER, K. (1975). Adsorption of water on well characterised solid surfaces. In *Water, a comprehensive treatise*, vol. 5, ed. F. Franks, pp. 249–251, New York: Plenum.

MOLECULAR STRUCTURE AND MECHANICAL PROPERTIES OF KERATINS

R. D. B. FRASER AND *T. P. MACRAE*

CSIRO Division of Protein Chemistry, 343 Royal Parade, Parkville, Victoria 3052, Australia

Introduction

Keratin is a fibrous, proteinaceous material that is produced in the integument, or outer covering, of the vertebrate body. The integument comprises two structural entities: the dermis, which is derived from the embryonic mesoderm, and the superficial epidermis, which is derived from the ectoderm. The principal fibrous proteins of the dermis are collagen and elastin; in association with inorganic salts and an organic matrix they constitute the connective tissue of the integument. The fibrous proteins produced by the epidermis are many and varied and are collectively termed keratin proteins. Keratin proteins characteristically contain cysteinyl residues which are oxidized in the final stages of biosynthesis, leading to a network of intramolecular and intermolecular disulphide cross-linkages.

In the dermis the collagen fibrils and elastic fibres are laid down extracellularly but remain under the influence of cells termed fibroblasts. The connective tissue may undergo substantial changes in structure with age, the resorption of tadpole-tail collagen being an extreme example. In contrast, keratin is laid down intracellularly and the cells responsible for keratin synthesis, termed keratinocytes, die in the terminal stage of keratin synthesis when the intermolecular cross-linkages are formed. Some remnants of the cellular apparatus of the keratinocyte are incorporated in the final keratin. The epidermis is constantly being renewed and this is a protective adaptation which compensates for wear and damage at the surface of the organism. Two patterns of activity can be distinguished: in the skin there is continuous exfoliation of dead cells from the surface which is balanced by continuous production of new cells in the underlying germinal layer (Fig. 1), but many appendages are shed at regular intervals and periods of intense mitotic activity in the germinal layers are interspersed with quiescent periods during which the appendage has a stable form.

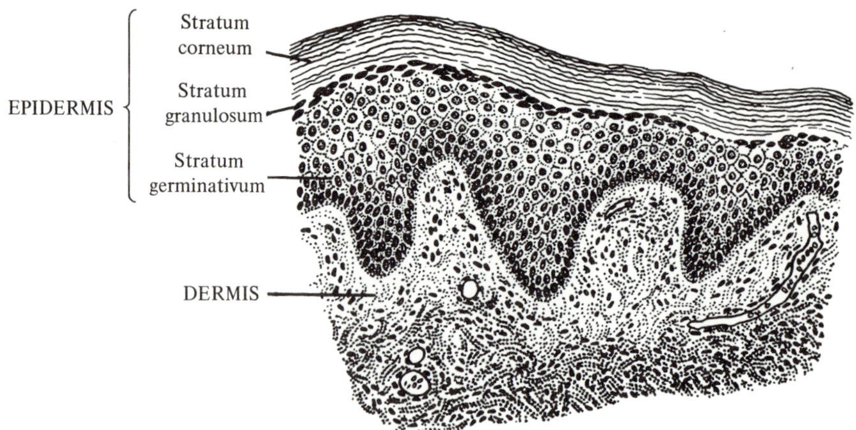

Fig. 1. Diagram of the structure of mammalian skin (human shoulder). Reproduced by permission from Romer (1970).

The integument constitutes the interface between the organism and its environment and its primary function is that of a barrier. In aquatic vertebrates the epidermis is often thin and plays only a minor role in determining the mechanical properties of the integument; in land-dwelling vertebrates however, the epidermis is generally thicker and an outer horny layer of cells may clearly be distinguished from germinal and differentiating layers of cells. It is this outer horny layer together with a profusion of specialized epidermal appendages that make a significant contribution to the mechanical properties of the integument of the land-dwelling vertebrates.

Before discussing the composition, structure and mechanical properties of keratins it is of interest to summarize the functions of keratinized structures since the mechanical properties are of necessity intimately related to these functions. Although not comprehensive, the following list includes the principal functions recognized by biologists.

(a). Prevention of egress of tissue fluids.
(b). Prevention of ingress of environmental fluids.
(c). Prevention of ingress of micro-organisms, parasites and foreign matter.
(d). Protection against mechanical injury and attack by predators.
(e). Aggressive and defensive actions both within and between species.
(f). Food gathering.
(g). Temperature regulation.
(h). Locomotion, including flight, climbing and floating.

Composition and molecular structure

Classification of Keratins

Two distinct patterns of differentiation have been observed in keratinocytes (Giroud & Leblond, 1951; Mercer, 1961), the product of the first being so-called 'soft' keratins such as the stratum corneum of skin and of the second being so-called 'hard' keratins such as hair and nail. Although the terms 'soft' and 'hard' are not much used at present no alternative succinct description has emerged and these terms will be retained in the present contribution. The classification is reinforced by studies which show that the protein compositions of 'soft' and 'hard' keratin are basically different. There is a great need for an alternative nomenclature, probably one which makes no presumptions about mechanical properties.

Keratin-containing tissues have also been classified on the basis of their high-angle X-ray diffraction patterns (Astbury & Street, 1931; Astbury & Marwick, 1932; Rudall, 1947; Fraser, MacRae & Rogers, 1972). Two types of pattern were distinguished depending on whether the dominant secondary structure of the polypeptide chains was the α-helix or the β-sheet conformation. An important fact to emerge was that the structure of the 'hard' keratin of mammals is based on the α-helix whilst the structures of the 'hard' keratins of birds and reptiles are based on the β-sheet. Again, the differences have been shown to have a molecular basis. The inter-relationship between the histological and conformational classifications is summarized in Fig. 2.

'Soft' keratins

The cells in the stratum corneum of the epidermis of both mammals, and birds and reptiles are densely packed with keratin which yields an X-ray diffraction pattern of the α-type (Rudall, 1947) and has a filament-matrix texture (Brody, 1960; Parakkal & Alexander, 1972). The filaments have a uniform diameter of around 7.0 nm and are composed, in mammalian epidermis, of three closely related types of protein chain of molecular weight around 50 000, combined together in a three-chain unit of precise stoicheiometry (Steinert, Idler & Zimmerman, 1976). The α-helix content of the unit is about 50% and the helical and non-helical portions of the chains have markedly different amino-acid compositions (Baden, 1970; Skerrow, Matoltsy & Matoltsy, 1973). It has been suggested that the structural unit of the filament consists of a dimer of three-chain units (Matoltsy, 1975; Steinert, 1978). Each chain

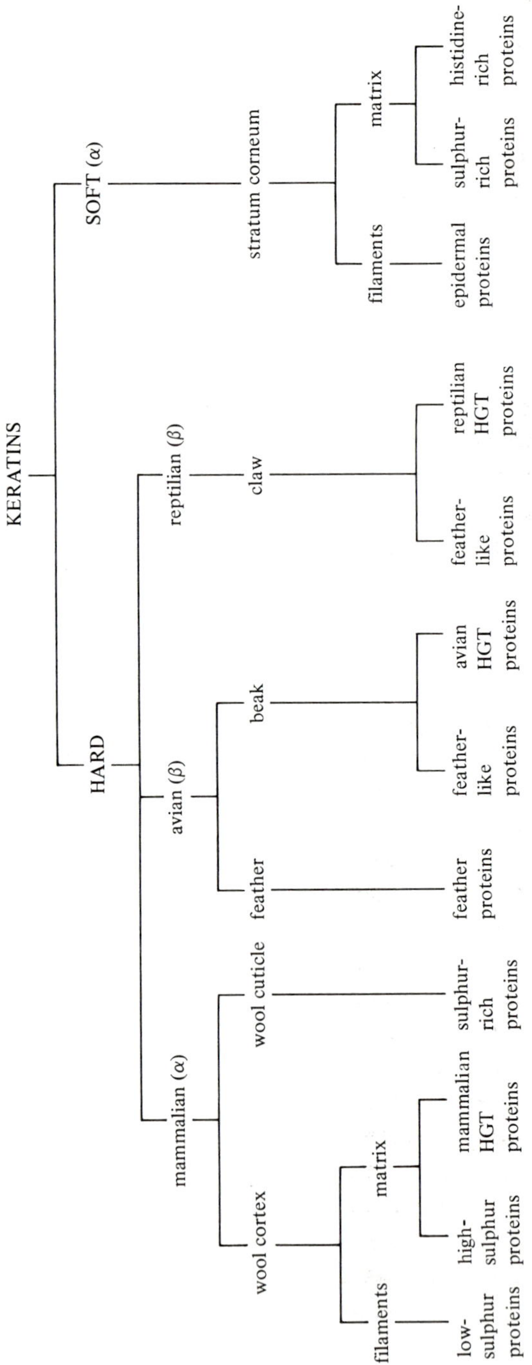

Fig. 2. Classification of keratins and keratin proteins.

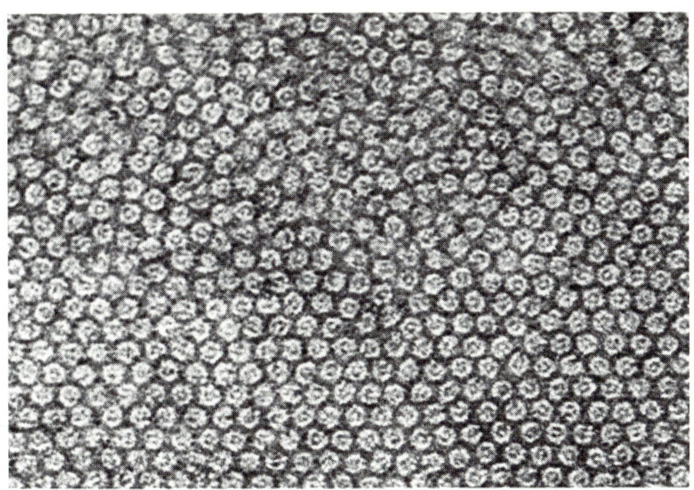

Fig. 3. Electron micrograph of a cross-section of a wool fibre showing the filament-matrix texture; the filaments are about 7.0 nm in diameter. (Courtesy Professor G. E. Rogers, University of Adelaide.)

is supposed to contain two segments of α-helix and corresponding segments in the three-chain units to be in register. The X-ray diffraction pattern indicates that the α-helical segments have a coiled-coil conformation (Crick, 1953) and it is further supposed that the three segments of α-helix that are in register are coiled around a common axis to form a three-strand rope. No detailed studies of the filament proteins of 'soft' keratins from birds or reptiles have been reported so far.

The matrix proteins of 'soft' keratins are less well characterized than the filament proteins. Two types of protein appear to be present, the first being rich in cysteinyl residues and in residues which inhibit α-helix formation (Matoltsy, 1975; Jones, 1980) and the second being rich in histidinyl residues (Dale, 1977; Jones, 1980).

Mammalian 'hard' keratins

Mammalian 'hard' keratins generally yield an α-type X-ray diffraction pattern (Rudall, 1947) and have a filament-matrix texture (Fig. 3) with filaments around 7.0 nm in diameter (Rogers, 1959; Fraser *et al.*, 1972). Superficially, the filaments resemble those found in soft keratins but the proteins that constitute the filaments are substantially different as regards the composition of the non-helical segments.

Filaments

The filaments, or microfibrils, in 'hard' mammalian keratins are composed of proteins that contain a lower proportion of cysteinyl residues than the whole complex (Jones, 1976) and the chemical nature of these so-called low-sulphur proteins has been summarized by Crewther (1976). As with the filament proteins of 'soft' mammalian keratins, three types of protein chain with molecular weights around 50 000 are present, but the stoicheiometry has not so far been determined. The helix content is similar to that of the filaments from 'soft' keratins but there is a substantial difference in the composition of the non-helical portion (Crewther, Dobb, Dowling & Harrap, 1968). The non-helical portions of the filament proteins from 'hard' keratins are rich in cysteinyl residues, whereas the non-helical segments of the filament proteins of soft keratins are rich in glycyl residues. In other respects there are many similarities.

Sequence studies of the α-helical segments of the mammalian filament proteins (Crewther, Inglis & McKern, 1978; Gough, Inglis & Crewther, 1978) reveal a repeating pattern of hydrophobic residues which favours the formation of coiled-coil rope structures (Crick, 1953). The rope segments are around 15 nm in length and a detailed model for their distribution in the filament has been suggested (Fraser, MacRae & Suzuki, 1976). This model is based on low-angle X-ray diffraction data which suggests that the filament has a helical symmetry (Fig. 4). At present there is no conclusive evidence concerning the number of α-helical strands in the rope. The data obtained by Crewther and coworkers (Crewther, 1976) favour a three-strand rope but do not exclude a two-strand rope. McLachlan (1978) has suggested that the available sequence data support a two-stranded structure.

Matrix

The matrix in mammalian 'hard' keratins is composed of two distinct types of protein. Proteins of the first type, the so-called high-sulphur (HS) proteins, have more cysteinyl residues per cent than the whole keratin. Proteins of the second type, the so-called high-glycine–tyrosine (HGT) proteins have very high contents of glycyl residues and appreciably higher contents of aromatic residues than whole keratin. The principal properties of these two classes of protein have recently been summarized (Fraser & MacRae, 1979). A wide spectrum of matrix proteins is produced in a single individual, and these may be classified into families on the basis of molecular weight and of composition.

The amino-acid sequences of a number of high-sulphur proteins have been determined. In some instances no obvious regularity is present but others exhibit periodic patterns of residues (Ycas, 1972; Swart, 1973;

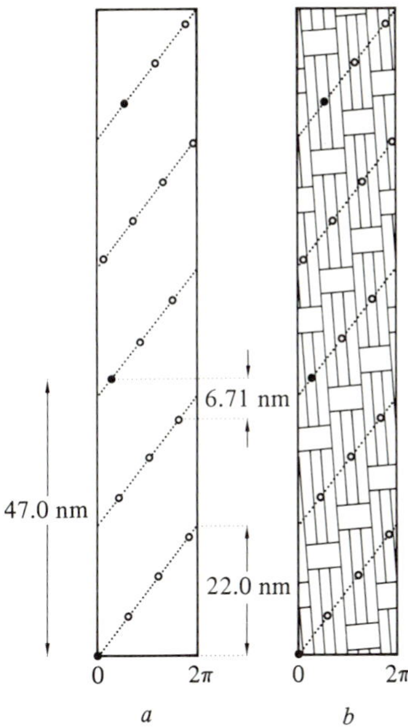

Fig. 4.a. Radial projection showing the helical symmetry of the filaments in mammalian hard keratin deduced from X-ray diffraction studies. The filled circles define the true helical symmetry and the open circles define the pseudo-symmetry. b. Radial projection of a model for the distribution of the segments of coiled-coil rope. Each pseudo-unit is associated with three segments of coiled-coil rope. (Adapted from Fraser, MacRae, Rowlands & Tulloch, 1976.)

Elleman, Lindley & Rowlands, 1973). The sequences have been analyzed using a Fourier transform technique and a favoured pentapeptide sequence Cys–Gln–Pro–(Ser, Thr)–Cys–detected (Parry, Fraser & MacRae, 1979). The significance of this repeat has not yet been established but it has been suggested that the sequence Gln–Pro–(Ser, Thr)–Cys may form a β-bend conformation which is stabilized by a disulphide linkage when both the Cys-favoured positions are occupied by Cys residues. The polypeptide chain would then have the form of a flexible but regularly convoluted chain. Sequence studies of a HGT protein indicate that in part of the molecule there is repetition of a dipeptide sequence Gly–X (Dopheide, 1973).

Cuticle

Animal hairs have an outer sheath, termed the cuticle, which is composed of specialized keratin-containing cells. The mature cells contain little or no filamentous material and X-ray diffraction patterns contain only diffuse rings indicating an amorphous texture (Woods,

1938). Amino-acid analyses of cuticle reveal a high content of Cys residues indicating the presence of HS proteins, but these have not been sufficiently well characterized to establish whether they are cuticle-specific.

Avian and reptilian 'hard' keratins

The molecular structures and protein compositions of avian and reptilian 'hard' keratins are basically different from those of mammalian 'hard' keratins. The X-ray pattern is of the β-type (Astbury & Marwick, 1932; Rudall, 1947) and, after suitable staining procedures, cross sections exhibit a filament-matrix texture when examined in the electron microscope. The 'filaments' have a diameter which is only about half that of the mammalian filaments (Filshie & Rogers, 1962; Roth & Jones, 1970). The range of keratin proteins found in a single individual is appreciably less than is the case with mammals, indeed separate filament and matrix proteins have not been identified. It must be concluded therefore that the 'filament' and 'matrix' components visualized in electron micrographs correspond to different parts of a single molecular type (Fraser, MacRae, Parry & Suzuki, 1971).

Feather proteins

The principal constituent of feathers is a single family of closely related proteins with molecular weights around 10 500 (Woodin, 1954; Harrap & Woods, 1964*b*). The complete amino-acid sequences of feather proteins from two species have been determined (O'Donnell, 1973; O'Donnell & Inglis, 1974) and they exhibit a high degree of homology. About one quarter of the protein has a β-conformation (Fraser *et al.*, 1971) and it has been shown that this regular secondary structure is concentrated in the central portion of the molecule which is rich in hydrophobic residues (Suzuki, 1973). Analysis of the sequence using Fourier transforms revealed an eight-residue periodicity in the distribution of β-favouring residues, and residues favouring reversal of chain direction (Fraser & MacRae, 1976). There was a phase difference of 180° between the two periodicities, suggesting that the chain is regularly folded to give a compact molecule.

The X-ray diffraction pattern of feather rachis can be interpreted in terms of a helical arrangement of molecules and it has been suggested (Fraser *et al.*, 1971) that the β-sheet portions of the molecules are arranged in the manner depicted in Fig. 5. If this model is correct the 'filaments' observed in electron micrographs correspond to the β-sheet portions of the molecules and the 'matrix' to the remainders of the

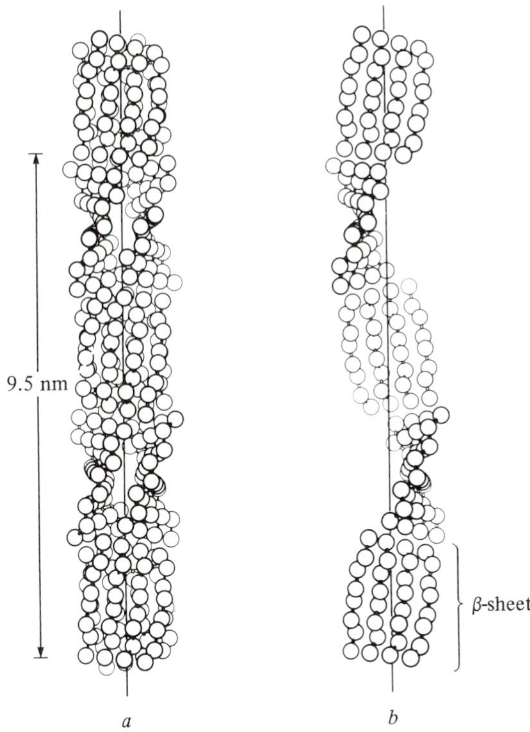

Fig. 5.a. Model for the arrangement of the β-sheet portions of the protein molecules in the filaments of feather keratin. Each sheet contains four segments of polypeptide chain and the sheets are twisted. The circles represent individual amino-acid residues. b. One of the pair of symmetry-related strands of β-sheets (Fraser & MacRae, 1976).

molecules. This would be consistent with the concentration of cysteinyl and charged residues in the non-β segments of the chain (O'Donnell, 1973; Suzuki, 1973).

Avian scale, beak and claw

The predominant proteins in the beaks and scales of birds have molecular weights around 14 000 (Walker & Bridgen, 1976; Frenkel & Gillespie, 1976). Small amounts of feather-like proteins are also present. The presently available evidence suggests that the scale proteins have a composition similar to that of feather proteins but that an additional segment approximately 40 residues in length is present. The additional segment has a repeating pattern of residues of the type Gly–Gly–X– where X is Tyr, Phe or Leu (Walker & Bridgen, 1976). Avian scale has a filament-matrix texture which resembles that observed in feathers (Stewart, 1977) and the X-ray patterns of avian scale, beak and claw are very similar to those of feather. The nature of the 'filament' is therefore presumed to be very similar in all avian 'hard' keratins whilst the 'matrix' contains an additional component in beak, claw and scale which is a counterpart of the HGT matrix proteins of mammalian 'hard' keratins.

Reptilian hard keratins
Proteins resembling those present in avian beak and scale have been isolated from lizard claw and turtle scute (Frenkel & Gillespie, 1976) but the compositions and molecular weights differ slightly. In particular the proportions of Cys, Tyr and Leu show significant variations; the lizard claw yielded proteins of molecular weight 11 500, 13 500 and 15 000. It remains to be established whether the lower molecular weight component can be regarded as the counterpart of a feather protein. Reptilian 'hard' keratins have a filament-matrix texture which resembles that found in avian 'hard' keratins (Roth & Jones, 1970; Parakkal & Alexander, 1972) and the X-ray diffraction pattern has many features in common with avian 'hard' keratins (Astury & Marwick, 1932; Rudall, 1947; Baden, Roth & Bonar, 1966; Fraser *et al.*, 1972). The structure of the filaments must therefore be of a type similar to that depicted in Fig. 5.

Mechanical properties

The mechanical properties of a biological tissue are a function of both its geometrical form and of the intrinsic properties of the constituent materials. The general principles of mechanical design in organisms have been reviewed by Wainwright, Biggs, Currey & Gosline (1976) and detailed descriptions of the intrinsic mechanical properties of keratins have been given by Bendit & Feughelman (1968), Chapman (1969) and Morton & Hearle (1975). With the possible exceptions of the cuticle of hair, keratins have a filament-matrix texture and thus belong to the class of materials termed composites. All epidermal appendages contain some non-keratinous material (Bradbury, 1973; Zahn, 1977) but this is usually neglected, without justification, in the interpretation of mechanical properties. The mechanical design and the character of the composites in specific tissues in relation to their mechanical properties have just been considered and now their general character will be described.

Keratins as composites

Composites in which linearly elastic, high modulus fibres are embedded in a viscoelastic matrix of lower modulus have important properties not possessed by the individual components (Slayter, 1962; Kelly, 1965; Wainwright *et al.*, 1976). Of particular biological value is the fact that the incorporation of the matrix tends to distribute any applied stress evenly over the filaments thus preventing the propagation of cracks from local imperfections or points of rupture. The main factors which determine the mechanical properties of filament-matrix composites are as follows.

(1). Mechanical properties of filaments.
(2). Filament length.
(3). Filament orientation and packing.
(4). Mechanical properties of matrix material.
(5). Proportion of matrix.
(6). Adhesion of matrix to filaments.

Mechanical properties of filaments

In mammalian keratins the filaments contain α-helices which are aligned almost parallel with the length of the filament. The individual segments of α-helix are quite short but they are twisted together in rope-like structures in which the twist of the minor and major helices are of opposite sense. The ropes, in turn, are arranged in a helical assembly (Fig. 4). For small strains (up to about 1%) it seems likely that the filaments will behave essentially as linearly elastic elements and that the stress developed in the filaments will result from the resistance to stretching offered by the α-helices (Enomoto & Krimm, 1962; Feughelman, 1971). Direct evidence for a stretching of the coiled-coils at small strains has been obtained from X-ray diffraction studies (Astbury & Haggith, 1953). At larger strains the molecular organization of the filament is progressively disrupted (Astbury & Woods, 1933; Bendit & Feughelman, 1968) but such strains are generally outside the normal physiological range encountered.

The filaments present in avian and reptilian keratins are, as discussed earlier, quite different from those in mammalian keratin. The central, hydrophobic portion of the structural filament is composed of a two-strand rope of crystallites having a β-conformation (Fig. 5) and this probably acts mechanically as a high-modulus, essentially linearly elastic element. The stress developed in extension will result from the resistance to stretching offered by the β-sheets and the predicted modulus is an order of magnitude greater than for α-helical structures (Enomoto & Krimm, 1962). Direct evidence has been obtained from X-ray diffraction studies indicating that the β-crystallites are stretched when feather is extended (Astbury & Marwick, 1932; Schor & Krimm, 1961).

Filament length

All keratins are assembled and cross-linked intracellularly and no evidence has been obtained to suggest that filaments pass through the modified cell membrane complex. This places a limitation on the possible length of a straight filament. As far as can be determined from electron micrographs of thin sections, keratin filaments are extremely long in relation to their width, that is, they have a very high aspect ratio. Under these circumstances the mechanical behaviour of the keratin will be close to that of a material with continuous filaments (Wainwright *et al.*, 1976) provided that the intercellular substance is not a limiting factor. Microscopic examination of strained keratins suggests that in 'hard' keratins the cellular structure conforms to the strain and that the cell boundaries are not points of weakness (Woods, 1938). This implies that the intercellular material is both elastic and adhesive. The important part played by the intercellular material in maintaining the mechanical continuity of hair was demonstrated by Elöd & Zahn (1946) who found that the tensile strength fell rapidly when the hair was treated with proteolytic enzymes which preferentially hydrolyzed the intercellular material. In 'soft' keratins the intercellular material is less durable and the cells flake apart at the skin surface (Montagna & Parakkal, 1974).

Filament orientation and packing

The contribution of the filament component to the elastic moduli of a composite depends upon the orientation density function of filament directions (Wainwright *et al.*, 1976). For tensile stress in certain specified directions the modulus of a composite may be approximated by the expression

$$E = \eta E_f V_f + E_m(1 - V_f) \quad (1)$$

where E_f and E_m are the moduli of the filament and matrix respectively and V_f is the volume fraction of filaments. The coefficient η is termed 'the efficiency of reinforcement' (Krenchel, 1964; see also Harris, 1980). For a random distribution of filament directions η has a value of 0.2 rising to a value of 1.0 for perfect alignment of the filaments parallel with the direction of stress. In materials such as hair the filaments have a strongly preferred orientation parallel to the fibre axis and the value of η must be close to unity. In the stratum corneum of skin and in reptile scales the filaments have preferred orientations parallel to the plane of the surface. For perfect planar orientation the coefficient η has a value of 0.375 for in-plane stresses and so the value for in-plane stresses in these materials will lie between 0.2 and 0.375, depending upon the form of the orientation density function for filament directions.

Although the consequences are not easy to predict there are two further aspects of filament arrangement in keratins that may introduce differences in mechanical properties. These are variations in the sizes of the filament bundles or macrofibrils, which are formed during biosynthesis, and the packing arrangement within the bundles. In the absence of complicating factors, cylindrical objects tend to pack hexagonally and this type of packing is observed to varying degrees of perfection in many keratinized tissues (Fig. 3). A second type of packing is characterized by a macrofibril composed of concentric layers in which inclination of the filaments to the axis of the macrofibril increases with each successive layer (Birbeck & Mercer, 1957).

Mechanical properties of matrix

The filament-matrix composite in keratins is assembled intracellularly in an aqueous environment with the cysteinyl residues in the reduced form. In the final stages of synthesis the cysteinyl residues are oxidized in pairs to yield disulphide cross-linkages which stabilize the matrix and, in the case of mammalian 'hard' keratins, probably link the filament and matrix proteins through covalent linkages. When the keratin is eventually exposed to the environment, equilibration of the water content takes place, accompanied by shrinkage. X-ray diffraction studies of porcupine quill suggest that the matrix shrinks much more than the filaments (Fraser *et al.*, 1972).

An insight into the mechanical properties of the matrix in wool was obtained from the recognition that the rigidity-modulus (G) in wool increases by a factor of about 15 when a hydrated fibre is dried, whilst the tensile modulus (E) increases by only 2.7 over the same range (Feughelman, 1959). For a mechanically isotropic material the two moduli are related by the expression

$$E/G = 2(1 + v) \tag{2}$$

where v is the ratio of the fractional decrease in diameter to the fractional increase in length accompanying a small extension (Poisson's ratio). Very little change in volume occurs when animal fibres are stretched (Banky & Slen, 1956; Haly & Feughelman, 1960) so the value of v must be close to 0.5. Thus from eqn (2) the value of E/G would be around 3.0 if the fibre were mechanically isotropic. In fact, the value measured for dry wool is 2.7 (Speakman, 1929) suggesting that the matrix has very similar mechanical properties to the filaments in the absence of water. In the fully hydrated state the ratio of E/G has a much higher value than that predicted by eqn (2) and the matrix must therefore be very compliant. The profound change in the mechanical properties of the matrix upon dehydration is also shown in the sensitivity of the viscoelastic properties of keratins to water content (Bendit & Feughelman, 1968; Chapman, 1969). In dry conditions 'soft' keratins tend to become stiff and brittle, whilst in wet conditions they tend to become pliant and highly extensible. No obvious biological advantages accrue from these variations in property with environmental conditions.

A side-effect of the fact that the keratin composite is assembled and cross-linked in the hydrated form is that subsequent equilibration with an unsaturated environment will result in contraction of the matrix and lead to small longitudinal compression of the filaments (Bendit & Feughelman, 1968). In an unsaturated environment the composite is therefore 'pre-stressed' to a small extent.

Proportion of matrix

Little is known about the proportion of matrix proteins in 'soft' keratins. It has been estimated that the sulphur-rich protein constitutes around 17% of the residues remaining after the extraction of bovine nose epidermis with formic acid to remove the living layers (O'Donnell, 1971; Fraser & MacRae, 1979). 'Soft' keratin obtained from the posterior heel of bovine hoof has been shown to contain around 22% matrix protein with approximately equal proportions of sulphur-rich and histidine-rich components (Jones, 1980). From the mechanical standpoint the proportion of matrix proteins cannot be equated directly to the matrix component of the filament-matrix composite since the filament proteins are only around 50% α-helical and it is possible that part of the non-helical material protrudes into the matrix and acts mechanically as part of the matrix.

The content of HS, HGT proteins in a wide range of mammalian 'hard' keratins has been determined by Gillespie & Frenkel (1974) and some representative values are given in Table 1. Wide variations in the proportions of both HS and HGT proteins are apparent as well as the total proportion of both types. The compositions of the HS and HGT proteins vary from appendage to appendage; the content of cysteinyl residues in the HS fraction for example increases by a factor of about 10 from rhinoceros horn to racoon hair. Some wools are almost devoid of

Table 1. *Protein compositions of various mammalian hard keratin appendages (Gillespie & Frenkel, 1974)*

	High-Sulphur (%)	High-Glycine–Tyrosine (%)	Total Matrix Proteins (%)
Rhinoceros horn	8	<1	8
Lincoln wool	20	<1	20
Merino wool	26	8	34
Porcupine quill	18	20	38
Mouse hair	25	18	43
Racoon hair	45	<1	45
Cat claw	30	25	55
Echidna quill	20	35	55

HGT proteins suggesting that this type of protein is not an essential matrix constituent (Frenkel, Gillespie & Reis, 1974). This is further supported by the observation that the proportion of HGT protein in other wools can be greatly reduced by dietary means (Frenkel *et al.*, 1974; Gillespie, Frenkel & Reis, 1980). The proportion and nature of the HS proteins is also subject to genetic and dietary influences (Gillespie, 1965, 1967). As with 'soft' keratins a small additional component of the 'mechanical' matrix may be derived from the non-helical component of the proteins that form the filaments.

In feather keratin it is likely that the non-β portion of the molecule will act mechanically as a matrix component and so the proportion of matrix would appear to be fixed at about 70% (Fraser *et al.*, 1971). The molecular weight of the scale and beak protein is around 40% greater than the feather proteins so if the additional material acts mechanically as matrix then the proportion of matrix in these materials will be correspondingly greater than in feather.

Adhesion of matrix to filaments

Since keratin is in effect a 'molecular composite', adhesion between the filaments and matrix will occur through a variety of secondary bonds including van der Waals' forces, hydrogen bonds and ionic interactions. In 'hard' mammalian keratins it has been pointed out that the net charge on the high-sulphur proteins and on the low-sulphur proteins is of opposite sign (Crewther & Dowling, 1960; Gillespie & Simmonds, 1960) and this must contribute to the adhesion between the filaments and the matrix. In addition to secondary bonds it is generally assumed, although not yet proved directly, that covalent disulphide bonds are formed between the matrix and filament proteins. In avian and reptilian 'hard' keratins the mechanical filament and matrix components are linked covalently by peptide bonds.

Stratum corneum

In the fur-bearing areas of mammalian skin and in the feather-bearing areas of avian skin the stratum corneum is generally very thin and is only around ten or less cells in thickness. In these areas the stratum corneum is too thin to make any significant contribution to the mechanical properties of the skin but the mechanical properties must be such that the stratum corneum can conform elastically to the strains to which the skin is subject in day-to-day encounters with the external environment. The primary function of the stratum corneum in fur-bearing and feather-bearing areas is that of a barrier to loss of body fluids and to the ingress of external fluids and microorganisms. The barrier function appears to be associated with the presence of a complex mixture of lipids, sterols and highly saturated long-chain fatty acids which are deposited intercellularly. The total content of such materials is around 3–4% (Elias, Goerke & Friend, 1977). In mammals such as man, where the pelage is vestigial, the stratum corneum is more prominent and may be up to around 50 cells in thickness (Brody, 1977). The contribution to the mechanical properties of the skin is probably significant in these instances.

The stratum corneum can be separated from the rest of the skin and its mechanical properties have been studied. In many instances chemical or thermal treatments have been used and the possibility must be borne in mind that the mechanical properties may have been modified by the treatment. This may account for the apparently contradictory results reported by different investigators. Blank (1952) found that at moisture contents above about 10%, corresponding to a relative humidity of around 60%, the material was soft and pliable but when dried it became brittle and inflexible. In contrast Kligman (1964) measured the tensile properties of the stratum corneum and found a large difference in behaviour between specimens equilibrated at 65% relative humidity and specimens saturated with water (Fig. 6*a*). In addition to reducing the tensile modulus, saturation with water had a plasticizing effect. At 65% relative humidity rupture occurred after about 5% extension whilst a tenfold increase in extension to rupture was found with saturated specimens. The influence of relative humidity on stratum corneum has

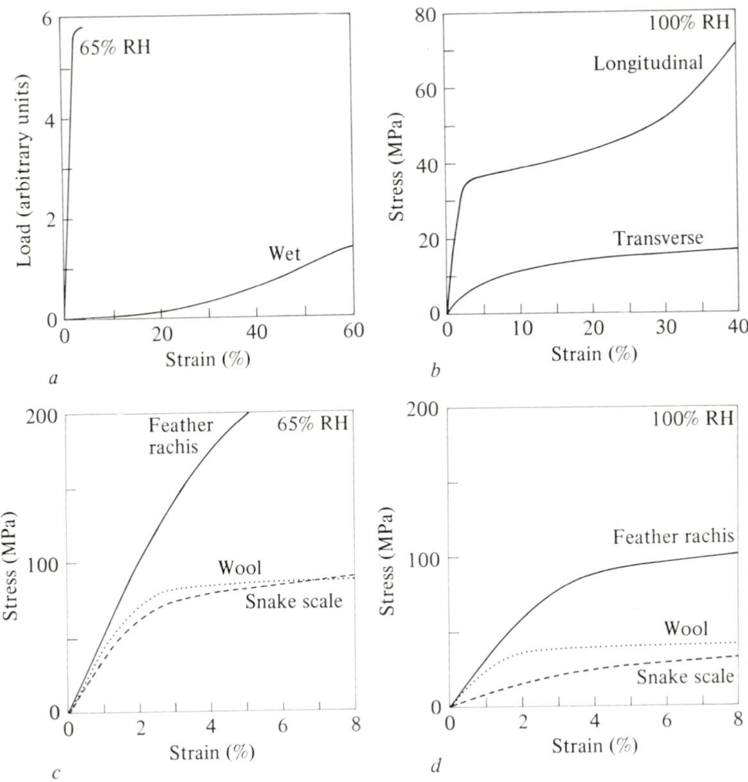

Fig. 6.a. Load–extension curves for human stratum corneum in water and at 65% relative humidity (RH). (Kligman, 1964.) b. Stress–strain curves for porcupine quill in water measured in the longitudinal and transverse directions. Redrawn from Feughelman & Druhala, 1976. c. Stress–strain curves for feather rachis, snake scale, and Lincoln wool measured at 65% relative humidity. Bendit, unpublished data. d. Stress–strain curves for feather rachis, snake scale, and Lincoln wool measured in water. Bendit, unpublished data.

also been studied (Wildnauer, Bothwell & Douglass, 1971). They found that over a 0–100% range of relative humidity the breaking strength decreased by 85% and extension at fracture increased from 20% to 190%. Baden, Goldsmith & Lee (1972) measured the initial tensile modulus of stratum corneum and obtained values of 0.19 and 0.13 GPa at relative humidities of 70% and 100% respectively. As will be seen from Table 2 these values are an order of magnitude lower than those reported for 'hard' keratins such as nail and hair.

The orientation of filaments in the greater part of the stratum corneum is such that there is a preferred direction parallel to the surface of the stratum but random orientation in the plane perpendicular to the

Table 2. *Elastic moduli of keratinized structures. Moduli are expressed in Giga Pascals (GPa)*[a]

Material	Relative humidity			Reference[c]
	0%	Intermediate[b]	100%	
		Longitudinal Extension		
Human nail	–	2.6 (70%)	1.8	1
hair	–	2.3 (70%)	1.5	1
stratum corneum	–	0.19 (70%)	0.13	1
Wool (Cotswold)	5.6	–	2.0	2
Horsehair	6.8	5.1 (60%)	2.4	3
Porcupine quill[d]	–	–	3.5	3
Wool (Lincoln)	–	4.5 (65%)	2.5	4
Feather rachis[e]	–	5.2 (65%)	3.4	4
Snake scale[f]	–	3.5 (65%)	0.86	4
Human stratum corneum	–	~1.0 (65%)	~0.004	4
		Transverse Extension		
Porcupine quill[d]	–	–	0.30	5
		Longitudinal Compression		
Horsehair	6.5	4.0 (60%)	1.8[g]	3
Porcupine quill[d]	6.7	5.8 (60%)	3.4	3
		Transverse Compression[h]		
Horsehair	3.5	2.0 (60%)	0.095[i]	3
Porcupine quill[d]	4.0	2.5 (60%)	0.32	3
		Torsion		
Wool (Cotswold)[j]	1.76	–	0.109	6

[a] 1 GPa = 1 $GNm^{-2} = 10^9 Nm^{-2}$. It should be noted that the values obtained for moduli vary according to: (i) the history of the specimen; (ii) the rate of strain; (iii) the particular part of the appendage used to prepare the specimen; and (iv) whether or not allowance is made for alignment and straightening of the specimen (Bendit, 1978). Significant variations are also found from animal to animal.

[b] Relative humidity given in parentheses.

[c] 1, Baden, Goldsmith & Lee (1974); 2, Meredith (1956); 3, Bendit (1976); 4, Bendit (unpublished observations); 5, Feughelman & Druhala (1976); 6, Speakman (1929).

[d] African porcupine (*Hystrix cristata*).

[e] Laysan albatross (*Diomedia immutabilis*).

[f] Australian copperhead (*Australaps superba*).

[g] Range 1.4–2.2 GPa.

[h] See also Table 3.

[i] Range 0.06–0.13 GPa.

[j] Mean values for a range of fibre diameters.

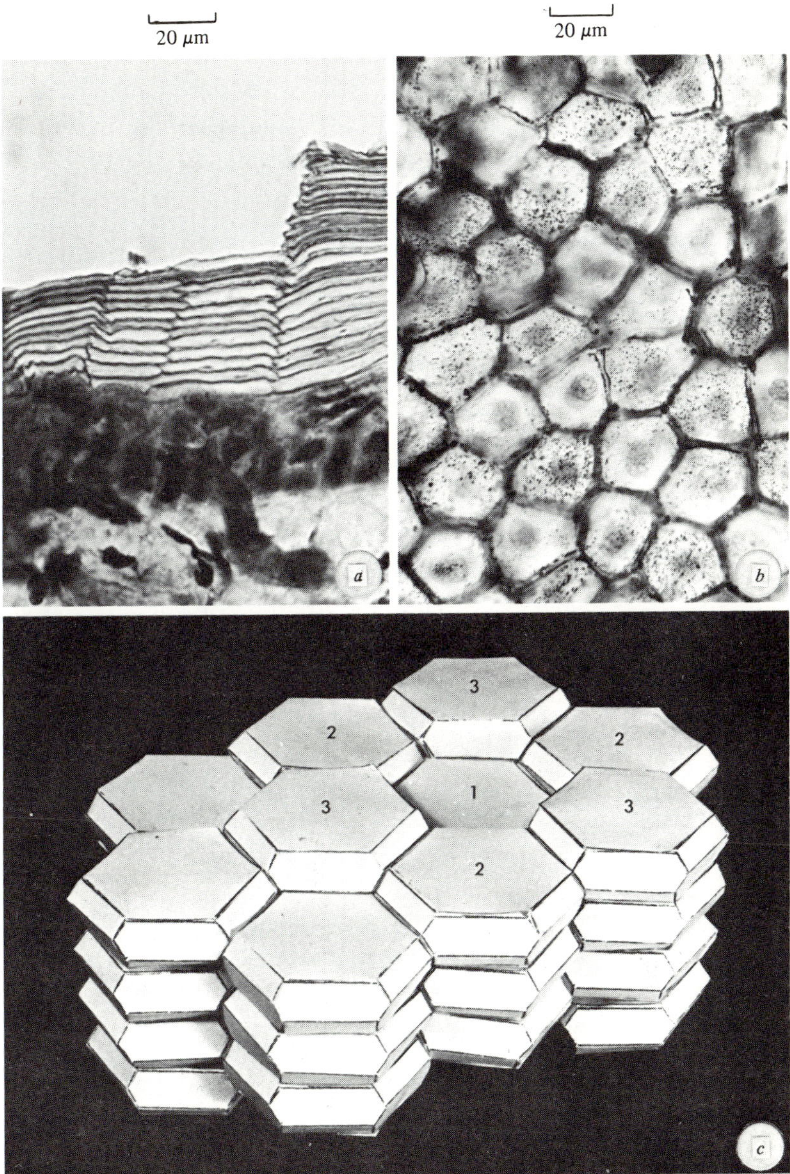

Fig. 7.*a*. Cell columns in the stratum corneum of mouse ear expanded with alkali; *b*, sheet of mouse-ear epidermis stained with Feulgen and silver; *c*, close-packed truncated tetrakaidecahedra; the upper cells in columns 2 and 3 must be released before a cell in column 1 can be released (Menton, 1976). Reproduced by permission.

stratum (Rudall, 1946; Brody, 1959, 1977) and the filaments are packed in bundles which interleave. As discussed earlier the in-plane tensile modulus for stratum corneum will be reduced, by this random orientation by a factor of between 0.2 and 0.375 compared with that of an axially orientated structure such as a hair. A second source of difference is likely to be the very much lower cystine content of stratum corneum compared with hair, and the corresponding reduction in the number of disulphide cross-linkages.

The cells of the stratum corneum are flattened and, over much of the body surface, are arranged in a regular manner in columns (Mackenzie, 1975). The shape of the cells approximates to that of a compressed orthic tetrakaidecahedron (Fig. 7). A regular array of orthic tetrakaidecahedra produces the greatest economy of surface in a homogeneous partition of space (Kelvin, 1887, 1894) and the relevance of this finding to the stratum corneum has been discussed by Potten (1976) and by Allen & Potten (1976). The cells are interlocked in such a way that at any one time only one third of the surface cells are free to desquamate.

Over most of the body surface the stratum corneum is typically membranous and is adapted for flexibility and impermeability. Most epidermal appendages are composed of 'hard' keratin but some specialization is also seen in structures composed of 'soft' keratins. For example, plantar pads and certain portions of hooves have a construction adapted to load-bearing and in higher primates palmar ridges are present which are an adaptation for gripping. Thickening of the stratum corneum also occurs in response to continued pressure and rubbing leading to the formation of calluses.

The chemical composition and physical properties of specialized 'soft' keratin structures differ somewhat from those of the stratum corneum (Montagna & Parakkal, 1974). The fine structure of human plantar epidermis has been studied by Brody (1972) and he noted that whilst the filaments still occur in bundles, the orientation of the bundles does not show the preferred planar orientation of the rest of the stratum corneum. The regular arrangement of cells into columns, which is common in the remainder of the stratum corneum, does not appear to be present in thickened structures (Mackenzie, 1975).

In sparsely haired mammals such as the seal, sealion, elephant, rhinoceros and hippopotamus a very thick stratum corneum is produced and the term hyperkeratosis is used to describe this process (Spearman, 1973). Little is known about the composition or mechanical properties of these materials. Similarly rodent tail scales and pangolin body scales have not been studied in any detail.

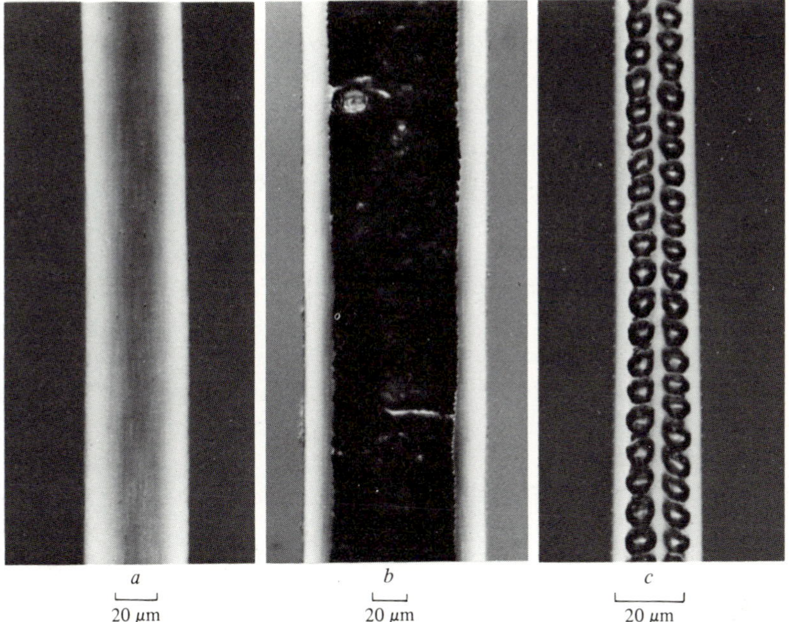

Fig. 8. Photomicrographs of mammalian fibres: *a*, infant human hair; *b*, cow hair; *c*, rabbit guard hair. (Courtesy Mr E. Suzuki, CSIRO Division of Protein Chemistry, Melbourne.)

Mammalian 'hard' keratins

A great variety of epidermal appendages is found in mammals (Spearman, 1966, 1973) and the mechanical properties vary from appendage to appendage depending upon geometrical form, molecular texture and chemical composition. In the majority of species much of the body is covered by a thick pelt, which provides mechanical protection and plays an important role in temperature regulation. In addition plate-like structures are found in nails, claws and hooves.

Hairs and spines

The gross structure of hairs has been described in detail by Wildman (1954) and by Ryder (1973). The simplest type of hair has the form of a solid cylinder, generally with an elliptical cross-section (Fig. 8*a*). Coarse hairs usually have hollow cores which may be continuous, giving a tubular structure (Fig. 8*b*), or may be elaborately subdivided (Fig. 8*c*). Medullation permits economy in weight and reduced thermal conductivity with little loss in bending stiffness. Most mammals have an outer coat of long, thick, medullated fibres termed guard hairs (Fig. 8*c*) and short, fine under-hair. The spines of hedgehogs and the quills of porcupines and echidnas may be regarded as specialized guard hairs. Since porcupines and echidnas are so distantly related the development of quills is an instance of convergent evolution.

In mammals with dense pelts a common feature of the under-hair is a waviness or crimp that entraps large volumes of air and provides an additional source of thermal insulation. In fine Merino wool the crimped fibres have a bilateral structure in which two strands of cells of different chemical composition and physical properties are arranged side-by-side in such a manner that one component, the paracortex, always follows the inside of the crimp wave, whilst the other, the orthocortex, is formed on the outside of the crimp wave (Ryder, 1973). It follows that the length of the paracortex must be less than that of the orthocortex.

Most hairs are enveloped in a cuticle which is composed of flattened cells that are rich in high-sulphur proteins and contain very few filaments. The cuticle usually represents only a small fraction of the cross-sectional area and does not make any significant contribution to the mechanical properties. It does, however, have a profound effect on the frictional properties since the arrangement of scale cells (Fig. 9*a*) gives the fibre a ratchet-like profile. This results in a difference between the coefficient of friction in the root-to-tip direction and in the tip-to-root direction and is termed the 'directional friction effect' (Moncrieff, 1953; Morton & Hearle, 1975). The difference between the two coefficients can be quite substantial and individual measurements and theories of the mechanism have been reviewed by Makinson (1972). The biological function of the ratchet-like profile is believed to be twofold. Its primary purpose is to anchor the fibre in the follicle but it also serves to expel desquamated material derived from the stratum corneum and foreign particles that penetrate the pelt (Moncrieff, 1953). When groups of hairs flex as the animal moves about the directional friction effect causes material in contact with the hairs to migrate towards the surface. In some porcupine quills there is also a ratchet-like profile (Fig. 9*b*) but the protrusions are oppositely directed to those in hair and serve as barbs to anchor the quill into the flesh of predators.

In both hairs and quills the filaments in the main body of the structure have a preferred orientation parallel to the long (fibre) axis and as a result the mechanical properties are anisotropic with reference to this axis (Table 2). The tensile stress–strain curves for porcupine quill are shown in Fig. 6*b* (Feughelman & Druhala, 1976). The material is stiffer for tensile stresses applied parallel to the fibre axis, and up to about $1\frac{1}{2}\%$

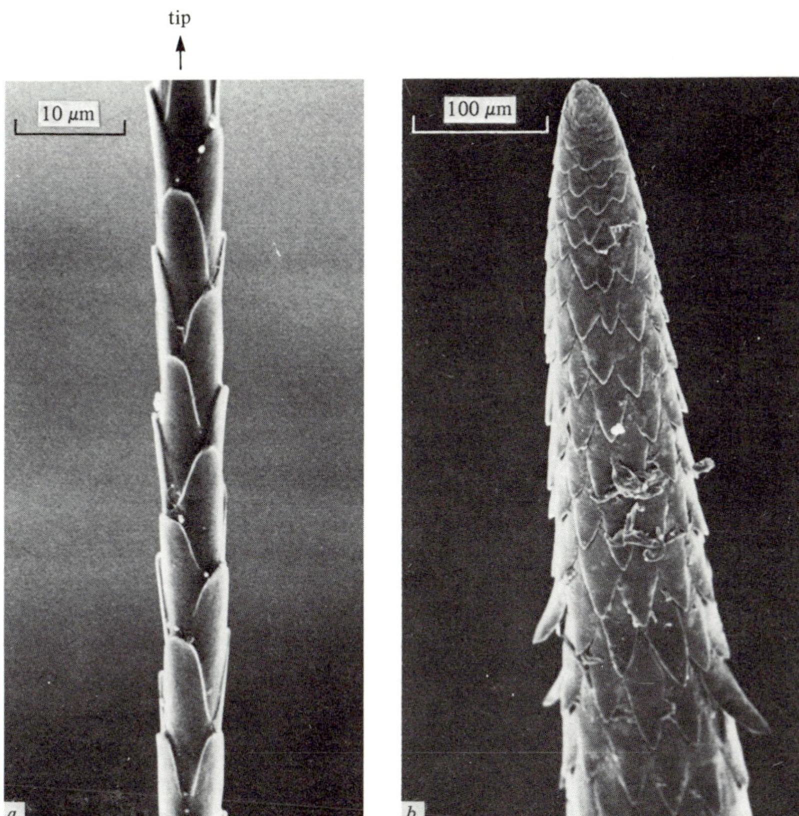

Fig. 9. Scanning electron micrographs: *a*, mink fibre; *b*, North American porcupine quill in tip region. (Courtesy Mr A. Van Donkelaar, CSIRO Division of Protein Chemistry, Melbourne.)

strain there is an approximately linear relationship (so-called Hookean) between stress and strain. However, on close examination the curve shows appreciable departures from ideality (Bendit, 1978). At about 2% extension the quill begins to yield and appreciable further extension yield takes place with only a small increase in stress. A great deal of information is available on the tensile properties of animal fibres and this has been reviewed by Bendit & Feughelman (1968) and by Chapman (1969). The transition from the Hookean to the yield region in longitudinal extension is associated with breakdown in the structure of the filament, and this continues in a progressive manner throughout the yield region. The α-helices are disrupted in the process but the polypeptide chains recrystallize in the β-conformation (Astbury & Street, 1931; Bendit & Feughelman, 1968) and the entropy contribution to the stress is small (Woods, 1946).

Table 3. *Transverse compressional moduli of mammalian hard keratins (measured in water) compared with the contents of HGT proteins (Bendit & Gillespie, 1978; Bendit, 1980).*

Material	E_T (GPa)	HS Content (%)	HGT Content (%)	Total (%)
Rhinoceros horn	0.030	7.3	5.3	12.6
Horse hair	0.095[a]	22.7	1.3	24.0
Rabbit claw	0.15	20.8	5.0	25.8
Finger nail	0.15	30[b]	0[b]	30[b]
Human hair	0.20[b]	31.5	0	31.5
Porcupine quill	0.32	11.2	21.6	32.8
Echidna quill	0.90	20.8	43.0	63.8

[a] Range 0.06–0.13 GPa.
[b] Approximate values.

The longitudinal mechanical properties in the Hookean region are not greatly dependent on rate of extension because of the predominant influence of the filaments which are essentially elastic (Bendit & Feughelman, 1968; Chapman, 1969: Feughelman, 1971; Wainwright *et al.*, 1976). The relatively small time-dependent effects in this region are due to the matrix, which constitutes the main viscous component in the structure (Chapman, 1975). In the yield region the equilibrium between the α and β phases predominates and results in an essentially constant stress. However, the shape of the curve in this region is again affected by the presence of the viscous matrix; both the level of stress and the slope increase slightly with increasing rates of strain (Rigby, 1955). This behaviour is observed at all relative humidities, but is most clearly demonstrated with the fully hydrated material, as the matrix, which contains the major proportion of the water in the composite, is then mechanically much weaker than the filaments. At low and intermediate relative humidities the stiffness of the matrix is high and the behaviour of the material approaches that of a uniformly hydrogen-bonded network (Feughelman, 1971).

Both the composition and the proportions of matrix show wide variation in hairs and quills (Gillespie & Frenkel, 1974) and this is believed to have a functional basis. A correlation has been established between the proportion of matrix and the transverse compressional modulus (Bendit & Gillespie, 1978; Bendit, 1980) which was shown to increase with the proportion of matrix (Table 3).

The variability of the matrix composition and proportion between hairs and quills from different animals, and its dependence on nutrition, raises the problem of how the orderly process of self-assembly can proceed during synthesis with so much variability in the constituent proteins. We believe that the answer may lie in a process whereby the filaments assemble and become organized into a lattice in which the interfilament distance is regulated either by projections from the filaments, derived from the low-sulphur proteins, or by interaction with an essential 'spacer' matrix protein. Since some fibres, for example human hair and Lincoln wool, contain virtually no HGT proteins, the 'spacer' matrix protein, if it is a single substance, is presumably of the HS type. Such a structure would permit wide variations in the amount and nature of the matrix protein deposited in the remaining interfilament spaces and the shrinkage of the structure upon drying would be predicted to be inversely related to matrix content. Such a relationship has, in fact, been observed by Bendit (1980). Direct evidence for ordering in the spaces between the microfibrils has been obtained from X-ray diffraction studies (Fraser & MacRae, 1973; Fraser *et al.*, 1976). Recent estimates of the intermicrofibrillar distance in various keratins in the hydrated state suggest that it is relatively constant from appendage to appendage (Spei & Zahn, 1979).

Although the filaments are preferentially orientated parallel to the fibre axis in hairs it has been shown (Earland, Blakey & Stell, 1962*a, b*) that in certain medullated structures, such as the tactile whiskers of lions and tigers, the filament orientation density function has a significant value normal to the fibre axis in a thin layer around the medulla. It was suggested that the function of this layer was to prevent splitting during flexing. Complex local distributions of filament orientation in the walls of quills may be inferred from measurements of mechanical anisotropy (Makinson, 1954; Bendit & Gillespie, 1978; Bendit & Kelly, 1978) and X-ray diffraction patterns (Bendit, personal communication).

There is evidence that small amounts of mineral salts are present in some hairs (Pautard, 1963, 1964; Blakey, Guy, Happey & Lockwood, 1971) but their effect on the mechanical properties has not been determined.

Nails, claws, hooves and horns

Very little work has been done on the mechanical properties of nails, claws and hooves. Their general form is similar but the filament orientation is adapted to function (Baden, 1970; Spearman, 1973). In human nail the filaments are preferentially orientated perpendicular to the direction of growth (Astbury & Sisson, 1935; Forslind, 1970) but in the claws of primates that have true claws, there is preferential orientation parallel to the growth direction (Baden, 1970). In transitional types an intermediate pattern of orientation is present. The degree of parallelism of the filaments in nails, claws and hooves is notably less than in hairs and quills. The mechanical properties of human nail have been studied by Baden (1970) who showed that the Young's modulus measured parallel to the growth direction was less than that measured parallel to the free edge of the nail. Baden, Goldsmith & Lee (1974) reported values of Young's modulus perpendicular to the growth direction for human nail at 70 and 100% relative humidity and these are given in Table 2. The measured values are very similar to those that they obtained from human hair.

Crystals of hydroxyapatite are found in some claws and this is believed to increase their hardness; primate nails are not strongly calcified (Spearman, 1973).

Horns exhibit anisotropic mechanical properties. According to Makinson (1954, 1955) rhinoceros horn is approximately isotropic transverse to the growth direction, and in this property resembles hairs, whilst ram's horn is transversely isotropic about a radial direction. These results apply to bulk specimens and it is to be expected that detailed studies would reveal local variations in the filament orientation density function such as has been found in quills (Bendit & Kelly, 1978; Bendit, personal communication).

Warburton (1948) measured the tangential Young's modulus of strips of sheep's horn cut from the inner portion and found a substantial decrease (about 7:1) in modulus in going from 0% to 100% relative humidity. This is appreciably less than the decrease in transverse modulus measured for materials with a high degree of axial orientation, as for example horsehair (Table 2), and is consistent with a wide dispersion of filament directions as suggested by Makinson's observations.

Avian and reptilian 'hard' keratin

Feathers represent the most specialized development of epidermal appendages and vary in complexity from hair-like phyloplumes to the elaborately structured primary flight feathers (Fig. 10*a*). The mechanical properties of the shaft of a primary feather were analysed by Purslow & Vincent (1978) who showed that the behaviour in bending was determined by the outer tube of keratin forming the so-called cortex. The forces to which the feather shaft is subjected in flight are primarily

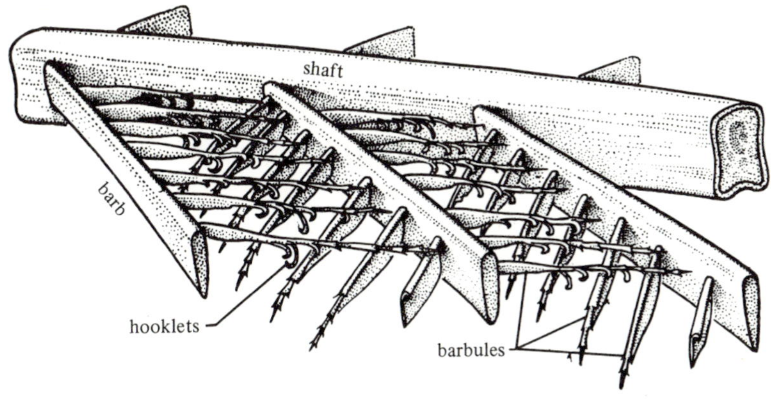

a

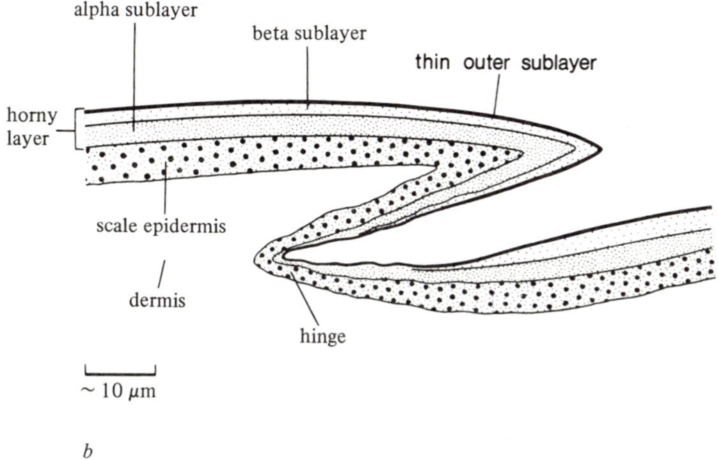

b

Fig. 10.*a*. Structure of primary flight feather (Storer, 1943). Used with permission. *b*. Structure of snake scale (Spearman, 1973). Reproduced by permission.

bending stresses and the tubular form of the shaft gives a high resistance to bending with economy in weight (Galileo, 1638). If a tube is to resist bending in all planes containing the axis of the tube the most efficient structure is one with a circular cross-section. Such tubes fail catastrophically, however, when the bending moment exceeds a critical value, the integrity of cross-sectional shape is lost and the tube buckles. Characteristically the circle is deformed into an elliptical shape and buckling occurs in the portion of the tube that is in compression. In a primary feather the bending stresses are confined to a small range of azimuth about the axis of the tube and the cross-section is close to that of a box girder (Fig. 10*a*), with an orientation appropriate to resist the type of buckling just described. The shaft of the feather is clamped at one end in the body of the bird and so the bending stress decreases with distance from the body. Correspondingly, it is found (Purslow & Vincent, 1978) that the shaft tapers with distance from the body thus affording a further economy in weight.

The predominant filament orientation in the distal portion of the shaft (the rachis) is parallel to the long direction; the orientation, as judged by X-ray diffraction patterns, is very high. In the proximal portion of the shaft (the calamus) the filaments in the inner part of the cortex are orientated in the same manner as in the rachis but there is an outer layer in which the filament orientation is perpendicular to the length of the shaft and circumferential (Earland, Blakey & Stell, 1962*a, b*).

As explained earlier the proportion of non β-sheet material, presumed to act as a mechanical matrix in feather, is relatively constant but it has been found that the protein compositions of the rachis and calamus are somewhat different from those of the barbs and medulla (Harrap & Woods, 1964*a*, 1967) and this may have a functional origin.

The modulus of feather rachis for longitudinal extension at 100% relative humidity is very similar to that of porcupine quill (Table 2). As with mammalian 'hard' keratins there is a substantial increase in modulus when the feather is equilibrated at 65% relative humidity, which may be attributed to a stiffening of that component of the structure which is acting mechanically as the matrix. The stress–strain curve of feather rachis for longitudinal extension at 65% relative humidity has a substantially different character from that of a mammalian 'hard' keratin such as Lincoln wool (Fig. 6*c*). The slope of the curve decreases slightly as extension proceeds but there is no counterpart of the yield region observed with wool, hairs and quills. The extensibility of feather rachis (around 12%) is much lower than hair (Bendit, personal communication). These differences are a reflection of the different molecular structures of the filamentous component of the mechanical composite. The β-sheet in the feather rachis filament (Fig. 5) would be expected to rupture at a comparatively low strain whilst the α-helices in the filaments of hair and quill can open out progressively when the strain exceeds that which can be accommodated by the α-helix. At 100% relative humidity (Fig. 6*d*) there is a pronounced decrease in modulus for strains over about 3% but the extension at break is not significantly different.

Scales are found on the tarsus of birds and, as mentioned earlier, the molecular weights of the main proteins are greater than those of feather keratin. The filament component resembles that present in feather keratin, being based on the β-sheet but the role of the additional material has not yet been established and it is not clear at present whether it behaves mechanically as an addition to the filament or to the matrix component. No mechanical measurements on avian scales have been reported so far but it has been established that the filaments are preferentially orientated in the plane of the scales although the alignment is poor (Rudall, 1947; Stewart, 1977). In avian claws the filaments are preferentially aligned parallel to the growth direction (Baden, 1970).

Most reptiles have a body covering which consists of scales having the general form shown in Fig. 10*b*. The scale consists of an outer layer of 'hard' keratin-containing filaments which resemble those present in feather keratin, and an inner layer of soft keratin-containing filaments of the epidermal type; the scales are hinged together by folds of 'soft' keratin. The filaments in the outer layer are preferentially orientated in the plane of the layer (Rudall, 1947). As with avian scales the proteins of the 'hard' outer layer have a higher molecular weight than those found in feather keratin and similar doubt exists about the structural role of the additional material.

The stress–strain curve of snake scale for in-plane extension, measured at 100% relative humidity, exhibits a substantially lower initial slope than that of feather rachis (Fig. 6*d*). This may be attributed in part to the presence of the soft, inner layer but the different type of filament orientation must also have a pronounced effect. In feather rachis the value of the coefficient of reinforcement η in eqn (1) must be close to 1 for axial stress whilst in the outer layer of snake scale the preferred in-plane orientation would suggest a value between 0.2 and 0.375 for in-plane stress.

At 65% relative humidity the modulus for in-plane stress increases to a value approaching that of feather rachis (Table 2) and this may be attributed to a concomitant increase in the stiffness of the matrix component, bringing the value of E_m in eqn (1) up to a value comparable with E_f. The value of η will then have a much smaller effect on the modulus of the composite, particularly if V_f is equated with the proportion of β-sheet material (about 0.3).

A further difference between the stress–strain curves of feather and snake scale is the presence of a yield region in the latter at 65% relative humidity which is similar to the yield region in Lincoln wool (Fig. 6*c*). Whether this has its origin in the different type of filament orientation or in the presence of a soft inner layer is not clear.

The claws of birds and reptiles are superficially similar to those of mammals but, as explained earlier, the filament structures are basically different. The filament orientation varies in different claws and in different parts of the same claw but generally shows a preferred alignment in the plane of the local surface and also a preferred alignment parallel to the direction of growth. No measurements of the mechanical properties of claws are currently available.

Summary

Keratinized epidermal appendages have mechanical properties which are typical of those expected for composite materials with a filament-matrix texture. At 100% relative humidity the matrix is highly hydrated, mechanically weak and exhibits viscoelastic behaviour. As the water content is reduced the matrix becomes progressively stiffer until at 0% relative humidity its properties approach those of the filaments. In the normally encountered range of relative humidities the advantages of such a texture are that stress is evenly distributed over the filaments, which constitute the load-bearing elements, thus preventing the propagation of cracks from local imperfections.

Based on studies of synthetic filament-matrix composites potentialities for the adaptation of the mechanical properties to specific functions are to be found in variation of the properties of the filaments and the matrix and also of the cross-linking. Since keratin is a filament-matrix composite at the molecular level the great diversity of molecular species presumably has its origin in such adaptation. Superimposed on filament and matrix composition and properties are the variables of filament orientation and proportion and there is abundant evidence that many of the subtler properties of keratinized appendages stem from variations in these factors.

We are indebted to our colleagues Dr E. G. Bendit, Professor M. Feughelman, Dr J. M. Gillespie, Mr E. Suzuki and Mr A. Van Donkelaar for their valuable advice and assistance in the preparation of this review. We are especially grateful to Dr Bendit for making the results of his studies on the mechanical properties of a number of keratins available to us prior to publication.

References

Allen, T. D. & Potten, C. S. (1976). Significance of cell shape in tissue architecture. *Nature, London*, **264,** 545–7.

Astbury, W. T. & Haggith, J. W. (1953). Pre-transformation stretching of the so-called 5.1 Å and 1.5 Å spacings in α-keratin. *Biochimica et Biophysica Acta*, **10,** 483–4.

Astbury, W. T. & Marwick, T. C. (1932). X-ray interpretation of the molecular structure of feather keratin. *Nature, London*, **130,** 309–10.

Astbury, W. T. & Sisson, W. A. (1935). X-ray studies of the structure of hair, wool, and related fibres. III. The configuration of the keratin molecule and its orientation in the biological cell. *Proceedings of the Royal Society of London, Series A*, **150,** 533–51.

Astbury, W. T. & Street, A. (1931). X-ray studies of the structure of hair, wool and related fibres I. General. *Philosophical Transactions of the Royal Society of London Series A*, **230,** 75–101.

Astbury, W. T. & Woods, H. J. (1933). X-ray studies of the structure of wool and related fibres. II. The molecular structure and elastic properties of hair keratin. *Philosophical Transactions of the Royal Society of London Series A*, **232,** 333–94.

Baden, H. P. (1970). The physical properties of nail. *Journal of Investigative Dermatology*, **55,** 115–22.

Baden, H. P., Goldsmith, L. A. & Lee, L. (1974). The importance of understanding the comparative properties of hair and other keratinized tissues in studying disorders of hair. In *The first human hair symposium*, ed. A. C. Brown, pp. 388–98. New York: Medcom Press.

Baden, H. P., Roth, S. I. & Bonar, L. C. (1966). Fibrous proteins of snake scale. *Nature, London*, **212,** 498–9.

Banky, E. C. & Slen, S. B. (1956). Dimensional changes and related phenomena in wool fibers under stress. *Textile Research Journal*, **26,** 204–10.

Bendit, E. G. & Feughelman, M. (1968). Keratin. In *Encyclopedia of Polymer Science and Technology*, Vol. 8, pp. 1–44. New York: Wiley.

Bendit, E. G. & Gillespie, J. M. (1978). The probable role and location of high-glycine–tyrosine proteins in the structure of keratins. *Biopolymers*, **17,** 2743–5.

Bendit, E. G. & Kelly, M. (1978). Properties of the matrix in keratins. Part I: The compression testing technique. *Journal of the Textile Institute*, **48,** 674–9.

Bendit, E. G. (1976). Longitudinal and transverse mechanical properties of keratin in compression. In *Proceedings of the 5th International Wool Textile Research Conference, Aachen*, 1975, ed. K. Zeigler, Vol. 2, pp. 351–60. Aachen: Deutches Wollforschungsinstitut.

Bendit, E. G. (1978). Properties of the matrix in keratins Part II: The Hookean region in the stress-strain curve of keratins. *Textile Research Journal*, **48,** 717–22.

Bendit, E. G. (1980). The location and function of the high-glycine-tyrosine proteins in keratins. In *Fibrous Proteins: Scientific, Industrial and Medical Aspects* ed. D. A. D. Parry & L. K. Creamer, Vol. 2, pp. 185–94. New York: Academic Press.

Birbeck, M. S. C. & Mercer, E. H. (1957). The electron microscopy of the human hair follicle I. Introduction and the hair cortex. *Journal of Biophysical and Biochemical Cytology*, **3,** 203–14.

Blakey, P. R., Guy, R., Happey, F. & Lockwood, P. (1971). The effect of chemical modifications on the morphological structure of keratin. *Applied Polymer Symposium, No. 18*, 193–200.

Blank, I. H. (1952). Factors which influence the water content of the stratum corneum. *Journal of Investigative Dermatology*, **18,** 433–40.

Bradbury, J. H. (1973). The structure and chemistry of keratin fibres. *Advances in Protein Chemistry*, **27,** 111–211.

Brody, I. (1959). The keratinization of epidermal cells of normal guinea pig skin as revealed by electron microscopy. *Journal of Ultrastructural Research*, **2,** 482–511.

Brody, I. (1960). The ultrastructure of the tonofibrils in the keratinization process of normal human epidermis. *Journal of Ultrastructural Research*, **4,** 264–97.

Brody, I. (1972). Electron microscopy of normal human plantar epidermis. I. Strata lucidum and corneum. *Z. Mikrosk – Anat. Forsch., Leipzig*, **86,** 305–31.

Brody, I. (1977). Ultrastructure of the stratum corneum. *Dermatology*, **16,** 245–56.

Chapman, B. M. (1969). A review of the mechanical properties of keratin fibres. *Journal of the Textile Institute*, **60,** 181–207.

CHAPMAN, B. M. (1975). The ageing of wool II: The ageing of disorganized fibres. *Journal of the Textile Institute*, **66,** 343–6.

CREWTHER, W. G. (1976). Primary structure and chemical properties of wool. In *Proceedings of the 5th International Wool Textile Research Conference, Aachen, 1975*, ed. K. Ziegler, Vol. 1, pp. 1–101. Aachen: Deutches Wollforschungsinstitut.

CREWTHER, W. G., DOBB, M. G., DOWLING, L. M. & HARRAP, B. S. (1968). The structure and aggregation of low-sulphur proteins derived from alpha keratins. In *Symposium on fibrous proteins, Australia 1967*, ed. W. G. Crewther, pp. 329–40. Sydney: Butterworths.

CREWTHER, W. G. & DOWLING, L. M. (1960). Effects of chemical modifications on the physical properties of wool: A model of the wool fibre. *Journal of the Textile Institute*, **51,** T775–91.

CREWTHER, W. G., INGLIS, A. S. & MCKERN, N. M. (1978). Amino acid sequences of α-helical segments from S-carboxymethylkerateine A. Complete sequence of a type-II segment. *Biochemical Journal*, **173,** 365–71.

CRICK, F. H. C. (1953). The packing of α-helices. Simple coiled-coils. *Acta Crystallographica*, **6,** 689–97.

DALE, B. A. (1977). Purification and characterization of a basic protein from the stratum corneum of mammalian epidermis. *Biochimica et Biophysica Acta,* **491,** 193–204.

DOPHEIDE, T. A. A. (1973). The primary structure of a protein, component 0.62, rich in glycine and aromatic residues, obtained from wool keratin. *European Journal of Biochemistry*, **34,** 120–4.

EARLAND, C., BLAKEY, P. R. & STELL, J. G. P. (1962*a*). Molecular orientation of some keratins. *Nature, London*, **196,** 1287–91.

EARLAND, C., BLAKEY, P. R. & STELL, J. G. P. (1962*b*). Studies on the structure of keratin IV. The molecular structure of some morphological components of keratins. *Biochimica et Biophysica Acta*, **56,** 268–74.

ELIAS, P. M., GOERKE, J. & FRIEND, D. S. (1977). Mammalian epidermal barrier layer lipids: Composition and influence on structure. *Journal of Investigative Dermatology*, **69,** 535–46.

ELLEMAN, T. C., LINDLEY, H. & ROWLANDS, R. J. (1973). Periodicity in high-sulphur proteins from wool. *Nature, London*, **246,** 530–1.

ELÖD, E. & ZAHN, H. (1946). Die löslichkeit chemisch Behandelter Schafwolle in Pankreatin. *Melliand Textilber*, **27,** 68–70.

ENOMOTO, S. & KRIMM, S. (1961). Elastic moduli of helical polypeptide chain structures. *Biophysical Journal*, **2,** 317–26.

FEUGHELMAN, M. (1959). A two-phase structure for keratin fibers. *Textile Research Journal*, **29,** 223–8.

FEUGHELMAN, M. (1971). The relationship between structure and the mechanical properties of keratin fibers. *Applied Polymer Symposium, No. 18*, 757–74.

FEUGHELMAN, M. & DRUHALA, M. (1976). The lateral mechanical properties of alpha-keratin. In *Proceedings of the 5th International Wool Textile Research Conference Aachen, 1975*, ed. K. Ziegler, Vol. 2, pp. 340–9. Aachen, Deutches Wollforschungsinstitut.

FILSHIE, B. K. & ROGERS, G. E. (1962). An electron microscope study of the fine structure of feather keratin. *Journal of Cell Biology*, **13,** 1–12.

FORSLIND, B. (1970). Biophysical studies of the normal nail. *Acta Dermato-Venereologica*, **50,** 161–8.

FRASER, R. D. B. & MACRAE, T. P. (1973). The structure of α-keratin. *Polymer*, **14,** 61–7.

FRASER, R. D. B. & MACRAE, T. P. (1976). The molecular structure of feather keratin. In *Proceedings of the 16th International ornithological Congress*, ed. H. J. Frith & J. H. Calaby, pp. 443–51. Canberra: Australian Academy of Sciences.

FRASER, R. D. B. & MACRAE, T. P. (1979). Current views on the keratin complex. In *The*

Skin of Vertebrates, London: Linnean Society. (In press.)

FRASER, R. D. B., MACRAE, T. P., PARRY, D. A. D. & SUZUKI, E. (1971). The structure of feather keratin. *Polymer*, **12**, 35–56.

FRASER, R. D. B., MACRAE, T. P. & ROGERS, G. E. (1972). *Keratins – Their Composition, Structure and Biosynthesis*, Springfield, Illinois: Thomas.

FRASER, R. D. B., MACRAE, T. P., ROWLANDS, R. J. & TULLOCH, P. A. (1976). Molecular structure of keratin. In *Proceedings of the 5th International Wool Textile Research Conference Aachen 1975*, ed. K. Ziegler, Vol. 2, pp. 80–8. Aachen, Deutches Wollforschungsinstitut.

FRASER, R. D. B., MACRAE, T. P. & SUZUKI, E. (1976). Structure of the α-keratin microfibril. *Journal of Molecular Biology*, **108**, 435–52.

FRENKEL, M. J. & GILLESPIE, J. M. (1976). The proteins of the keratin component of bird's beaks. *Australian Journal of Biological Science*, **29**, 467–79.

FRENKEL, M. J., GILLESPIE, J. M. & REIS, P. (1974). Factors influencing the biosynthesis of the tyrosine-rich proteins of wool. *Australian Journal of Biological Science*, **27**, 31–8.

GALILEO, G. (1638). *Dialogues concerning two new sciences*. Trans. H. Crew & A. de Salvio (1914), pp. 149–51. New York: MacMillan.

GILLESPIE, J. M. (1965). The high-sulphur proteins of normal and aberrant keratins. In *Biology of skin and hair growth*, ed. A. G. Lyne & B. F. Short, pp. 377–98. Sydney: Angus & Robertson.

GILLESPIE, J. M. (1967). The high-sulphur proteins of α-keratins: Their relation to fiber structure and properties. *Journal of Polymer Science*, (C) **20**, 201–14.

GILLESPIE, J. M. & FRENKEL, M. J. (1974). The macroheterogeneity of type I tyrosine-rich proteins of Merino wool. *Australian Journal of Biological Science*, **27**, 617–27.

GILLESPIE, J. M. & SIMMONDS, D. H. (1960). Amino acid composition of a sulphur-rich protein from wool. *Biochimica et Biophysica Acta*, **39**, 538–9.

GILLESPIE, J. M., FRENKEL, M. J. & REIS, P. (1980). Changes in the matrix proteins of wool and mouse hair following the administration of depilatory compounds. *Australian Journal of Biological Science*, **33**, 125–36.

GIROUD, A. & LEBLOND, C. P. (1951). The keratinization of epidermis and its derivatives, especially the hair, as shown by X-ray diffraction and histochemical studies. *Annals of the New York Academy of Sciences*, **53**, 613–26.

GOUGH, K. H., INGLIS, A. S. & CREWTHER, W. G. (1978). Amino acid sequences of α-helical segments from S-carboxymethyl-kerateine-A. Complete sequence of a type-1 segment. *Biochemical Journal*, **173**, 373–85.

HALY, A. R. & FEUGHELMAN, M. (1960). Stress changes at constant strain and hydrogen bonding in keratin fibres. *Journal of the Textile Institute*, **51**, T573–88.

HARRAP, B. S. & WOODS, E. F. (1964*a*). Soluble derivatives of feather keratin I. Isolation, fractionation and amino acid composition. *Biochemical Journal*, **92**, 8–18.

HARRAP, B. S. & WOODS, E. F. (1964*b*). Soluble derivatives of feather keratin II. Molecular weight and conformation. *Biochemical Journal*, **92**, 19–26.

HARRAP, B. S. & WOODS, E. F. (1967). Species differences in the proteins of feathers. *Comparative Biochemistry and Physiology*, **20**, 449–60.

HARRIS, B. (1980). The mechanical behaviour of composite materials. In *The Mechanical Properties of Biological Materials* (34th Symposium of the Society for Experimental Biology) ed. J. F. V. Vincent & J. D. Currey, pp. 37–74. Cambridge University Press.

JONES, L. N. (1976). Studies on microfibrils from α-keratin. *Biochimica et Biophysica Acta*, **446**, 515–24.

JONES, L. N. (1980). Protein composition of mammalian stratum corneum. In *Fibrous Proteins: Scientific, Industrial and Medical Aspects*, Vol. 2, ed. D. A. D. Parry & L. K. Creamer, pp. 167–75. New York: Academic Press.

KELLY, A. (1965). Fiber-reinforced metals. *Scientific American*, **212**, Feb Issue, pp. 28–37.

KELVIN, W. T. (1887). On the division of space with minimal partitional area. *Philosophic-*

al Transactions of the Royal Society of London, **24,** 503–14.

KELVIN, W. T. (1894). On homogeneous division of space. *Proceedings of the Royal Society of London*, **55,** 1–16.

KLIGMAN, A. M. (1964). The biology of the stratum corneum. In *The Epidermis*, ed. W. Montagna & W. C. Lobitz, pp. 387–433. New York: Academic Press.

KRENCHEL, H. (1964). *Fiber Reinforcement*. Stockholm: Akademisk Vorlag.

MACKENZIE, I. C. (1975). Ordered structure of the epidermis. *Journal of Investigative Dermatology*, **65,** 45–51.

MCLACHLAN, A. D. (1978). Coiled-coil formation and sequence regulations in the helical regions of α-keratin. *Journal of molecular Biology*, **124,** 297–304.

MAKINSON, K. R. (1954). The elastic anisotropy of keratinous solids I. The dilational elastic constants. *Australian Journal of Biological Science*, **7,** 336–47.

MAKINSON, K. R. (1955). The elastic anisotropy of keratinous solids II. The rigidity constants of ram's horn. *Australian Journal of Biological Science*, **8,** 278–87.

MAKINSON, K. R. (1972). The role of the scales of wool fibres in felting and in shrinkproofing. *Wool Science Review, No. 42*, 2–16.

MATOLTSY, A. G. (1975). Desmosomes, filaments, and keratohyaline granules: their role in the stabilization and keratinization of the epidermis. *Journal of Investigative Dermatology*, **65,** 127–42.

MENTON, D. N. (1976). A liquid film model of tetrakaidecahedral packing to account for the establishment of epidermal cell columns. *Journal of Investigative Dermatology*, **66,** 283–91.

MERCER, E. H. (1961). *Keratin and keratinization*, Oxford: Pergamon Press.

MEREDITH, R. (1956). *The Mechanical Properties of Textile Fibres*. Amsterdam: North-Holland.

MONCRIEFF, R. W. (1953). *Wool Shrinkage and Its Prevention*. London: National Trade Press.

MONTAGNA, W. & PARAKKAL, P. F. (1974). *The Structure and Function of Skin*, 3rd edn, New York: Academic Press.

MORTON, W. E. & HEARLE, J. W. S. (1975). *Physical Properties of Textile Fibres*, London: Heinemann.

O'DONNELL, I. J. (1971). The search for a simple keratin – the precursor keratins from cow's lip epidermis. *Australian Journal of Biological Science*, **24,** 1219–34.

O'DONNELL, I. J. (1973). The complete amino acid sequence of a feather keratin from emu (*Dromaius novae-hollandiae*). *Australian Journal of Biological Science*, **26,** 415–37.

O'DONNELL, I. J. & INGLIS, A. S. (1974). Amino acid sequence of a feather keratin from silver gull (*Larus novae – hollandiae*) and comparison with one from emu (*Dromaius novae – hollandiae*). *Australian Journal of Biological Science*, **27,** 369–82.

PARAKKAL, P. F. & ALEXANDER, N. J. (1972). *Keratinization. A Survey of Vertebrate Epithelia*, New York: Academic Press.

PARRY, D. A. D., FRASER, R. D. B. & MACRAE, T. P. (1979). Repeating patterns of amino acid residues in the sequences of some high-sulphur proteins from α-keratin. *International Journal of Biological Macromolecules*, **1,** 17–22.

PAUTARD, F. G. E. (1963). Mineralisation of keratin and its comparison with the enamel matrix. *Nature, London*, **199,** 531–5.

PAUTARD, F. G. E. (1964). Calcification of keratin. In *Progress in the biological sciences in relation to dermatology*, ed. A. Rook & R. H. Champion, Vol. 2, pp. 227–40. Cambridge University Press.

POTTEN, C. S. (1976). Epidermal cell production rates. *Journal of Investigative Dermatology*, **65,** 488–500.

PURSLOW, P. P. & VINCENT, J. F. V. (1978). Mechanical properties of primary feathers from the pigeon. *Journal of experimental Biology*, **72,** 251–60.

RIGBY, B. J. (1955). Stress-relaxation of wool fibres in water at strains of 5–20%

extension. *Australian Journal of Physics*, **8,** 176–83.

Rogers, G. E. (1959). Electron microscope studies of hair and wool. *Annals of the New York Academy of Sciences*, **83,** 378–99.

Romer, A. S. (1970). *The Vertebrate Body*. Philadelphia: W. B. Saunders.

Roth, S. I. & Jones, W. A. (1970). The ultrastructure of epidermal maturation in the skin of the boa constrictor (*Constrictor constrictor*). *Journal of Ultrastructural Research*, **32,** 69–93.

Rudall, K. M. (1946). The structure of epidermal protein. In *Fibrous Proteins Symposium of the Society of Dyers and Colourers*, ed. C. L. Bird, pp. 15–22. Leeds: Chorley & Pickersgill.

Rudall, K. M. (1947). X-ray studies of the distribution of protein chain types in the vertebrate epidermis. *Biochimica et Biophysica Acta*, **1,** 549–62.

Ryder, M. L. (1973). *Hair*, London: Edward Arnold.

Schor, R. & Krimm, S. (1961). Studies on the structure of feather keratin I. X-ray diffraction studies and other experimental data. *Biophysical Journal*, **1,** 467–87.

Skerrow, D., Matoltsy, A. G. & Matoltsy, M. N. (1973). Isolation and characterization of the α-helical regions of epidermal prekeratin. *Journal of Biological Chemistry*, **248,** 4820–6.

Slayter, G. (1962). Two-phase materials. *Scientific American*, **206,** Jan Issue, pp. 124–34.

Speakman, J. B. (1929). The rigidity of wool and its change with adsorption of water vapour. *Transactions of the Faraday Society*, **25,** 92–103.

Spearman, R. I. C. (1966). The keratinization of epidermal scales, feathers and hairs. *Biological Reviews*, **41,** 59–96.

Spearman, R. I. C. (1973). *The integument*. Cambridge University Press.

Spei, M. & Zahn, H. (1979). Röntgenkleinwinkeluntersuchungen von gedehnten Faserkeratinen mit verschiedenem Cystingehalt. *Progress in Colloid and Polymer Science,* **66,** 387–91.

Steinert, P. M. (1978). Structure of the three-chain unit of the bovine epidermal keratin filament. *Journal of molecular Biology*, **123,** 49–70.

Steinert, P. M., Idler, W. W. & Zimmerman, S. B. (1976). Self-assembly of bovine epidermal keratin filaments *in vitro*. *Journal of molecular Biology,* **108,** 547–67.

Stewart, M. (1977). The structure of chicken scale keratin. *Journal of Ultrastructural Research*, **60,** 27–33.

Storer, T. I. (1943). *General Zoology*. New York: McGraw Hill.

Suzuki, E. (1973). Localization of β-conformation in feather keratin. *Australian Journal of Biological Science*, **26,** 435–7.

Swart, L. S. (1973). Homology in the amino-acid sequences of the high-sulphur proteins from wool. *Nature, New Biology*, **243,** 27–9.

Wainwright, S. A., Biggs, W. D., Currey, J. D. & Gosline, J. M. (1976). *Mechanical Design in Organisms*, New York: Wiley.

Walker, I. D. & Bridgen, J. (1976). The keratin chains of avian scale tissue. Sequence heterogeneity and the number of scale keratin genes. *European Journal of Biochemistry*, **67,** 283–93.

Warburton, F. L. (1948). Determination of the elastic properties of horn keratin. *Journal of the Textile Institute*, **39,** 297–307.

Wildman, A. B. (1954). *The Microscopy of Animal Textile Fibres*. Leeds: Wool Industries Research Association.

Wildnauer, R. H., Bothwell, J. W. & Douglass, A. B. (1971). Stratum corneum biomechanical properties I. Influence of relative humidity on normal and extracted human stratum corneum. *Journal of Investigative Dermatology*, **56,** 72–8.

Woodin, A. M. (1954). Molecular size, shape and aggregation of soluble feather keratin. *Biochemical Journal*, **57,** 99–109.

Woods, H. J. (1938). X-ray studies of the structure of hair, wool and related fibres IV.

The molecular structure and elastic properties of the biological cells. *Proceedings of the Royal Society of London, Series A*, **166,** 76–96.

WOODS, H. J. (1946). The contribution of entropy to the elastic properties of keratin, myosin and some other high polymers. *Journal of Colloid Science*, **1,** 407–19.

YCAS, M. (1972). *De novo* origin of periodic proteins. *Journal of Molecular Evolution*, **2,** 17–27.

ZAHN, H. (1977). Die Fasern in der makromolecularen Chemie. *Lenzinger Ber*. **42,** 19–34.

SILKS – THEIR PROPERTIES AND FUNCTIONS

MARK W. DENNY

Department of Zoology, University of Washington, Seattle, WA 98195, USA

Introduction

The term 'silk' is generally associated with the protein fibre produced by the larvae of the silkworm, *Bombyx mori*. The material is drawn in continuous strands of considerable length from glands opening beneath the larva's head and is used to construct the larval cocoon. Over 4000 years ago it was discovered in China that this protein fibre could be unwound from the cocoon and used to weave fabrics that were sheer, yet strong. This property of *B. mori* silk has resulted in the formation of a large textile industry and concomitant political and historical consequences for international trade. Secondary consequences have included the complete domestication of the silkworm and considerable study into the physical and chemical properties of *B. mori* silk fibres. Despite the economic importance of *B. mori* silk, it is far from being the only such protein fibre produced in nature. Many arthropods, in particular the arachnids and insects, produce protein fibres for use outside their bodies, and the term silk is applied to these fibres as well. In general any protein fibre produced by drawing a viscous liquid through a fine orifice is a silk, regardless of its precise chemical composition or physical properties. Silks form a functionally diverse lot. They are widely used (by moths, for example) in the construction of cocoons and egg cases where they provide physical support, protection, and thermal insulation. Further examples include burrowing spiders, which use silk fibres as structural elements in the construction of their burrows, and caddis fly larvae, which spin silk underwater to build about themselves a camouflaged armour. The boleadora, or bolo spider, uses silk as a sticky lariat when catching flying insects, and some tropical ants use the silk of their larvae to fashion leaves into nests. Other examples abound. In many of these animals each silk destined for a specific purpose is produced by a separate gland and each has its own particular chemical structure and physical properties. Perhaps the best examples of this specialization are the spiders that spin orb-webs. *Araneus diadematus*, a common garden spider, produces five separate types of silk (Lucas, 1964; Peakall, 1964). One silk is used as a safety line and also forms the framework of the web. Another silk forms the viscid catching net of the web and still another the fibrous attachment discs which secure the web to surrounding structures. Prey, once captured is swathed in a fourth type of silk, and finally one more silk is used to construct egg cases. Each type of silk is formed in a separate gland or set of glands and is extruded from a separate set of silk orifices or spinnerets.

Given these varied functions how can the mechanical properties of the silks be accounted for by their macromolecular structure and chemical composition? And how are the mechanical properties of silks suited to their functions in biological structures? The purpose of this review is to offer some insight into these questions.

Mechanical properties

While many of the mechanical properties of *B. mori* silk have been known for some time, only recently has sufficient information been available for other types of silk to allow general statements to be made concerning the properties of silks as a class of materials. The first fact which is now emerging is that unlike some other biological fibres (e.g. cellulose), the silks are *not* restricted as regards their mechanical properties. Rather, as can be seen from Fig. 1, these properties vary over a wide range. Despite this diversity, several generalizations can be made about the mechanical properties of silks. In order to ease this task, I have divided silks into three groups. The characteristic properties of each of these groups will be discussed in turn, and specific examples cited.

Group I

The silks of this first group are characterized by high tensile strength and extensibility which is low relative to other silks. The most thoroughly studied examples of this group are the cocoon silk of *B. mori* and the dragline silks of the closely related spiders *A. diadematus* and *A. sericatus*.

The physical dimensions of silkworm and dragline silks are typical for most silks. The individual fibres are 1 to 3 μm in diameter and roughly circular in cross-section. Both silks are drawn from the animal in pairs of fibres. The individual strands of a spider's dragline are twisted about each other and the threads may be easily separated with forceps. In

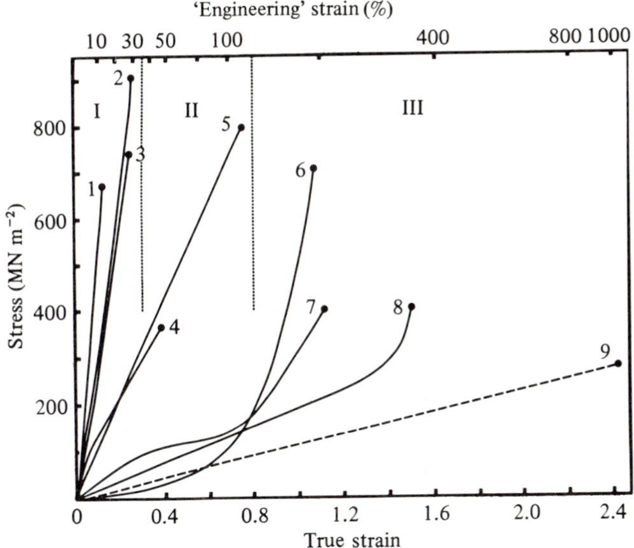

Fig. 1. Silks show a wide range of stress–strain properties. The examples shown here are divided into three groups on the basis of their extensibility. 1, *Anaphe moloneyi*; 2, *Araneus sericatus* dragline; 3, *Bombyx mori* cocoon; 4, *A. diadematus* cocoon; 5, *Galleria mellonella* cocoon; 6, *A. sericatus* viscid; 7, *Apis mellifora* larval; 8, *Chrysopa carnea* egg stalk; 9, *Meta reticulata* viscid. 1 and 3 redrawn from Lucas *et al.* (1955). 2 and 7 redrawn from Denny (1976). 4 redrawn from Lucas (1964). 5, 7 and 8 redrawn from Hepburn *et al.* (1979). 9 inferred from de Wilde (1943).

contrast, the two silk fibres of the silkworm are coated and held together by a protein gum – sericin. In the process of producing practical textile fibres, *B. mori* silk is separated from the sericin by boiling it in water, and it is the properties of this cleaned silk which are discussed here.

Typical stress–strain curves for silkworm and dragline silks are shown in Fig. 2. In this and all other plots of stress and strain presented here, stress and true strain, ε_t, are used, as suggested by Hepburn, Chandler & Davidoff (1979). While ε_t is the measure of choice, it is not an intuitive measure of extension and stress–strain plots include calibration of the strain axis in per cent extension ('engineering' strain).

The curves shown in Fig. 2 are notable in a number of respects. First, the tensile strength of these silks is quite high, approaching 1 GNm^{-2}. This tensile strength is comparable to that of other high tensile strength biological fibres such as cellulose, collagen and chitin, and is considerably greater than that of keratin (Wainwright, Biggs, Currey & Gosline, 1976). The stress–strain curves of *A. sericatus* silk show a shape characteristic of many Group I silks. At low strains ($\varepsilon < 0.05$) the material is very stiff, with an initial modulus that may be as high as

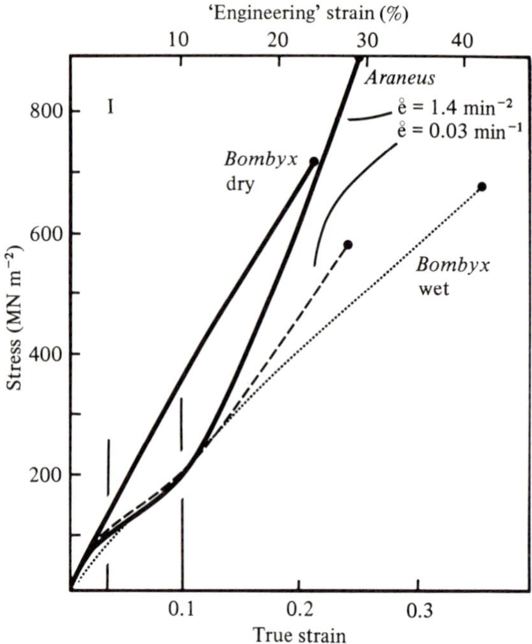

Fig. 2. Examples of Group I silks. The vertical bars indicate the yield plateau typical of Group I silks. Note that these silks are affected by strain rate and humidity. *B. mori* curves redrawn from Hepburn *et al.* (1979). *A. sericatus* dragline curves redrawn from Denny (1976).

20 GNm^{-2}. At strains of about 0.05 to 0.10 the modulus decreases, forming a 'plateau' or yield region as the material deforms plastically. Above a strain of 0.10 the modulus rises and then remains roughly constant until failure. While this shape for the stress–strain curve is typical of the silks of Group I it is not universal. For example, the cocoon silks of both *B. mori* and *Anaphe moloneyi* exhibit a stress–strain relationship that is very nearly linear (Lucas, Shaw & Smith, 1955).

Failure occurs in the silks of Group I at a strain ranging from 0.10 to 0.35, though for most silks in the group failure occurs between strains of 0.20 and 0.30. The extensibility of these silks is five to ten times greater than that of chitin and cellulose, and equal to or twice that of keratin and collagen (Wainwright *et al.*, 1976). Thus, while the silks of this group are characterized in part by extensibilities less than those of other silks their extensibilities are none the less very high compared to other fibrous materials. This unusually high extensibility coupled with high tensile strength gives these silks a breaking energy that is greater than that of any other biological material. A rough estimate of the breaking

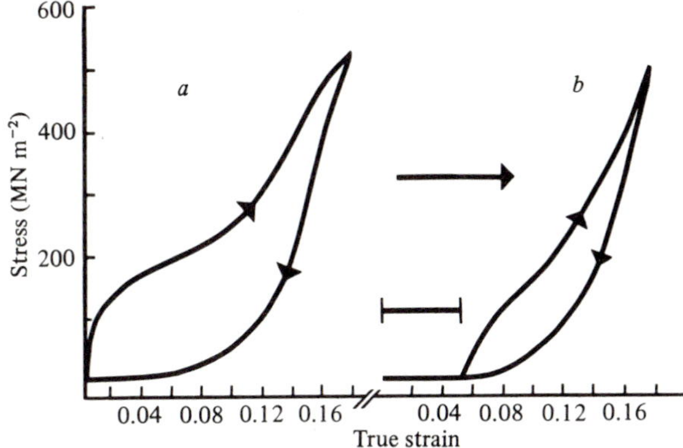

Fig. 3. A cyclic loading test on *A. sericatus* dragline silk. The initial cycle, *a*, shows a hysteresis of about 65%, associated with plastic deformation. The horizontal bar shows the extent of this plastic deformation in *b*, the second cycle. Hysteresis in *b* is about 45%.

energy of materials may be calculated from their tensile strengths and extensibilities by assuming a linear stress–strain curve. Using values drawn from Wainwright *et al.* (1976) the breaking energy of collagen and chitin is estimated to be of the order of 2 to 3 MJm^{-3}. Keratin and cellulose yield higher values, about 23 to 30 MJm^{-3}. The breaking energy of *Araneus sericatus* dragline silk is 158 MJm^{-3} (Denny, 1976). The significance of this high breaking energy will be discussed below.

The characteristics of the stress–strain relationship for the silks of Group 1 are dependent upon a number of factors. At high relative humidities, or when the silk is actually immersed in water, the tensile strength and modulus of the silk are decreased and the extensibility is increased (Fig. 2). This phenomenon has been described for *B. mori* and other silks by Lucas *et al.* (1955) and for *A. diadematus* dragline by Work (1977*a*). The shape, and even the presence, of the yield plateau is affected by the water content of the silk, the plateau generally being less pronounced at high relative humidities.

The stiffness and tensile strength of these silks are also a function of the strain rate. An example is shown in Fig. 2. The tensile strength of *A. sericatus* dragline increases by 75% as the strain rate is increased from 0.03 s^{-1} to 1.41 s^{-1} (Denny, 1976). At the same time the initial modulus rises from 9.8 GNm^{-2} to 20.5 GNm^{-2}. Similar results have been reported for *B. mori* silk. (Iizuka, 1965.) This strain rate dependence indicates the presence of a viscous component in this primarily elastic material, a fact which is borne out by other tests. When the *A. sericatus* dragline silk is cyclically loaded it shows pronounced hysteresis (Fig. 3)

(Denny, 1976). This hysteresis is found in all silks and is highest during the first loading cycle as a result of including the energy required to deform the silk plastically. In this case the hysteresis is about 65%. Further extensions to the same strain do not deform the material plastically further and the hysteresis is lower, about 45% (see also Mullins, 1980). In stress relaxation, *A. sericatus* dragline silk will flow to a limited extent under the action of a load (Denny, 1976). The stress at infinite time is about 75% of that at time zero. Again this indicates the presence of a viscous component in the silk.

In summary, the silks of Group I are very strong ($\sigma_f = 1\,\text{GNm}^{-2}$) yet surprisingly extensible ($\varepsilon_f = 0.2$) viscoelastic materials and as a consequence their breaking energy is very high. In many cases the modulus of these silks varies widely with strain. Without knowledge of the range of strains at which the material functions it is of little use to cite a single modulus value for a silk. The best compromise available is to use the overall modulus, E_{ave}, the tensile strength divided by the breaking strain. E_{ave} of Group I silks ranges from about 1 to $6\,\text{GNm}^{-2}$.

Group II

A second general class of silks is that exemplified by the cocoon silk of *A. diadematus* and the cocoon silk of the greater waxmoth, *Galleria mellonella*. These silks have received far less attention than those of Group I and for the most part only their stress–strain characteristics have been measured. The stress–strain curves for these silks are shown in Fig. 4. The primary difference between these silks and those of Group I lies in their greater extensibility: failure does not occur until the fibres have been extended by 50 to 100%. When compared with other biological fibres these silks are extremely extensible, roughly 50 times as extensible as cellulose or chitin and 2–10 times as extensible as keratin or collagen. While the spider cocoon silk shows a lower tensile strength than dragline silk the high tensile strength of *G. mellonella* silk (Hepburn *et al.*, 1979) shows that a decrease in tensile strength is not a necessary result of the increase in extensibility. E_{ave} of silks in this group is about $1\,\text{GNm}^{-2}$. As with the silks of Group I, those of Group II have a high breaking energy; about $350\,\text{MJm}^{-3}$ for *G. mellonella*. The shape of the stress–strain curves for Group II is variable. The cocoon silk shows a yield plateau as do dragline silks but *G. mellonella* silk does not. Hepburn *et al.* (1979) report that *G. mellonella* silk shows hysteresis when cyclically loaded but they do not report the magnitude. The tensile strength of *G. mellonella* silk decreases and its extensibility increases in high relative humidity or in water (Hepburn *et al.*, 1979; see Fig. 4).

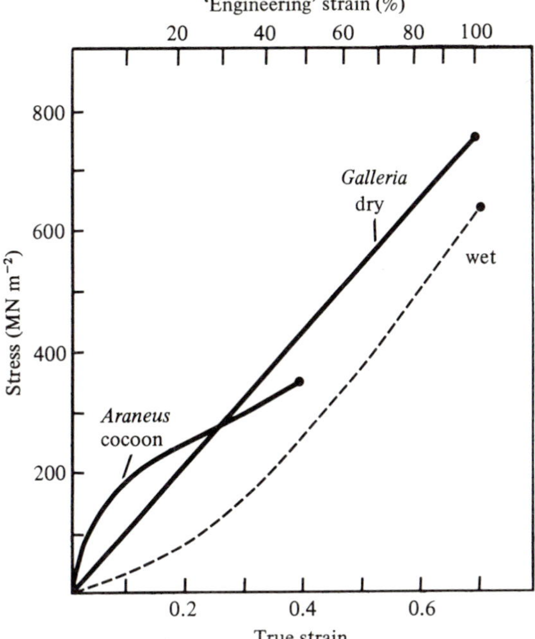

Fig. 4. Examples of the silks of Group II. Note the effect of hydration on *Galleria mellonella* silk. *Galleria* curves redrawn from Hepburn *et al.* (1979). *Araneus diadematus* curve redrawn from Lucas (1964).

In summary, the silks of Group II are considerably more extensible than those of Group I and generally, though not always, have a lower tensile strength. Their overall modulus is lower than silks of Group I. In other respects, especially in having a high breaking energy, they are quite similar to the silks of Group I.

Group III

The final class of silks to be examined here consists of a small number of silks, again characterized by their extensibility. To date this class includes the viscid silks of two species of spiders, the larval silk of the honey bee, *Apis mellifera*, and the egg-stalk silk of the green lacewing, *Chrysopa carnea* (Fig. 5). The extensibilities of these silks range from the viscid silk of *Araneus sericatus* which fails at an extension of 200% (Denny, 1976) to the viscid silk of the spider *Meta reticulata* for which de Wilde (1943) reports a breaking extension of 600–1600%. These high extensibilities are very unusual in biological fibrous materials, matching or even exceeding those of such non-fibrous materials as the protein rubbers (Wainwright *et al.*, 1976). In general the tensile strength of

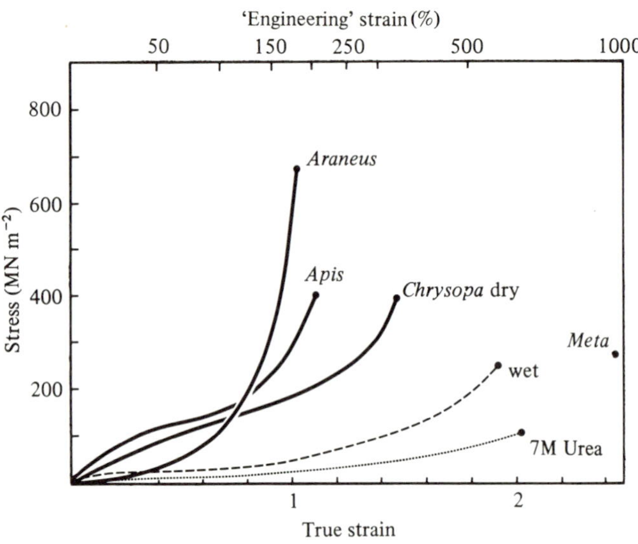

Fig. 5. Examples of the silks of Group III. Note how the modulus increases with increasing strain in *Araneus sericatus* viscid silk, and the dependence of *Chrysopa carnea* egg-stalk silk upon solvent properties. *Araneus* curve redrawn from Denny (1976). *Apis mellifera* and *Chrysopa* curves redrawn from Hepburn *et al.* (1979). The breaking point for *Meta reticulata* is inferred from de Wilde (1943).

these silks is lower than that of less extensible silks, though the high breaking strength of *A. sericatus* viscid silk shows that, again, this is not a necessary concomitant of greater extensibility. The overall modulus of these silks varies from about 100 to 600 MNm^{-2}. The breaking energy of Group III silks is quite high. Even *A. sericatus* viscid silk, with its highly nonlinear upwardly concave (J-shaped) stress–strain curve, has a breaking energy of about 150 MJm^{-3}.

The tensile strength of *Chrysopa* silk decreases and its extensibility increases (Hepburn *et al.*, 1979) in high relative humidity or in water (see Fig. 5). Relative humidity appears to make little difference to the properties of the silk of *Apis* (Hepburn *et al.*, 1979). The effect of humidity has not been studied in the viscid spider silks. Denny (1976) performed a number of tests on *A. sericatus* viscid silk which indicated the presence of a considerable viscous component. When cyclically loaded the silk showed a hysteresis of about 40%. Further, *A. sericatus* viscid silk stress-relaxes at infinite time to a stress only 51% of that at time zero.

In summary, the silks of Group III are very highly extensible, with tensile strength generally, though not necessarily, less than that of silks of both Groups I and II. In all other respects these silks are similar to Group I and II silks.

It should be emphasized here that the grouping of silks followed in this review is a matter of convenience. The grouping has been made on the basis of extensibility and it is probable that when more silks have been tested the range of extensibilities will prove to be a continuum. If so, separating any group of silks on the basis of a physical property such as extensibility will be seen to be purely arbitrary.

The broad range of mechanical properties of silks may be summarized as follows (see Table 1). All silks studied so far have a high tensile strength, varying from 200 to 1500 MNm^{-2}. The extensibility of silks varies to a much larger degree, from *Anaphe moloneyi* cocoon silk, which breaks down when extended by 10% to *M. reticulata* viscid silk which can be extended to 1600%. Often the tensile strength decreases as extensibility increases and this has been thought to be a general characteristic of silks (Lucas *et al.*, 1955). However, using the tensile strengths and extensibilities for the 18 silks listed in Table 1, no significant correlation ($r = -0.31$) is found between these variables. However, the tensile strengths used in this calculation were obtained using various strain rates, so until further testing has been carried out this point remains uncertain. The silks are unique among natural fibres in combining very high tensile strength with relatively large extensibility. As a consequence it takes more energy to break a given volume of silk than it takes to break the same volume of any other biological material. This energy absorbing capacity may be the factor of primary importance in the functioning of silk and will be discussed below.

Macromolecular and chemical structure

As with any material, the mechanical properties of silks are determined by their molecular structure. The macromolecular chemistry of many silks is well described and it is possible to account (at least in a preliminary fashion) for the observed mechanical properties of silks. It is not possible or appropriate here to review in any detail the sizeable literature concerning the chemistry and macromolecular structure of silks. Rather, this review will be limited to those aspects of silk structure that bear directly on the explanation of mechanical properties. Further information on the macromolecular chemistry of silks can be found in reviews such as Lucas & Rudall (1968).

The first clue to the molecular structure of silk is found in the nature of silk production. When contained within the silk gland, silk is a

Table 1. *Properties of silks produced by various insects and spiders*

Species	Tensile strength ($Nm^{-2} \times 10^8$)	Extensibility (%)	Crystal type	Function	Reference
Bombyx mori	7.4	24	parallel β	cocoon	Lucas, Shaw & Smith, 1955
Anaphe moloneyi	6.7	13	parallel β	cocoon	Lucas *et al.*, 1955
A. infracta	6.2	14	parallel β	cocoon	Lucas *et al.*, 1955
Antheraea mylitta	6.6	35	parallel β	cocoon	Lucas *et al.*, 1955
A. pernyi	5.8	35	parallel β	cocoon	Lucas *et al.*, 1955
Caligula japonica	2.8	20	parallel β	cocoon	Lucas *et al.*, 1955
Galleria mellonella	7.5	101	parallel β	cocoon	Hepburn, Chandler & Davidoff, 1979
Araneus sericatus	14.2	27	parallel β	dragline	Denny, 1976
A. sericatus	12.7	229		viscid	Denny, 1976
A. diadematus	14.2	30	parallel β	dragline	de Wilde, 1943
A. diadematus	5.4	24		cocoon	de Wilde, 1943
A. diadematus	11.9	31	parallel β	dragline	Lucas, 1964
A. diadematus	3.6	46		cocoon	Lucas, 1964
A. diadematus	14.0	31.5	parallel β	dragline	Work, 1977*a*
various spiders	18.0	35		dragline	Work, 1977*a*
Nephila madagascariensis	4.6	31	parallel β	dragline	Lucas *et al.*, 1955
Meta reticulata	2.8	1100		viscid	de Wilde, 1943
Apis mellifera	4.0	200	α helix		Hepburn *et al.*, 1979
Chrysopa carnea	3.8	310	cross β	egg stalk	Hepburn *et al.*, 1979

concentrated aqueous protein solution. In *B. mori* silk the solution contains 15–30% protein by weight (Iizuka, 1966). The silk is extruded from the gland into the outside world through the fine opening of the spinneret. The force necessary to extrude the silk is applied not as a pressure on the fluid in the gland but as tension on the silk that has already passed through the spinneret. Thus the viscous protein solution is pulled, rather than pushed, through the spinneret orifice and this tension brings about a structural change in the material: the viscous protein solution of the gland becomes the strong, primarily elastic, silk fibre. This change is not easily reversed. For example, even though its properties change to a certain extent as a function of hydration, silk is not readily dissolved.

This transformation in mechanical properties as a result of an applied force is a well-known phenomenon in man-made crystalline polymers (Work, 1977*b*). The basis of the process lies in the ordering of the protein chains as they are sheared in their passage through the silk gland and the spinneret. Within the silk gland the protein chains are randomly arranged (Magoshi *et al.*, 1977). As the silk passes through the spinneret, shearing of the fluid compels the protein chains to become extended in the direction of flow and in so doing to lie parallel to one another. Once the chains are properly positioned, bonds form between adjacent chains and the material crystallizes. This tightly bound crystalline form is energetically favourable, and therefore not easily disrupted. The percentage of chains which become sufficiently aligned to crystallize should be proportional to the shear rate of the fluid and this is inversely proportional to the diameter of the spinneret orifice and directly proportional to the rate of drawing. Iizuka (1966) has found this to be so for *B. mori* silk.

The crystalline structure of silk fibres has received considerable attention and is well understood. The primary tool used to examine this structure has been the analysis of X-ray diffraction patterns, supported by amino-acid analysis and sequencing. Owing to its availability and economic importance, the silk of *B. mori* was among the first silks studied and its structure is the best documented. This structure is also one of the simplest found among the silks and will be used as an example to introduce the β pleated sheet, the crystalline configuration found in the vast majority of the silks (all of Groups I and II) so far studied.

In *B. mori* silk, those protein chains which lie in a crystalline region of the material are fully extended with their long axes parallel to the fibre axis (Fig. 6). At intervals chains will fold back upon themselves so that adjacent chains run in opposite directions (i.e. antiparallel). In this

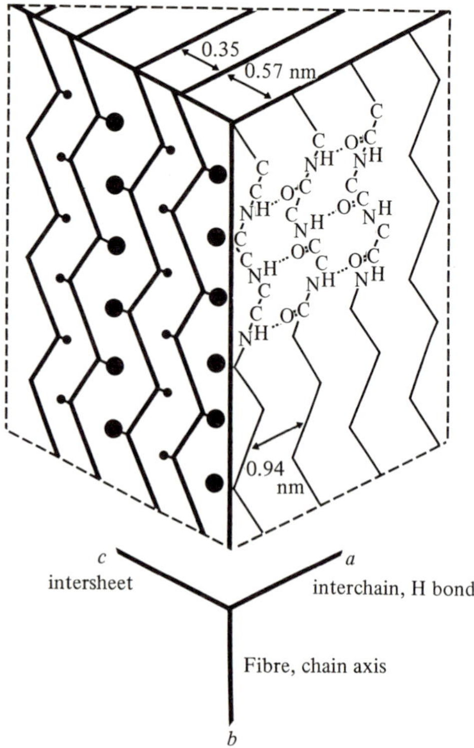

Fig. 6. The parallel β pleated sheet configuration of *Bombyx mori* silk. The chain axis is parallel to the fibre axis. After Marsh *et al.* (1955*a*) courtesy of *Biochim. Biophys. Acta.*

configuration hydrogen bonds form easily between the peptide groups of adjacent chains, and several antiparallel chains may be bound together to form a 'pleated sheet'. These sheets then stack one atop another to form a crystallite. The dimensions of the crystallite depend on a number of factors. On the smallest scale the repeat distance along the fibre axis (*b* in Fig. 6) is determined by the dimensions of amino acids and is constant at 0.695 nm for all parallel β silks (Warwicker, 1960). Similarly the interchain repeat distance (*a* in Fig. 6) is constant at 0.944 nm. The intersheet repeat distance (*c* in Fig. 6) is determined by the dimensions of the amino-acid side-chains. Marsh, Corey & Pauling (1955*a*) found that the intersheet repeat distance in *B. mori* silk is 0.93 nm and proposed that this distance is the sum of alternate spacings of 0.35 nm and 0.57 nm. This alternation in intersheet spacing is explained with evidence drawn from the amino-acid composition of the silk (Table 2). The glycine content of *B. mori* silk is very nearly equal to the sum of the alanine and serine contents. Thus glycine with its small

Table 2. *Amino-acid compositions of silks*

	Bombyx mori[a]	*Antheraea pernyi*[b]	*Araneus diadematus* dragline[c]	*Araneus diadematus* viscid[d]	*Galleria*[e] *melonella*	Vespidae and Apidae[a]	*Chrysopa flava*[f]	*Anaphe molonyi*[a]
Glycine	445	265	320	442	318	46–65	246	424
Alanine	293	441	363	83	249	161–367	212	530
Valine	22	7	16	67	52	14–65	–	21
Leucine	5	–	16	14	67	47–95	–	2
Isoleucine	7	–	13	10	35	10–51	–	1
Serine	121	118	59	31	166	87–227	427	3
Threonine	9	1	17	25	18	28–53	31	2
Aspartic acid	3	47	8	27	21	82–127	60	5
Glutamic acid	10	8	121	29	5	83–140	8	3
Lysine	3	1	12	13	2	23–44	–	1
Arginine	5	26	18	11	13	1–53	5	1
Histidine	2	8	–	7	1	2–13	–	2
Tyrosine	52	49	10	26	3	1–25	8	1
Phenylalanine	6	6	–	11	2	1–22	–	1
Proline	3	3	27	205	47	13–31	–	2
Tryptophan	2	11	–	–	–	–	3	–
Methionine	1	–	–	–	–	–	–	–
Cystine (half)	2	–	–	–	–	some	–	–
crystal type	anti-parallel β-pleated sheet	anti-parallel β-pleated sheet	anti-parallel β-pleated sheet	not determined	anti-parallel β-pleated sheet	α helix	cross β	anti-parallel β-pleated sheet

[a] Lucas & Rudall, 1968
[b] Schroeder & Kay, 1955
[c] Peakall, 1964
[d] Andersen, 1970
[e] Lucas *et al.*, 1960
[f] Lucas *et al.*, 1955

side-chain (–H) could alternate with either alanine ($-CH_3$) or serine (–OH). This strict alternation of small side-chains with large side-chains assures that glycine side-chains in one sheet will oppose glycine side-chains in the next sheet, forming the 0.35 nm spacing and that serine and alanine side-chains will oppose like side-chains, forming the 0.57 nm repeat. This proposed structure has been well supported by X-ray diffraction evidence from man-made protein crystals. Fraser, MacRae & Stewart (1966) found that crystalline poly-L-alanylglycyl-L-alanylglycyl-L-serylglycine produced a diffraction pattern very similar to that of *B. mori* silk. Further, various amino-acid sequencing studies reveal considerable fractions of *B. mori* silk with sequences compatible with this structure (Lucas & Rudall, 1968). While large portions of *B. mori* silk may be bound in crystalline regions, the amino-acid composition of the silk is not confined solely to glycine, alanine, and serine. The other amino acids which occur in the silk generally have bulky side-chains and therefore will not fit into the observed crystalline structure. Thus, in addition to crystalline regions, the silk must contain regions of noncrystalline, and presumably randomly arranged, protein chains. These are generally referred to as amorphous regions.

The overall size of crystallites in *B. mori* silk has been examined by using both electron microscopy and X-ray diffraction. Dobb, Fraser & MacRae (1967) reported microfibrils that are approximately rectangular in cross-section with dimensions of 2.0 nm in the intersheet direction and 6.0 nm in the interchain direction. Koratky *et al.* (1964) using low angle X-ray diffraction, observed particles consistent with the dimensions reported by Dobb *et al.* (1967) extending about 21 nm along the fibre axis. Thus it seems likely that *B. mori* silk is formed of a large number of small crystalline regions, $2 \times 6 \times 21$ nm, separated by amorphous regions containing randomly arranged chains. Iizuka (1966) reports that *B. mori* silk is 40–45% crystalline.

Structures differing only slightly from that of *B. mori* silk occur in a large number of other silks (Warwicker, 1960). The main difference in structure concerns the size of the side-chains of the amino acids incorporated into the crystalline regions. This difference is exemplified by the silk of the moth *Antheraea pernyi*, first examined by Marsh *et al.* (1955*b*). The combined content of alanine and serine in *A. pernyi* silk is much larger than the content of glycine, rendering a glycine–(serine/alanine) alternation unlikely. As a result the intersheet spacing does not alternate and is found to be a constant 0.53 nm (see Fig. 7). Other silks with considerable variation in amino-acid composition are found to have crystal dimensions similar to *A. pernyi* silk, differing only in the

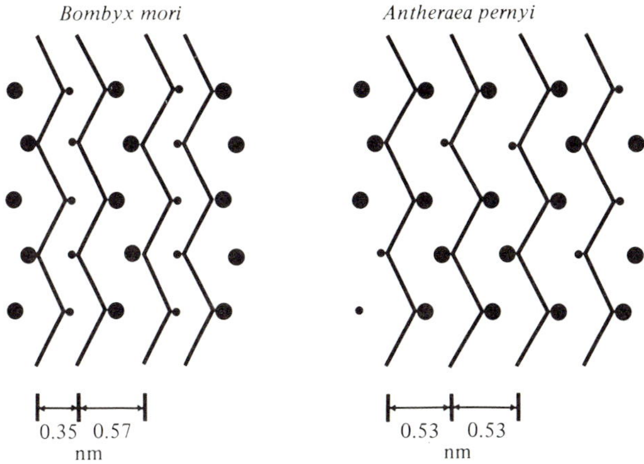

Fig. 7. A comparison of the intersheet packing configurations for *Bombyx mori* and *Antheraea pernyi* silks. After Marsh *et al.* (1955*a*) courtesy of *Biochem. Biophys. Acta.*

intersheet spacing and Warwicker (1960) divides the parallel β silks into five categories on the basis of this spacing. Despite these variations in crystal dimensions all parallel β silks are thought to consist of a combination of crystalline and amorphous regions.

While the vast majority of silks contain parallel β pleated sheets other crystalline forms are present. Parker & Rudall (1957) and more recently Geddes, Parker & Beighton (1968) have described a crystalline configuration found in the egg-stalk silk of the green lacewing *Chrysopa carnea*. This configuration, the cross β (or $\times\beta$) configuration, is similar to that of the parallel β pleated sheet except that the axes of the extended chain lie perpendicular to the fibre axis instead of parallel to it (Fig. 8). This requires each protein chain to fold back on itself every eight amino acids, a length of 2.5 nm. Tension along the fibre axis is placed directly on the bonds holding chains together and can cause this $\times\beta$ configuration to 'unfold' into a typical parallel β pleated sheet (Geddes *et al.*, 1968). This unfolding process involves a considerable extension of the crystallite along the fibre axis and this is reflected in the high extensibility of the material as a whole (Fig. 5). The X-ray data predict that the $\times\beta$ crystallite must be extended by six times to be transformed into a parallel β structure and this correlates well with the observed extensibility. As with the parallel β silk the material will consist of both crystalline and amorphous regions. The mechanism by which a $\times\beta$ configuration could be formed if it were to be extruded through a spinneret is not clear. *Chrysopa* silk is formed, however, by

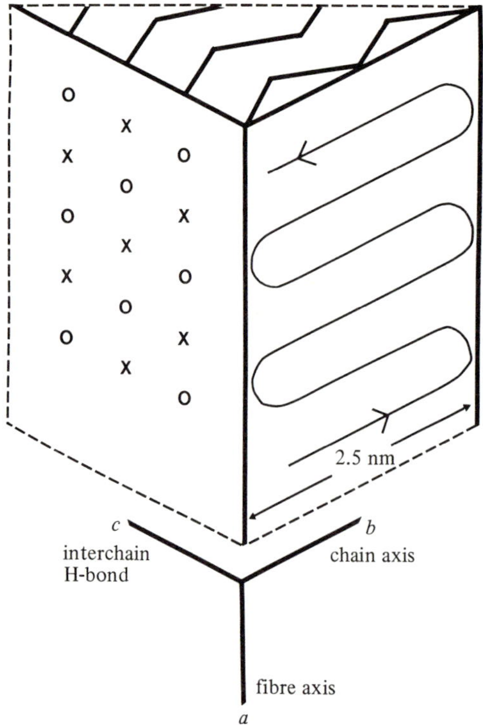

Fig. 8. The $\times\beta$ configuration. The chain axes are perpendicular to the fibre axis. After Geddes *et al.* (1968) courtesy of *J. molec. Biol.*

being drawn from an externally deposited fluid drop and the fibres are considerably larger (15–20 μm) than those of other silks (Parker & Rudall, 1957). Presumably this mode of formation contributes to the difference in crystal structure.

Several silks, for example the larval silk of the honey bee *Apis mellifera*, have been found to contain regions with chains arranged in various α helical configurations (Lucas & Rudall, 1968). As with the $\times\beta$ configuration these α helical structures may be transformed into parallel β structures upon extension, a process much like the α to parallel β transformation in stretched keratin (Fraser & MacRae, 1980). Again this transformation involves an extension of the crystallite, a fact which is reflected in the high overall extensibility of the silk (Fig. 5).

Other crystalline configurations have been identified in silks (e.g. the collagen configuration in the silk of the common gooseberry sawfly (Lucas & Rudall, 1968)). The consequence of these configurations as regards the mechanical properties of the material is not known. Conversely, the mechanical properties of some silks are known (e.g.

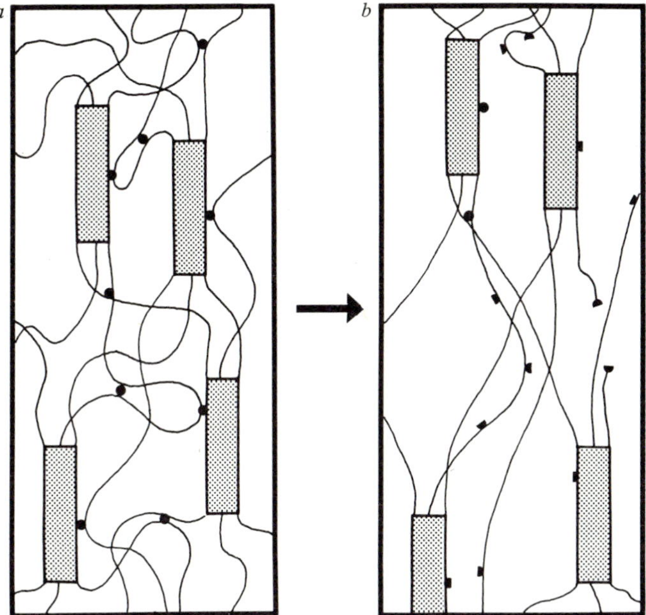

Fig. 9. A schematic representation of the molecular events accompanying extension in a crystalline polymer such as silk. The fibre axis is vertical. Crystalline regions (shaded) are connected by random chains. In the unstressed silks, *a*, the random chains are bound to themselves (solid dots) and to crystalline regions. As the silk is extended, *b*, the random chains are extended and bonds are broken.

spider viscid silks), but their crystalline configurations have not been determined.

From these molecular studies a reasonably clear picture of the basis for the mechanical properties of silks can be drawn: silks are partially crystalline polymeric materials. The crystalline regions are accurately aligned with the fibre axis and are interspersed with regions of randomly arranged chains (see Fig. 9). As tension is applied to a silk fibre the material deforms. On a molecular scale, this deformation occurs primarily in the amorphous regions rather than the highly aligned (and therefore stiff) crystallites. The elastic modulus and viscosity of the material during this initial extension will depend upon the extent to which the amorphous chains are bound to each other, the interaction of amorphous chains with the crystallites and the proportion of amorphous material in the fibre. While the amorphous chains are randomly arranged, it is unlikely that they are kinetically free (as are the chains in a rubber). Instead bonds between chains and 'short' chains connecting crystallites must be broken before the bulk of the amorphous chains may extend. The strength of these bonds will account for the high initial modulus of silks such as *A. sericatus* dragline silk (Lucas *et al.*, 1955). Once these bonds have been broken, however, the amorphous regions will deform easily, producing the lower modulus yield plateau. In the process of deformation, chains must slide past each other and past the crystallites. Any interaction between chains, and between chains and crystallites, will result in viscous loss of energy as the material is extended. If, in the process of extension, amorphous chains become sufficiently aligned it may be possible for them to crystallize, producing permanent plastic deformation in the material.

As the silk is extended further an increasing percentage of the amorphous chains will become fully extended. As a result more of the load will be borne by the chains of the crystalline regions. In parallel β silks considerable lengths of these chains are already fully extended along the fibre axis and the only possible accommodation by the material to further extension is breakage of the bonds between chains or breakage of the protein chains themselves. Either results in failure of the silk. The strength of the material at breaking strain will generally depend on three factors: 1, the backbone strength of the protein chains; 2, the proportion of chains which are fully extended and bearing the load; and 3, the strength of the bonds that keep chains from sliding past each other. The backbones of all proteins are the same, so it is unlikely that the tensile strength of these chains will vary from one silk to the next. Thus, differences in tensile strength should reflect differences in the percentage of chains bearing the load and differences in the combined strength of the bonds tying chains together. This last factor can be tested by placing a silk in water. The crystalline regions are too closely packed to allow entry of water, but the more open structures of the amorphous regions will become hydrated. The imbibed water will compete for hydrogen bonds between amorphous chains and facilitate the sliding of chains past each other, resulting in lower modulus and greater extensibility (Lucas *et al.*, 1955; Hepburn *et al.*, 1979). If the silk is soaked in a more potent hydrogen bond competitor, such as lithium thiocyanate, 8-M urea, or formamide, the hydrogen bonds of the crystalline regions may be broken and the tensile strength will be lower even than that in water, and the extensibility greater (see Fig. 5, Hepburn *et al.*, 1979).

Silks with other than parallel β crystallites will follow a different sequence of events during the final stage of extension. As the amorphous regions become fully extended the load will be transferred to the crystallites. In this case the chains of the crystallites are not fully extended along the fibre axis and the tension will be borne by the interchain hydrogen bonds. At some critical extension the stress on the crystallites will be sufficient to transform the crystallite chains from either $\times\beta$ or α helix to parallel β. As explained earlier, this transformation involves extension of the crystallite and of the silk as a whole. As an increasing percentage of crystallites becomes fully extended, more stress will be placed on the chain axis itself and, as with the parallel silks, at some extension the chains will break or slip past each other and the material will fail. Again variations in tensile strength should reflect differences in the percentage of chains bearing the load and in the combined strength of interchain bonds.

Thus the physical properties of silks will be determined on the molecular scale by a number of factors. 1, The crystalline configuration will, to a large extent, determine the overall extensibility. 2, Within the range set by crystalline configuration the relative proportions of crystalline and noncrystalline regions will determine the extensibility. 3, The nature of the bonding between amorphous chains, and between amorphous chains and crystallites, will determine both the tensile strength of the material and its viscous and plastic nature.

The unstressed macromolecular arrangement of a silk is largely determined by the amino-acid composition and sequence of the silk. Little work has been directed to this problem, apart from the studies mentioned above regarding the sequences of *B. mori* silk and man-made analogues, so few generalizations can be drawn. One fact of interest is the generally high proline content of silks. It has been shown (Szent-Gyorgi & Cohen, 1957) that proline produces a 'kink' in protein chains. Such a kink is incompatible with either the parallel β or $\times\beta$ configurations. Thus the position and amount of proline in a silk may be a mechanism for controlling the extent of the amorphous regions (Wainwright *et al.*, 1976). The high proline content of silks may also play a role in ensuring that the silk does not crystallize while in the silk gland.

The relationship of mechanical properties to function

The many functions of silks have been well described and a few examples were given in the introduction to this review. However, in all but a few cases, the mechanical forces placed on silks during their functioning have not been studied. Indeed, in many cases the unique mechanical properties of silks seem to be incidental to their use in a particular structure. A prime example is the cocoon of *B. mori*. In a typical cocoon 500 to 1200 m of double stranded silk is used to form a fibre-wound capsule which encloses and protects the pupa (Encyclopaedia Britannica, 1971). It can be calculated that the pupa would have to exert a pressure of about 6 MNm^{-2} on the walls of the cocoon in order to break the silk fibres. It seems unlikely that the metamorphosing moth will stress the cocoon to anything approaching this value, though these forces have not been measured. Protection from predators and the elements seems likely to come more from the sheer bulk of the cocoon than from the strength of its fibres. Thus in cases such as this it appears that factors other than mechanical properties have governed the evolution and use of silks.

Cases exist, however, where the mechanical properties of a silk play a direct and important role in its functioning. The best documented are the silks produced by the spiders and two examples are discussed here.

It is a near universal characteristic of spiders that as they move they trail behind themselves a silk dragline. This dragline serves the same function as a mountain climber's safety rope: if the spider falls the dragline will stop its descent and allow the spider to climb back to the spot from which it fell. The ingenious mechanism used by spiders to brake the dragline during a fall is described in Work (1978). Lucas (1964) found that a female *A. diadematus* weighing 0.65 g produced a dragline capable of supporting a static load of about 1 g. This provides an adequate safety margin when supporting a stationary spider. For a dragline consisting of two threads each only 1 μm in diameter this is a tribute to the high tensile strength of this material. However, this story is more complex if, as Lucas notes, the kinetic energy of a falling spider is considered. A spider of mass M, falling in a vacuum at velocity v, will have kinetic energy equal to $Mv^2/2$. If the spider is to be brought to a halt by its dragline, the dragline must absorb this kinetic energy. The breaking energy of *A. sericatus* dragline silk has been measured by Denny (1976). In this case the spider weighs 0.1 g and we will assume that it has fallen 1 m. At this point *in vacuo* its velocity will be about 4.5 ms^{-1} and its kinetic energy will be 10^{-4} J. Fig. 10 shows the force–extension curve for a typical *A. sericatus* dragline 1.0 m in length and $6.9 \times 10^{-12} m^2$ cross-sectional area. The area beneath the curve is a measure of the energy required to break the dragline, in this case 1.1×10^{-4} J. This dragline would be able to break the spider's fall. If, however, the thread had the same tensile strength, but were less extensible, the energy required to break it would be less. For example the dashed line in Fig. 10 shows a hypothetical force–extension curve for a silk with the same high tensile strength as dragline silk, but with the extensibility of collagen. The energy required to break this thread would

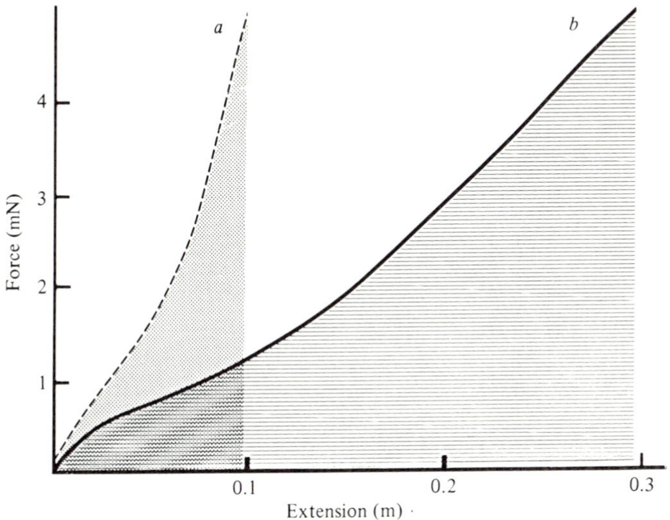

Fig. 10. The energy absorbing capacity of spiders' dragline silk. The total area beneath a force elongation curve is the breaking energy. The high extensibility of *Araneus sericatus* dragline (*b*) gives it a large breaking energy (horizontal hatching). A hypothetical silk with the same tensile strength, but less extensibility (*a*) shows a lower breaking energy.

be only 3×10^{-5} J and this would not necessarily stop the fall described above. In air, drag will slow the spider's fall, and its kinetic energy will be considerably lower than that calculated here. The same principles will apply, however. Thus a high breaking energy such as is shown by the spider's silk is a necessity for proper biological functioning with sufficient margin of safety.

This close relationship between mechanical properties and biological function is further exemplified by the orb-web silks of spiders such as *A. sericatus*. Here requirements in addition to high breaking energy are placed on the silks by the functions they perform. The webs of orb-weavers are the sole means by which these animals obtain food. Thus the energy gained from captured prey must be weighed against the energy expended in producing a functional web silk. Indeed the cost of producing the web is one of the largest energy expenditures encountered by these animals (Lubin, 1973; Peakall & Witt, 1976). Consequently it is an advantage to the spider to produce a web with a minimum volume of silk and thereby minimum expenditure in the form of protein secreted. The volume of the web is most easily minimized by constructing it of very fine fibres. The practical lower limit to the size of fibres is set by their tensile strength and breaking energy for, fine as the fibres are, they must be capable of withstanding the forces placed on the web by impacting insects. As seen in Fig. 1, *A. sericatus* dragline (the

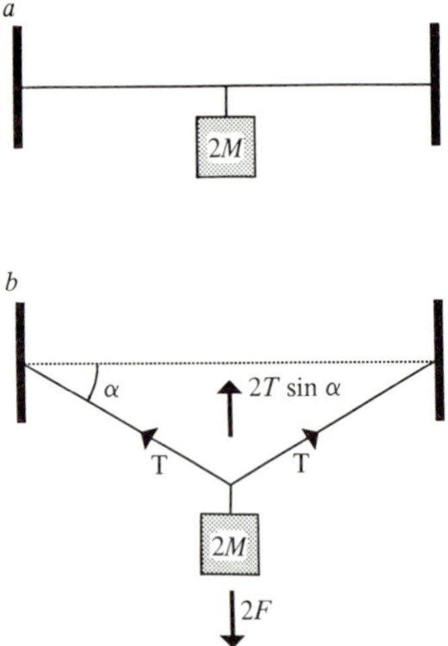

Fig. 11. The forces acting on horizontal threads. Two threads each supporting a mass M (total mass $2M$) are stretched. The tension in each thread is $F(\sin \alpha)^{-1}$.

silk used to form the framework of the orb web) has both a very high tensile strength and high breaking energy.

Closer examination of the role of dragline silk in an orb web reveals a further correlation between mechanical properties and function. The radial fibres of the web must support the weight of the spider plus the weight of any prey captured. In the orb web many of these fibres will be nearly horizontal when unstressed as in Fig. 11*a*. If a force $2F$, due to spider and prey, is applied, each of the two silk threads will deform to some angle α. The tension T which must be developed in each fibre if the weight is to be held stationary is $T = F/\sin \alpha$. Thus the larger the angle α, the less tension is required to support a given weight. However, for a fibre held horizontally the only way to increase α is by stretching the fibre. As the fibre is extended its cross-sectional area decreases and the stress placed on the fibre by a given tension increases. Beyond a certain optimum point this increase in stress will offset any decrease in tension due to the increase in α. Thus for a fibre of a given tensile strength, σ_{max}, and initial cross-sectional area, A_0, the maximum force that can be resisted, F_{max}, (when loaded as in Fig. 11*b*) is a function of the breaking extension ratio L ($L = \exp \varepsilon_t$).

$$F_{max} = \sigma_{max} (A_0/L) \sin \alpha. \quad (1)$$

$$F_{max} = \sigma_{max} (A_0/L) (L^2 - 1)^{1/2}/L. \quad (2)$$

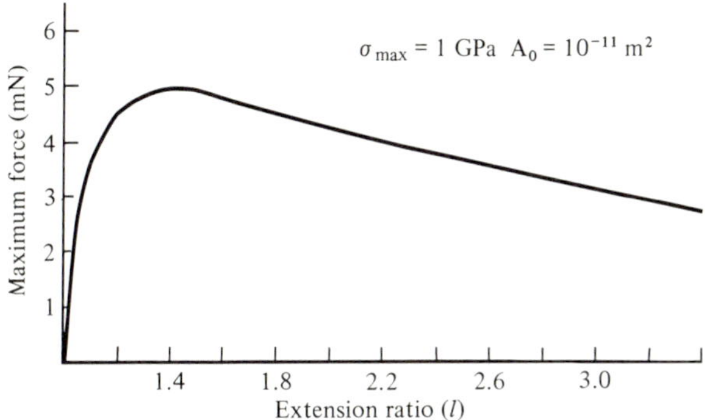

Fig. 12. The maximum force horizontal threads can support is a function of their extensibility. For a thread of given dimensions and tensile strength, an extensibility of 42% is optimal (from Denny, 1976).

This function is plotted in Fig. 12 and reaches a maximum at $L = 1.42$, an extension of 42%. Fibres with extensibilities much less than 20% are relatively weak under such a loading regime, as are fibres with L much beyond 2.0. Dragline silks with extensibilities of 20 to 30% ($L = 1.2$ to 1.3) are quite near the theoretical maximum. In this respect dragline silks are well adapted to their function as a scaffolding material, being maximally strong in the resistance of static loads.

The viscid silk of orb webs, with an extensibility of 200 to 1600%, is well beyond the point of maximum strength calculated above. The usefulness of this greater extensibility is again related to the function of the silk in the web. The viscid silk acts as the primary catching net of the web. In order for a flying insect to be caught its speed relative to the web must be brought to zero. The magnitude of deceleration the insect must undergo in being caught is a function of the distance over which the deceleration is accomplished; the smaller the distance the higher the deceleration. For an insect of mass M, a force equal to Ma (where a is the rate of deceleration) must be applied by the viscid silk to the insect and this force will be smaller the greater the distance over which the force is applied. The great extensibility of viscid silk allows a flying insect to be slowed over a large distance relative to that offered by dragline silk, thereby decreasing the force placed on the web. As a result a web formed with viscid silk can catch either larger insects or insects flying faster than a web formed solely of dragline silk.

Further, an insect trying to struggle free from a web is held by the highly extensible viscid threads. In the web these threads lie in the initial, low modulus, portion of their stress–strain range. Thus any thrust with a leg or wing is met not with an elastic force that would provide something to push against, but rather with the near forceless extension of the viscid thread to no avail. It is interesting to note that the stress–strain curve of *A. sericatus* viscid silk (see Fig. 5) resembles that of many soft composite membrane materials (e.g. blood vessels). This shape of curve with a low modulus at low strains and a rapidly increasing modulus at higher strains has been shown by Gordon (1978) to be a general characteristic of materials used in membranes. Indeed an orb web may be thought of as a porous membrane which is deformed by insects hitting it.

One final adaptation of orb-web silks should be mentioned here. If the web silks were perfectly elastic an orb web would act much like a trampoline; the energy of an insect hitting the web would be elastically stored and the insect would be catapulted whence it came. The high initial hysteresis of both the dragline and viscid silks (65%) ensures that much of the energy of impact is dissipated viscously and the web consequently functions as a proper net.

The spiders' silks discussed here are but one example of the manner in which the physical properties of silks are tied to their biological functioning. Further research into the role of silks in nature is certain to bring to light other such examples, and will help to elucidate not only the structure of this one class of fibres, but the nature of the evolutionary mechanisms by which these and all other biological materials are designed.

References

Andersen, S. O. (1970). Amino acid composition of spider silks. *Comparative Biochemistry and Physiology*, **35,** 705–11.

Denny, M. W. (1976). The physical properties of spiders' silk and their role in the design of orb-webs. *Journal of experimental Biology*, **65,** 483–506.

Dobb, M. G., Fraser, R. D. B. & MacRae, T. P. (1967). The fine structure of silk fibroin. *Journal of Cell Biology*, **32,** 289–95.

Encyclopaedia Britannica (1971). Volume 20, pp. 519–26. Chicago: Encyclopaedia Britannica Inc.

Fraser, R. D. B., MacRae, T. P. & Stewart, F. H. C. (1966). Poly-L-alanylglycyl-L-alanylglycyl-L-serylglycine: A model for the crystalline regions of silk fibroin. *Journal of molecular Biology*, **19,** 580–2.

Fraser, R. D. B. & MacRae, T. P. (1980). Molecular structure and mechanical properties of keratin. In *The Mechanical Properties of Biological Materials* (34th Symposium of the Society for Experimental Biology) ed. J. F. V. Vincent & J. D. Currey, pp. 211–46. Cambridge University Press.

Geddes, A. J., Parker, K. D. & Beighton, E. (1968). Cross-β conformation in proteins. *Journal of molecular Biology*, **32,** 343–58.

Gordon, J. E. (1978). *Structures or Why Things Don't Fall Down*. London: Penguin Books.

Hepburn, H. R., Chandler, H. D. & Davidoff, M. R. (1979). Extensometric properties of insect fibroins: The green lacewing cross β, Honey Bee helical and greater waxmoth parallel β conformations. *Insect Biochemistry*, **9,** 69–77.

Iizuka, E. (1965). Degree of crystallinity and modulus relationship of silk threads from *Bombyx mori* and other moths. *Biorheology*, **3,** 1–8.

Iizuka, E. (1966). Mechanism of fiber formation in the silkworm, *Bombyx mori*. *Biorheology*, **3,** 141-52.

Koratky, O., Wawra, H., Pily, I., Sekora, A. & van Deinse, A. (1964). Röntgen Kleinwinkeluntersuchungen an Lösungen von der nativer Seide. *Monatshefte für Chemie*, **95,** 359–72.

Lubin, Y. D. (1973). Web structure and function: The non-adhesive orb-web of *Cyrtophora atricola* Forskal. *Forma et Functio*, **6,** 337–58.

Lucas, F. (1964). Spiders and their silk. *Discovery*, **25,** 20–6.

Lucas, F. & Rudall, K. M. (1968). Extracellular fibrous proteins: The silks. In *Comparative Biochemistry*, ed. M. Florkin & E. H. Stotz, **26B,** 475–558. Amsterdam: Elsevier.

Lucas, F., Shaw, J. T. B. & Smith, S. G. (1955). The chemical constitution of some silk fibroins and its bearing on their physical properties. *Journal of the Textile Institute*, **46,** T440–2.

Lucas, F., Shaw, J. T. B. & Smith, S. G. (1960). Comparative studies of fibroins. 1. The amino acid composition of various fibroins and its significance in relation to their crystal structure and taxonomy. *Journal of molecular Biology*, **2,** 339–49.

Magoshi, J., Magoshi, Y., Nakamura, S., Kasai, N. & Kakudo, M. (1977). Physical properties and structure of silk. V. Thermal behaviour of silk fibroins in the random coil conformation. *Journal of Polymer Science*, **15,** 1675–83.

Marsh, R. E., Corey, R. B. & Pauling, L. (1955*a*). An investigation of the structure of silk fibroin. *Biochimica et Biophysica Acta*, **16,** 1–34.

Marsh, R. E., Corey, R. B. & Pauling, L. (1955*b*). Structure of Tussah silk fibroin. *Acta Crystallographica*, **8,** 710–15.

Mullins, L. (1980). Theories of rubber-like elasticity and the behaviour of filled rubbers. In *The Mechanical Properties of Biological Materials* (34th Symposium of the Society for Experimental Biology) ed. J. F. V. Vincent & J. D. Currey, pp. 273–88. Cambridge University Press.

Parker, K. D. & Rudall, K. M. (1957). The structure of the silk of Chrysopa egg-stalks. *Nature, London*, **179,** 905–7.

Peakall, D. B. (1964). Composition, function, and glandular origin of the silk fibroins of the spider *Araneus diadematus* Cl. *Journal of experimental Zoology*, **156,** 345–50.

Peakall, D. B. & Witt, P. N. (1976). The energy budget of an orb-web building spider. *Comparative Biochemistry and Physiology*, **54A**, 187–90.

Schroeder, W. A. & Kay, L. M. (1955). The amino acid composition of *Bombyx mori* silk fibroin and Tussah silk fibroin. *Journal of the American Chemical Society*, **77,** 3908–13.

Szent-Gyorgi, A. G. & Cohen, C. (1957). Role of proline in polypeptide chain configuration of proteins. *Science*, **126,** 697–8.

Wainwright, S. A., Biggs, W. D., Currey, J. D. & Gosline, J. M. (1976). *Mechanical Design in Organisms*. London: Edward Arnold Ltd.

Warwicker, J. O. (1960). Comparative studies of fibroins II. The crystal structure of various fibroins. *Journal of molecular Biology*, **2,** 350–62.

de Wilde, J. (1943). Some physical properties of the spinning threads of *Aranea diadema* L. *Archives néerlandaises Sciences*, **27,** 118–32.

Work, R. W. (1977*a*). Dimensions, birefringence, and force-elongation behaviour of major and minor ampullate silk fibers from orb-web-spinning spiders – the effects of wetting on these properties. *Textile Research Journal*, **47,** 650–62.

Work, R. W. (1977*b*). Mechanism of major ampullate silk fiber formation by orb-web-spinning spiders. *Transactions of the American Microscopical Society*, **96,** 170–89.
Work, R. W. (1978). Mechanism for the deceleration and support of spiders on draglines. *Transactions of the American Microscopical Society*, **97,** 180–91.

THEORIES OF RUBBER-LIKE ELASTICITY AND THE BEHAVIOUR OF FILLED RUBBER

LEONARD MULLINS
Malaysian Rubber Producers' Research Association,
Brickendonbury, Hertford, SG13 8NL, UK

A number of biological materials show highly-elastic behaviour, that is, they can be subject to large deformations and recover almost completely on removal of the deforming force. There is no need for me to do more than list a few examples of these materials since many pioneering studies of their behaviour have been made. In vertebrates (i) ligaments, such as are attached to bones where they meet at joints, restrict motion provided for by the joint and hence prevent dislocation, and (ii) tissues in arterial walls, which accommodate the rapid systolic stretch, both contain the highly-elastic protein elastin. In insects, cuticle, as in wing hinges, plays an important role in flight as it stores energy in the upstroke of the wing for release in the downstroke. Similarly the energy storing pads in the legs of fleas release energy much more quickly than muscles can. These cuticular structures contain the highly-elastic protein resilin. In scallops the ligaments in the hinge between the valves, which enable the mollusc to swim by rapid abductions of the valves, contain the highly-elastic protein abductin. In most of these applications it is important not only that recovery of the deformation is complete, but also that the cycle of deformation and recovery should occur in a most efficient way with minimum loss of energy.

Kinetic theory of rubber-elasticity

This behaviour is characteristic of rubbers and it is to rubbers that we must look for the most advanced theoretical description and understanding of the principal elastic qualities of highly-elastic materials. The unusual elastic qualities of rubber have attracted interest for a long time, and as early as 1805 Gough described its anomalous elastic behaviour; he recorded that, unlike other known materials, heat was evolved on stretching, that when held extended under constant load it became shorter as the temperature was raised, and that on removal of the load retraction was more rapid and more complete the higher the temperature. But it was not until 1932 that Meyer, von Susich & Valko were able fully to appreciate the significance of these observations, to put forward the essential molecular basis for rubber-like elasticity, and to show that the retractive force in stretched rubber resulted from the thermal motion of the rubber molecules as distinct from interatomic forces, just as the pressure due to a gas is a result of the kinetic energy of its molecules. The problem of rubber elasticity was thus chiefly concerned with an account of the origin of entropy changes accompanying deformation, changes in internal energy being relatively small and unimportant.

Meyer *et al.* (1932) showed that the decrease in entropy on elongation was a consequence of the orientation and rearrangement of the long molecular chains of rubber, and that the restoring force originated from the tendency of these chains rearranged by deformation to return to their original configurations. This picture of the kinetic origin of rubber-like elasticity has provided a sound basis for the development by statistical methods of a theory for the elastic behaviour of rubber.

Central to this theory was the idea that rubber-like elasticity was an attribute of a general type of flexible long-chain molecular structure rather than of the specific constitution of the natural rubber polyisoprene molecule. This was an idea which was to be amply confirmed with the advent of a wide range of synthetic rubbers.

Similarities between the behaviour of natural rubber and many biological materials of widely diverse chemical nature, such as gelatin, cellulose, muscle, silk, etc., have not gone unnoticed and have helped to provide insight into their molecular structure. It is now clear that all of these materials comprise long polymeric chain molecules. Structural formulae for the repeating molecular units of some typical rubbers and related materials are given in Table 1.

General conditions for rubber-like elasticity

A characteristic of these long molecular chains is their ability to assume a very large array of configurations as a result of relatively free rotation about single bonds in the chain. Large deformations are accommodated by rearrangement of the chains but, of course, they can only be achieved if there is also adequate internal mobility between the chains to permit these rearrangements to take place. For reversible elastic behaviour the long chain molecules must also be linked together into a permanent network structure so that there is no plastic flow.

To complete our picture of the molecular structure of rubber-like

Table 1. *Structural formulae for some typical rubbers and related materials*

Structural formula	Material
$-CH_2-C(CH_3)=CH-CH_2-$	polyisoprene (natural rubber)
$-CH_2-CH=CH-CH_2-$	polybutadiene
$-CH_2-C(Cl)=CH-CH_2-$	polychloroprene (Neoprene rubber)
$-CH_2-CH=CH-CH_2-$ ⁞ $-CH_2-CH(C_6H_5)-$	butadiene–styrene copolymer
$-CH_2-CH=CH-CH_2-$ ⁞ $-CH_2-CH(CN)-$	butadiene–acrylonitrile copolymer
$-O-Si(CH_3)_2-$	polydimethyl siloxane (Silicone rubber)
$-CH(R)-C(=O)-NH-$	polypeptide (protein rubbers)

materials, each of the rubber chains is considered to be in constant thermal motion (Brownian type) and in the course of this motion each chain twists itself into a very large number of shapes or configurations consistent with the restrictions imposed by the network crosslinks.

When a stress is applied the random molecular motions are perturbed and the network chains seek new configurations commensurate with the stress. The rate of the stress-biased diffusion depends upon the viscous resistance encountered by the chain segments as they move among their neighbours. At elevated temperatures the viscous resistance is low, the chains in the network rearrange rapidly and equilibrium stress–strain data can be obtained readily, while at lower temperatures the rearrangements occur more slowly due to increased viscous resistance.

The properties dependent on the rate of segmental diffusion are beyond the scope of this chapter and will be ignored.

Statistical theory of rubber-like elasticity (Treloar, 1975, Chapter 4)

The derivation of stress–strain relationships for rubber from a quantitative analysis of the changes in entropy with deformation demands further simplifying assumptions regarding the molecular model. A model which has proved extremely powerful involves the replacement of the real molecular chains by equivalent hypothetical chains composed of a large number of freely jointed links in which complete freedom of rotation exists between neighbouring links. The lengths of the chains between junction points in the network are all assumed the same and it is further assumed that no interaction occurs between elements of the chain other than at junction points which move affinely during deformation. (A small volume of a material is said to deform affinely if its strains are the same as those in the larger volume in which it is embedded.)

Even with this simplified model it is still hopeless to attempt any detailed analysis of molecular configuration but the model is ideal for the application of statistical methods. The basic problem is to determine the probability that the average distance between the ends of a chain lying between network junction points is changed in proportion to the dimensions of the deformed rubber. With the probability P thus derived it is possible to associate an entropy S according to Boltzmann's relation

$$S = k \log_n P$$

Then from considerations of changes in the thermodynamic quantities, internal energy (U) and entropy (S) and the work done (W) during deformation we have for a reversible process

$$\mathrm{d}U = T\,\mathrm{d}S + \mathrm{d}W$$

and for simple extension

$$\mathrm{d}W = f\,\mathrm{d}l.$$

It has been shown that for rubbers at moderate elongations changes in internal energy are relatively small compared with changes in entropy and that thus to a first approximation, where f is force and l is length is

$$f = T\frac{\mathrm{d}S}{\mathrm{d}l}.$$

The equation of state derived from this simple molecular model can be expressed in a general form in terms of the energy W stored during deformation

$$W = \tfrac{1}{2}NkT(\lambda_1^2 + \lambda_2^2 + \lambda_3^2 - 3)$$

where N is the density of network chain segments per unit volume; λ_1, λ_2, and λ_3, the three principal extension ratios, k is Boltzmann's constant and T is the absolute temperature. In principle this equation permits the calculation of stress–strain relationships for any type of deformation in terms of the single parameter N which is the basic variable characterizing the structure of the molecular network. Thus in simple extension the force f required to stretch a sample of rubber to an extension ratio of λ is given by

$$f = NkTA_0(\lambda - \lambda^{-2})$$

where A_0 is the unstrained cross-sectional area.

The stress required to extend rubber is thus given by the product of the three following factors: (i), a deformation factor $(\lambda - \lambda^{-2})$; (ii), the temperature T; (iii), a network structure factor.

In considering the extent to which the theory has been confirmed it is convenient to consider separately the dependence of stress upon strain, temperature, and network structure.

The predicted dependence of stress upon strain can only be tested adequately by determining the behaviour of rubbers under various pure homogeneous deformations. It is natural that most investigations have been in simple extension because of the relative ease of the experimental techniques but measurements have also been carried out in pure shear, two dimensional extension, and equivalently simple compression. The results show that although the theory provides a basis for understanding the relation between stress–strain curves, for different types of strain there are appreciable departures from theoretical predictions, and the theory thus only provides a first approximation to a description of the stress–strain behaviour of rubber. This is illustrated in Fig. 1 by results obtained by Treloar in simple extension. Appreciable departures occur at moderate extensions and the theoretical curve does not follow the rapid rise in stress at high extensions.

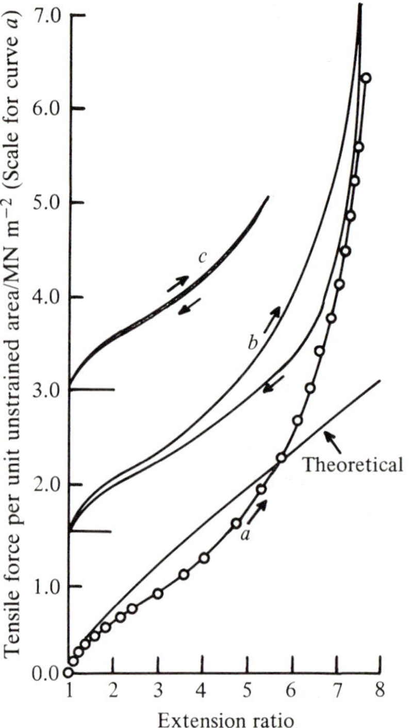

Fig. 1. Comparison of experimental simple extension stress–strain curves with theoretical form (Treloar, 1975).

Phenomenological theory of large elastic deformations

Considerably better description of experimental stress–strain data can be obtained by the use of a phenomenological theory developed by Rivlin (1948). In this generalized theory of large elastic deformations the energy stored during deformation is described in terms of a double infinite power series in terms of two strain invariants I_1 and I_2 defined in terms of the principal extension ratios λ_1, λ_2 and λ_3 where $I_1 = \lambda_1^2 + \lambda_2^2 + \lambda_3^2 - 3$ and $I_2 = \lambda_1^{-2} + \lambda_2^{-2} + \lambda_3^{-2} - 3$. In its most general form the stored energy function for an incompressible isotropic material is

$$W = \sum_{i=0,\, j=0}^{\infty} C_{ij}(I_1)^i(I_2)^j$$

where C_{ij} is a constant. The first two simplest terms of this series correspond to $i = 1$, $j = 0$ and $i = 0$, $j = 1$ and the expression containing these two terms has been found to be of particular interest

$$W = C_{10}(\lambda_1^2 + \lambda_2^2 + \lambda_3^2 - 3) + C_{01}(\lambda_1^{-2} + \lambda_2^{-2} + \lambda_3^{-2} - 3)$$

For simple extension this gives the stress–strain relation

$$f = 2A_0(\lambda - \lambda^{-2})(C_{10} + C_{01}\lambda^{-2})$$

and gives a very satisfactory description of the stress–strain behaviour at low and moderate extensions. It will be noticed that the term involving C_{10} in this phenomenological theory is identical to that given in the statistical theory.

Although no physical significance can be assigned to the constants C_{10} and C_{01} in the phenomenological theory, investigations on the influence of swelling and temperature in simple extension have shown that the dependence of the first term of this theory are in close accord with the predictions of the statistical theory. Further, when corrections have been made for the effect of network frames due to chain ends and chain entanglements, there appears to be an absolute correspondence between its magnitude and that determined from the molecular network structure and using the statistical theory.

The second term of the generalized theory thus describes departures from the theory. These are found to be large in dry rubbers but to decrease progressively with increase in strain and degree of swelling, and highly swollen rubbers show behaviour closely corresponding to that predicted by the statistical theory (Gee, 1946; Gumbrell, Mullins & Rivlin, 1953).

However, data obtained on other forms of deformation show that the situation is more complex and that the behaviour cannot be described by a single additional term to the stored energy function of the statistical theory. Nevertheless the results of numerous investigations confirm that the basic tenets of the molecular theory provide a correct basis for describing the equilibrium elastic properties of rubber and also provide a molecular model for considering the viscoelastic properties and other properties such as strength, extensibility and diffusion. The model involves a network of interpenetrating molecular chains coupled at sparsely spaced points through chemical linkages; as a consequence of interpenetration of chains the network contains entanglements which may or may not be trapped during the formation of the network.

No clear picture of the physical source of the second term of the theory has yet been established. Many hypotheses have been advanced in terms of its molecular origin but criticism can be levelled at most or all of them. However it appears probable that entanglements play an important role and contribute in large measure to deviations from theory. At large deformations, when the individual network chains become highly extended, further departures from theoretical predictions occur due to the finite extensibility of the network chains. At large strains the use of the Gaussian distribution of chain segment displacement length which is adopted in the derivation of simple statistical theory becomes quite unacceptable but this deficiency can be dealt with by the introduction of more suitable distributions (Treloar, 1946).

Swelling behaviour

Perhaps the application of the statistical theory to the swelling of rubbers is of more interest to readers of this book. The property of swelling in suitable low-molecular-weight liquids is an essential of all living matter, and is also characteristic of rubbers and other organic high polymers. In many respects swelling is akin to solution, and like solution is markedly dependent on the nature of the swelling liquid. Just as materials may be divided in respect to solubility into two broad classes, those which are soluble in water (hydrophilic), and those which are soluble in organic solvents (hydrophobic), high polymers may similarly be divided in respect of swelling into water swelling and organic-liquid swelling classes. The first group includes cellulose, proteins, etc., and the second rubbers and other organic high polymers. Protein rubbers swell in water.

In rubbers, swelling behaviour can be described in terms of a rather simple statistical thermodynamic model and there has been considerable success in accounting for observed behaviour. Here again we are concerned with equilibrium, in this case between a mixed polymer plus liquid phase and a single phase of pure liquid. The condition for equilibrium is that Gibbs' free energy should be a minimum in respect to changes in the composition of the phases

$$\Delta G = 0$$
$$\Delta G = \Delta H - T\Delta S.$$

This is closely analogous to the treatment of elastic behaviour. ΔH can be obtained from the heat of mixing and ΔS from a statistical treatment of the increase in entropy with diffusion of liquid molecules into the polymer. At equilibrium we get the well-known Flory–Huggins equation (Flory, 1942) which relates swelling to the degree of crosslinking of the network

$$-(\log_n(1-v_2)+v_2+\chi v_2^2)=\frac{V_1}{M_c}\left(v_2^{1/3}-\frac{v_2}{2}\right)$$

where v_2 is the volume fraction of polymer in the mixture, V_1 the molar volume of the liquid and χ a solvent–polymer interaction parameter.

A straightforward extension gives a theory of the solubility of any crystalline material in the rubber. This requires the introduction of an extra term involving the latent heat of fusion (L) and the melting temperature (T_0). The resulting relation

$$\frac{1}{T}=\frac{1}{T_0}-\frac{R}{L}\left(\log_n(1-v_2)+v_2+v_2^2+\frac{V_1}{M_c}\left(v_2^{1/3}-\frac{v_2}{2}\right)\right)$$

expresses the solubility in terms of the volume fraction of rubber.

This simple molecular theory has application in such processes as diffusion and transport of sparingly soluble substances through membranes and other polymeric materials. For example in rubber much success has recently been achieved in describing the blooming on the surface of sparingly soluble waxes incorporated in the rubber (Nah & Thomas, 1978). Here the driving force leading to the transport of waxes through the rubber results from elastic stresses developed around particles of wax precipitated from a supersaturated solution at imperfections in the rubber, leading to redissolution and reappearance on the surface in an unstressed state. The mechanism has been put on a satisfactory quantitative basis with good agreement between theory and experiment.

The theory has also been applied to the inflation of balloons and to the deformation of thin- and thick-walled tubes under extension, compression and inflation and provides criteria describing the conditions for instability. These aspects of rubbery behaviour have relevance to the performance of various biological systems such as lungs, veins etc. (Charrier & Gent, 1971; Charrier, 1974).

Composite highly-elastic materials

All of the considerations developed to account for the behaviour of rubbers apply with equal force to other materials which contain flexible long chain-like molecules and in which the molecules are free to move relative to each other. Such materials will be amorphous, homogeneous and isotropic. However there are many highly-elastic materials of technological or biological importance to which the theory is not directly applicable. These are ordinarily composite materials with a rubber-like matrix but contain dispersed particulate fillers or plastic domains which are either indigenous in the material or result from crystallization induced by strain or low temperature. The plastic domains are oriented structural units or fibres containing parallel and associated sections of the long chain molecules. Such materials will show anisotropic mechanical properties. Examples of biological substances of this type include stroma of red blood corpuscles, tendons and elastic ligaments.

The elastic behaviour of such composite materials reflects in varying degree the behaviour of the rubber-like amorphous matrix. Here again the most extensive investigations have been carried out on rubbers in which a wide variety of fillers is used to increase stiffness, hardness and strength and in which crystallites or orientated molecular domains may be present.

Elastic behaviour of filled rubbers

The primary sources of the increased stiffness (higher elastic modulus) of filled rubbers are (i) the absence of deformation within the rigid filler particles, and (ii) immobilization of the rubber matrix at the surface of the filler particles.

This is a formally identical problem to the increase in viscosity of a liquid caused by a suspension of solid particles.

The modulus thus depends on the concentration and shape of the filler particles as described, for example, by Guth–Einstein equations (Guth, 1945) for spherical particles

$$E = E_0(1 + 2.5c + 14.1c^2)$$

and asymmetric particles

$$E = E_0(1 + 0.67c + 1.62f^2e^2)$$

where c is the volume concentration of the filler, f is a factor describing the shape of the asymmetric particles as expressed by the ratio of their length to diameter, E_0 is Young's Modulus of the unfilled rubber and E is Young's Modulus of the filled rubber.

Good agreement with the predicted dependence on concentration has been obtained with rubbers containing carbon black with essentially spherical particles (Fig. 2). With finer carbon blacks containing chain-like agglomerates or clusters of particles fair agreement has been obtained using appropriate shape factors. (Mullins & Tobin, 1965.)

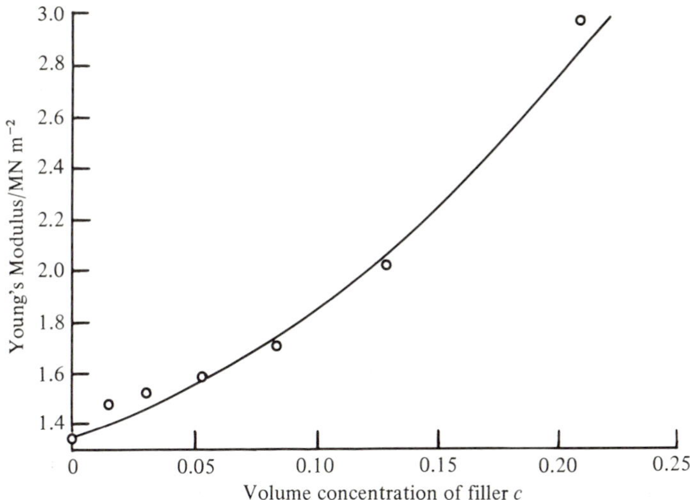

Fig. 2. Young's Modulus of natural rubber–MT black vulcanizate as a function of filler concentration (Mullins & Tobin, 1965).

Strain amplification

The simplest interpretation of these observations is that the ratio of stress to strain is increased by the presence of filler by a factor X and for small deformations and for spherical particles

$$X = \frac{E}{E_0} = 1 + 2.5c + 14.1c^2$$

E is the ratio of stress to strain for small deformations, but for larger deformations, stress and strain are no longer directly proportional and it is necessary to consider whether it is the stress, or the strain, or a function of both stress and strain which is influenced by the presence of filler. It has been found that the stress–strain properties of filled rubber can be described in terms of those of the corresponding rubber matrix without filler if the effective deformation of the rubber matrix is taken as equivalent to x times the measured overall deformation where x was determined experimentally as the ratio of the Young's moduli (Mullins & Tobin, 1965).

This concept of strain amplification produced by fillers permits results of the statistical theory already discussed to be applied to the elastic behaviour of filled rubbers, and has been found to have considerable relevance to other properties such as breaking elongation and creep (Mullins, 1978).

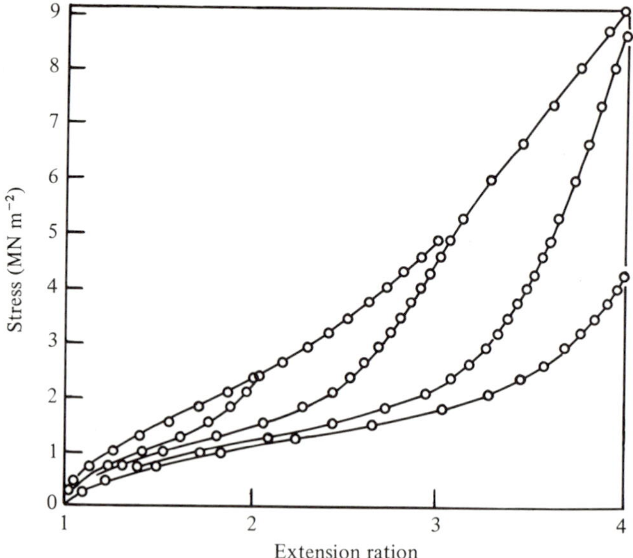

Fig. 3. Effect of previous stretching on stress–strain curves of natural rubber–MPC black vulcanizate (Mullins, 1978).

The use of a strain amplification factor to describe the increase in modulus resulting from incorporation of fillers is an important unifying concept. The concept should be equally applicable to the behaviour of other similar composite highly-elastic materials and the filler may be plastic domains or crystallites formed by portions of oriented and aligned long chain molecules.

Stress softening

It has been observed for many years that deformation results in the softening of rubber and that the initial stress–strain curve of a new sample of rubber is unique (Mullins, 1947). Most of the softening occurs in the first deformation and after a few cycles of deformation the rubber approaches a steady state with a constant stress–strain curve. The softening is usually only present at deformations smaller than the previous maximum (Fig. 3). Softening in this way occurs not only in rubbers but also in all highly-elastic materials.

Various qualitative theories have been advanced to describe the effect, and as it appears to be more pronounced in materials containing fillers discussion has mostly concentrated on breakdown of filler structure. Thus the softening has been attributed to breakage or slippage of linkages between filler and rubber, breakdown of filler aggregates, and

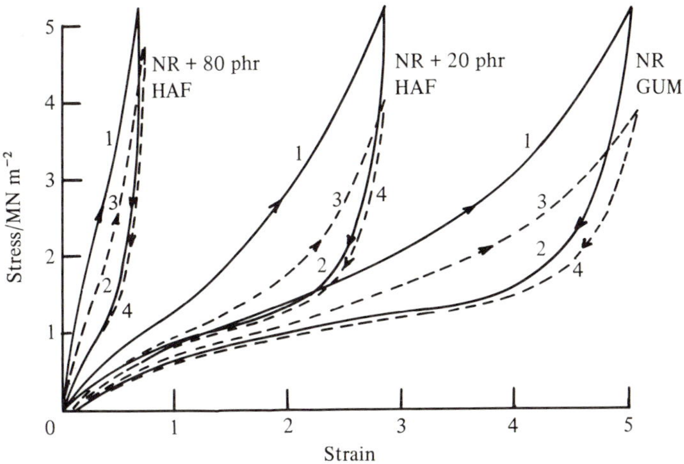

Fig. 4. Stress–strain curves of natural rubber vulcanizates. First extension 1; first retraction 2; second extension 3; second retraction 3. (Harwood, Mullins & Payne, 1965.)

breakage of network chains. Although each mechanism will lead to softening there is an absence of direct evidence to identify their individual contributions to the softening process.

New light was thrown on the source of softening resulting from previous stretching by the observation that when allowance was made for the strain amplification resulting from the presence of fillers the degree of softening was similar in magnitude for both filled and unfilled rubbers (Harwood, Mullins & Payne, 1965). Typical stress–strain curves obtained during repeated extension and retraction of natural rubbers, both with and without fillers, are shown in Fig. 4. The close correspondence between the softening of filled and unfilled rubbers when compared at the same stress is demonstrated in Fig. 5 and has led to the use of the term 'stress softening' and the conclusion that the softening was primarily due to the rubber phase itself.

It appears that stress softening is primarily a result of the rearrangement of the configuration of the molecules in the molecular network. The rearrangement involves displacement of network junctions and chain entanglements from their original random position – owing for example to localized non-affine deformation as shorter network chains become fully extended – and results in less stress being required to produce deformations that are less than the previous largest deformation. This picture is immediately evident to all who have played with a tangled mass of string.

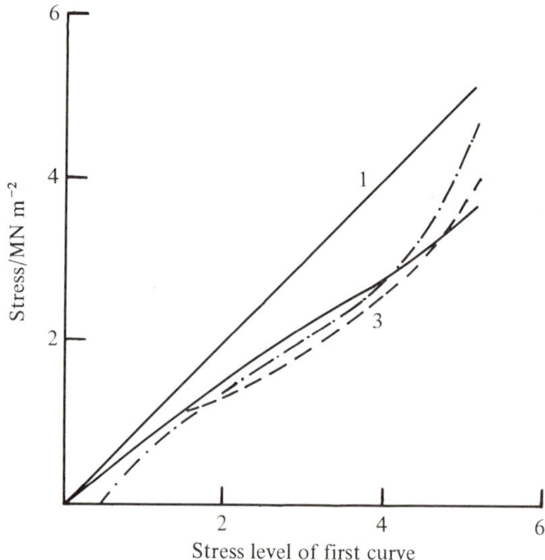

Fig. 5. Stress–strain curves in Fig. 4 obtained in first extension (1) and second extension (3) replotted as stress against stress in first extension. Full line, natural rubber; dashed and dotted line, natural rubber + 20 p.h.r. HAF black; dashed line, natural rubber + 80 p.h.r. HAF black (Harwood, Mullins & Payne, 1965).

Other mechanisms, involving breakdown of filler–filler and filler–rubber linkages, are normally relatively unimportant and only responsible for a minor part of the softening. However filled rubber filler particles and aggregates are also displaced and orientated in the network during deformation and account for the slower and less complete recovery of stress softening in filled rubbers compared with its rapid and almost complete recovery in unfilled rubbers.

Stress softening resulting from previous deformation is a frequent and widespread phenomenon. It is observed in polymeric materials with and without fillers and reflects configurational changes within the fine structure of the material which permits subsequent deformation to take place more readily.

In filled rubbers pronounced softening may also take place on very small deformations owing to the breakdown of weak aggregates of filler particles. This breakdown is essentially complete at small deformations and at larger deformations softening due to this cause is relatively small (Fig. 6).

Although the main sources of stress softening in polymers now appear to be correctly identified the problem of characterization of stress–strain behaviour remains complex. It appears that the stress related to each

THE THEORY OF VISCOELASTICITY IN BIOMATERIALS

K. L. DORRINGTON

Hertford College, Oxford, UK

Introduction

The first experiments

If it is appropriate to attribute the discovery of viscoelasticity to any one person, then it seems that the honour must go to Wilhelm Weber. It is apposite to this meeting that the discovery appears to have been made during the course of experiments on a biological material, for it was in describing his studies on silk threads that Weber (1835) fabricated a name for the phenomenon we now call viscoelasticity, which rapidly became widespread in the German scientific literature to describe similar behaviour discovered in silver wire, glass filaments, rubber and other materials (Braun, 1891).

Weber observed that his silk threads obeyed the law of proportionality between a stretching load and the resulting elongation (Hooke's Law), but only for a short time after applying the load. If the load were left applied for a long time then the elongation would continually increase. To this observation he gave the name *elastische Nachwirkung* (see Meyer, 1874) which, in 1880, the distinguished yet sadly short-lived pathologist Charles Roy (1854–97) translated, if a little imprecisely, into English as the 'elasticity after-action'. Roy undertook tensile experiments on a wide range of animal tissues and found the time-dependent elongation to be universal, noting that 'of all the animal tissues, the wall of the arteries presents the elasticity after-action in the most marked degree' (Roy, 1880).

As has so often been the case with later workers viscoelastic behaviour was for Roy primarily merely a nuisance because it tended to introduce errors into his thermoelastic measurements. 'Care must constantly be taken', he warns, 'that error is not introduced owing to the disturbing influence' – advice we do well to heed and which is pertinent coming from a man of whom Sir Charles Sherrington said: 'As an

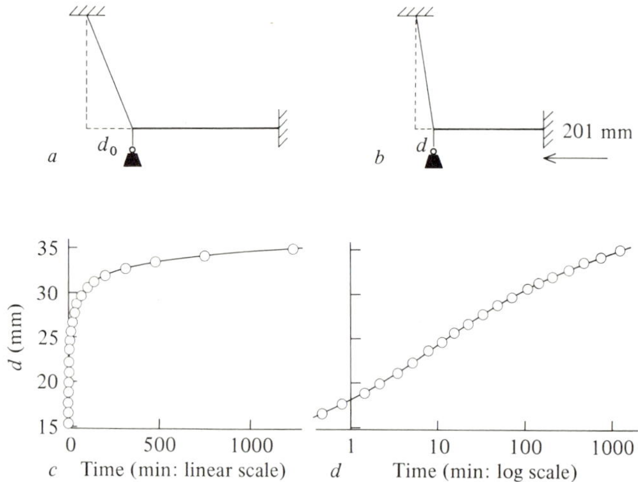

Fig. 1. Time-dependent elongation of silk thread; *a*, diagram of Weber's apparatus in resting position; *b*, initiation of experiment; *c*, *d* recorded in linear time; *d*, the same data recorded in log time. From Weber (1841).

operator in the laboratory he had no equal in this country' (Granit, 1966). Despite these difficulties, Roy succeeded in being the first to discover that pliant tissues from the body share the same thermoelastic idiosyncrasy as natural rubber, a fact we are now beginning to take for granted.

Weber refined his experimental technique and then published some results for silk threads which well illustrate time-dependent elastic behaviour (Weber, 1841). Fig. 1*a* depicts Weber's apparatus in which the thread under test lay horizontally under tension attached on the right to a moveable support and on the left to a tie by which a weight hung from the ceiling. (The weight was immersed in a bowl of water to damp its oscillations.) In the experiment illustrated in Fig. 1 the silk thread was left for 24 h to reach a steady state in which the left end lay a distance d_0 from the vertical passing through the ceiling support. At a time designated zero the right end of the thread was displaced rapidly by 201 mm as indicated in Fig. 1*b*. Subsequently the distance d was recorded as a function of time. The effect of the displacement is a rapid unloading of the silk initiating a recovery which appears as the increasing distance d plotted in Figs. 1*c*, *d*.

In Weber's second experiment (Fig. 2) the end support was displaced to the right, this time causing a sudden rise in tension in the silk. In this case the thread slowly elongates with time and d decreases.

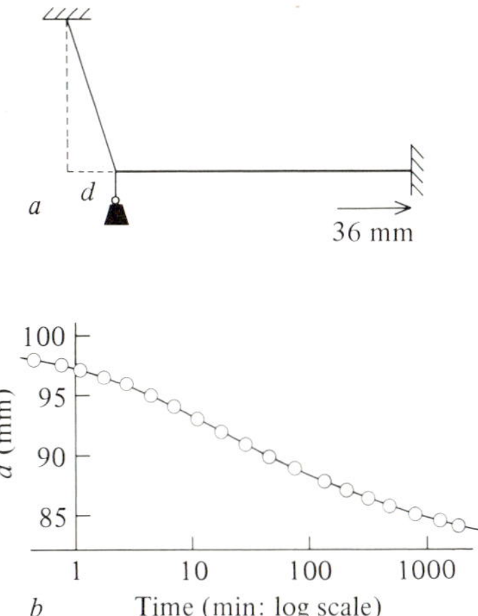

Fig. 2. Time-dependent elongation of silk thread: *a*, initiation of the experiment *b*, *d* recorded in log time. From Weber (1841).

It is important to note that these drifts with time are not once-and-for-all irreversible flow-like events but are reproducible. To reproduce them each experiment must be preceded by a long period under steady conditions during which the specimen's memory of past events is erased. This is typical of viscoelastic behaviour in general.

A second feature to note is the time scale of the changes. When plotted as a linear function of time, as in Fig. 1*c*, there is a rapid drift for short times and then an asymptotic approach to equilibrium. It is only when data of this kind are represented on a logarithmic scale in time (Figs. 1*d*, 2*b*) that they become more manageable. This too is typical of viscoelastic behaviour.

(An unusual property of Weber's apparatus can be noted. With the geometry adopted the tension applied to the horizontal thread is proportional to the distance *d* (for small angles of inclination of the ceiling tie to the vertical). The tension is thus a linear function of specimen length. This method of loading has not been favoured in subsequent tensile experiments because of the difficulties it usually introduces into a theoretical description of the experiment. There are occasions, however, on which a known linear relationship between applied tension and length can simplify the analysis. Such would be the case, for example, if the viscoelasticity were governed by a single relaxation time (see Fig. 7).)

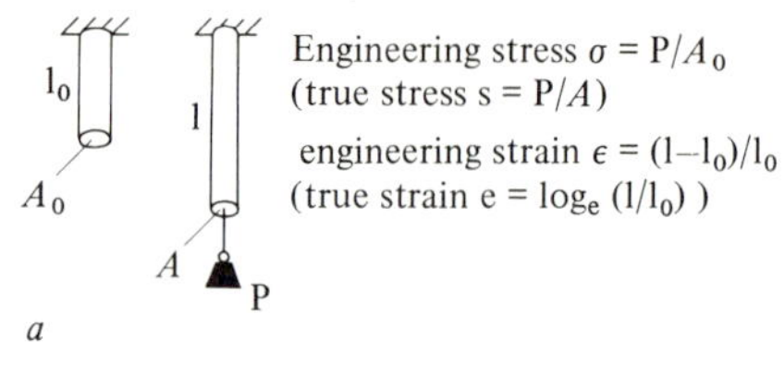

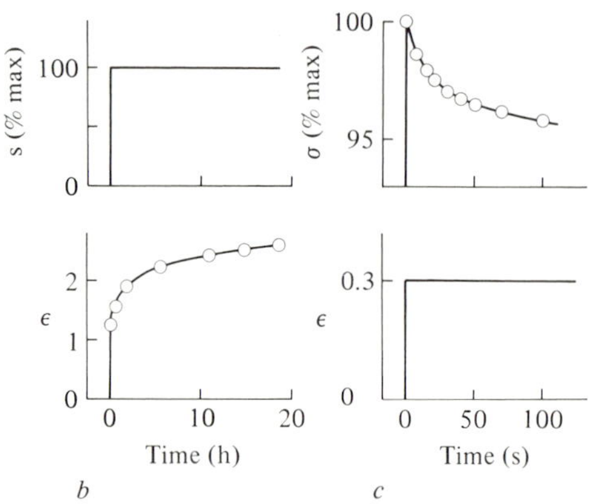

Fig. 3. The methods of creep and stress relaxation: *a*, parameters for quantifying loading (stress) and elongation (strain); *b*, creep of mesogloea from the jellyfish *Cyanea capillata* (data from Alexander, 1964); *c*, stress relaxation of isolated elastin from porcine aortic wall.

Creep and stress relaxation

A more convenient mathematical description is afforded by the experiments of creep and stress relaxation than by the method of Weber. These experiments are illustrated in Fig. 3.

Fig. 3*a* depicts a tensile specimen in the rested and extended state and defines parameters in common use for the normalized quantification of loading (stress) and elongation (strain). For materials in which the elongations are a small fraction of the original length it is common to define an engineering stress in terms of the cross-sectional area of the *undeformed* specimen A_0. For larger elongations, in which the cross-sectional area of the deformed specimen A differs markedly from the original area, a true stress may permit a better interpretation of the observed behaviour. The choice of a definition of strain tends not greatly to influence ease of interpretation. Engineering strain is widely used for both highly extensible and less extensible materials. This indifference to choice between engineering and 'true' strain rests on the fact that they are directly related and yield equivalent information. This is not the case for the two definitions of stress.

In a creep experiment a rested specimen is rapidly loaded so as to bear a constant stress and the strain is monitored with time. The example of creep given in Fig. 3*b* is taken from the data of Alexander (1964) and is obtained from a specimen of body wall of jellyfish. The strains being very large in this experiment, a constant true stress was imposed. It should be noted that attainment of constant true stress is achieved experimentally only by making assumptions about the specimen volume. It is common to assume a constant volume during deformation and this may give rise to substantial errors when testing water-swollen biological materials in which large volume changes can occur. (For a brief discussion of volume changes in elastin see McCrum & Dorrington (1976).)

In a stress relaxation experiment it is the strain of the specimen which is rapidly imposed and held constant whilst the stress falls as a function of time. The example of stress relaxation in Fig. 3*c* is obtained from a ring of porcine aortic wall from which all components except elastin had been extracted.

These two examples illustrate two marked differences between mesogloea and elastin. First, the approach to equilibrium is much slower in mesogloea than in elastin. The time scales on the graphs differ by a factor of around 600. Second, not only does elastin in this example reach equilibrium more speedily but the change in stress is a mere 5% or so of the maximum, whereas the change in strain of the mesogloea during creep is around 200% of the initial value. We might speak of mesogloea as being more highly viscoelastic than elastin under the conditions represented.

It will be clear from these examples that viscoelasticity can be a major disturbing influence in any attempt to measure the *equilibrium* mechanical properties of a body tissue. In a stress relaxation experiment, for example, the asymptotic approach to equilibrium is such that we can more closely approximate to a final stress by waiting for a longer time, but such a wait is invariably limited by practical considerations, not least of which with biological materials is irreversible degradation of the tissue. The possibility of using changes of temperature to help overcome this difficulty will become apparent later in this paper.

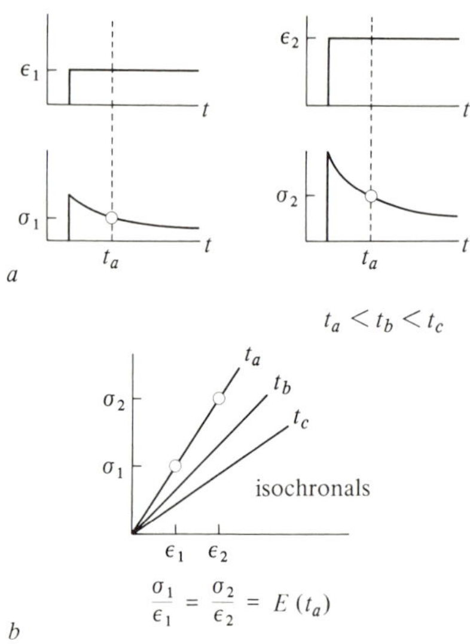

Fig. 4. Boltzmann's assumption of linearity: *a*, separate hypothetical stress relaxation experiments on the same specimen for different imposed strains; *b*, linear isochronals (one from *a*) with gradients equal to the relaxation modulus $E(t)$.

Phenomenology

The principle of Boltzmann

For forty years after the experiments of Weber in the 1830s there was lacking a coherent rational basis for the quantitative description of viscoelasticity. Many investigators, like Weber himself, fitted their data with empirical formulae which were of no wider use than for the individual experiments for which they were computed. It was in 1874 that Ludwig Boltzmann put forward his superposition principle (*Princip der Superposition*), which has formed a sound basis for subsequent theory. In the following the principle is considered as it applies to the stress relaxation experiment (Boltzmann, 1874, 1876). It can be analogously applied to creep.

Fig. 4 illustrates the first of two assumptions made by Boltzmann, the assumption of *linearity*. If two separate stress relaxation experiments were to be performed on the same rested specimen, in the first one a strain ε_1 being imposed and in the second one a strain ε_2 being imposed, then in each experiment, it is assumed, the resulting stress will bear the same ratio to strain if the stress is measured *at the same time* after application of the strain (Fig. 4*a*). This ratio of stress to strain is termed

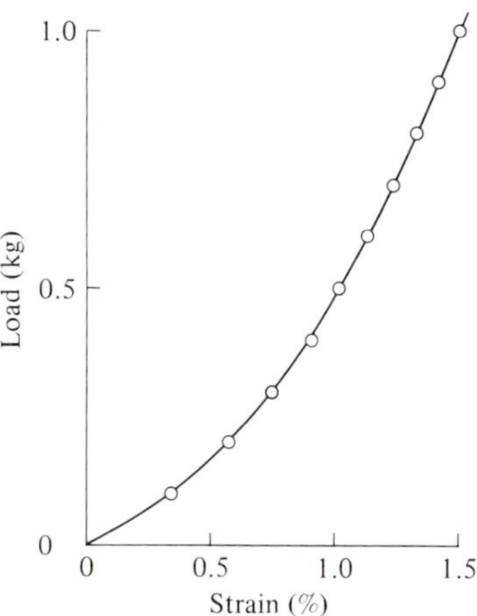

Fig. 5. A markedly non-linear creep load–strain isochronal for time 20 s from human flexor digitorum tendon. From Cohen *et al.* (1976).

the relaxation modulus $E(t)$. It is clearly a function of time. It follows that, if a series of stress relaxation experiments is completed with the modulus being measured at different times, then the resulting stress–strain *isochronals* will be straight lines with gradients decreasing with time (Fig. 4*b*).

The importance of this isochronal check on linearity can hardly be overemphasized because it has been so infrequently undertaken when biological materials have been studied. And yet the application of *linear* viscoelasticity theory can only be wholly justified when the system is linear in this sense. Non-linear stress–strain curves are a feature of many tissues and for such tissues the linear theory is at best an approximation. It should be emphasized here that stress–strain curves measured at a constant strain rate, a procedure commonly employed, are not isochronals as defined above. Fig. 5 is an isochronal load–strain curve taken from creep experiments of Cohen, Hooley & McCrum (1976) on human finger tendon and exemplifies the marked non-linearity of this particular tissue. Indeed, the observation of a highly non-linear isochronal for collagenous tissue has led Hooley & Cohen (1979) to propose a new viscoelastic model of the straightening of crimps in the collagen molecule. In this model the low strain region of the isochronal curve is dictated by shear in the matrix of mucopolysaccharide gel and only the high strain region by the collagen fibres in direct tension. The degree of success for this model may go some way to establishing whether or not the sharply angled crimps observed by microscopy are merely artifacts as has been claimed (Viidik, 1978).

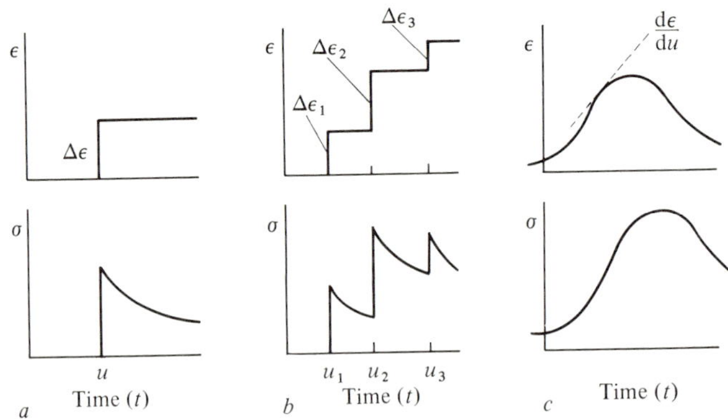

Fig. 6. Boltzmann's Principle of Superposition: *a*, time course in a single stress relaxation experiment; *b*, assumed linear superposition of multiple experiments running concurrently; *c*, generalization using the calculus to an arbitrary strain history. Full explanation in text.

The second of Boltzmann's assumptions is that there is a linear superposition of stresses on a material for *concurrently running* stress relaxation experiments which have *begun at different times*. This is illustrated in Fig. 6. For a single relaxation with a strain $\Delta\varepsilon$ imposed at time $t = u$ the stress for $t > u$ is given by the product of relaxation modulus and strain (Fig. 6*a*)

$$\sigma = E(t - u)\Delta\varepsilon. \tag{1}$$

It is then assumed that, for a strain history comprising a series of strain steps $\Delta\varepsilon_i$ at times u_i, the stress for times after the last step is the sum of stresses of individual stress relaxation experiments (Fig. 6*b*)

$$\sigma = \sum_i E(t - u_i)\,\Delta\varepsilon_i. \tag{2}$$

In the limit, as the number of steps tends to infinity and their height becomes limitingly small the calculus expresses the stress for any strain history as an integral summation (Fig. 6*c*)

$$\sigma = \int_{u=-\infty}^{u=t} E(t - u)\ \mathrm{d}\varepsilon/\mathrm{d}u\ \mathrm{d}u. \tag{3}$$

In this integral the strain history is represented by $\mathrm{d}\varepsilon/\mathrm{d}u$, which is actually the gradient of its tangent (see Fig. 6*c*).

Using the Boltzmann Superposition Principle we see, therefore, that the stress response of a specimen experiencing an arbitrary strain history

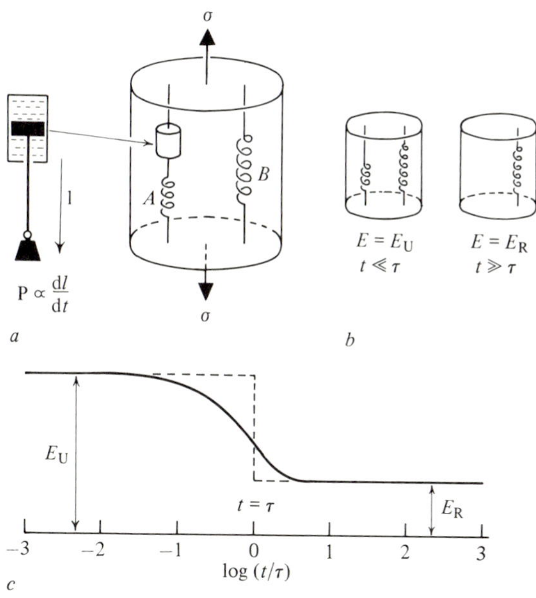

Fig. 7. The Standard Linear Solid: *a*, property of the Newtonian dashpot and its incorporation with two spring elements to form the model; *b*, behaviour of the model for rapid and slow loadings; *c*, the relaxation modulus displayed by the model and the approximate step relaxation modulus (interrupted line).

can be found if the relaxation modulus is known. It is this generalization which has formed the foundation for the theory of viscoelasticity.

The relaxation modulus $E(t)$

It has been shown that a stress response to a quite general strain history can be predicted using eqn (3) if the relaxation modulus is known. The form of $E(t)$ is now considered and how, in the most general case, it can be expressed in terms of a distribution of relaxation times representing physical processes in the material.

In order that the form of $E(t)$ be interpreted we must ask what kind of physical process in a material can give rise to the relaxation behaviour. In Roy's discussion and subsequently it has been customary to think of the action within a tissue of a 'molecular friction' giving rise to a time-dependent flow. The most simple model of friction is the first of those proposed by Newton (1684, 1687) and subsequently termed Newtonian. In a system displaying Newtonian friction the force resisting a motion is proportional to the speed of the motion. Such a system is often called a simple dashpot and is depicted in the left of Fig. 7*a*. A model solid incorporating such a frictional element has two further requirements. First, it must display a measure of instantaneous elasticity. Second, it must give reversible rather than flow deformation. To incorporate these requirements in the simplest form we envisage one linear elastic element, or spring, A, in series with the dashpot and one in parallel, B. The model thus formed is known as the Standard Linear Solid (Maxwell, 1868; Zener, 1948). The differential equation describing its behaviour is

$$\mathrm{d}\sigma/\mathrm{d}t + \sigma/\tau = E_{\mathrm{U}}\mathrm{d}\varepsilon/\mathrm{d}t + E_{\mathrm{R}}\varepsilon/\tau \tag{4}$$

where E_{U}, E_{R} and τ are constants.

For very rapid deformation of the Standard Linear Solid the derivative terms dominate and eqn (4) yields

$$\mathrm{d}\sigma/\mathrm{d}t \simeq E_{\mathrm{U}}\mathrm{d}\varepsilon/\mathrm{d}t \tag{5}$$

and hence

$$\sigma \simeq E_{\mathrm{U}}\varepsilon. \tag{6}$$

The solid displays an *unrelaxed modulus* E_{U} under such loading. The behaviour is as though the dashpot were replaced by an inextensible tie (Fig. 7*b*).

For very slow deformations the derivative terms become small and eqn (4) yields directly

$$\sigma \simeq E_{\mathrm{R}}\varepsilon. \tag{7}$$

The solid displays a *relaxed modulus* E_{R} under such loading. The behaviour is then as though only spring B remained (Fig. 7*b*).

Consider now the response of this Standard Linear Solid in a stress relaxation experiment. The relaxation modulus is obtained from eqn (4) by setting $\mathrm{d}\varepsilon/\mathrm{d}t = 0$ and integrating

$$E(t) = E_{\mathrm{R}} + (E_{\mathrm{U}} - E_{\mathrm{R}})\exp(-t/\tau). \tag{8}$$

The function is plotted in Fig. 7*c*. Note that the horizontal scale is logarithmic in time. When time t is in the region of τ the modulus decays from the unrelaxed value to the relaxed value and τ is known as the *relaxation time*. It should further be noted that the decay of the modulus occurs mainly within a total range of about 1.5 decades along the time axis. This implies that for a relaxation under way at a time $t = 1\,\mathrm{s}$, for example, we would expect it to be near completion at $t = 1 \times 10^{1.5} \simeq 30\,\mathrm{s}$.

The dotted function in Fig. 7*c* is a step approximation to the relaxation modulus which may sometimes be used as a mathematical

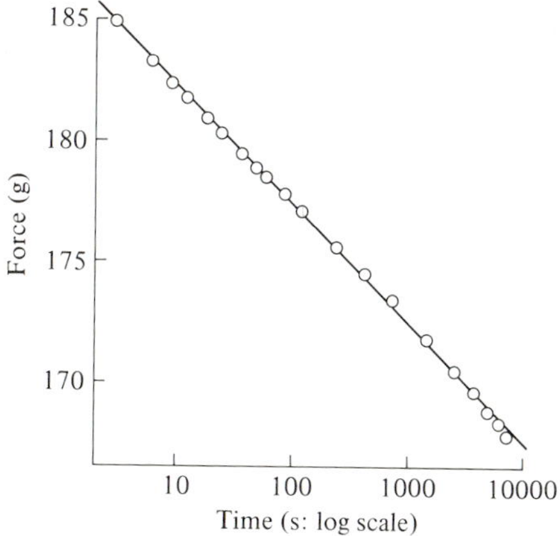

Fig. 8. Force–log time data for isolated porcine aortic elastin in a stress relaxation experiment. The applied strain was 30% with the specimen immersed in water at 70°C. (Some of the data are presented on a linear time scale in Fig. 3*c*.) Being nearly linear over approximately four decades these data show that the relaxation of elastin cannot be represented by a Standard Linear Solid in which relaxation is complete within about 1.5 decades.

simplification and will be referred to as a 'step relaxation modulus'. Physically it represents the absence of a dashpot in the model plus a sudden removal of spring A at $t = \tau$.

The simple model of the Standard Linear Solid has been found to describe well some materials. Alexander (1962) found good agreement for the body wall of sea anemones (Anthozoa) using creep, though the body wall of the quite closely related jellyfish (Scyphozoa) showed relaxation over a much broader range of time than can be accounted for by the single relaxation time model (Alexander, 1964). The data of Weber (Figs. 1, 2) showed relaxations in silk extending to three decades and beyond. Fig. 8 shows typical stress relaxation data for elastin extending nearly linearly over almost four decades in time. It is clear that the majority of biological materials can only be represented by a broader spread of relaxation times than one alone. This is considered in the following section.

The distribution of relaxation times $\phi(\ln \tau)$

Let it be considered that a small fraction only, $\phi(\ln \tau)\ \Delta(\ln \tau)$ (Fig. 9), of the total modulus to be decayed away $(E_U - E_R)$ does so with a relaxation time whose natural logarithm lies between $\ln \tau$ and $\ln \tau + \Delta$ $(\ln \tau)$. (It has become customary to write ϕ as a function of the natural logarithm of τ $(\ln \tau)$ rather than the logarithm to base 10 $(\log \tau)$ or even τ

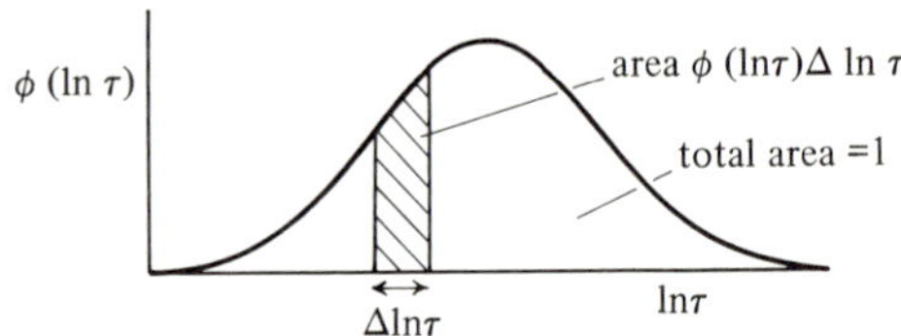

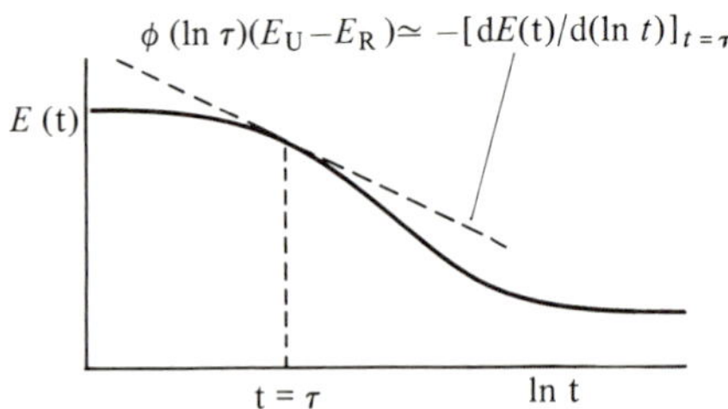

Fig. 9. The distribution of relaxation times and its approximate derivation from the relaxation modulus curve by use of the Alfrey approximation which gives $\phi(\ln \tau)$ $(E_U - E_R)$ as the negative gradient of the tangent to $E(t)$ for $t = \tau$.

itself. This convention is continued here.) From eqn (8) we see that this fraction contributes to the total modulus

$$\Delta(E(t) - E_R) = \phi(\ln \tau)(E_U - E_R)\exp(-t/\tau)\Delta(\ln \tau). \qquad (9)$$

Upon summing all contributions of this kind to the modulus over all possible relaxation times from 0 to ∞ we obtain

$$E(t) - E_R = \int_{-\infty}^{\infty} \phi(\ln \tau)(E_U - E_R)\exp(-t/\tau)\mathrm{d}(\ln \tau). \qquad (10)$$

As it stands this integral equation is usually difficult to solve for $\phi(\ln \tau)$ given a measured $E(t)$. If however we use the step relaxation modulus approximation to which reference was made earlier we may simplify to obtain an approximate solution. We therefore set $\exp(-t/\tau) = 1$ for $t < \tau$ and $\exp(-t/\tau) = 0$ for $t > \tau$ and eqn (10) becomes

$$E(t) - E_R \simeq \int_{\ln t}^{\infty} \phi(\ln \tau)(E_U - E_R)\mathrm{d}(\ln \tau). \qquad (11)$$

Upon differentiation eqn (11) yields

$$\phi(\ln \tau)(E_U - E_R) \simeq -[\mathrm{d}E(t)/\mathrm{d}(\ln t)]_{t=\tau}. \qquad (12)$$

This is the simple result known as the Alfrey approximation (Alfrey & Doty, 1945), which states that $\phi(\ln \tau)(E_U - E_R)$ is measured for relaxation time τ by taking the negative of the gradient of the relaxation modulus curve at a time $t = \tau$ (Fig. 9). From Fig. 7*c* we may estimate that this approximation, if applied to the relaxation modulus for a Standard Linear Solid would suggest a distribution $\phi(\ln \tau)$ of width ~1.5 decades whereas $\phi(\ln \tau)$ is in fact singular for this solid. The approximation will consequently only be applicable if $E(t)$ decays over substantially more than 1.5 decades in time.

Consider again Fig. 8. For this linear graph the Alfrey approximation gives a $\phi(\ln \tau)$ which is constant from $\tau \simeq 1$ s to $\tau \simeq 10^4$ s. This may be termed a 'box-like' distribution of relaxation times. Zatzman, Stacy, Randall & Eberstein (1954) used such a box-like distribution to describe stress relaxation in arterial segments. In their work on carotid and umbilical arteries the relaxation was attributed to smooth muscle cells.

The distribution $\phi(\ln \tau)$ shown in Fig. 9 is in fact a Gaussian error curve and hence its integral, giving $E(t)$, is the normal probability function. There are theoretical grounds for taking a curve of this kind (Feltham, 1955). By assuming a normal distribution in the specimen of lengths of parts of molecules which move to induce relaxation Feltham has derived a distribution of relaxation times which is gaussian in the logarithm of τ. Such a Gaussian distribution was first proposed by Wiechert (1893) on empirical grounds (to describe stress relaxation data for torsion of glass threads) and has been found to occur in many materials, e.g. in wheat dough (Grogg & Melms, 1958).

The Alfrey approximation has been used by Haughton, Sellen & Preston (1968) to derive a relaxation time distribution from stress relaxation experiments on the cell wall of the plant *Nitella opaca*. In Fig. 10 are plotted the relaxation modulus at 0°C (derived in fact from relaxation data taken at four temperatures) and the distribution function measured as the negative gradient of the modulus curve. We can envisage a very broad Gaussian-like curve of which the calculated curve is a mere fraction on the right side of the maximum. If such a curve is indeed part of a Gaussian-like distribution then it is clear that the whole distribution spreads over many (~10) decades of time and that the relaxation modulus decays over that same large number of decades. Such a conclusion is also apparent for the relaxation time spectrum for wood published by Moriizumi, Fushitani & Kaburagi (1973) of which the experimental range in time spans only the *left* tail or low-time region of the whole distribution.

The data of Fig. 10 serve to illustrate one very great weakness of the

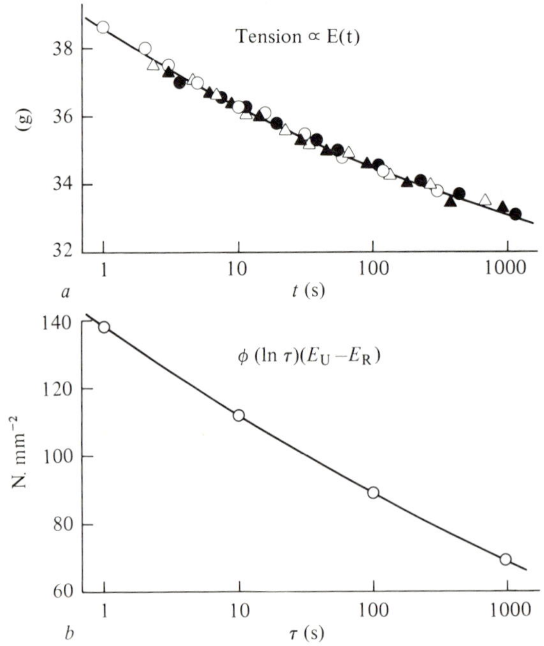

Fig. 10. The use of the Alfrey approximation: *a*, relaxation tension (proportional to $E(t)$) at 0°C for cell wall of *Nitella opaca* (derived from experiments at 0°C (open circles), 20°C (open triangles), 35°C (filled triangles) and 50°C (filled circles) using time-temperature superposition); *b*, the distribution of relaxation times $\phi(\ln \tau)(E_U - E_R)$ derived from the negative gradient in *a*. Modified from Haughton *et al.* (1968).

stress relaxation experiment in the study of viscoelasticity. If there is indeed a Gaussian-like distribution of relaxation times in *Nitella*, then most of the curve lies along the log time scale for times well below 1 s. Moreover, quite a large part of the distribution lies at times much longer than 1000 s. In practice the short times cannot be studied because of the finite time needed to strain the specimen and the long times cannot be studied because life is too short. It is necessary to resort to other methods to study more of the whole distribution.

The problem is represented in Fig. 11, together with two possible solutions. In Fig. 11*a* is shown a hypothetical (Gaussian) distribution of relaxation times centred at $\tau = 1$ s and spreading over about 12 decades in time. The range of time suitable for study in a stress relaxation (or creep) experiment is labelled, extending over four decades in time.

One method for studying more of the distribution at the short time end is to use the technique of forced vibration in which a specimen is strained with a simple harmonic motion for frequencies above about

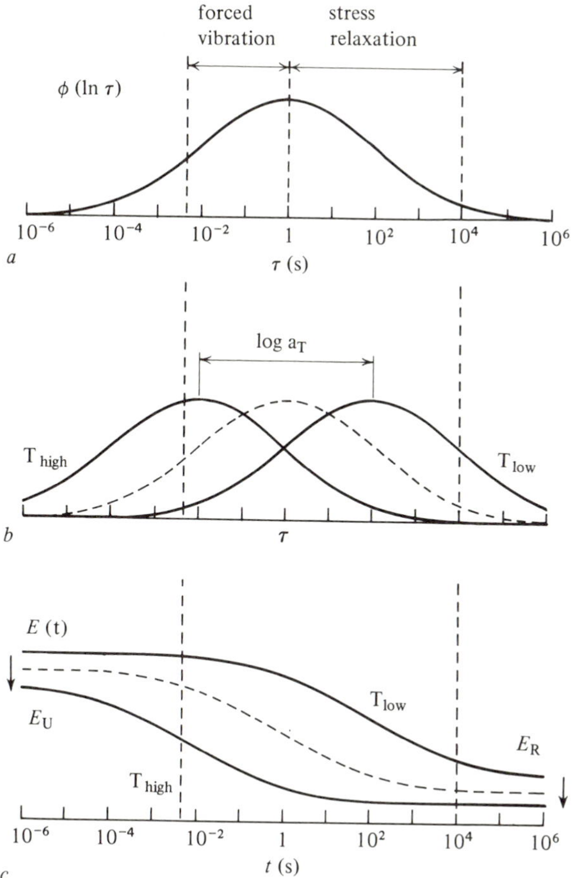

Fig. 11. *a*, The experimental range in time superimposed on a hypothetical Gaussian distribution of relaxation times centred at $\tau = 1$ s and spreading over approximately 12 decades in time; *b*, shifts with temperature of the distribution function; *c*, shifts with temperature of the corresponding relaxation modulus. Arrows indicate decreasing limiting moduli with temperature.

1 Hz. The permissible range on the τ scale for such studies is shown in Fig. 11*a* and further considered below.

The second method for studying a broader range of the distribution than is enclosed within the experimental range is to make use of the displacement of the distribution along the ln τ axis which occurs as the temperature of the specimen is changed. This shift is depicted in Fig. 11*b* for changes to both high and low temperatures. At high temperatures the right tail of the distribution enters the experimental region and for low temperatures the left tail enters the region. Corresponding shifts

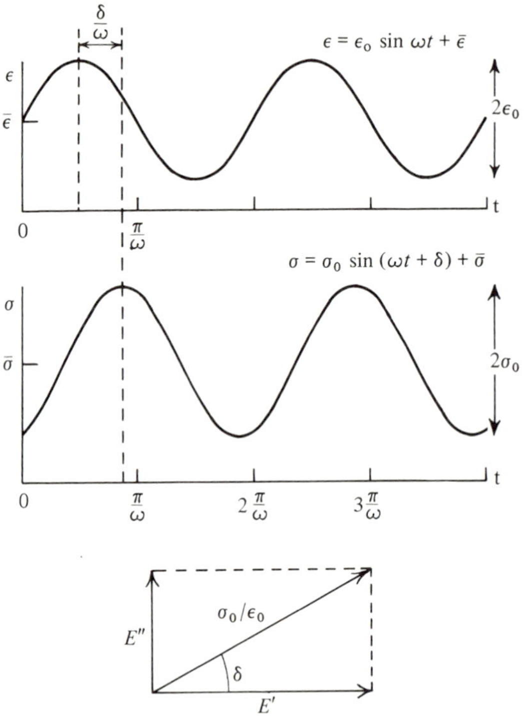

Fig. 12. The response of a linear viscoelastic solid to a sinusoidally oscillating strain. Full explanation in text.

of the relaxation modulus curve are shown in Fig. 11*c* and are further considered below.

The method of forced vibration

In the forced vibration experiment continuously oscillating sinusoidal strain is applied to a specimen and the resulting stress oscillations are monitored (Fig. 12). For specimens which buckle in compression, such as the body connective tissues, the oscillating strain of amplitude ε_0 is applied about a mean tensile strain $\bar{\varepsilon}$ where $\varepsilon_0 < \bar{\varepsilon}$. The strain may be represented by

$$\varepsilon = \varepsilon_0 \sin \omega t + \bar{\varepsilon} \tag{13}$$

where ω is the frequency of oscillation in radians per second. (The relation between frequency f in Hertz and ω is given by $\omega = 2\pi f$.) For a linearly viscoelastic material the stress response is also sinusoidal but lags the strain by a phase angle δ or a time δ/ω

$$\sigma = \sigma_0 \sin(\omega t + \delta) + \bar{\sigma}. \tag{14}$$

Expanding the sine function in eqn (14) we obtain

$$\sigma = \sigma_0 \cos\delta \cdot \sin\omega t + \sigma_0 \sin\delta \cdot \cos\omega t + \bar{\sigma}. \tag{15}$$

Division through by ε_0 obtains

$$\sigma/\varepsilon_0 = E' \sin\omega t + E'' \cos\omega t + \text{constant}, \tag{16}$$

where $E' = \sigma_0/\varepsilon_0 \cos\delta$ and $E'' = \sigma_0/\varepsilon_0 \sin\delta$. A geometric relationship between E', E'' and δ is shown in Fig. 12. We note that

$$\tan\delta = E''/E'. \tag{17}$$

E' provides a measure of the stress which lies in phase with the applied strain. It is often termed the *storage modulus* because it defines the energy stored in the specimen. E'' provides a measure of the stress which lies out of phase with the applied strain. It is often termed the *loss modulus* because it defines the dissipation of energy (Ward, 1971). In a perfectly elastic Hookean solid E'' would be zero and the stress and strain would remain perfectly in phase.

The significance of the parameters E', E'' and $\tan\delta$ becomes apparent when the response of the Standard Linear Solid is considered (Fig. 7, eqn (4)). For this model

$$E' = E_R + (E_U - E_R)\,\omega^2\tau^2/(1 + \omega^2\tau^2), \tag{18}$$

$$E'' = (E_U - E_R)\,\omega\tau/(1 + \omega^2\tau^2) \tag{19}$$

and

$$\tan\delta = \omega\tau(E_U/E_R - 1)/(1 + \omega^2\tau^2 E_U/E_R). \tag{20}$$

These functions are plotted in Fig. 13*a* with E_U/E_R taken to be 3 (as in Fig. 7). Comparison can be made with data for water swollen elastin in Fig. 13*b* from Gosline & French (1979). It was noted earlier, with reference to Fig. 7*c* and Fig. 8, that the relaxation of elastin extends over a much broader time scale than can be accounted for by the single relaxation time model. This conclusion is confirmed by Fig. 13, in particular by comparison of the $\tan\delta$ responses.

For a distribution of relaxation times eqn (18) and eqn (19) become generalized to:

$$E' = E_R + \int_{-\infty}^{\infty} (E_U - E_R)\phi(\ln\tau)\omega^2\tau^2/(1 + \omega^2\tau^2)\,\mathrm{d}(\ln\tau) \tag{21}$$

$$E'' = \int_{-\infty}^{\infty} (E_U - E_R)\phi(\ln\tau)\omega\tau/(1 + \omega^2\tau^2)\,\mathrm{d}(\ln\tau), \tag{22}$$

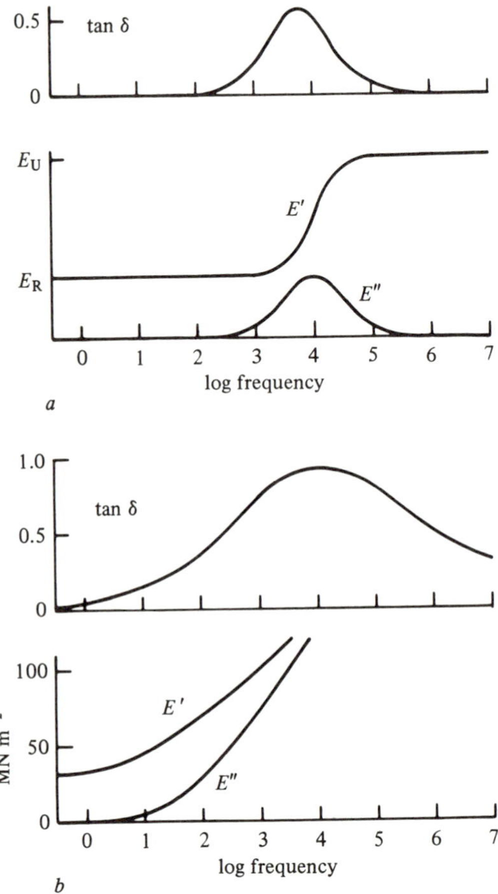

Fig. 13. Responses to forced vibration: *a*, for a Standard Linear Solid ($E_U/E_R = 3$, as in Fig. 7*c*; $\tau = 10^{-4}/2\pi$); *b*, for elastin hydrated at 55°C with reference temperature 36°C, derived from data over a three-decade frequency range using time-temperature superposition (From Gosline & French, 1979). Note that for the glass transition of elastin represented here $E_U/E_R \simeq 10^3$ and therefore little of the modulus response can be presented on the linear scale of *b*.

and to a similar degree of approximation to that of eqn (12) $\phi(\ln \tau)$ may be expressed in terms of E' and E'' as follows

$$\phi(\ln \tau)(E_U - E_R) \simeq [dE'/d(\ln \omega)]_{\tau=1/\omega} \tag{23}$$

$$\simeq (2/\pi)[E'']_{\tau=1/\omega}. \tag{24}$$

The loss modulus therefore gives an approximate immediate measure of the distribution of relaxation times.

The maximum frequency for forced vibration experiments must be below the frequencies of resonances in specimen and apparatus (Ward,

1971). In the experiments of Gosline & French (1979) a maximum frequency of 200 Hz was found convenient. This value corresponds to a relaxation time of 1/200 = 0.005s and is shown in Fig. 11 as the lower limit in time of the whole experimental range. The range can be further extended using resonance and wave-propagation methods. The latter has been a popular method for study of the whole systemic vasculature. Neither technique is considered further here.

The method of forced vibration has been used to study viscoelasticity of whole arteries (Bergel, 1961; Learoyd & Taylor, 1966; Apter & Marquez, 1968*a*), isolated elastin (Gotte, Mammi & Pezzin, 1968, Pezzin & Scandola, 1976; Scandola & Pezzin, 1978) and mesogloea (Gosline, 1971). The relation between the forced vibration method and stress relaxation in biomaterials has been discussed by Apter & Marquez (1968*b*).

Causative mechanism

Time–temperature equivalence

The effect of temperature on relaxation modulus has been depicted in Fig. 11*c*. Upon raising the temperature the process of relaxation is accelerated and the modulus curve shifts to lower times; the decay from E_U to E_R occurs more rapidly. Upon lowering the temperature the relaxation is slowed and the curve shifts to the right. It has been widely found that this shift with temperature can be accounted for by a corresponding shift of the relaxation time distribution (Fig. 11*b*). There is a need, furthermore, to allow for changes with temperature in the limiting moduli E_U and E_R (McCrum & Morris, 1964). For certain relaxations it is adequate to assume that both moduli show the same temperature dependence of the form

$$E_U^T/E_U^{T_0} = E_R^T/E_R^{T_0} = b_T. \qquad (25)$$

If this is the case then we may expect two relaxation modulus curves taken at temperatures T and T_0 to coincide when there is a horizontal shift of one of $\ln a_T$ along the $\ln t$ axis and a vertical shift of $\ln b_T$ up the modulus axis *when the modulus is plotted along a logarithmic scale.* Usually $\ln a_T \gg \ln b_T$ and for approximate work the vertical shift may be ignored.

A consequence of the shift with temperature of the modulus curves along the log time axis is that a relaxation modulus measured at some constant reference time t_r shows a similar dependence on temperature to

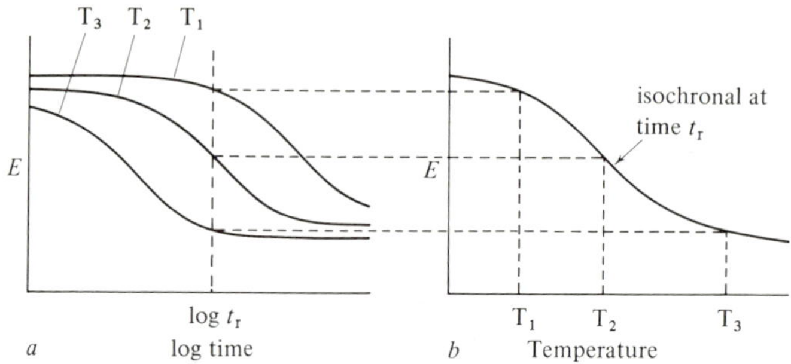

Fig. 14. 'Equivalence' of time and temperature in determining modulus: *a*, isothermal relaxation modulus curves for three temperatures $T_3 > T_2 > T_1$; *b*, isochronal relaxation modulus for time t_r constructed from *a*. It can be seen that a relaxation appears as a decay of modulus with increase of temperature in a manner similar to its decay with time.

that shown by the modulus plotted as a function of log time for a constant temperature. In short the isochronal modulus shows a similar functional dependence on temperature to that of the isothermal modulus on logarithmic time. The 'time–temperature equivalence' is depicted in Fig. 14. It can be seen that the shape of a modulus isochronal depends upon two main factors: the shape of the isothermal modulus and the degree of shift with temperature of the curves along the log time axis. The shift is given by the parameter log a_T (Fig. 11*b*). It should be noted that similar time–temperature equivalence is found for E', E'' and tan δ in forced vibration experiments.

The glass transition and secondary transitions

It has been tacitly assumed so far that any specimen will display only one relaxation. The existence of a plurality of relaxations may become apparent when a specimen (isochronal) modulus is plotted over a wide range of temperature. Fig. 15*a* shows a typical plot for an amorphous polymer. Individual relaxation mechanisms can be distinguished by the separate step reductions in modulus to which they give rise as the temperature increases (cf. Fig. 14*b*).

For an amorphous polymer by far the largest relaxation is the glass–rubber transition (often termed just the 'glass transition' or α-relaxation). In this transition the modulus falls by a factor of around 10^4 as a consequence of which the modulus must be plotted logarithmically if it is to be readily presented graphically (Fig. 15). The glass transition represents the setting-in of a molecular mobility which gives the solid a liquid-like deformability constrained only by cross-links between long chains.

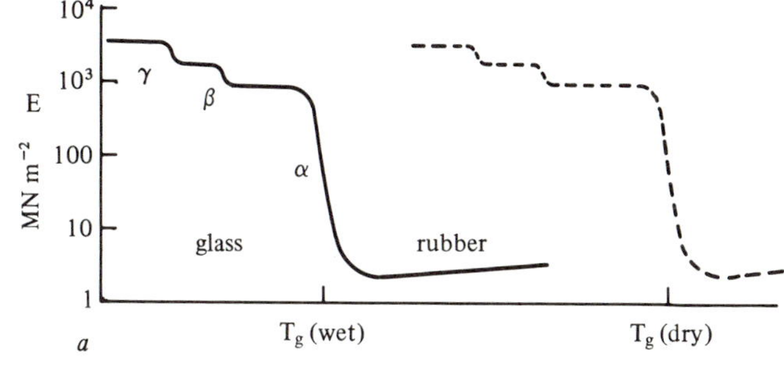

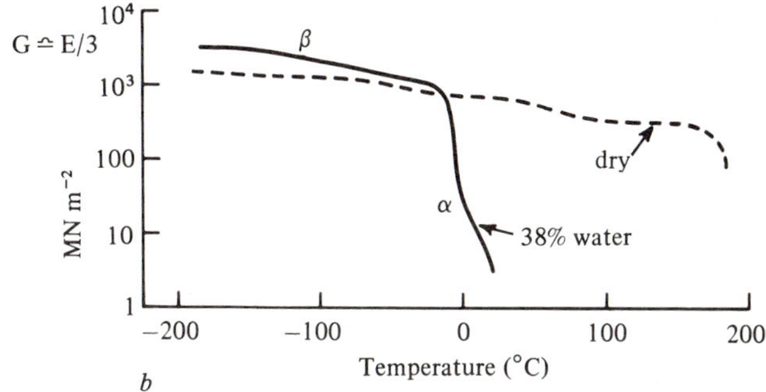

Fig. 15. Temperature dependence of isochronal modulus showing multiple transitions α (glass transition), β, γ etc: *a*, hypothetical typical amorphous water-swollen polymer with distinct transitions, showing also the shift of the modulus upon drying (interrupted line); *b*, shear modulus from forced vibration experiments (approximately equal to *E*/3) for elastin (Gotte *et al.*, 1968). The solid line is for a water content of 38% and shows a distinct α and a broad β transition. The curve for dry elastin (interrupted line) is shown for comparison. Note the 200°C change in T_g for elastin upon drying.

Other, secondary, transitions exist at low temperature involving relatively small changes in modulus. These are termed β- and γ-relaxations etc. They are attributable to more localized arrangements within polymer chains than those occurring in the glass transition and are most similar to relaxations in crystalline polymers such as collagen, which can show no glass transition.

It has been found that the shapes of the glass transition and secondary transitions are described by quite different models of molecular behaviour. These models predict shift factors given respectively by the equation derived by Williams, Landel & Ferry (1955) (WLF equation) and the Arrhenius equation.

Williams, Landel & Ferry proposed that the rate of the relaxation process in the glass transition *depends on temperature primarily through its dependence on free volume*, that is the volume of the solid not occupied by atoms directly. It was furthermore assumed that this free volume has associated with it a constant thermal expansion coefficient above the glass transition temperature T_g. Inserting then an exponential dependence on volume of molecular friction as found empirically to obtain in liquid paraffins by Doolittle (1951), Williams *et al.* (1955) obtained the following expression for the shift factor (relative to T_g):

$$\log a_T = C_1(T - T_g)/[C_2 + (T - T_g)] \quad (26)$$

where C_1 and C_2 are constants. This WLF equation has recently been found applicable to the glass transition of elastin by Gosline & French (1979).

The WLF equation is not successful in describing secondary transitions. For such transitions it is commonly assumed that the viscoelastic behaviour can be related to a simple molecular rate process and described consequently by the equation first put forward by Svante Arrhenius (1889) to model the rates of various chemical reactions, notably the optical inversion of sucrose under acid hydrolysis. The Arrhenius equation predicts a shift factor (relative to an arbitrary T_0) of

$$\ln a_T = \Delta H/R\{1/T - 1/T_0\}. \quad (27)$$

Here ΔH is the activation energy (strictly an enthalpy) representing a barrier height to a molecular process, and R is the universal gas constant in units appropriate to ΔH. Eqn (27) has been used by Cohen *et al.* (1976) to account for the relaxation of collagenous tissue using the Temperature-Jump technique.

The plasticizing action of water in tissues

Dry elastin shows a T_g of about 200°C (Fig. 15*b*). The addition of a mere 10% by weight of water to dry elastin lowers T_g to about 70°C (Kakivaya & Hoeve, 1975) and a further increase to 38% water brings T_g to about 0°C (Fig. 15*b*).

All available information points to a high sensitivity of the viscoelastic properties of biological materials on water content. There remains to be elucidated a quantitative model for this sensitivity but to a first approximation we may represent the effect of water uptake as a marked shift along the temperature axis of the isochronal modulus curve (Fig. 15*a*). This is exactly analogous to the *temperature* shift of $\phi(\ln \tau)$ along the log τ axis. In Fig. 15*b* are presented the data of Gotte *et al.* (1968) showing the shear modulus ($\simeq E/3$) of elastin in dry and wet states for comparison.

In the body, elastin has a water content which places it precariously close to the rising step of the glass transition (Dorrington, Grut & McCrum, 1975). According to Fig. 15 a small decrease in water content may be expected to lead to a 'hardening of the arteries' as an onset of glassiness occurs in the isochronal modulus appropriate to the cardiac pulse frequency. That such an onset of glassiness might be caused by a displacement of water by lipids and be related to atherosclerotic disease has been suggested by Gosline (1976).

There is, moreover, a further complication to be considered. This is that the water content of a material is itself a function of the stresses applied to it. This is because all stress systems contain a hydrostatic or pure 'pressure' component which acts either to increase or decrease the tissue volume. (For the case of simple tension under a stress σ (Fig. 3*a*) this pressure component is equal to $\sigma/3$ and acts to increase volume.) Indeed a feature of the literature on the aetiology of atherosclerosis is that much attention has been given to, and success claimed in, correlating the location of plaques with local peculiarities of stresses in artery walls.

It should be noted that the fluid exchanges induced by an application of stresses to a solvated polymer will themselves produce a time-dependence in deformational response which must be clearly distinguished from that viscoelasticity which occurs in the closed system. This problem has been given some consideration by Hayes & Mockros (1971) whose work on cartilage clearly indicates a water flux contribution to viscoelasticity in that system.

Closing remarks

Viscous elasticity is elasticity with losses of work energy in the form of heat. It seems difficult not to conclude, therefore, that viscoelasticity is something which nature could well do without, a property of solids which introduces design snags for the living organism, as well as making an unwelcome appearance in some of Roy's and our experiments. If it is true that nature has to design against viscoelasticity then we might expect to learn something from its products.

In the structural protein resilin, for example, the insect phylum has inherited a highly non-viscous rubber which serves efficiently to store energy during high frequency wing oscillations (but see Gosline, 1980). In contrast to resilin, suggested Weis-Fogh (1972), elastin is a material 'unsuited for the construction of oscillating systems working at high rates of deformation. Since elastin is found in all chordates it must have evolved at an early stage of evolution and long before fast running or flying became part of the vertebrate repertoire'.

Hubbard (1958) has drawn attention to the interesting property of the Pacinian nerve ending that 'adaptation of the corpuscle is a direct consequence of its mechanical properties' suggesting, therefore, that viscoelasticity is used to good effect in the receptor to process the input stimulus before passing it to the central nervous system.

In this paper I have attempted to mention some of the most important aspects of viscoelasticity theory as they apply to biology. The traffic of valuable insights has so far tended to be in the direction from polymer science to biology though I expect the coming years to show us some reversal of this state of affairs with biology contributing much to the physical science, as it has done in so many other fields.

References

Alexander, R. McN. (1962). Viscoelastic properties of the body wall of sea anemones. *Journal of experimental Biology*, **39**, 373–86.

Alexander, R. McN. (1964). Viscoelastic properties of the mesogloea of jellyfish. *Journal of experimental Biology*, **41**, 363–9.

Alfrey, T. & Doty, P. (1945). The methods of specifying the properties of viscoelastic materials. *Journal of Applied Physics*, **16**, 700–13.

Apter, J. T. & Marquez, E. (1968*a*). Correlation of viscoelastic properties of large arteries with microscopic structure. *Circulation Research*, **22**, 393–404.

Apter, J. T. & Marquez, E. (1968*b*). A relation between hysteresis and other viscoelastic properties of some biomaterials *Biorheology*, **5**, 285–301.

Arrhenius, S. (1889). Über die Reaktionsgeschwindigkeit bei der Inversion von Rohrzucker durch Säuren. *Zeitschrift für Physik und Chemie*, **4**, 226–48.

Bergel, D. H. (1961). The dynamic elastic properties of the arterial wall. *Journal of Physiology*, **156**, 458–69.

Boltzmann, L. (1874). Zur Theorie der elastischen Nachwirkung. *Sitzungsberichte der Akademie der Wissenschaften II*, **70**, 275–306.

Boltzmann, L. (1876). Zur Theorie der elastischen Nachwirkung. *Annalen der Physik und Chemie*, **7**, (*Ergebnisband*), 624–54.

Braun, F. (1891). Elastische Nachwirkung. In *Handbuch der Physik I*, ed. A. Winkelmann, 321–42. Breslau.

Cohen, R. E., Hooley, C. J. & McCrum, N. G. (1976). Viscoelastic creep of collagenous tissue. *Journal of Biomechanics*, **9**, 175–84.

Doolittle, A. K. (1951). Studies in Newtonian flow II. The dependence of the viscosity of liquids on free-space. *Journal of Applied Physics*, **22**, 1471–5.

Dorrington, K., Grut, W. & McCrum, N. G. (1975). Mechanical state of elastin. *Nature, London*, **255**, 476–8.

Feltham, P. F. (1955). On the representation of rheological results with special reference to creep and stress relaxation. *British Journal of Applied Physics*, **6**, 26–31.

Gosline, J. M. (1971). Connective tissue mechanics of *Metridium senile*. *Journal of experimental Biology*, **55**, 775–95.

Gosline, J. M. (1976). The physical properties of elastic tissue. *International Review of Connective Tissue Research*, **7**, 211–49.

GOSLINE, J. M. (1980). The elastic properties of rubber-like proteins and highly extensible tissues. In *The Mechanical Properties of Biological Materials* (34th Symposium of the Society for Experimental Biology) ed. J. F. V. Vincent & J. D. Currey, pp. 331–57. Cambridge University Press.

GOSLINE, J. M. & FRENCH, C. J. (1979). Dynamic Mechanical Properties of Elastin. *Biopolymers*, **18,** 2091–103.

GOTTE, L., MAMMI, M. & PEZZIN, G. (1968). Some structural aspects of elastin revealed by X-ray diffraction and other physical methods. In *Symposium of fibrous proteins, Australia, 1967*, ed. W. G. Crewther, pp. 236–45. London: Butterworths.

GRANIT, R. (1966). *Charles Scott Sherrington: An appraisal*. pp. 14, 15. London: Nelson.

GROGG, B. & MELMS, D. (1958). A modification of the extensograph for study of the relaxation of externally applied stress in wheat dough. *Cereal Chemistry,* **36,** 189–95.

HAUGHTON, P. M., SELLEN, D. B. & PRESTON, R. D. (1968). Dynamic mechanical properties of the cell wall of *Nitella opaca*. *Journal of experimental Biology*, **19,** 1–12.

HAYES, W. C. & MOCKROS, L. F. (1971). Viscoelastic properties of human articular cartilage. *Journal of Applied Physiology*, **31,** 562–8.

HOOLEY, C. J. & COHEN, R. E. (1979). A model for the creep behaviour of tendon. *International Journal of Biological Macromolecules*, **1,** 123–32.

HUBBARD, S. J. (1958). A study of rapid mechanical events in a mechanoreceptor. *Journal of Physiology*, **141,** 198–218.

KAKIVAYA, S. R. & HOEVE, C. A. J. (1975). The glass point of elastin. *Proceedings of the National Academy of Sciences of the U.S.A.*, **72,** 3505–7.

LEAROYD, B. M. & TAYLOR, M. G. (1966). Alterations with age in the viscoelastic properties of human arterial walls. *Circulation Research*, **18,** 278–92.

MCCRUM, N. G. & DORRINGTON, K. L. (1976). The bulk modulus of solvated elastin. *Journal of Materials Science*, **11,** 1367–8.

MCCRUM, N. G. & MORRIS, E. L. (1964). On the measurement of the activation energies for creep and stress relaxation. *Proceedings of the Royal Society of London, Series A*, **281,** 258–73.

MAXWELL, J. C. (1868). On the dynamic theory of gases. *Philosophical Magazine*, **35,** 129–45.

MEYER, O. E. (1874). Theorie der elastischen Nachwirkung. *Annalen der Physik und Chemie*, **151,** 108–19.

MORIIZUMI, S., FUSHITANI, M. & KABURAGI, J. (1973). Viscoelasticity and Structure of Wood. *Journal of the Japan Wood Research Society (Mokuzai Gakkaiski)*, **19,** 81–8.

NEWTON, I. (1684). *De motu corporum in gyrum*, definition 3, hypothesis 1, translated in *The mathematical papers of Isaac Newton*, ed. D. T. Whiteside, 1967, vol. 6, pp. 33, 65. Cambridge University Press.

NEWTON, I. (1687). *Philosophiae Naturalis Principia Mathematica*, p. 307. Jussu Societatis Regiae ac typis Josephi Streatii; Londini.

PEZZIN, G. & SCANDOLA, M. (1976). The low-temperature mechanical relaxation of elastin, I. The dry protein. *Biopolymers*, **15,** 283–92.

ROY, C. S. (1880). The elastic properties of the arterial wall. *Journal of Physiology*, **3,** 125–59.

SCANDOLA, M. & PEZZIN, G. (1978). The low-temperature mechanical relaxation of elastin. II. The solvated protein. *Biopolymers*, **17,** 213–23.

VIIDIK, A. (1978). On the correlation between structure and mechanical function of the soft connective tissues. *Verh. Anat. Ges.*, **72,** 75–89.

WARD, I. M. (1971). *Mechanical Properties of solid Polymers*. New York: John Wiley & Sons Ltd.

WEBER, W. (1835). Über die Elasticität der Seidenfäden. *Annalen der Physik und Chemie*, **34,** 247–57.

WEBER, W. (1841). Über die Elasticität fester Körper. *Annalen der Physik und Chemie*, **54,** 1–18.

WEIS-FOGH, T. (1972). Energetics of hovering flight in hummingbirds and in *Drosophila. Journal of experimental Biology*, **56,** 79–104.

WIECHERT, E. (1893). Gesetze der elastischen Nachwirkung für constante Temperatur. *Annalen der Physik und Chemie*, **50,** 335–48; 546–70.

WILLIAMS, M. L., LANDEL, R. F. & FERRY, J. D. (1955). The temperature dependence of relaxation mechanisms in amorphous polymers and other glass-forming liquids. *Journal of the American Chemical Society*, **77,** 3701–7.

ZATZMAN, M., STACY, R. W., RANDALL, J. & EBERSTEIN, A. (1954). Time course of stress relaxation in isolated arterial segments. *American Journal of Physics*, **177,** 299–302.

ZENER, C. (1948). *Elasticity and Anelasticity of Metals*. The University of Chicago Press.

THE MECHANICAL PROPERTIES OF PLANT CELL WALLS

D. B. SELLEN

Astbury Department of Biophysics, University of Leeds, Leeds LS2 9JT, UK

Introduction

There are two principal, but inter-related, reasons for studying the mechanical properties of plant cell walls. First, cell growth by elongation is thought to arise from mechanical deformation resulting from the osmotic pressure acting from within the cell. Several experimental studies have been made in order to correlate observed growth rates with mechanical properties. There is an extensive review of this kind of work by Preston (1974), and a paper dealing with cell enlargement has been read at a recent symposium of this society by Cleland (1977). Nevertheless a brief review of this aspect of the subject will be given in the present paper. In order to correlate mechanical properties with growth rates it is not necessary to define rigorously the mechanical parameters actually being measured. The second reason for studying the mechanical properties of the plant cell wall is to try to explain them in terms of the ultrastructure and the molecular structure of the ultrastructural components, together with the associated dynamic molecular processes. In this case more rigorous measurements have to be carried out over a wider range of experimental conditions, and studies of this sort have been fewer. A review of the results of these experiments and their interpretation will be given in the present paper.

Mechanical properties and cell wall growth

Many types of growing cell wall such as that of the internodal cell of the alga *Nitella* (Probine & Preston, 1961) may be thought of as a cylindrical structure consisting of several lamellae. Each lamella is made up of a mesh of microfibrils embedded in a matrix. On the inside of the cell the microfibrils are predominantly transverse with respect to the cell axis, and on the outside they are predominantly longitudinal. Roelofson (1951) observed a structure of this type for growing cotton hairs and proposed a 'multi-net growth hypothesis'. This proposes that the innermost lamella is initially laid down with the microfibrils predominantly transverse, but that as it extends the microfibrils first become progressively more disordered and then eventually take up a predominantly longitudinal orientation. Meanwhile successive new lamellae are laid down with their initial microfibrillar direction predominantly transverse. Cell walls such as those of the algae *Cladophora* and *Chaetomorpha* (Frei & Preston, 1961) have a helically organized crossed microfibrillar structure with alternate lamellae having microfibrils nearly transverse and nearly longitudinal with respect to the cell axis. In the case of these algae, individual lamellae may be separated and examined. On passing from the inside to the outside of the wall the nearly transverse microfibrils become less transverse and more disordered, a phenomenon similar to that predicted for multinet growth. The nearly longitudinal microfibrils become more ordered as would be expected, but also become less longitudinal, the average angle between microfibrils in alternate lamellae being maintained. The latter fact has been explained by Frei & Preston (1961) in terms of twisting of the cell during growth and they have established that the amount of twisting observed experimentally is consistent with this idea. According to Roland, Vian & Reis (1975, 1977), however, the inner and outer walls of the roots and stems of mung beans and peas show little detectable structural differences and they have proposed, for some higher plants, an 'ordered growth hypothesis'. This envisages a crossed microfibrillar structure with the transverse microfibrils assembled into annular rings, thus preventing growth in that direction, whilst the longitudinal microfibrils are free to slide over each other.

Whatever structural changes take place during growth, the question arises as to whether the lamellae extend due to the osmotic turgor pressure within the cell, or whether they extend by intussusception of new material between the microfibrils. Probine & Preston (1962) have investigated the mechanical properties of strips of cell wall from the internodal cells of *Nitella opaca* in various stages of growth, and compared them with the growth rates averaged over the 24 h prior to harvest. They showed that growth rates fell with time and could be correlated with E_T/E_L which decreased from ~5 in young cells to ~2 in mature cells. E_L and E_T here represent the dynamic moduli (Dorrington, 1980) in the longitudinal and transverse directions one minute after loading. It is inconceivable however that the elastic moduli represent the factor determining growth as this would require a continuous increase in turgor pressure for growth to continue. Probine & Preston

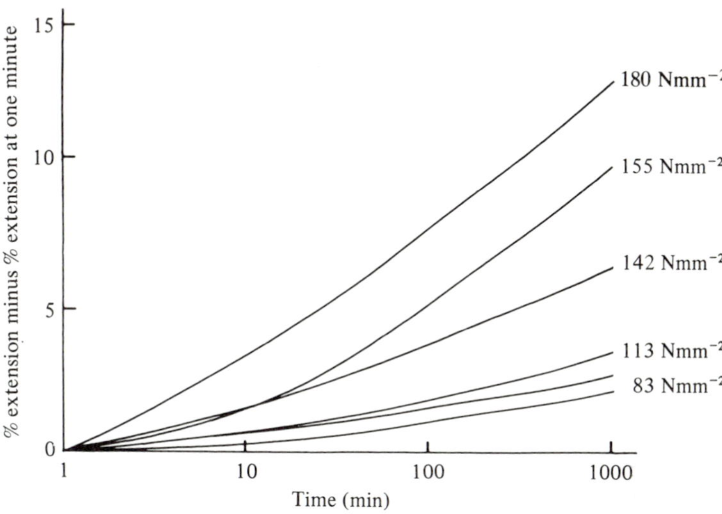

Fig. 1. Creep curves for *Nitella opaca* (Probine & Preston, 1961).

(1962) also investigated the creep properties of *Nitella*. Fig. 1 shows the longitudinal creep rate relative to the extension at one minute for various loads. It would appear that creep is a linear function of the logarithm of time and increases markedly for stresses above 100 Nmm^{-2} (corresponding to an initial strain ~10%). This has recently been confirmed by Metraux & Taiz (1978) who also showed that creep in the transverse direction is logarithmic with time, the rates of creep per unit stress in the longitudinal and transverse directions being proportional to the compliances. Probine & Preston (1961) showed that rates of growth increased monotonically with longitudinal rates of creep under given stresses provided that the wall stresses were measured relative to that which the cell would experience under a given turgor pressure (i.e. after correcting for cell diameter and cell wall thickness). This too has been confirmed by Metraux & Taiz (1978). Whereas a correlation between rates of creep and rates of growth does not necessarily establish cause and effect, if growth were not to result from creep it would be necessary that in the living cell the creep be counteracted in some way in order that the same effect could be achieved by some other means, a somewhat unlikely situation. The mechanical properties of the living cell wall must however be different from those of the isolated wall for the following reasons:

1. Cell wall growth is far from being a logarithmic function of time and over the time scale of mechanical experiments is linear.

2. The relative average rates of creep observed by Probine & Preston (1962) over the first 100 min of test corresponding to a typical turgor pressure (0.8 MNm^{-2}), were an order of magnitude greater than the relative growth rates over the preceding 24 h. The stress above which there was a marked increase in creep corresponded to a turgor pressure of only 0.2 MNm^{-2}. According to Kamiya, Tazawa & Takata (1963), the strains produced by a multiaxial turgor pressure are some three to fourfold less than those produced by the corresponding uniaxial stress. This might partially account for the above discrepancies, since creep rates appear to be proportional to compliances, but a discrepancy still remains.
3. The transverse modulus (Probine & Preston, 1962) and the transverse rate of creep under given stress (Metraux & Taiz, 1978) of *Nitella* cell walls isolated at various stages of growth is approximately constant. The relative growth rate in the transverse direction is, however, proportional to that in the longitudinal direction (albeit five times less (Probine & Barber, 1966)) and decreases with time.
4. The work of Green, Erickson & Buggy (1971) and Green & Cummins (1974) suggests that in the living cell the mechanical properties of the cell wall respond to turgor pressure in a compensatory manner. Whereas small changes in turgor pressure produce large transient changes in growth rate, the change in growth rate then proceeds to diminish until a steady-state rate of growth is achieved which is far less turgor sensitive. There is a turgor pressure below which cell growth stops permanently, which in the case of *Nitella* is about 0.2 MNm^{-2}, approximately equal to the yield stress observed by Probine & Preston (1962) in the isolated wall.

All of the above phenomena suggest that cell wall growth is a biochemically controlled creep, and at the time of writing this is the accepted point of view. However in spite of the amount of work which has been done on cell wall growth, the mechanisms involved are not yet understood. The pros and cons of the various suggestions that have been made have been discussed by Preston (1974) and Cleland (1977).

Mechanical properties and cell wall structure

On the basis of the multinet growth hypothesis, the structure of the cell wall does not change as the cell extends (Probine & Barber, 1966), orientation of microfibrils towards the longitudinal direction being compensated by the laying down of new lamellae with their microfibrillar direction predominantly transverse, in such a way that the distribution of microfibrillar orientations remains constant. On this basis it is difficult to understand why the longitudinal modulus and the longitudinal creep rate, under given stress, fall as the cell extends. However, it has been shown in *Nitella* (Probine & Preston, 1961), that although the microfibrillar direction on the inside of the cell is predominantly transverse, there is a minority of longitudinal microfibrils present. It has been suggested that the proportion of these increases as the cell grows so as to limit growth. In higher plants the limitation of growth is thought to be achieved by the deposition of the secondary wall.

Probine (1959) has investigated the mechanical anisotropy of young *Nitella* cell walls by measuring the modulus of strips cut at various angles to the cell axis. He has shown that the modulus possesses orthorhombic symmetry, i.e. the major and minor axes are at right angles and approximately (but not quite) transverse and longitudinal with respect to the cell axis. In this case the compliance, S, measured at an angle θ with respect to the major axis is given by Cowdrey (1965):

$$S = a + b\sin^2\theta + c\sin^4\theta$$

where

$$\begin{aligned} a &= S_{22} \\ b &= 2S_{12} + S_{66} - S_{22} \\ c &= S_{11} + S_{22} - 2S_{12} - S_{66} \end{aligned} \qquad (1)$$

S_{22} and S_{11} are the compliances in the direction of the major and minor axis of symmetry, S_{12} is the compliance determining the strain in the direction of the major axis resulting from the stress in the direction of the minor axis and *vice versa*, and S_{66} is the torsional compliance about an axis perpendicular to the wall. Fig. 2 shows a plot of S against $\sin^2\theta$ using Probine's data. It can be seen that within experimental error this is a straight line, i.e. $S_{11} + S_{22} \simeq 2S_{12} + S_{66}$. Similar plots were made by Cowdrey & Preston (1966) using data obtained from measurements with pieces of wood taken from a series of annual rings of samples of *Picea sitchensis*. Moduli were measured parallel to the cell axis. The S_2 layer of the secondary wall makes the major contribution to the modulus of each cell in this case, and it consists of helically organized microfibrils whose helical angle varies across the tree. Thus the cells may be regarded as cylinders whose major mechanical axis of symmetry lies at a variable angle θ (the helical angle) to the cylinder axis. Here again the plots were straight lines within experimental error, although it is not suggested that this is necessarily of any fundamental significance. In the work of Cowdrey & Preston (1966) it has been assumed that there is no

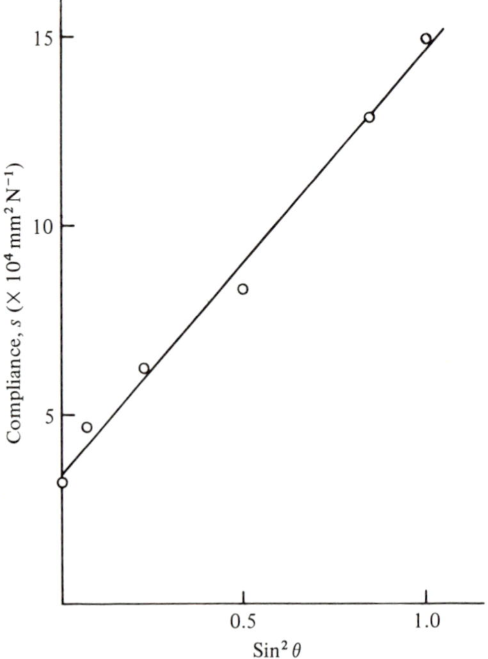

Fig. 2. Variation of compliance of *Nitella* cell wall with direction of stress. θ is the angle between the direction of stress and the mechanical major axis of symmetry. From data obtained by Probine (1959).

variation in the mechanical properties of the cell wall material across the tree, apart from its orientation.

If a rectangular strip of wall material is stretched in any direction other than along one of the axes of mechanical symmetry it will, in the absence of an opposing couple, deform into a parallelogram. Thus if a cylindrical cell is stretched along its axis and the cell axis does not correspond to an axis of mechanical symmetry, one end will twist with respect to the other. If S_θ is the angular deformation per axial stress then:

$$S_\theta = \mathrm{Sin}\ 2\theta\left(\frac{b}{2} + c\sin^2\theta\right) \qquad (2)$$

where S_θ is positive if the major axis of mechanical symmetry, which forms a helix around the cylinder, untwists. In higher plants this twisting effect is inhibited by the presence of neighbouring cells. In *Nitella* the major axis of symmetry as determined by polarization microscopy is not quite transverse, deviating by up to 10°. In young cells it forms a left hand helix and in older cells a right hand helix around the cell (Probine,

1963). Probine has carried out some experiments with *Nitella* cells in which the turgor pressure was reduced to zero by plasmolysis and then the cell allowed to return to full turgor. On returning to full turgor twisting was observed in the opposite direction to the helical sense of the major axis of symmetry, i.e. the major axis of symmetry untwisted. The twisting was also in the same direction as that observed during growth with cells of similar length, thus supporting the idea that cell wall growth arises from a mechanical deformation, assuming once again that creep is proportional to compliance. The untwisting of the major axis of symmetry is consistent with eqn (2) since if $c = 0$, S_θ must always be positive. However it must be emphasized that eqn (2) refers to uniaxial stress along the cell axis, whilst turgor pressure is multiaxial. A general expression for mechanical deformation as a result of multiaxial stress has not been obtained, even for a thin walled cylinder. In general a cylinder subjected to multiaxial stress would not retain its exact cylindrical shape.

It is difficult to reconcile the above observations with the proposed growth mechanism of *Cladophora* and *Chaetomorpha* (Frei & Preston, 1961) where the nearly longitudinal microfibrils in the crossed microfibrillar structure twist up in order to become less longitudinal. However the mechanical properties of individual lamellae of these algae have not been investigated. Also the nearly transverse microfibrils untwist during growth, and it is possible that there is some cooperative movement between the various layers.

Viscoelastic properties of the cell wall

Experimentally all moduli of elasticity and elastic compliances are dynamic in character and have to be obtained by some specific, but arbitrary, time-dependent procedure (Dorrington, 1980). The dynamics of the mechanical properties are, however, themselves of interest not only because of the correlation between creep and cell wall growth but also because they are related to the relaxation processes which take place between and within the components of the cell wall. These are the relaxation processes which have to be modified during growth. They are best investigated by means of stress relaxation experiments rather than creep, as in this case the strain is constant, and may therefore be kept small and a very wide range of time of relaxation investigated. In order to investigate activation energies associated with relaxation times, measurements must be made over a range of temperatures. Experiments of this sort have been carried out by Haughton, Sellen & Preston

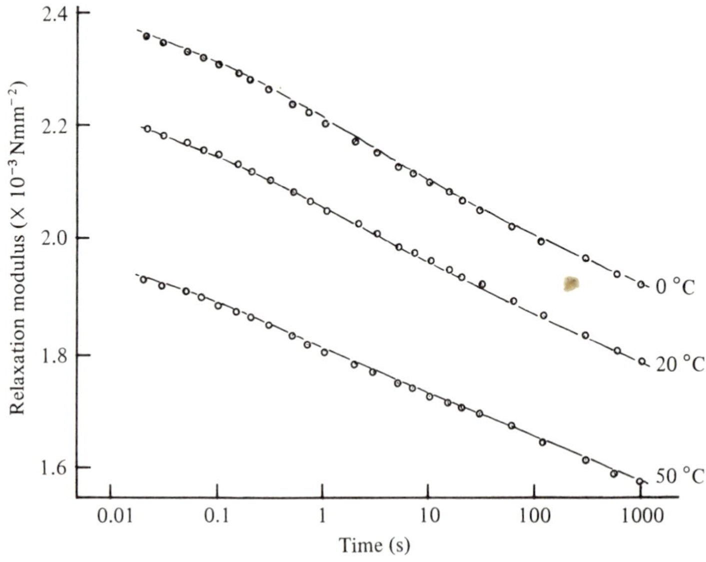

Fig. 3. Stress relaxation of *Nitella translucens*. Strain 1%. (Haughton & Sellen, 1969).

(1968) and Haughton & Sellen (1969), with strains in the range 1% to 3%, using a specially constructed apparatus. The cell walls of three species of algae were investigated: the internodal cells of *Nitella* (*opaca* and *translucens*), the stipes of *Acetabularia crenulata*, and single cells from filaments of *Penicillus dumetosus*. These have as their major structural polysaccharides cellulose, mannan and xylan respectively. All three types of cell wall have a microfibrillar structure (Probine & Preston, 1961; Frei & Preston, 1964; Mackie & Preston, 1968). Measurements were made on whole cells with the cytoplasm removed, strains being imposed in the longitudinal direction. Prior to making stress relaxation measurements specimens were conditioned by stretching them several times to a strain somewhat higher than that to be employed, so as to obtain reproducible results and ensure that only reversible dynamic processes were observed. Stress relaxation curves for *Nitella translucens* (curves for *Nitella opaca* were similar) and *Penicillus dumetosus* are shown in Figs. 3 and 4. Similar curves were obtained with *Acetabularia*, but it was not found possible to condition the specimens, successive relaxation curves at a given temperature lying lower on the plot. The dynamic moduli however, were of the same order as those of *Penicillus*, the results for which were highly reproducible. It is possible that in the case of *Acetabularia* the wall structure is comparatively unstable. Mannan microfibrils appear to be destroyed

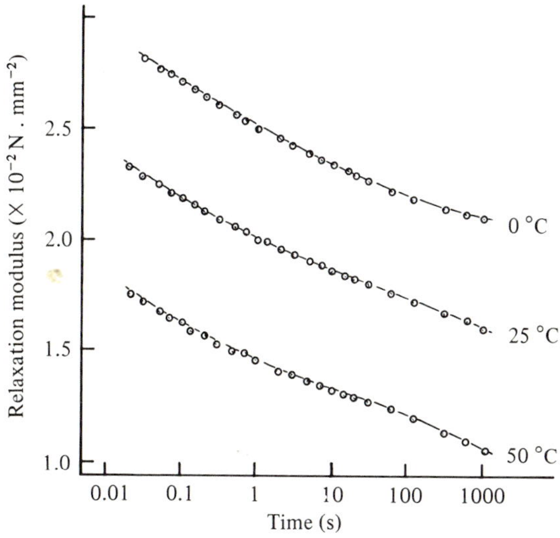

Fig. 4. Stress relaxation of *Penicillus dumetosus*. Strain 3%. (Haughton & Sellen, 1969).

under all but the mildest preparative procedures for electron microscopy (Mackie & Preston, 1968). Also mannan has a comparatively low degree of polymerization. Experiments with *Codium fragile*, whose cell wall also consists of mannan (Mackie & Sellen, 1969) yielded weight and number average degrees of polymerization of 1100 and 260 respectively. This compares with weight average values of >3000 for xylan (Mackie & Sellen, 1971) and 8000 for cellulose (Holt, Mackie & Sellen, 1973).

It is obvious from Figs. 3 and 4 that neither the instantaneous modulus nor the equilibrium modulus (if it exists) is an experimentally accessible quantity near room temperature. It is perhaps not surprising therefore that creep rates have been found to be proportional to compliances in the cell wall, as the compliances being measured are themselves mostly creep. Many other fibrous materials yield approximately linear relaxation curves on a logarithmic scale: in particular it has been shown with regenerated cellulose (Haughton & Sellen, 1973) that it is necessary to reduce the temperature to −180°C before stress relaxation ceases and the true instantaneous modulus is measured. This is thought to be the temperature at which hydrogen bonds acting singly are immobilized.

The fact that in *Nitella* both creep and stress relaxation are linear functions of the logarithm of time is consistent with the system being linear (Dorrington, 1980). However this is not quite so. Fig. 5 shows a stress–strain loop for *Nitella* obtained from a series of stress relaxations,

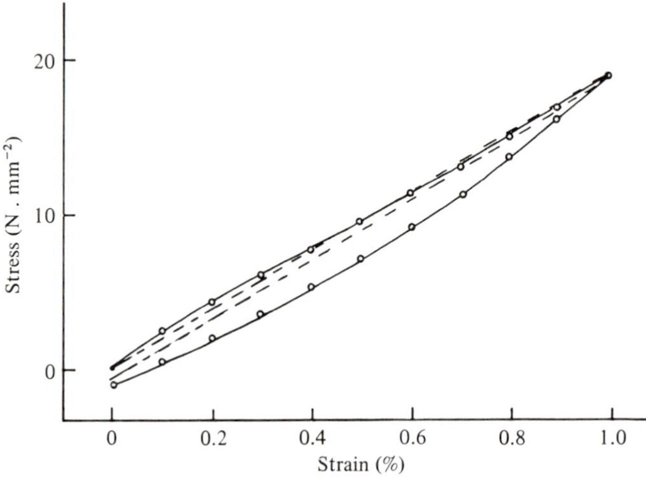

Fig. 5. Experimental stress–strain loop for *Nitella translucens* (full line), compared with that derived from Fig. 3 by means of the Boltzmann superposition principle (dashed line). (Haughton & Sellen, 1969.)

each point on the curve representing a strain increment. If *Nitella* is a linear viscoelastic system then it should be possible to predict this loop from the data of Fig. 3 by applying the Boltzmann superposition principle (*n.b.* a comparison of this sort would not be valid for material which had not previously been conditioned). In fact the predicted loop shows much less hysteresis. A similar result was obtained with *Penicillus*. The disparity is much less in regenerated cellulose (Haughton & Sellen, 1973). It is possible that the additional hysteresis arises from a small degree of reversible crystallization on stretching.

With fibres and amorphous polymers below their glass transition temperature it is usual to calculate activation energies by shifting curves horizontally so that they superimpose. The shift factor, log α_T, is then assumed to be given by the Arrhenius equation:

$$\text{Log}\,\alpha_T = \frac{1}{2.303}\frac{E_a}{R}\left[\frac{1}{T}-\frac{1}{T_s}\right] \tag{3}$$

where E_a is the activation energy and T_s an arbitrary reference temperature. If this procedure is applied to the data in Figs. 3 and 4 the detailed deviations from linearity do not superimpose, but activation energies may nevertheless be calculated. These are found to be 24 kcal/mol for *Nitella opaca*, 36 kcal/mol for *Nitella translucens* and 55 kcal/mol for *Penicillus*. (In Haughton & Sellen (1969) different values are quoted because the relaxation moduli were multiplied by T_s/T prior to superposition. This procedure was adopted because it was then thought that the straight line relaxation curves corresponded to a single Eyring element above its yield point. This is not now thought to be the case.)

The Arrhenius equation in the context of mechanical properties arises from the Eyring theory (Tobolski & Eyring, 1943). This however predicts, for a single activation energy and small stress, a single relaxation time and creep which is linear. At high stress, above a yield point in the predicted stress–strain curve, the Eyring model (which then becomes very non-linear), does indeed predict stress relaxation which is linear with the logarithm of time over a wide time scale, but it still predicts creep which is linear with time albeit increasing exponentially with stress. It seems unlikely that, at the small strains used by Haughton *et al.* (1968, 1969) the nature of the stress relaxation curves arises entirely from the non-linearity of the systems. It seems more likely that the stress relaxation curves represent a distribution of relaxation times and therefore distributions of activation energies. However in this case the Arrhenius equation is not strictly applicable since it assumes a single activation energy. A modified method of time–temperature superposition has been evolved by Haughton & Sellen (1973). They assume a distribution of relaxation times (τ) of the form

$$\tau = c\varepsilon^{Ea/RT} \tag{4}$$

The procedure adopted is to multiply the logarithm of time axis by T/T_s and then displace the curves horizontally to superimpose, thus obtaining a 'master' curve at the temperature T_s. A plot of the displacement required to do this against temperature yields a straight line of slope $-\log(C/T_s)$, from which log C may be calculated. A distribution of relaxation times may then be calculated from the master curve by means of a first order approximation and the corresponding activation energies calculated from eqn 4 by inserting T_s for T and the calculated value of C. Haughton & Sellen (1973) applied this procedure to stress relaxation data for regenerated cellulose. Figs. 6 and 7 show it applied to the data of Figs. 3 and 4. Here again the details of the relaxation curves do not superimpose. The distribution of activation energies in each case is but a small part of the complete spectrum and represents those activation energies that are relevant to the mechanical properties over the scale of time and temperature investigated. The values calculated by straightforward application of the Arrhenius equation lie within these distributions. Arrhenius plots yield straight lines because the temperature range is restricted and the range of relevant activation energies is small. This

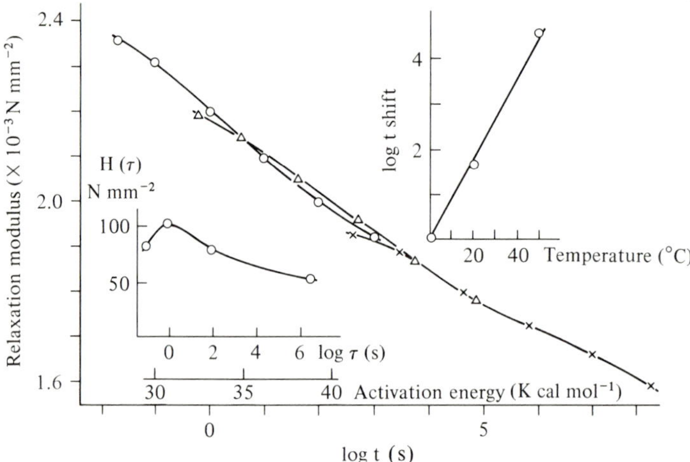

Fig. 6. Data of Fig. 3 treated according to the method of Haughton & Sellen (1973). Open circles, 0°C (reference temperature); triangles, 20°C; crosses, 50°C. Further details in the text.

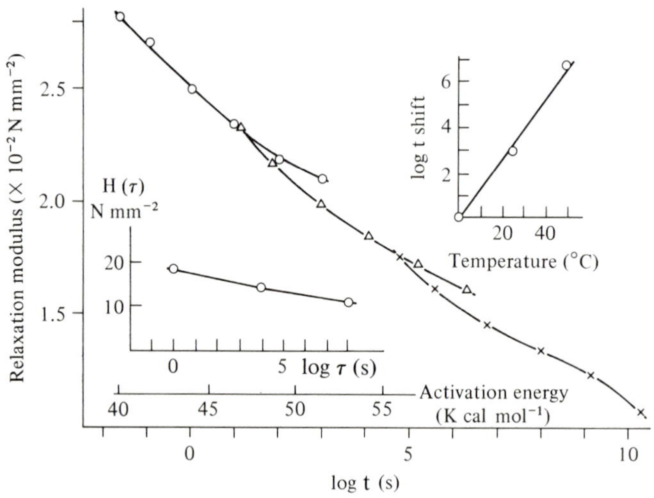

Fig. 7. Data of Fig. 4 treated according to the method of Haughton & Sellen (1973). Open circles, 0°C (reference temperature); triangles, 25°C; crosses, 50°C. Further details in the text.

narrow range of activation energies, however, represents a very broad range of relaxation times.

Values of log C were found to be -17, -25, -35 for *Nitella opaca*, *Nitella translucens*, and *Penicillus* respectively (C is in seconds). On the basis of the Eyring theory these values are related to the sizes of flow units, *Penicillus* in this case having the largest. Quantitative calculations of the sizes of flow units are, at the present time, dubious. (For a full discussion see Haughton & Sellen, 1973.) However, qualitatively it can be seen that a greater dependence of mechanical properties on temperature must involve larger flow units, as larger mechanical energies per flow unit need to be involved.

Conclusion

It is generally accepted that growth by cell elongation is some form of biochemically controlled creep, but the mechanisms involved are not understood. It is difficult to see what sort of experiments of a purely physical nature could be made in order to improve our understanding. There are however some gaps in the theory relating mechanical properties to structure which could usefully be filled. In particular the mechanical response of the cell to multiaxial turgor in the general anisotropic case has not been worked out.

Rigorous stress relaxation experiments at different temperatures over a wide range of time scale do not at present yield much specific information as to the relaxation processes involved and therefore the relaxation processes which have to be modified during growth. The reason for this is that there is at present no generally accepted theory for the mechanical properties of fibrous material and indeed the only experimental work which would be capable of testing such a theory is that of Haughton & Sellen (1973) on regenerated cellulose. Experiments over a very wide range of temperature, particularly very low temperatures, are not possible with plant material, but experiments on other fibrous material could help to establish some general principles of interpretation.

Activation energies calculated from stress relaxation experiments do, however, show that relaxation processes near room temperature do not involve large heavily bonded units, as they correspond to one or two covalent bonds or five to ten hydrogen bonds. In view of the spread of activation energies it is unlikely that covalent bonds are involved and more likely that the relaxation processes arise from hindrances to movement of lower activation energy acting cooperatively. These may or may not involve specific chemical bonding.

It is tantalizing to note that the Eyring model with a single activation energy predicts a yield point and a creep which is linear with time, i.e. properties similar to the response of cell wall growth to turgor pressure. It could be speculated that growth involves the narrowing of the activation energy spectrum by some biochemical agent. The present author can, however, make no suggestions as to how this could come about.

References

Cleland, R. E. (1977). The control of cell enlargement. In *Integration of Activity in the Higher Plant* (31st Symposium of the Society for Experimental Biology) ed. D. H. Jennings, pp. 101–16. Cambridge University Press.

Cowdrey, D. R. (1965). The structure and physical properties of timber. Ph.D. thesis, University of Leeds.

Cowdrey, D. R. & Preston, R. D. (1966). Elasticity and microfibrillar angle in the wood of Sitka spruce. *Proceedings of the Royal Society of London, Series B,* **166,** 245–72.

Dorrington, K. L. (1980). The theory of viscoelasticity in Biomaterials. In *The Mechanical Properties of Biological Materials* (34th Symposium of the Society for Experimental Biology) ed. J. F. V. Vincent & J. D. Currey, pp. 289–313. Cambridge University Press.

Frei, E. & Preston, R. D. (1961). Cell wall organization and wall growth in the filamentous green algae *Cladophora* and *Chaetomorpha*: I. The basic structure and its formation. *Proceedings of the Royal Society of London, Series B,* **154,** 70–94. II. Spiral structure and spiral growth. *Proceedings of the Royal Society of London, Series B,* **155,** 55–77.

Frei, E. & Preston, R. D. (1964). Non-cellulosic structural polysaccharides in algal cell walls. I. Xylan in siphoneous green algae. *Proceedings of the Royal Society of London, Series B,* **160,** 293–313.

Green, P. B. & Cummins, W. R. (1974). Growth rate and turgor pressure – Auxin effect studied with an automated apparatus for single coleoptiles. *Plant Physiology,* **54,** 863–69.

Green, P. B., Erickson, R. O. & Buggy, J. (1971). Metabolic and physical control of elongation rate – *in vivo* studies in *Nitella. Plant Physiology,* **47,** 423–30.

Haughton, P. M. (1968). Viscoelastic properties of the cell walls of some algae. Ph.D. thesis, University of Leeds.

Haughton, P. M., Sellen, D. B. & Preston, R. D. (1968). Dynamic mechanical properties of the cell wall *Nitella opaca. Journal of experimental Botany*, **19,** 1–12.

Haughton, P. M. & Sellen, D. B. (1969). Dynamic mechanical properties of the cell walls of some green algae. *Journal of experimental Botany,* **20,** 516–35.

Haughton, P. M. & Sellen, D. B. (1973). Stress relaxation in regenerated cellulose. *Journal of Physics, D. Applied Physics,* **6,** 1998–2011.

Holt, C., Mackie, W. & Sellen, D. B. (1973). Degree of polymerization and polydispersity of native cellulose. *Journal of Polymer Science,* **42,** 1505–12.

Kamiya, N., Tazawa, M. & Takata, T. (1963). The relation of turgor pressure to cell volume in *Nitella* with special reference to the mechanical properties of the cell wall. *Protoplasma,* **57,** 501–21.

Mackie, W. & Preston, R. D. (1968). The occurrence of mannan microfibrils in the green algae *Codium fragile* and *Acetabularia crenulata*. *Planta*, **79**, 249–53.

Mackie, W. & Sellen, D. B. (1969). The degree of polymerization and polydispersity of mannan from the cell wall of the green seaweed *Codium fragile*. *Polymer*, **10**, 621–32.

Mackie, W. & Sellen, D. B. (1971). Degree of polymerization and polydispersity of xylan from the cell wall of the green seaweed *Penicillus dumetosus*. *Biopolymers*, **10**, 1–9.

Metraux, J. P. & Taiz, Z. (1978). Transverse viscoelastic extension in *Nitella*. *Plant Physiology*, **61**, 135–8.

Preston, R. D. (1974). *The physical biology of plant cell walls*. London: Chapman and Hall.

Probine, M. C. (1959). Molecular structure and mechanical properties of plant cell walls in relation to growth. Ph.D. thesis, University of Leeds.

Probine, M. C. (1963). Cell growth and the structure and mechanical properties of the wall in internodal cells of *Nitella opaca*. III. Spiral growth and cell wall structure. *Journal of experimental Botany*, **14**, 101–13.

Probine, M. C. & Barber, N. F. (1966). The structure and plastic properties of the cell wall of *Nitella* in relation to extension growth. *Australian Journal of Biological Sciences*, **19**, 439–57.

Probine, M. C. & Preston, R. D. (1961). Cell growth and the structure and mechanical properties of the wall in the internodal cells of *Nitella opaca*. I. Wall structure and Growth. *Journal of experimental Botany*, **12**, 261–82.

Probine, M. C. & Preston, R. D. (1962). Cell growth and the structure and mechanical properties of the wall of the internodal cells of *Nitella opaca*. II. Mechanical properties of the wall. *Journal of experimental Botany*, **13**, 111–27.

Roelofson, P. A. (1951). Orientation of cellulose fibrils in the cell wall of growing cotton hairs and its bearing on the physiology and cell wall growth. *Biochimica et Biophysica Acta*, **7**, 43–53.

Roland, J. C., Vian, B. & Reis, D. (1975). Observations with cytochemistry and ultracryatomy on the fine structure of the expanding walls in actively elongating plant cells. *Journal of Cell Science*, **19**, 239–59.

Roland, J. C., Vian, B. & Reis, D. (1977). Further observations on cell wall morphogenesis and polysaccharide arrangements during plant growth. *Protoplasma*, **91**, 125–41.

Tobolsky, A. & Eyring, H. (1943). Mechanical properties of polymeric materials. *Journal of Chemical Physics*, **11**, 125–34.

THE ELASTIC PROPERTIES OF RUBBER-LIKE PROTEINS AND HIGHLY EXTENSIBLE TISSUES

JOHN M. GOSLINE

Department of Zoology, University of British Columbia, Vancouver, BC, Canada

Introduction

To many, the term skeleton brings to mind the numerous bony structures that constitute the most visible portion of the vertebrate support system. In reality, however, a skeleton is simply that structure which functions to resist external mechanical forces and to provide locomotion, and there are a number of different design strategies that can achieve these general requirements. For example, many of the 'lower' invertebrates lack stiff skeletal elements and are supported entirely by a pressurized fluid constrained within a deformable container (i.e. a hydrostatic skeleton). Other animals, like arthropods and vertebrates, employ a complex assembly of rigid elements articulated by elaborate muscular systems to achieve mechanical support and locomotory movements. However, even in skeletal systems dominated by rigid elements, there is an absolute need for deformable components (connective tissues) to attach the soft parts of the animal to the mobile framework, to provide protective coverings, etc. In general, the deformable components of skeletal systems exist in the form of thin sheets or membranes that can readily bend or fold in accordance with body movements, and they are constructed from pliant materials that have low tensile modulus and can withstand large tensile strains. Thus, both the shape and the materials determine the important functional properties.

At the microscopic level, most pliant biomaterials are composites of a stiff fibre (usually collagen) associated with a low modulus 'matrix'. The fibrous component is typically arranged to allow the composite as a whole to be deformed to large strains without the fibres themselves being strained. Two common fibre patterns found in pliant composites are: 1) crossed-fibre (crossed-helical) arrays, as in the body walls of many hydrostatic organisms, and 2) random feltworks of wavy or crimped fibres, as in mammalian skin (Wainwright, Biggs, Currey &

Gosline, 1976). With extension the fibres, which are linked together into a mechanically continuous mesh, become straightened and aligned in the direction of extension and at large strains the composite takes on the properties of the fibres alone. (See Kastelic & Baer (1980) for a description of the properties of parallel collagen fibres.) Thus, the fibres provide important limits to extension which prevent the rupture of low modulus components (Gordon, 1975). In some cases, however, the fibres are not linked, but are discontinuous and serve as reinforcing fillers. This arrangement allows for extremely large, biaxial, extensions, but there are no built-in high strain limits. Discontinuous fibre systems are quite rare, but *Metridium* mesogloea (Gosline, 1971) and locust inter-segmental membrane (Vincent, 1976) provide examples of this type of behaviour.

Although the fibres dominate the high strain properties and may be the most abundant constituent in the composite, the low modulus 'matrix' components are responsible for other important functional properties. Many pliant biomaterials show long-range elastic properties that are similar to those exhibited by lightly cross-linked rubbers. In some cases these elastic properties are associated with protein–polysaccharide complexes containing glycosaminoglycans (Gosline, 1971), but in the great majority of cases the long-range elastic properties can be attributed to cross-linked protein rubbers. The distribution and diversity of such protein rubbers is probably quite broad (Elder & Owen, 1967). In fact, there is now evidence for the presence of an intracellular protein rubber in muscle cells (Maruyama *et al.*, 1977). However, in most tissues the protein rubber is found as extremely fine fibres that are difficult to isolate and test. For this reason most of what is known about protein rubbers and their functional role in pliant biomaterials is derived from the study of only three protein rubbers. This paper attempts to develop a general picture of the protein rubbers by discussing the properties of these three materials. In addition, the properties of vertebrate arteries are discussed to provide a well documented example of the subtle interactions that exist between the protein rubber and the fibre meshwork in an elegantly designed pliant composite.

Distribution and function of protein rubbers

To date only elastin, resilin and abductin have been isolated and investigated in any great detail. Elastin is found in the vertebrates (Sage & Gray, 1976), and is quite widely distributed in such pliant composites as skin, elastic ligaments like the *ligamentum nuchae*, arteries, veins, mesenteries, elastic cartilage, etc. Elastin functions both in statically loaded tissues, such as skin, to resist long-term tensile forces, and in dynamically loaded tissues, such as artery, to provide an efficient energy storage system. Resilin is a component of insect cuticle, usually found in association with chitin fibres (Andersen, 1971). The first description of resilin, by Weis-Fogh (1960), was of elastic ligaments associated with the wings of the locust and elastic tendons in the flight musculature of the dragonfly. Resilin has been found in the jumping mechanism of fleas (Bennet-Clark & Lucey, 1967), a number of other insect structures and in some crustaceans (Andersen & Weis-Fogh, 1964). Resilin probably evolved as a component of the insect flight system, where it functions as an energy store that helps to overcome the inertia of the rapidly reciprocating wings (Weis-Fogh, 1973). Abductin is found in the inner hinge-ligament of the bivalve mollusc shell, where it acts as an elastic pivot that antagonizes the muscles that close the shell. The rubbery region of the ligament is commonly filled by mineralized inclusions, except in the Pectinidae where relatively homogeneous deposits of the protein abductin are found (Trueman, 1953; Kahler, Fisher & Sass, 1976). Abductin serves as an energy store that opens the shell when the abductor muscles relax. This system is particularly well developed in *Pecten* and related genera that are able to swim by rapid cyclical abductions of the shell.

The chemistry of protein rubbers

Protein rubbers appear to conform to the predictions of the kinetic theory of rubber elasticity (see Mullins, 1980). The two major requirements for this theory are that the material be constructed from flexible, kinetically-free, linear polymer molecules, and that these molecules be linked together through permanent cross-links to form a random network of essentially infinite molecular weight. The requirement for kinetically free protein chains goes against the Central Dogma of molecular biology, that each protein sequence corresponds to a unique functional conformation. Indeed, the strong appeal of this Dogma has led some workers to consider unique structures to explain the properties of elastin. However, rubber-like behaviour arises only when network chains are mobile and have no stable conformations, but have instead the conformation of a totally disordered molecule (i.e. a random-coil). Of the enormous number of proteins that have been identified and studied from different organisms, all appear, under cellular conditions, to have a single or very limited number of stable conformations that correspond to some highly specific function. The protein rubbers appear to be the only exception to this generalization. At present, very little is known about the design of sequences that give rise to random structures, but it is clear that careful design is essential. A consideration of amino-acid composition and of the limited sequence data may give some clues, but much is yet to be learned. The requirement for stable cross-links is rather better understood. Careful chemical analyses of insoluble elastin and resilin have shown that, as with other structural proteins, the protein rubbers utilize unique amino-acid derivatives and elaborate biosynthetic mechanisms to create the stable linkages needed to form a network.

Amino-acid composition

Table 1 summarizes the available data on amino-acid composition for the three protein rubbers. As with all data about amino-acid composition, it is very difficult to make any meaningful deductions about sequence–conformation relations. Perhaps the only real generalization that can be drawn is that protein rubbers are rich in glycine. This presumably reflects the fact that glycine, having a very small side-group, will impose the least restrictions to rotational freedom of the backbone chain (Flory, 1969). However, the amount of glycine is quite variable, being about 325 residues per 1000 in elastin, 400 in resilin and about 600 in the abductins. It should be noted that a glycine content of about 330 per 1000 (i.e. the same as that for elastin) is very characteristic of all collagens, and collagen is structurally and mechanically very different from a protein rubber. The protein rubbers are all rich in small amino acids, with (gly + ala + ser) making up 55% to 70% of the total composition. Again, these small amino acids are probably associated with reducing steric restrictions to the rotational freedom of the backbone. However, these three amino acids make up more than 85% of silkworm silk (see Denny, 1980), and silk is highly crystalline and not at all like a protein rubber. Clearly, composition is not enough to determine conformation. Highly regular repeating sequences, like poly(gly–ala) for silk or poly(gly–pro–X) for collagen, produce crystalline structures (Wainwright *et al.*, 1976), and therefore such regular repeat sequences are inappropriate for a protein rubber. The proline content is quite high in most of the protein rubbers and since proline is known to terminate α-helical segments in globular proteins, it may be argued that prolines help to keep the protein chains kinetically free. However, the abductin from *Pecten irradians* has very little proline.

Mammalian elastin is noted for being very hydrophobic. With over

Table 1. *The amino-acid composition of protein rubbers. The data for the abductins and the elastins are from Sage & Gray (1976). The data for the resilins are from Andersen (1971)*

	Abductins		Elastins		Resilins	
	Pecten irradians	*Spondylus varians*	pig artery	cod artery	locust wing-hinge	dragonfly tendon
asx	39	6	6	12	102	94
thr	7	2	15	60	29	20
ser	51	16	12	29	77	128
glx	21	3	19	27	43	42
pro	8	72	113	98	75	75
gly	584	544	313	449	411	422
ala	67	75	224	136	112	70
½ cys	2	tr	–	1	–	–
val	6	116	128	47	23	12
met	125	2	–	4	–	–
ileu	3	2	18	8	13	9
leu	4	2	54	29	22	30
tyr	1	20	19	36	21	12
phe	69	108	33	33	25	12
lys	9	5	5	5	4	8
his	tr	tr	1	2	8	12
arg	4	30	8	17	36	46
OHpro	–	–	9	8	–	–
des	–	–	4	1	–	–
di-tyr	–	–	–	–	8	–
tri-tyr	–	–	–	–	4	–
small amino acids	702	635	549	614	600	620
gly	584	544	313	449	411	422
polar	131	62	67	152	299	350
non-polar	285	397	602	392	295	229

60% of its amino acids carrying non-polar side-chains, mammalian elastin is one of the most hydrophobic proteins known (Gosline, 1976). This composition has some important implications for the elastic mechanism and the mechanical properties, but it is unclear how it relates to the problem of generating protein chains that remain kinetically free under normal cellular conditions. Interestingly, elastins isolated from the lower vertebrates are considerably less hydrophobic (Sage & Gray, 1976). There are also some striking differences in the compositions of the abductins. As mentioned above, there are large differences in the proline content of abductin from the scallop *Pecten* and the oyster *Spondylus*. In addition, note the large differences in the content of methionine and valine in abductin from these two animals and, although it is not shown in Table 1, there are differences in methionine and valine of almost equal magnitude in the abductin in animals from a single genus (Sage & Gray, 1976; Kahler *et al.*, 1976). The composition of the resilins seems to be more consistent, but there are too few data available to establish the variation.

Cross-links

In addition to the normal amino acids listed in Table 1, hydrolysates of purified, insoluble, elastin and resilin yield specialized amino acids that are thought to be derived from the network cross-links. Elastin hydrolysates contain desmosine and isodesmosine (Fig. 1A, B), that are derived from the oxidation and condensation of four lysine side chains, and a small amount of lysinonorleucine (Fig. 1C) that is derived from two lysine side chains (Sandberg, 1976). The desmosines and the lysinonorleucine are present in sufficient quantity to account for the network properties, and are probably the only cross-links present in elastin. Cystine is absent from elastin, and disulphide-bond breaking agents have no effect on cross-linking. Similarly, disulphide bonds do not contribute to the cross-linking of resilin. However, di- and tertyrosine (Fig. 1D, E) can be isolated from hydrolysates, and these compounds are present in sufficient quantity to account for the properties of resilin (Andersen, 1971). The abductin cross-links are still unknown. Andersen (1967) isolated 3, 3′-methylene-bistyrosine from hydrolysates of *Mytilus edulis* hinge, which he thought could provide a bifunctional cross-link like dityrosine. However, it has been suggested recently that this compound is created during acid hydrolysis (Price & Hunt, 1974).

Amino-acid sequence data

The only sequence data available for a protein rubber are for elastin, and unfortunately no simple sequence pattern can explain the fact that the protein chains of elastin remain kinetically free under conditions where virtually all other proteins have unique, stable conformations. Several repeat sequences have been found, but attempts to construct helical secondary structures based on these sequences have somewhat confused the issue. Gray, Sandberg & Foster (1973) published the first extensive sequence data on tryptic fragments of soluble elastin. Soluble elastin, the uncross-linked precursor of insoluble elastin, contains about 850 residues, and the first sequences were for about 350 residues, or

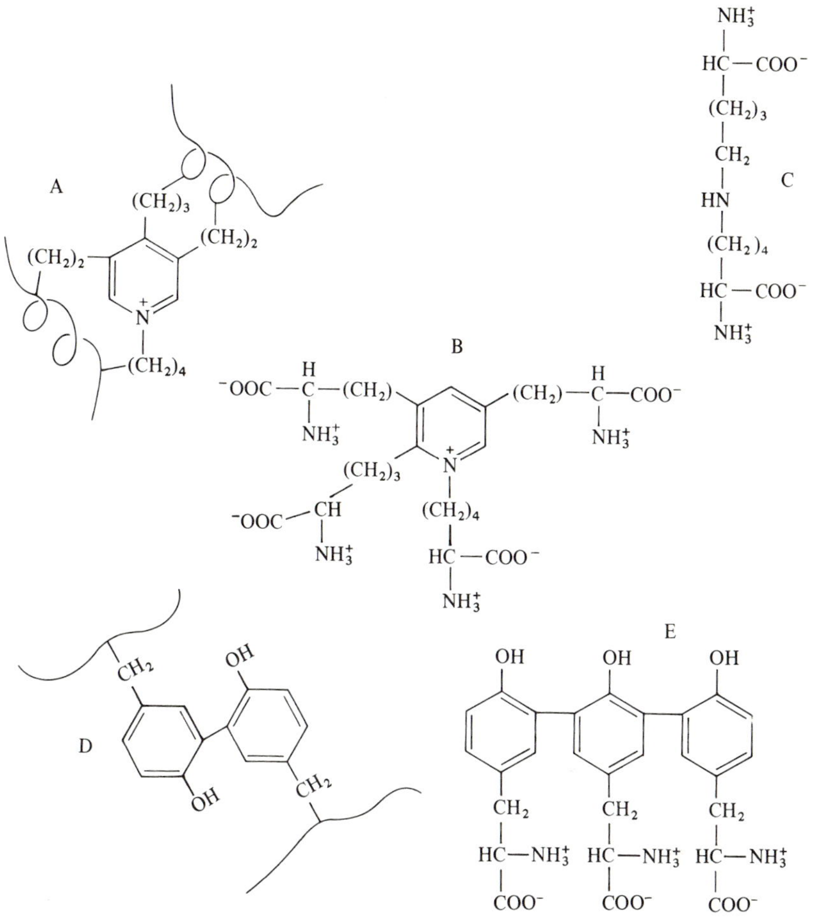

Fig. 1. The cross-links of elastin and resilin. A, desmosine as it probably exists in the elastin network, linking two chains; B, isodesmosine and C, lysinonorleucine, as they exist after the hydrolysis of insoluble elastin; D, dityrosine, as it probably exists in insoluble resilin; E, tertyrosine, after the hydrolysis of resilin.

about 40% of the total. A number of large peptides were isolated corresponding to the random chains between cross-links. (Trypsin cleaves at lysine residues, and lysines are involved in the cross-links.) The following three repeat sequences were observed in these large peptides: a tetrapeptide, val–pro–gly–gly; a pentapeptide, val–pro–gly–val–gly; and a hexapeptide, ala–pro–gly–val–gly–val. Gray *et al.* (1973) suggested that the tetrapeptide would form β-turns, and they discussed the possibility of a network of helical structures based on a sequence of these β-turns, called oiled-coils (Fig. 2). The elasticity was seen to arise

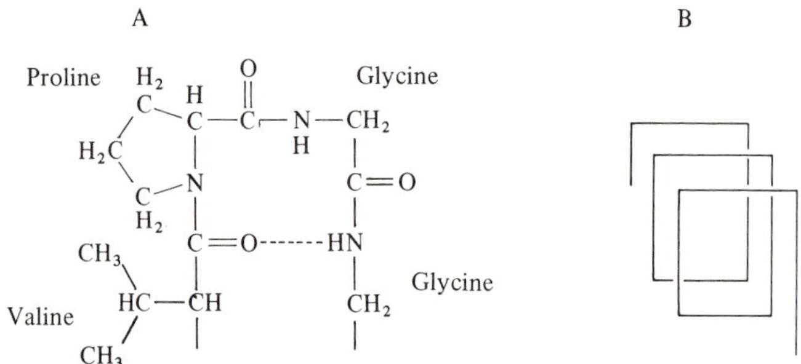

Fig. 2. Proposed β-helical structure based on the repeat tetrapeptide, val–pro–gly–gly (after Gray *et al.*, 1973). A, a β-turn formed from one of the tetrapeptide units. B, a β-helix (oiled-coil) formed from a number of β-turns in succession.

from the opening of the coils and exposure of non-polar amino-acid side-chains to water. Urry & Long (1976*a, b*) have carried out extensive analyses of the pentapeptide and hexapeptide. They find that synthetic peptides with these repeat sequences will also adopt extended coils as preferred conformations, and they also suggest that helical structures form the basis for the elasticity of elastin. However, only about 25% of the residues that were sequenced in these non-cross-link regions were present as the three repeat sequences (Gray *et al.*, 1973; Sandberg, 1976) and more recent data suggest that the repeat sequences make an even smaller portion of the total (L. Sandberg, personal communication). In addition, several nuclear magnetic resonance studies of insoluble elastin clearly demonstrate that the carbon atoms in the backbone chain are highly mobile and are able to tumble in three dimensions on a time scale of the order of 10^{-6} s (Torchia & Piez, 1973; Ellis & Packer, 1976). Therefore, the proposed helical structures cannot exist as stable structures, but it is not possible to say that these repeat sequences are not important features of the overall sequence design. At present, no other sequence patterns have been proposed to account for the kinetic mobility of the protein chains of elastin.

The cross-linking regions of elastin are somewhat better understood. Extensive proteolytic action on insoluble elastin yields desmosine-containing peptides that are very rich in lysine and alanine (Fig. 3). The important feature of these peptides is the positioning of two lysines within a stretch of poly-alanine, with the lysines separated either by two or three alanines (Sandberg, 1976). It has been suggested that these regions form short α-helical segments with the two lysine side-chains sticking out from the same side of the helix and thus the lysines are prestructured into reactive pairs. In cross-link formation one desmosine

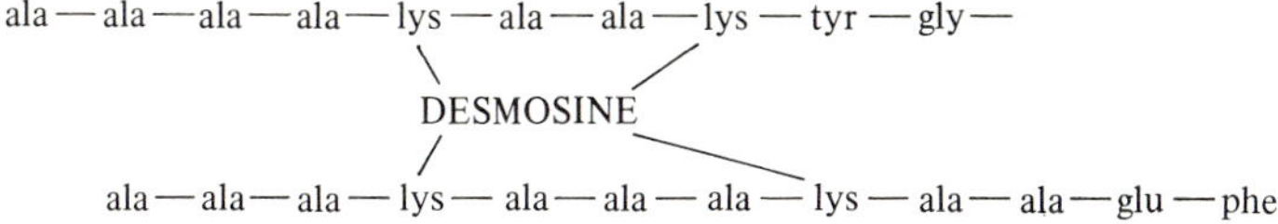

Fig. 3. A desmosine-containing peptide derived from the extensive proteolytic digestion of insoluble elastin (see Sandberg, 1976). Note the long poly-alanine peptides with two lysines inserted.

(or isodesmosine) forms between two of these cross-link regions to connect two network chains, rather than form a complex between four chains if each chain provides one of the lysines.

Elastic mechanism and static elastic properties of protein rubbers

Contribution of the swelling solvent to the elastic mechanism

As discussed in Mullins (1980), the kinetic theory of rubber elasticity provides an excellent molecular model to explain the long-range elastic properties of polymeric materials. This theory is also quite adequate to explain all aspects of the elastic mechanism of the protein rubbers as well, but one important feature of protein rubbers has made the interpretation of their properties more complex and has led to the formation of 'novel' elastic mechanisms based on the interaction of the protein chains and their swelling solvent. Protein rubbers are only rubber-like when they are swollen with a polar solvent like water. For this reason much of the experimental work on the elastic mechanism has been carried out with specimens that are in swelling equilibrium with water or some similar aqueous solvent. Analysis of the thermodynamics of such open systems is quite complex and this is where the problems have arisen.

No attempt will be made here to derive the details of the kinetic theory (see Mullins (1980), Flory (1953) and Treloar (1975) for excellent reviews); however, a brief discussion of open systems will be presented. The basic premise of the kinetic theory is that the mechanical distortion of kinetically free, random polymer chains will decrease their conformational entropy, S_{conf}, by causing the chains to shift from a most probable set of conformations to a less probable set. Elastic recoil arises because the 'deformed' chains will spontaneously diffuse back to the most probable set of shapes, the driving force being provided by an increase

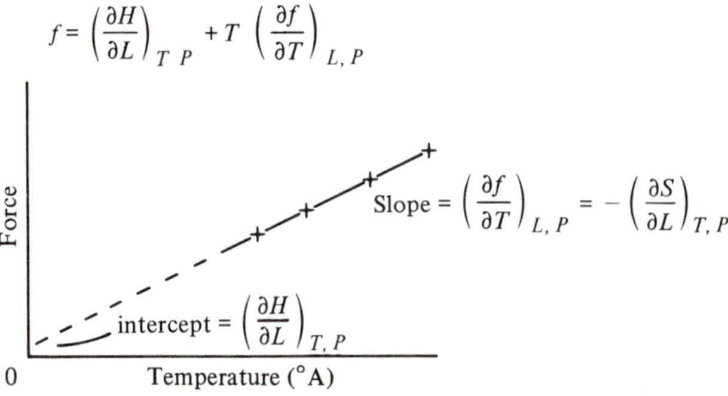

Fig. 4. Thermoelastic measurements for determining the enthalpic and entropic contribution to the stored energy of an elastic solid. If the elastic solid is held at constant length and the force measured at different temperatures, the slope of the force–temperature curve will give the entropic contribution to the stored energy, and the intercept will give the enthalpic contribution. The plot shown is characteristic of that seen for most rubbers.

in entropy. The elasticity of stiff solids (e.g. steel) is derived from the direct distortion of chemical bonds (i.e. changes in internal energy, E), rather than in changes in conformational entropy as in compliant, rubbery solids. The Wiegand-Schneider (1934) equation, shown below, provides the basis for an experimental protocol, namely thermoelastic measurements, that enables us to determine the energy and entropy contributions to the stored elastic energy, W, for any elastic solid. The equation is:

$$f = (\partial H/\partial L)_{T,P} + T(\partial f/\partial T)_{L,P} \tag{1}$$

where f is the elastic force, H the enthalpy, L the length of the sample, P the pressure, and T the absolute temperature. The first term on the right gives the enthalpic contribution to the elastic force, and the second term gives the entropic contribution. That is, $(\partial f/\partial T)_{L,P} = -(\partial S/\partial L)_{T,P}$. The kinetic theory of rubber elasticity predicts that all of the elastic energy will be stored as a change in entropy (the second term) and that the contribution of the first term will be zero. Thus, if a sample of rubber is held stretched at constant length and the temperature is varied, the elastic force should be directly proportional to absolute temperature, and the intercept of the elastic force at zero degrees absolute should be zero, or very close to it (Fig. 4).

Most rubbers meet these criteria quite well. However, the use of this relation involves an approximation that is not always valid. The kinetic theory predicts specifically that $(\partial E/\partial L)_{L,V} = 0$, or that there is no change in *internal energy* with stretching at constant *volume* (Flory, 1953). The Wiegand-Schneider relation provides a means of measuring

$(\partial H/\partial L)_{T,P}$, the change in enthalpy at constant pressure. For most rubbers tested as closed systems (i.e. if they cannot absorb solvent during stretching), $(\partial E/\partial L)_{T,V} \simeq (\partial H/\partial L)_{T,P}$, and no problems are encountered. This approximation also holds for rubbers tested as open systems, but only if the swelling of the rubber does not change with temperature (Hoeve & Flory, 1958). However, for the open systems the energy and entropy terms relate to the entire system consisting of the network, absorbed solvent and free solvent. Thus, these terms cannot be unequivocally attributed to the conformational state of the chains in the network, but also include a contribution from the mixing of the solvent and the polymer.

Both resilin (Weis-Fogh, 1961*a*) and abductin (Alexander, 1966) show very small changes in swelling with temperature, and thermoelastic analysis indicates that $(\partial H/\partial L)_{T,P} \simeq 0$. Thus, to a first approximation, one can conclude that for these protein rubbers the elastic energy is stored primarily as a change in entropy. Presumably this entropy change is associated with changes in the conformation of network chains, in accordance with the kinetic theory.

Water-swollen elastin, however, shows large changes in swelling with temperature; to the extent that the volume is decreased by more than 40% as temperature is increased from 0 to about 50°C. Thus, $(\partial E/\partial L)_{T,V} \neq (\partial H/\partial L)_{T,P}$ because changes in enthalpy associated with the mixing of the solvent and the network chains contribute to the total enthalpy change. Indeed, thermoelastic measurements (Meyer & Ferri, 1937; Wöhlisch *et al.*, 1943) and calorimetric measurements (Weis-Fogh & Andersen, 1970) of water-swollen elastin indicate that $(\partial H/\partial L)_{T,P}$ is extremely large and unexpectedly *negative*. Since the swelling changes with temperature are apparently due to the extreme hydrophobic character of the elastin protein (Gosline, 1978*a*), and since the enthalpy changes for the mixing of water and non-polar compounds are large and negative (Nemethy & Scheraga, 1962), it has been suggested that part of the elastic force of elastin arises from a 'hydrophobic mechanism' involving the exposure of non-polar groups to water (i.e. breaking hydrophobic bonds) when the network is deformed (Partridge, 1968; Robert, Robert & Robert, 1970; Weis-Fogh & Andersen, 1970; Hoffman, 1971; Gray *et al.*, 1973; Urry & Long, 1976*a*). In fact, a detailed analysis of the thermodynamics of elastin (Gosline, 1978*b*) shows that such a hydrophobic mechanism does make a very large contribution to the stored elastic energy. Fig. 5*a* shows the changes in free energy associated with the absorption of water on to hydrophobic groups, ΔF_{h}, and with changes in the conformational state of the backbone chains,

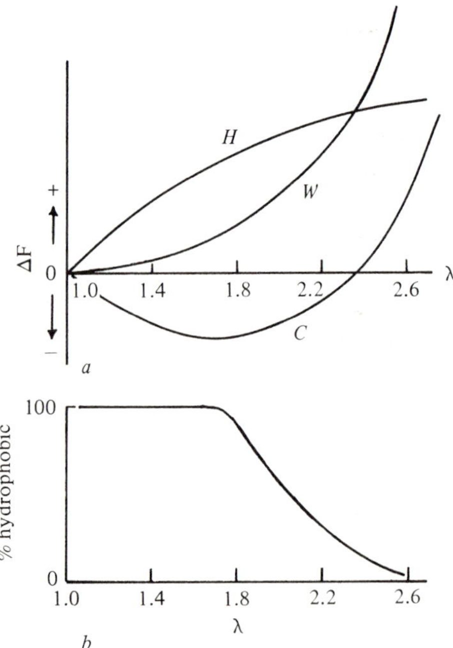

Fig. 5. The thermodynamics of elastin stretched as an open system, after the data of Gosline (1978*b*). *a*. The free-energy change associated with the stress-induced absorption of water on to hydrophobic groups, *H*, added to the free-energy change associated with changes in the backbone conformational state, *C*, is equal to the stored elastic energy, *W*. *b*. The relative importance of the hydrophobic mechanism to the stored elastic energy as a function of extension ratio. Since $dC/d\lambda$ is negative at extension ratios below about 1.7, the chain conformation mechanism makes no contribution below this value.

ΔF_{conf}, as a function of extension. The hydrophobic free energy is positive because work must be done to cause water to hydrate non-polar groups in the network. The conformational free energy is negative at low extensions but becomes positive at high extensions. This shift occurs because at low extensions the dilution of the network by the absorbed solvent increases entropy (i.e. an ideal entropy of mixing) faster than the uniaxial extension decreases entropy by 'pulling-out' the coiled chains. At large extensions the situation is reversed. The slope of these free energy–extension curves provides an indication of the contribution of each mechanism to the stored elastic energy, *W*, and Fig. 5*b* shows the relative importance of the hydrophobic mechanism, calculated from these data. These curves are not exact, but it is clear that the hydrophobic mechanism predominates at low and medium extensions, while the chain conformation mechanism dominates at large extensions.

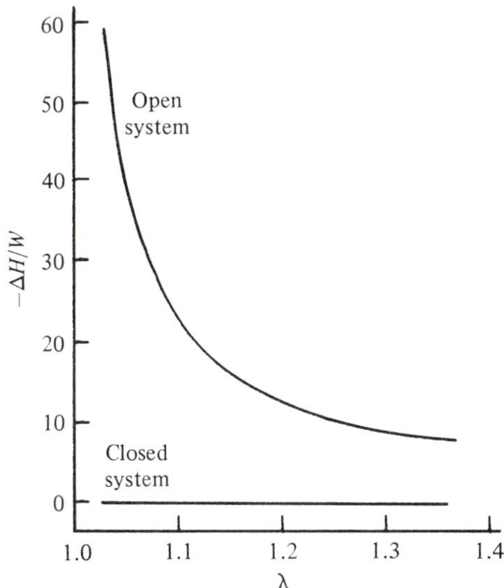

Fig. 6. The ratio of the calorimetrically measured change in enthalpy, ΔH, to the stored elastic energy, W, for elastin tested as a closed system with constant water content and as an open system that can absorb water as the sample is stretched.

The importance of hydrophobic interactions to the elasticity of elastin provides perhaps the major stimulus for the formulation of novel network structures, such as the oiled-coil and the helical fibrillar models discussed previously and the liquid-drop model of Weis-Fogh & Andersen (1970). These models were proposed to provide structures that could accommodate a hydrophobic mechanism. However, a completely random network structure can easily account for this phenomenon, since the extension of any system that is in equilibrium with its swelling solvent will result in the absorption of solvent (Hoeve & Flory, 1974; Grut & McCrum, 1974). If the network contains non-polar groups, then the absorbed solvent will increase the average hydration of these groups and, in effect, disrupt hydrophobic interactions. Thus, there is no need to specify an elaborate network structure to explain the hydrophobic contribution. Furthermore, analysis of elastin tested as a closed thermodynamic system will allow us to exclude these other network structures, since they all *require* that hydrophobic interactions contribute to the elastic mechanism.

Fig. 6 shows the results of some unpublished calorimetric data for elastin tested both as an open system and as a closed system. The closed system was created by immersing hydrated elastin in mineral oil, so that no water could be absorbed when the sample was stretched. The static mechanical properties of elastin as an open and as a closed system are virtually identical, but the energetics of the elastic mechanism are dramatically different. The figure shows the ratio of the change in enthalpy, ΔH, to the stored elastic energy, W, as a function of extension. The presence of the hydrophobic component for the open system is indicated by the large negative ratio, particularly obvious at low extensions. (Remember that large negative enthalpy changes are characteristic of the mixing of water and non-polar compounds.) For the closed system, however, this ratio is zero. All the elastic energy is stored as a change in entropy and, since water cannot enter the network with stretching, this entropy change must be associated with conformational changes of the network chains in accordance with the kinetic theory. The random network model provides the only structure that can explain the properties of elastin both as a closed and as an open system. In fact, the quantitative analysis of random networks as open systems have been worked out in some detail (see Treloar, 1975), and the mechanical and thermodynamic properties of elastin fit the predictions quite well.

Network properties of protein rubbers

According to the kinetic theory, the force–extension curve for the uniaxial extension of a swollen rubber should follow

$$f = NkTv_2^{1/3}(\lambda - 1/\lambda^2) \quad (2)$$

where f is the force per unit swollen, unstressed cross-sectional area, N is the number of chains per unit volume of the unswollen sample, k is the Boltzmann constant, T is the absolute temperature, v is the volume fraction of polymer and λ is the extension ratio related to the dimensions of the swollen sample (L/L_0). The number of chains, N, provides the only description of the network structure that is necessary, and the term NkT provides a measure of the stiffness or modulus of the rubber. The symbol G is commonly used for the elastic modulus,

$$G = NkT = \varrho RT/M_c \quad (3)$$

where ϱ is the density of the dry polymer, R the gas constant and M_c the average molecular weight of random chains between cross-links. Eqns (2) and (3) provide useful, although not exact, descriptions of the static properties of many rubbers.

Figure 7 shows a plot of eqn (2) for which $G = 6.5 \times 10^5\,\mathrm{Nm}^{-2}$ and $v_2 = 0.45$, typical values for resilin tendons from a dragonfly at pH 6.7 (Weis-Fogh, 1961*b*). In addition, one of the experimental curves for a resilin tendon has been redrawn from Weis-Fogh (1961*b*). Note that the

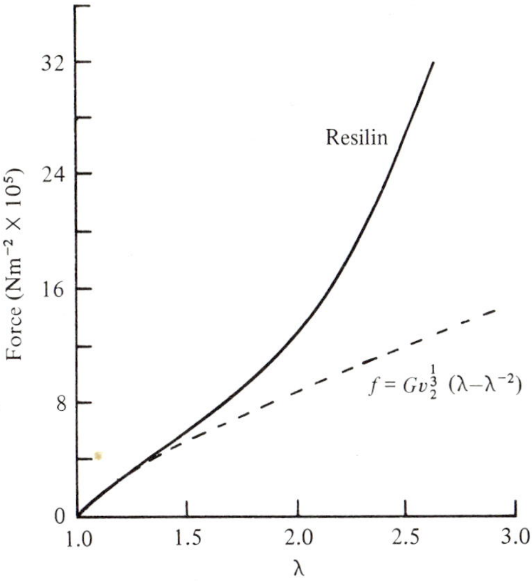

Fig. 7. The force-extension curve for dragonfly tendon resilin (after Weis-Fogh, 1961*b*) compared to the elastic properties of a Gaussian network according to the kinetic theory of rubber elasticity.

experimental curve follows the theoretical curve only to about $\lambda = 1.5$, and beyond this extension ratio the force becomes increasingly larger than the predictions. This deviation from eqn (2) is seen for all rubbers. It arises because this equation is based on a Gaussian distribution function to describe the chain dimensions in the network. This approximation works at small extensions, where the network chains are highly coiled, but real networks become non-Gaussian and get increasingly stiff as the chains are pulled out. More realistic ways of describing the large strain behaviour of rubbers are described by Mullins (1980). The fact that resilin shows non-Gaussian properties at relatively low extensions suggests that its protein chains are quite stiff. Indeed, Weis-Fogh's (1961*b*) elegant analysis of the mechanical and photoelastic properties of resilin suggests that the random chain contains only 5 to 25 random segments, with a best guess at about 10 to 15.

From eqn (3), we can calculate that M_c for resilin is about 5100, and since the average residue weight is 88.5, there are about 55 to 60 amino acids per random chain (Weis-Fogh, 1961*b*). If we assume, from above, that the random chain contains 10 to 15 random segments, then each segment is about 4 to 6 amino-acid residues in length and contains 8 to 12 single bonds allowing rotational freedom. This suggests a rather inflexible chain, and is probably quite characteristic of other protein rubbers as well.

In comparison to resilin, less is known about the detailed network properties of elastin than abductin. This is probably because it is difficult to obtain suitable test samples. Abductin forms a small, irregular piece of the shell hinge, and elastin is always found as bundles of very fine fibres or very thin sheets. In both cases it is difficult to obtain convenient specimens. For example, many experiments with purified elastin from bovine *ligamentum nuchae* show increasing stiffness at low extensions, in contrast to the decreasing slope predicted by eqn (2) (see Fig. 7). This shape of force–extension curve probably reflects non-uniform loading of the fibres and the recruitment of fibres with extension (Hoeve & Flory, 1958; Gosline, 1978*b*). The only way to avoid this sort of problem is to test individual 5 μm thick fibres, and this is experimentally quite difficult. The following are some unpublished data for single elastin fibres obtained in my laboratory by Mr Ben-Meyer Aaron. The elastic modulus, G, is about $4.5 \times 10^5\,\mathrm{Nm}^{-2}$, and the corresponding M_c is 7100. The average residue weight for elastin is 85, making the average number of amino acids in a random chain about 85. The analysis of the mechanical and photoelastic properties suggests that non-Gaussian effects become important at extension ratios around 1.8, or at slightly higher extensions than for resilin. Thus, we should expect that the number of random segments per chain is greater in elastin than in resilin, perhaps in the range of 15 to 20. This means that the random segment in elastin is of the same order of size as in resilin.

Alexander (1966) reports that E (Young's modulus) for abductin is about $4 \times 10^6\,\mathrm{Nm}^{-2}$ and that v_2 is about 0.5. Since $G \simeq E/3$ for rubbers at low extensions, G for abductin is about $1.5 \times 10^6\,\mathrm{Nm}^{-2}$, making M_c about 2000. Virtually nothing is known about the details of the force-extension or the photoelastic properties.

Dynamic elastic properties

As with all polymeric materials, protein rubbers demonstrate viscous as well as elastic properties. The analysis of the elastic mechanism and network properties in the previous section assumes that the rubber is deformed very slowly, so that viscous processes are negligible. However, if reasonable rates of deformation are used, viscous properties can become extremely important. The viscoelastic–glass transition (see Dorrington, 1980) provides the major viscous process that affects the functional properties of protein rubbers. This transition, over which pliant rubbery solids are converted into rigid glasses, is associated with a dramatic increase in frictional forces between segments in the network, and ultimately with the total loss of segmental mobility. As already mentioned (p. 333), a major design problem for protein rubbers is the construction of peptide sequences that allow segmental mobility in aqueous environments. Protein rubbers appear to have solved this problem but, as with all polymeric materials, variables such as temperature, strain rate and swelling will affect the ability of the protein chains to remain mobile. A reasonably complete analysis of the glass transition of elastin is now available and this will be presented to provide an indication of the properties to be expected for protein rubbers in general. The limited dynamic data available for resilin and abductin will be included as the discussion proceeds.

At normal temperatures all protein rubbers are rigid glasses when dry (Weis-Fogh, 1960; Gosline, 1976). This lack of segmental mobility undoubtedly arises from the polar character of the peptide group in the backbone chain of all proteins. In the absence of water or some other polar solvent, peptide–peptide hydrogen bonds will hold the folded chains into stable, static conformations. Increased temperature should disrupt these hydrogen bonds and allow dry proteins to become rubbery, but the strength of the hydrogen bonds is so great, and the glass structure so stable, that dry protein rubbers like elastin decompose before they become rubbery (Gotte *et al.*, 1968). Polar solvents can compete for peptide sites and reduce the peptide–peptide interactions that limit segmental mobility, but the analysis of the network properties of water-swollen protein rubbers (p. 316) indicates that even hydrated peptides are relatively inflexible. Thus, it is to be expected that protein rubbers will be quite close to their glass transition and may show viscoelastic properties that could limit their use for high frequency energy storage systems.

The recent study by Gosline & French (1979) provides an analysis of the high frequency properties of elastin. Because elastin is very hydrophobic its water content is quite low at physiological temperatures. Further, the hydrophobic properties make the swelling of elastin very temperature-dependent. Since the glass transition of an amorphous polymer is affected by both temperature and swelling, the analysis of the properties of elastin can be quite complex. However, if elastin is isolated as a closed system with fixed water content (see p. 343), it is possible to apply standard time-temperature shifting procedures (see Dorrington, 1980) and investigate the dynamic elastic properties over a broad range of frequency. Fig. 8 shows the results of such dynamic

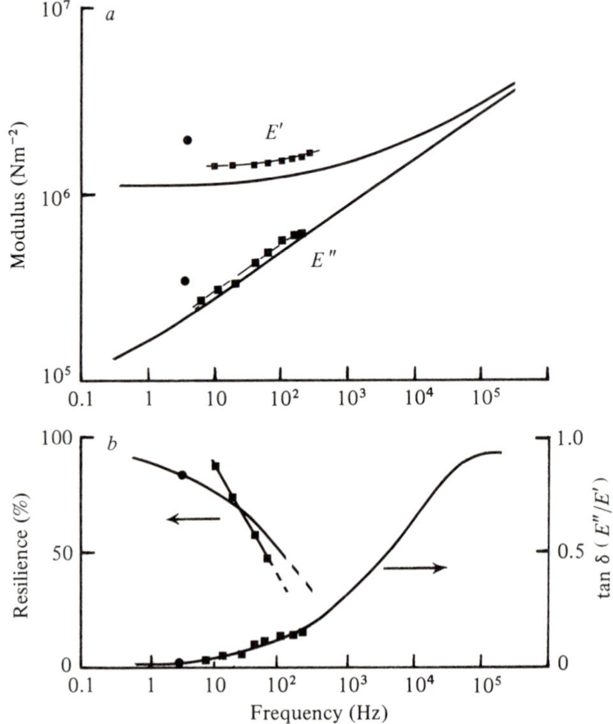

Fig. 8. The dynamic elastic properties of protein rubbers. The smooth solid lines are for elastin, the lines connecting solid squares are for resilin, and the single solid circles are for abductin. *a*, The storage modulus, E' and the loss modulus E'' as a function of frequency. *b*. The loss tangent, $\tan\delta = E''/E'$ and the resilience, $R = e^{-2\pi\tan\delta}$ as a function of frequency.

measurements for purified *ligamentum nuchae* elastin hydrated at 36°C (water content is 0.46 g water per g protein) and temperature shifted to 36°C. These data probably give a reasonable indication of elastin's properties as it exists *in vivo* in a mammal. Figure 8*a* shows the storage modulus, E', and the loss modulus, E'', as a function of frequency at a strain amplitude of about 0.01. Figure 8*b* shows the loss tangent ($\tan\delta = E''/E'$), which gives a measure of the relative amount of energy lost to the energy stored per cycle. In addition, Fig. 8*b* shows the resilience, R, calculated from these dynamic data. Resilience is defined as the ratio of the energy stored in two successive cycles of a damped free oscillation, and as long as the loss tangent is small ($\tan\delta < 0.1$) $R = e^{-2\pi\tan\delta}$. For $\tan\delta$ greater than 0.1 it is difficult to relate resilience to loss tangent (Bueche, 1962).

The properties are quite characteristic of an amorphous polymer close to its glass transition. Although the frequency range is not great enough to see the complete transition, the dramatic increase in the loss modulus, E'', and the corresponding peak in tan δ with increasing frequency are exactly as expected. Detailed analysis of the data reveals that the transition fits the WLF equation (Ferry, 1970), an empirical relation that provides an accurate description of the glass transition. The transition temperature is about -40°C. The calculated resilience values plotted in Fig. 8*b* show that elastin can function as an efficient energy storage system as long as frequency is kept reasonably low.

Fig. 8 also shows the data for abductin (Alexander, 1966) and resilin (Jensen & Weis-Fogh, 1962). Abductin is very similar to elastin at 3 Hz, although it is about three times as stiff. The strain amplitude of this measurement is not known. Resilin, surprisingly, has dynamic elastic properties that are also almost identical to those of elastin. This is surprising because resilin is normally regarded as being exceptionally resilient. This reputation, however, is based on the rather non-standard way that Jensen & Weis-Fogh (1962) presented their dynamic data (i.e. the loss factor, h). In a later publication (Andersen & Weis-Fogh, 1964) the data are presented as the ratio of the out-of-phase and in-phase components of the dynamic force, a measure that is equal to the loss tangent, tan δ. Therefore, the values presented here have been calculated from Jensen & Weis-Fogh's data for the per cent increase in dynamic stiffness, a static Young's modulus of 2×10^6 Nm^{-2}, and the tan δ values from Andersen & Weis-Fogh (1964). The resilience of resilin is greater than that of either elastin or abductin at low frequencies, but appears to be lower at high frequencies. This is surprising in light of the fact that resilin pads appear to be able to release large amounts of stored elastic energy over a time scale of less than 1 ms (a 1-ms time constant is approximately equivalent to cyclic loading at 160 Hz) in the jump of a flea (Bennet-Clark & Lucey, 1967). However, the locust prealar arm resilin used by Jensen & Weis-Fogh (1962) is probably not an ideal test specimen and the strain amplitudes, although not specified, were much higher than 0.01, which were used in the elastin studies presented there. Further, the swelling of a protein rubber has a dramatic effect on the dynamic properties, and the swelling of resilin is very pH-dependent (Weis-Fogh, 1961*b*). It is possible that the mechanical properties of resilin are somewhat different under *in vivo* conditions.

The effect of swelling is dramatically demonstrated in Fig. 9, which presents the dynamic properties of elastin hydrated at 97% relative

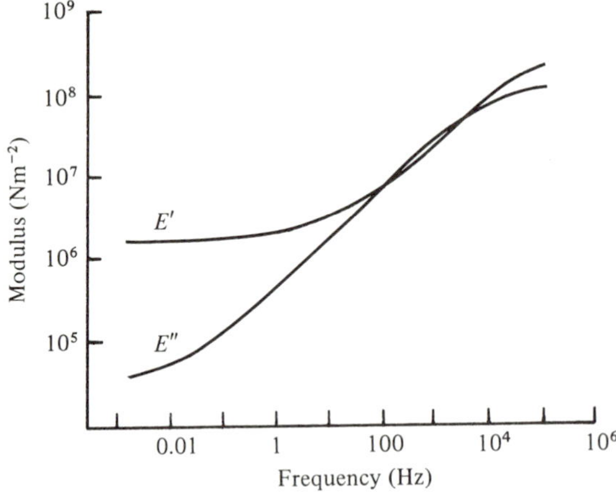

Fig. 9. The dynamic elastic properties of elastin hydrated at 97% relative humidity, having a water content of about 0.32 g/g.

humidity at 39°C. The reduced humidity lowers the water content from 0.46 $g g^{-1}$ (Fig. 8) to about 0.32 $g g^{-1}$. Almost the entire glass transition can be seen. The storage modulus increases about 200-fold, and the loss tangent goes through a peak and begins to decrease. These changes indicate that the material is becoming very glass-like. The form of the curves is virtually unchanged from Fig. 8*a*, with the exception that the data are shifted along the time scale by two to three decades of frequency. WLF analysis indicates that the glass transition temperature is increased to about 0°C. This strong dependence of the glass transition temperature on water content for elastin has been demonstrated by differential scanning calorimetry over a wide range of water contents (Kakivaya & Hoeve, 1975), and thus any consideration of the use of protein rubbers in high frequency energy storage systems must take this variable into account.

A useful example of how swelling behaviour may determine functional properties is presented in Fig. 10. This figure shows the calculated resilience at 1 Hz for elastin tested both as a closed system with a constant water content of 0.46 $g g^{-1}$, and as it probably exists in vivo as an open system. Elastin is present in the major arteries of all vertebrates, and it must function efficiently to store energy from the contraction of the heart and smooth the pulsatile flow of blood. However, lower vertebrates do not control their body temperature and animals like arctic fish must function at temperatures as low as −3°C. Since elastin functions close to its glass transition, a 40°C reduction in temperature could easily decrease resilience to the point where the

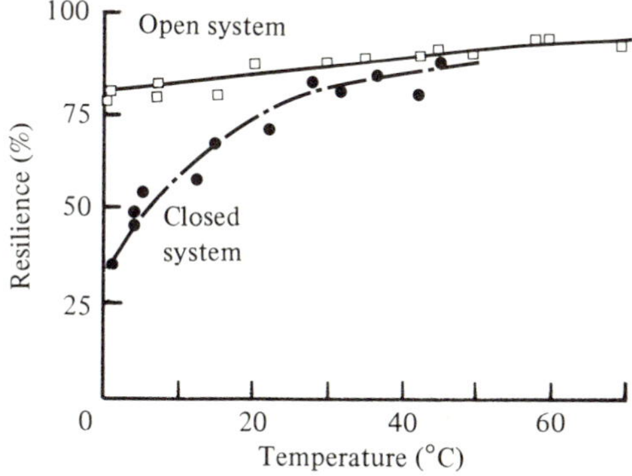

Fig. 10. The resilience of elastin tested as a closed system with a constant water content of about 0.46 g g^{-1} and as an open system in which the water content increases with decreasing temperature.

material was useless. This loss of resilience is demonstrated by elastin tested as a closed system. At 0°C the resilience of the closed system is in the range of 30 to 40%. (Note, this is only an approximation because tan δ is greater than 0.1 by this point.) However, elastin is unusual in its extremely high content of hydrophobic amino acids and the strong temperature-dependent swelling behaviour that arises because of this composition. If elastin is left in swelling equilibrium with water, it dramatically increases its water content at low temperatures (e.g. at 2°C the water content is about 0.76 gg^{-1}). As shown by the curve labelled 'open system' the increased water content balances the effect of decreased temperature yielding a material that shows virtually no decrease in resilience, even at 0°C. Gosline & French (1979) suggest that the evolution of elastin as a hydrophobic protein may well have taken place for this reason.

Artery wall – a pliant composite material

The large elastic arteries of mammals provide perhaps the best example of a pliant composite showing the interaction of a protein rubber, elastin, with a stiff fibre, collagen. The description presented here is by no means complete, and interested readers should consult Patel & Vaishnav (1972), Bergel (1972), or McDonald (1974). The major arteries of all animals that have closed, high pressure circulatory systems are compliant, rubber-like tubes. This type of construction is required to reduce the pressure peak during systole and maintain blood flow during diastole. The elastic arteries expand in radius during a pressure pulse

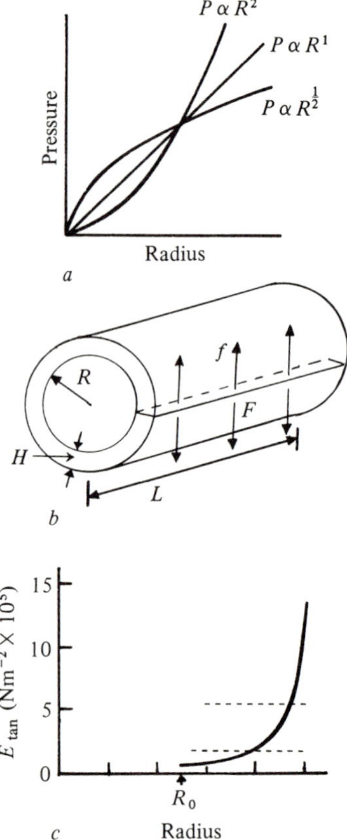

Fig. 11. The design of materials for the arterial wall. *a*, the shape of some pressure–radius curves; *b*, measurements required for estimating wall stress in a cylinder such as an artery: *R*, radius; *H*, wall thickness; *f*, force across cylinder wall; *L*, length of cylinder; *c*, the change in the tangential modulus, E_{tan}, of canine thoracic aorta (after Bergel, 1972). The interrupted lines indicate the approximate range of physiological values. See the text for a more complete analysis.

from the heart (the major arteries are generally tethered in vivo and do not change in length), and the elastic energy stored in the vessel wall provides the force that maintains the flow of blood.

The cylindrical geometry of an artery imposes a major structural problem that requires careful design of the materials in the tissue. Inflation experiments carried out on tethered arteries indicate that the pressure–radius curve can be reasonably described by a power function, $P \propto R^n$, where n is usually between 2 and 3. The implications of this kind of pressure-radius curve are described in Fig. 11*a*, which shows plots of $P \propto R^n$, for $n = 0.5$, 1.0 and 2.0. Consider first the case of an inflation for which $n = 0.5$. Since this curve has a continuously decreasing slope, it will be easier to inflate the tube if it has a larger radius. If the tube is not perfectly uniform, an area with slightly larger radius than the rest will 'balloon-out', leaving the rest of the tube unchanged. This, in fact, is just what happens when one inflates a cylindrical rubber balloon. The balloon does not burst at this 'aneurism' only because the rubber in this region is forced into its non-Gaussian region and gets very stiff. It is clear, however, that this sort of unstable behaviour is inappropriate for an artery. If $P \propto R^1$, the slope is constant at all values of R. The system should be stable, but only just stable. Animal design has provided a safety factor by making n considerably larger than unity.

In terms of structural design, how are these properties achieved? A linear pressure–radius curve might suggest a material with a constant modulus, but the geometry of a cylindrical tube makes the problem more complex. The force across the wall of a cylinder (assuming a thin-walled tube) can be calculated as, $f = PRL$, where P is pressure, R radius, and L length (see Fig. 11*b*). The wall stress $\sigma = f/A = PRL/LH$, where A is the cross-sectional area and H is the wall thickness. Taking an intuitive rather than an exact approach, let $H \propto 1/R$ (because the wall must get thinner as the circumference increases) and $P \propto R^n$. More accurately, $H \propto R/R_0$ and $P \propto (R - R_0)^n$, where R_0 is the unstressed radius, but this complicates the arithmetic greatly. Thus, $\sigma \propto R^{n+2}$.

The incremental or tangential modulus, E_{tan}, provides a convenient way of expressing the change in modulus with inflation, where,

$$E_{tan} = \mathrm{d}\sigma/\mathrm{d}\varepsilon = R\,\mathrm{d}\sigma/\mathrm{d}R$$

Thus, to a first approximation $E_{tan} \propto R^{n+2}$. Since n is 2 or greater for arteries, arterial stiffness must increase with radius to the fourth or higher power (see Fig. 11*c*). For example, the tangential modulus of canine thoracic aorta goes from about $2 \times 10^5\,\mathrm{Nm^{-2}}$ to $5 \times 10^5\,\mathrm{Nm^{-2}}$ over the biological pressure range (10.6 to 16.0 $\mathrm{kNm^{-2}}$), involving radial strains of the order of 8% (Bergel, 1972). The incremental modulus of a rubber varies only slightly at low strains, and this is one reason why arteries are not simply made from elastin.

Some sort of reinforcement is required, and the parallel arrangement of elastin and collagen, with stress being transferred from the elastin to the collagen in the biological range, is the normal explanation of the properties of arteries (Roach & Burton, 1957). The structural analysis of Wolinsky & Glagov (1964) suggests, however, that the situation is more complex, and that tissue architecture plays an important role. This aspect really goes beyond the present analysis of materials, but briefly the system can be described as follows. A three-component structure consisting of concentric elastin lamellae, fine inter-lamellar elastin fibres, and helically arranged, wavy collagen fibres, provides a reasonable explanation of the static elastic properties. At pressures below about 10 kNm^{-2} the elastin lamellae are unstrained and appear wavy in cross-section. In this range, the wall tension is carried by the fine inter-lamellar fibres. At the low end of the physiological range of pressure the elastin lamellae become straight and begin to carry some of the wall tension and throughout the biological range of pressures the increase in stiffness is largely attributable to the recruitment of elastin components. In addition, of course, the wavy collagen fibres become increasingly extended and start to carry more of the load, particularly at higher pressures. However, the limiting properties attributed to the collagen mesh alone are reached at pressures well above the normal range (i.e. at about 33 kNm^{-2}).

Summary

We are currently approaching a reasonable understanding of the macroscopic properties of protein rubbers, although much remains to be learned at the molecular level. Protein chains can form kinetically-free, random, networks and the kinetic theory of rubber elasticity does provide an accurate description of the elastic properties. Information on the network properties of other protein rubbers should be available in the near future, but we are still a long way from understanding the principles that govern the sequence design of kinetically-free peptides. Although we probably have a good general understanding of the dynamic elastic properties of protein rubbers, there are undoubtedly many interesting details that await discovery such as the relationship between amino acid sequence, swelling and the functional requirements of individual organisms. We also have a reasonable, general, understanding of how protein rubbers are incorporated into pliant composites, but again many of the subtle interactions that undoubtedly determine important functional properties in specific tissues are yet to be described.

References

Alexander, R. McN. (1966). Rubber-like properties of the inner hinge-ligament of *Pectinidae. Journal of experimental Biology*, **44,** 119–30.

Andersen, S. O. (1967). Isolation of a new type of crosslink from the hinge ligament protein of molluscs. *Nature, London,* **216,** 1029–30.

ANDERSEN, S. O. (1971). Resilin. In *Comprehensive Biochemistry*, ed. M. Florkin & E. H. Stotz, Vol. 26C, pp. 633–57. Amsterdam: Elsevier.

ANDERSEN, S. O. & WEIS-FOGH, T. (1964). Resilin, a rubber-like protein in arthropod cuticle. *Advances in Insect Physiology*, **2**, 1–65.

BENNET-CLARK, H. C. & LUCEY, E. C. A. (1967) The jump of the flea. *Journal of experimental Biology*, **47**, 59–76

BERGEL, D. H. (1972). The Properties of Arteries. In *Biomechanics*, ed. Y. C. Fung, N. Perrone & M. Anlicker, pp. 105–39. Englewood Cliffs, New Jersey: Prentice-Hall.

BUECHE, F. (1962). *Physical Properties of Polymers*. New York: John Wiley & Sons.

DENNY, M. W. (1980). Silks – their properties and functions. In *The Mechanical Properties of Biological Materials* (34th Symposium of the Society for Experimental Biology) ed. J. F. V. Vincent & J. D. Currey, pp. 247–71. Cambridge University Press.

DORRINGTON, K. L. (1980). The theory of viscoelasticity in biomaterials. In *The Mechanical Properties of Biological Materials* (34th Symposium of the Society for Experimental Biology) ed. J. F. V. Vincent & J. D. Currey, pp. 289–314. Cambridge University Press.

ELDER, H. Y. & OWEN, G. (1967). Occurrence of 'elastic' fibres in the invertebrates. *Journal of Zoology*, **152**, 1–8.

ELLIS, G. E. & PACKER, K. J. (1976). Nuclear spin-relaxation studies of hydrated elastin. *Biopolymers*, **15**, 813–32.

FERRY, J. D. (1970). *Viscoelastic properties of Polymers*, 2nd edition. New York: John Wiley & Sons.

FLORY, P. J. (1953). *Principles of Polymer Chemistry*. Ithaca, New York: Cornell University Press.

FLORY, P. J. (1969). *Statistical Mechanics of Chain Molecules*. New York: Wiley.

GORDON, J. E. (1975). Mechanical instabilities in biological membranes. In *Comparative Physiology – Functional Aspects of Structural Materials*, ed. L. Bolis, S. H. P. Maddrell & K. Schmidt-Nielsen. pp. 49–57. Amsterdam: North Holland.

GOSLINE, J. M. (1971). Connective tissue mechanics of *Metridium senile*. II. Viscoelastic properties and a macromolecular model. *Journal of experimental Biology*, **55**, 775–95.

GOSLINE, J. M. (1976). The physical properties of elastic tissues. *International Review of Connective Tissue Research*, **7**, 211–49.

GOSLINE, J. M. (1978*a*). The temperature-dependent swelling of elastin. *Biopolymers*, **17**, 697–707.

GOSLINE, J. M. (1978*b*). Hydrophobic interaction and a model for the elasticity of elastin. *Biopolymers*, **17**, 677–95.

GOSLINE, J. M. & FRENCH, C. (1979). Dynamic mechanical properties of elastin. *Biopolymers*, **18**, 2091–103.

GOTTE, L., MAMMI, M. & PEZZIN, G. (1968). Some structural aspects of elastin revealed by X-ray diffraction and other physical methods. In *Symposium of Fibrous Proteins*, ed. W. G. Crewther, pp. 236–245. Chatswood, N.S.W.: Butterworths.

GRAY, W. R., SANDBERG, L. B. & FOSTER, J. A. (1973). Molecular model for elastin structure and function. *Nature, London.*, **246**, 461–6.

GRUT, N. & MCCRUM, N. G. (1974). Liquid drop model of elastin, *Nature, London.*, **251**, 165.

HOEVE, C. A. J. & FLORY, P. J. (1958). The elastic properties of elastin. *Journal of the American Chemical Society*, **80**, 6523–6.

HOEVE, C. A. J. & FLORY, P. J. (1974). The elastic properties of elastin. *Biopolymers*, **13**, 677–86.

HOFFMAN, A. S. (1971). A critical evaluation of the application of rubber elasticity principles to the study of structural proteins such as elastin. In *Biomaterials*, ed. A. L. Bement, Jr. Seattle: University of Washington Press.

JENSEN, M. & WEIS-FOGH, T. (1962). Biology and physics of locust flight. V. Strength and elasticity of locust cuticle. *Philosophical Transactions of the Royal Society of London, Series B*, **245,** 137–69.

KAHLER, G. A., FISHER, F. M. & SASS, R. L. (1976). The chemical and mechanical properties of the hinge ligament in bivalve molluscs. *Biological Bulletin*, **151,** 161–81.

KAKIVAYA, S. R. & HOEVE, C. A. J. (1975). The glass point of elastin. *Proceedings of the National Academy of Sciences of the USA*, **72,** 3505–7.

KASTELIC, J. & BAER, E. (1980). Deformation in tendon collagen. In *The Mechanical Properties of Biological Materials* (34th Symposium of the Society for Experimental Biology) ed. J. F. V. Vincent & J. D. Currey, pp. 397–435. Cambridge University Press.

MCDONALD, D. A. (1974). *Blood Flow in Arteries*. London: Edward Arnold.

MARUYAMA, K., MATSUBARA, S., NATORI, R., NONOMURA, Y., KIMURA, S., OHASHI, K., MURAKAMI, F., HANDA, S. & EGUCHI, G. (1977). Connectin, an elastic protein of muscles. *Journal of Biochemistry*, **82,** 317–37.

MEYER, K. H. & FERRI, C. (1937). Die elastischen Eigenschaften der elastischen und der kollagenen Fasern und ihre molekulare Deutung. *Pflugers Archiv für die gesamte Physiologie*, **238,** 78–90.

MULLINS, L. (1980). Theories of rubber-like elasticity and the behaviour of filled rubbers. In *The Mechanical Properties of Biological Materials* (34th Symposium of the Society for Experimental Biology) ed. J. F. V. Vincent & J. D. Currey, pp. 273–88. Cambridge University Press.

NEMETHY, G. & SCHERAGA, H. A. (1962). The structure of water and hydrophobic bonding in proteins. III. The thermodynamic properties of hydrophobic bonds in proteins. *Journal of Physical Chemistry*, **66,** 1773–89.

PARTRIDGE, S. M. (1968). Elastin structure and biosynthesis. In *Symposium on Fibrous Protein*, ed. W. G. Crewther, pp. 246–64. Chatswood, N.S.W.: Butterworths.

PATEL, D. J. & VAISHNAV, R. N. (1972). The rheology of large blood vessels. In *Cardiovascular Fluid Dynamics*, ed. D. H. Bergel, Vol. 2, pp. 2–64. London: Academic Press.

PRICE, N. R. & HUNT, S. (1974). Fluorescent chromaphore components from the egg capsules of the gastropod mollusc *Buccinum undatum* L., and their relation to fluorescent compounds in other structural proteins. *Comparative Biochemistry and Physiology*, **47***B*, 601–16.

ROACH, M. & BURTON, A. (1957). Reason for the shape of the distensibility curve of arteries. *Canadian Journal of Biochemistry and Physiology*, **35,** 681–90.

SANDBERG, L. (1976). Elastin structure in health and disease. *International Review of Connective Tissue Research*, **7,** 160–210.

SAGE, E. H. & GRAY, W. R. (1976). Evolution of elastin structure. In *Elastin and Elastic tissues*, ed. L. B. Sandberg, W. R. Gray & C. Franzblau, pp. 291–312. New York: Plenum Press.

TORCHIA, D. A. & PIEZ, K. A. (1973). Mobility of elastin chains as determined by ^{13}C NMR. *Journal of molecular Biology*, **76,** 419–24.

ROBERT, L., ROBERT, B. & ROBERT, A. M. (1970). Molecular biology of elastin as related to aging and atherosclerosis. *Experimental Gerontology*, **5,** 339–56.

TRELOAR, L. R. G. (1975). *Physics of Rubber Elasticity*. Oxford: Clarendon Press.

TRUEMAN, E. R. (1953). Observations on certain mechanical properties of the ligament of *Pecten, Journal of experimental Biology*, **30,** 453–67.

URRY, D. W. & LONG, M. M. (1976*a*). On the conformation, coacervation and function of polymeric models of elastin. In *Elastin and Elastic Tissues*, ed. L. B. Sandberg, W. R. Gray & C. Franzblau, pp. 685–714. New York: Plenum Press.

URRY, D. W. & LONG, M. M. (1976*b*). Conformations of the repeat peptides of elastin in solution. *Critical Reviews in Biochemistry*, **4,** 1-45.

VINCENT, J. F. V. (1976). Design for living – the elastic-sided locust. In *The Insect Integument*, ed. H. R. Hepburn, pp. 401–19. Amsterdam: Elsevier.

Wainwright, S. A., Biggs, W. D., Currey, J. D. & Gosline, J. M. (1976). *Mechanical Design in Organisms*, London: Edward Arnold.

Weis-Fogh, T. (1960). A rubber-like protein in insect cuticle. *Journal of experimental Biology*, **37**, 889–907.

Weis-Fogh, T. (1961*a*). Thermodynamic properties of resilin, a rubber-like protein. *Journal of molecular Biology*, **3**, 520–31.

Weis-Fogh, T. (1961*b*). Molecular interpretation of the elasticity of resilin, a rubber-like protein. *Journal of molecular Biology*, **3**, 648–67.

Weis-Fogh, T. (1973). Energetics of hovering flight in hummingbirds and in *Drosophila. Journal of experimental Biology,* **56**, 79–104.

Weis-Fogh, T. & Andersen, S. O. (1970). New molecular model for the long-range elasticity of elastin. *Nature, London.,* **227**, 718–21.

Wiegand, W. B. & Schneider, J. W. (1934). The rubber pendulum, the Joule effect and the dynamic stress–strain curve. *Transactions of the Institution of the Rubber Industry*, **10**, 234–62.

Wöhlisch, E., Weitnauer, H., Gruning, W. & Rohrbach, R. (1943). Thermodynamische Analyse der Dehnung des elastischen Gewebe vom Standpunkt der statischkinetischen Theorie der Kautschukelastizität. *Kolloid Zeitschrift*, **104**, 14–24.

Wolinsky, H. & Glagov, S. (1964). Structural basis for the static mechanical properties of aortic media. *Circulation Research*, **20**, 99–111.

THE VISCOELASTICITY OF MUCUS: A MOLECULAR MODEL

ROGER H. PAIN

Department of Biochemistry, The University of Newcastle upon Tyne, Newcastle upon Tyne, NE1 7RU, UK

If it is love that makes the world go round, then it is surely mucus and slime which facilitate its translational motion. For there are few organisms which do not at some point make use of a mucous secretion to fulfil some vital and often intriguing biological function. Flexible and sometimes fragile surfaces which are in contact with the environment are frequently covered by a renewable film of mucus. The earthworm, for instance, carries a protective layer which lubricates its journeys through the earth but which is nevertheless permeable to oxygen and carbon dioxide. Fish and eels are covered, the latter rather liberally, with slime which protects against sudden osmotic shock, implying the ability to restrict mixing within its boundaries. The slug makes use of the unusual rheological properties of its pedal mucus to progress, with the mucus alternating between an adhesive gel and, under shear, a viscous fluid (M. Denny, personal communication). The frog palate is covered by a mucous layer which is motivated by the underlying cilia to transport food to the gastrointestinal tract. The slime is sufficiently coherent to allow the frog to keep its mouth open under water without the layer being dissolved away or the cells beneath becoming unduly hydrated (Silberberg, Meyer, Gilboa & Gelman, 1977).

Indian medicine has long recognised mucus or 'Kapha' as one of the vital components in its 'tri dosha' theory of human health and disease. One important location of mucus in the human is the respiratory system. This is a physiological cul-de-sac from which particulate material has to be swept out on a 'ciliary escalator'. The latter is composed, like the surface of the frog palate, of a moving belt of mucus supported and driven, directly or indirectly, by the co-ordinated beating of the cilia of the upper respiratory tract. In this way, particles of 2–10 μm diameter can be moved at speeds of up to 16 mm min^{-1}. The ability of the cilia to shift the mucus depends critically upon the water content which determines the rheological properties of the mucus. We shall return to the importance of the water content of mucus in relation to its physiological function.

The genito-urinary system is another flexible tract which makes use of the particular properties of mucus in performing its physiological function (Gibbons, 1978; Elstein, 1978). The most studied component is the cervical mucus which plays a spectacular role in fertility. It acts as a biological valve for the transport of sperm through the upper genital tract, its rheological properties being controlled synchronously with the menstrual cycle. At the time of ovulation its rheology changes to that of a generally watery fluid which allows the passage of motile sperm but which at the same time imposes some directional constraint on their flow. As other sperm swim up the cervix, so a selection process operates within the mucous secretion whereby only the stronger sperm reach their destination, the weaker going to the wall. This too may be a function of the complex structure of the cervical mucus.

Perhaps the most complex of the systems in which mucus plays a role is the gastrointestinal tract. The passage from mouth to anus is lined with mucous secretions which have differing properties and structures tailored, in ways not yet wholly understood, to the great variety of surfaces over which food or waste must pass. The stomach, for example, must be able to contain the acid and proteolytic components by which food is denatured and digested ready to be squeezed, at higher pH values, between the mobile and expandible walls of the intestine where much of the absorption occurs. The different types of mucus must be able to protect the cells at the surface of such organs against extremes of pH and proteolytic activity, but allow the diffusion of nutrients across this barrier. The coating must be highly flexible and be capable of being continually renewed by the secretion of mucous components from the underlying cells as the outer layers are swept away.

The approach in this paper is to show how it is possible to obtain some understanding of the characteristic viscoelastic properties of mucus by studying the molecular architecture of its essential components (cf. Allen, 1978; Pain, 1980). Most of the examples described will be taken from work on gastric mucus but a few comparisons and contrasts with mucus from other sources will be noted at appropriate points. An alternative approach is to study the rheological properties of the whole system and to relate these to the mucous function. This will not be dealt with here, and the reader is referred to reviews by Silberberg (1977) and by Litt, Khan & Wolf (1976) for information about this field.

Molecular structure

The rheological properties of mucus depend fundamentally on the physical and chemical properties of one class of macromolecule, the mucous glycoproteins. There is confusion about nomenclature in this field which has not been entirely avoided by recent attempts at rationalization (see, for example, Reid & Clamp, 1978). The main characteristics of the mucous glycoprotein molecule are as follows.

(i). High molecular weight (usually at least 10^6 daltons).

(ii). High carbohydrate content (>50%, the remainder being protein).

(iii). Rather short chains of carbohydrate (2–18 residues long) attached in large numbers to a central chain or core of polypeptide.

The most convincing evidence for the central role of glycoproteins in mucous structure is twofold. Firstly, glycoproteins have been isolated by a variety of chemically non-degradative methods and shown to be the major components in several types of mucus (see e.g., Robson, Allen & Pain (1975), Wolf, Sokoloski, Khan & Litt (1977), and Creeth (1978)). Second, the isolated and purified glycoprotein has been shown, in stomach and cervical mucus, to be able to reform a mucus-like gel exhibiting similar properties to the original gel. While the possible involvement of an 'aggregating factor' (cf. Creeth *et al.*, (1977) and Roberts, (1978)) is difficult to establish, all the evidence points to the absolute requirement of a high proportion of glycoprotein for the formation of mucous gel. In the gastric glycoprotein there seems no necessity to postulate other components for gel formation.

Other components than water, are ions, native and denatured proteins, and nucleic acid. These may all modify the rheological properties of the gel to a greater or lesser extent, but there is little quantitative information on their effect in vivo. An interesting component, which would be expected to contribute significantly to the viscosity, is a highly sulphated cellulose which occurs in the mucus of the whelk *Buccinum undatum* (Hunt & Oates, 1977). It would appear, however, that the major determinant of the viscoelasticity of a given mucous glycoprotein gel is the water content.

Chemical composition

There are five carbohydrate residues which occur most frequently, as shown in Table 1, although others, such as uronic acid in slug-pedal mucus (Denny, personal communication), and mannose in earthworm mucus (Clamp, 1977) are also found. It is probable that the non-polar methyl group adds to the non-polar character of some of the carbohy-

Table 1. *Common carbohydrate residues in mucous glycoproteins*

Residue type	Special characteristics	Content in pig gastric mucus glycoprotein (% by weight)
L-fucose	non-polar methyl	11
D-galactose	–	26
N-acetylglucosamine	may carry a negatively charged sulphate	20
N-acetylgalactosamine	non-polar methyl	8
Sialic acids	negatively charged carboxyl	0.2
Sulphate	–	3

drate residues (cf. Fig. 4 in Morris & Rees, 1978) making hydrophobic interactions a possibility. The charged groups on sialic acid and on sulphated N-acetylglucosamine provide the possibility of some degree of polyelectrolyte behaviour, although this is likely only when these groups form a large proportion of the residues such as in sheep salivary mucus (Gottschalk, Bhagarva & Murty, 1972; Litt, Khan, Shih & Wolf, 1977). Gibbons (1978) argues that the rheological properties of mucous glycoproteins are not sensitive to detailed carbohydrate composition, the argument being based on the considerable inter-species variation and also on chemical degradation studies. However, there is no doubt that the general hydrophilicity of the carbohydrates is important in determining the high degree of expansion of these glycoproteins in aqueous media.

The carbohydrate residues are linked in short chains, sometimes branched, of from 2 to 18 residues. Sheep salivary mucous glycoprotein molecules contain approximately 800 disaccharide chains attached to a polypeptide core while gastric mucous glycoprotein has a branched, 'bushy' side-chain of up to 18 sugar residues. In all examples the side-chains form the characteristic 'bottle-brush' structure with a 'bristle' connected to the polypeptide chain, frequently at every third or fourth residue.

The polypeptide or protein core contains a high proportion of serine and threonine residues, these amino acids forming the chemical links to the carbohydrate side-chains. In stomach, intestinal and colonic glycoproteins there is a similarly high proportion of proline, the one amino acid in which rotation of the backbone is specifically blocked. In the salivary glycoproteins, by contrast, there is a high proportion of the small side-chain residues, glycine and alanine, the former offering least

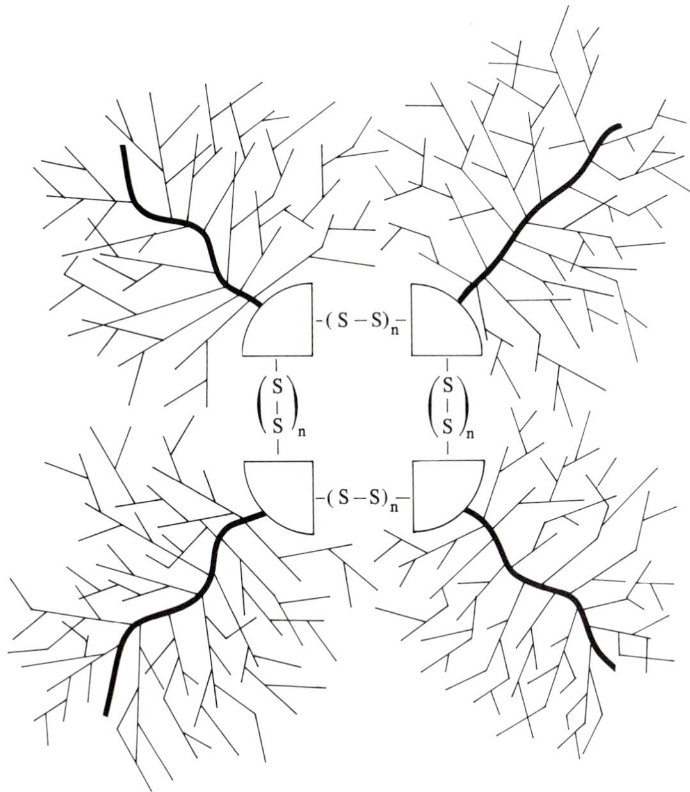

Fig. 1. Diagrammatic representation of the gastric mucous glycoprotein molecule. Each of the four disulphide bridged units has a molecular weight of 5×10^5 daltons. The non-glycosylated, half-cysteine containing portions of the protein chains may or may not be folded. There are many more oligosaccharide chains for a given length of protein than are shown.

constraint to rotation about the bonds forming the backbone of the molecule.

In cervical, respiratory and stomach mucous glycoproteins, the protein core has been shown to be glycosylated at one end and 'naked' or free from sugar at the other. The naked end has an amino-acid composition comparable to those of globular proteins, including cysteine residues capable of forming disulphide links within and between polypeptide chains. This offers the possibility for such glycoproteins to consist of more than one such bottle-brush unit and chemical reduction has been shown to lower the molecular weight for all the above classes of glycoprotein. In stomach mucous glycoprotein, for example, the molecule consists of four such units, each of 5×10^5 daltons linked by disulphide bands as shown diagrammatically in Fig. 1. Meyer &

Silberberg (1978) suggest that this is a rather general structure from their results on bovine cervical (oestrus), human ear and lung mucus.

It is not possible at the moment to generalize about the relation between function and disulphide-linked multimeric glycoprotein structure. For example, while pig and dog salivary mucous glycoproteins are reported to exist as disulphide bonded units, glycoprotein from the same organ in sheep does not (Holden, Yim, Griggs & Weisbach, 1971). There are clearly detailed functions specific to different species which are not yet understood; it may be, for instance, that the above difference in structure reflects a difference between carnivores and herbivores (see also Phelps & Young, 1977). It is interesting to note that the pedal mucus of the slug makes use of a disulphide-bonded glycoprotein structure (M. Denny, personal communication).

These broad structural features are summarized diagrammatically for the gastric mucous glycoprotein in Fig. 1. In relation to the structure of the viscoelastic mucous gel, it must be noted that cleavage of the disulphide bridges will solubilize the gel, and that the individual units, though unchanged in every other part of their structure, including all the carbohydrate moiety, will not now form a gel at the physiological concentrations. Proteolytic digestion of the naked polypeptide – which also results in separated subunits (see Fig. 1) – has the same effect on gel formation. Thus, the macromolecular integrity is essential for the characteristic mucous rheology.

Homogeneity of mucous glycoproteins

Proteins, whose synthesis is directly under gene control, are usually homogeneous. Glycosylation by contrast is a much more 'probabilistic' process, thus offering the possibility of a less homogeneous product. *In vivo*, the glycoproteins are also subject to breakdown and it is not easy to distinguish biosynthetic from degradative causes of any heterogeneity found in isolated materials. Further, heterogeneity in these large expanded molecules, whose flow properties exhibit a high degree of non-ideality, is difficult to quantitate. Gibbons (1978) argues strongly for considerable heterogeneity in the molecular weights of cervical mucous glycoprotein. Creeth & Horton (1977) have shown a heterogeneity with respect to density in bronchial mucous glycoprotein. The heterogeneity in preparations of gastric mucous glycoprotein can be reduced by careful removal of aggregated material, but the distribution of sedimentation coefficients shown in Fig. 2 suggests a heterogeneity spread around a mean value. It is not possible at the present time to account for the molecular basis for such heterogeneity, whether of mass,

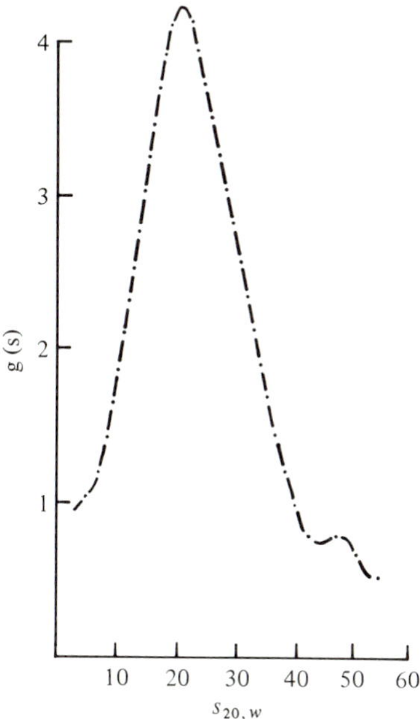

Fig. 2. Distribution of sedimentation coefficients (g(s)) for a purified sample of gastric mucous glycoprotein. A small amount of faster, aggregated, material is visible together with a shoulder corresponding to dimerized (?) glycoprotein. 1.1 g l^{-1} glycoprotein; 795s after start of centrifugation at 30 000 r.p.m. (D. Mantle & R. H. Pain, unpublished).

shape or density and indeed, the whole question of heterogeneity of the glycoprotein molecules *as secreted* must be treated with reserve. Nevertheless, the heterogeneity of carefully isolated glycoprotein does not prevent their reforming a native-like gel.

Molecular conformation

The way in which macromolecules perform their function and interact with other molecules in solution depends not only upon their chemical structure, but also upon the three dimensional distribution of that structure in space. This conformation is determined by the partial rotation of covalent bonds along the protein and oligosaccharide chains, by non-covalent interactions between different parts of the glycoprotein and, most important, by the nature of the interaction of the various parts of the molecule with water.

It follows from the earlier description of the chemistry of the glycoprotein that there is opportunity for hydrophilic interactions with solvent, particularly through hydroxyl groups, the degree to which the hydroxyls are bound to water molecules being limited by the stereochemistry of the sugar units. In the absence of other strong attractive forces between molecules, the saccharide moiety will tend to hydrogen-bond to water rather than to other like groups, owing to the large excess of water in the environment.

It has been mentioned that sugars can also offer non-polar, hydrophobic faces which will tend to associate with other like surfaces in order to exclude water. It is not known how significant these hydrophobic interactions are in determining glycoprotein conformation, but the expanded nature to be described later suggests that they are not very important.

Charged groups will interact favourably with water and offer the possibility of repelling other neighbouring charges if the ionic strength of the environment is sufficiently low to prevent charge-shielding by counter ions. The only evidence for their having an effect on conformation under physiological conditions appears to be that salivary mucous glycoprotein becomes less extended when its sialic acid carboxyl charges are neutralized (cf. Litt *et al.*, 1977).

The protein core acts as a rather extended support for the oligosaccharide chains and is not in general accessible to other molecular species. For example, this part of the protein is not cleaved by proteolytic enzymes. The naked part has a composition of amino acids which suggests the possibility of folding into a close packed conformation like a globular protein. Nothing, however, is known of these more detailed aspects of conformation. What has been determined has been at a lower level of information and concerns questions of flexibility and overall shape and expansion.

Flexibility

The gastric mucous glycoprotein is soluble in concentrated solutions both of guanidinium chloride, a standard denaturing agent, and caesium chloride. With each solvent the sedimentation coefficient increases markedly without any change in molecular weight, indicating that the molecule has contracted (Snary, Allen & Pain, 1974). Thus, the glycoprotein is partially expanded under normal conditions and has a certain flexibility so that a change of solvent only causes it to shrink. Conversely, reducing the ionic strength to lower than 10^{-4}M brings about a dramatic increase in expansion and intermolecular interaction. This change in shape is completely and rapidly reversible and demon-

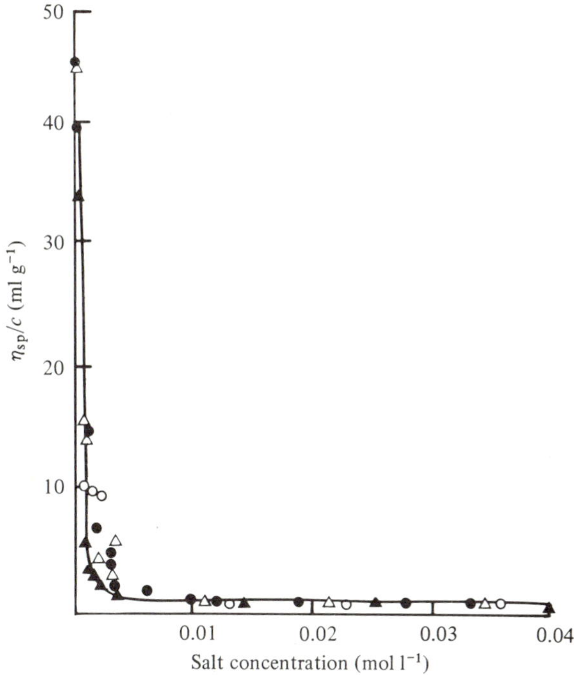

Fig. 3. The effect of a selection of ions on the reduced specific viscosity (η_{sp}/c) of gastric mucous glycoprotein. The curve may be approached from high or from low ionic strength. Glycoprotein 1.2 g l^{-1}; filled triangles, LiCl; open triangles, KCl; open circles, CsCl; filled circles, $(C_2H_5)_4NCl$. (D. Mantle & R. H. Pain, unpublished).

strates that the native molecule is not fully extended but occupies some intermediate distribution of structure and flexibility through space. The effect of a variety of ions (Fig. 3) shows the expansion to be a polyelectrolyte effect involving repulsion of ionic groups on the glycoprotein.

It is possible from flow properties to calculate an effective hydrodynamic volume V_e which may be thought of as indicating an average sphere of influence of the molecule. The gastric mucous glycoprotein is found to be approximately spherical in shape and highly expanded with $V_e \approx 40$ ml g^{-1}, indicating that approximately 97% of the volume within the molecule is occupied by solvent (Pain, 1980). Referring back to the model shown in Fig. 1, a value of $V_e = 20$ ml g^{-1} has been obtained for the subunits resulting from chemical reduction. This implies that, in the intact glycoprotein, these subunits are not closely packed together but that, linked together, they sweep out an additional volume, probably implying some flexibility about the central disulphide bonded protein region.

Formation of mucous gel

Three factors which govern the formation of mucous gel may be summarized as follows.

Requires non-covalent interactions only

Glycoproteins are now routinely isolated from mucous gels by dilution together with chemical or physical agents which break only non-covalent intermolecular interactions. In certain instances, such as the rather sloppy gastric mucous gel, some dissociation occurs spontaneously on dilution, which suggests that the forces of attraction are rather weak. However the barriers which prevent a compact bronchial plug from dissolving in water in this way may well be kinetic rather than thermodynamic. As has been mentioned above, these processes of dissociation are reversible, indicating that no irreversible change occurs on dissociating the gel by such means.

Interaction with water is low

The concentration of glycoprotein in mucous gel is low, although not as low as in agar gels, for example, where a gel can be stable with as little as 0.1% agar. The prodigality of the snail, suggested by the trails of slime, is more apparent than real for the interference patterns which draw attention to its peripatetic behaviour are a result of the extreme thinness of the films of glycoprotein. Even then, most of the snail's secretion is water. It would be interesting to know something of the role of this water. The interaction of water with biological materials can be measured by the technique of Kuntz (Kuntz & Kaufmann, 1974). Lowering the temperature of a solution to −25°C will result in all the water being frozen, except that which is bound to the solute and therefore unable to fit into the ice crystal lattice. The frozen water is immobile and the proton magnetic resonance peak is so broad as not to be seen. However, the bound water has a certain degree of rotational motion and a peak, whose area is proportional to this 'non-frozen' water, becomes evident. Results for gastric mucous glycoprotein in solution show that not more than 0.5 g water is bound per gram glycoprotein, that is, approximately 1% of the total water within the effective volume of the molecule. This is in accord with earlier findings on other gels and with the fact that molecules can diffuse through agar gels unhindered by any 'structured' water. There is no evidence for particular structure being induced in the water of the mucous gel by the glycoprotein.

Consideration of some of the physiological phenomena, however, suggests that the *content* of water in mucus is extremely important in determining its rheological and therefore physiological properties. Canine tracheal mucus is involved in the ciliary escalator described earlier and the ability of a sample of mucus to move particles can be assayed on the isolated frog palate preparation. Shih, Litt, Khan & Wolf (1977) have examined the rate of transport as a function of water content of this mucus and have shown that there is a critical optimum concentration for transport to operate effectively. Too little water, and the gel cannot be moved by the cilia; too much water and the mucus becomes mainly viscous in its properties and there is no reaction to the beat of the cilia. This property is important in the treatment of chronic obstruction lung disease, when a water spray may be used as a decongestant. The ideal would be to balance the final water content, not only to relieve congestion but also to enable the mucus to exert its protective function towards the lungs (Saketkhoo *et al.*, 1978).

Certain disease states frequently result in dramatic increases in the stiffness of mucus and this is correlated with a decrease in water content of the mucus. The processes which control the water content of the mucus as it is secreted are clearly vital to its proper state and function.

Intermolecular interaction is high

The viscous behaviour of gastric mucous glycoprotein is Newtonian and characteristic of a large, spherical, non-interacting molecule when measured in the concentration range 0–20 $g\,l^{-1}$ (Fig. 4). This behaviour is confirmed by the unexceptional sedimentation and diffusion properties (Snary, Allen & Pain, 1970). At higher concentrations, however, there is a marked change in behaviour, with the reduced specific viscosity at zero shear increasing very sharply with concentration and the viscosity becoming non-Newtonian. Further increase in concentration, effected by gradual lowering of the water content, leads to the formation of a gel. It is evident, therefore, that strong intermolecular interactions begin to occur in the concentration range 20–30 $g\,l^{-1}$, and that these are preliminary to and necessary for gel formation. It has already been shown that the glycoprotein is even more expanded in conditions of low ionic strength where its viscosity behaviour at finite concentrations shows strong interactions occurring at much lower concentrations than in higher salt concentrations. A gel formed at 10 $g\,l^{-1}$ in the absence of salt will liquefy on dialysis against 0.1 M sodium chloride and gel again on dialysis against water. This shows that the

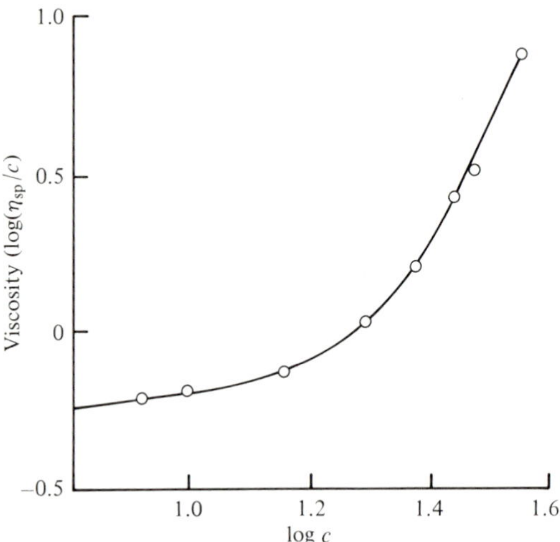

Fig. 4. The viscosity of gastric mucous glycoprotein as a function of glycoprotein concentration, $c(\mathrm{g\,l^{-1}})$. Buffer: 0.2 M KCl, 0.02 M potassium acetate, 0.02% (w/v) azide; pH 5.5. Viscosity values at zero shear. (Allen, Pain & Robson, 1976.)

interactions between molecules and subsequent gel formation are strongly dependent upon the degree of expansion of the glycoprotein.

A model for gastric mucous gel

Gel formation has been discussed for many years in terms of the formation of 'junction zones' which stabilize the gel (cf. Hermans, 1953). Work on polysaccharide gels has led to an understanding of the molecular basis of such junction zones for several different types of gel (Rees, 1969). Relatively strong, cooperative interactions are formed by the intertwining of helices (occurs in carrageenans), by metal-ion chelation-dependent crystallites (occurs in alginates) or by hydrophobic interactions (occurs in carboxymethyl celluloses). The model suggested for gel formation by gastric mucous glycoprotein differs markedly from any of these comparatively regular structures.

The evidence is as follows (Allen, Pain & Robson, 1976).

(i). The effective volume, obtained from flow property measurements at low concentrations, is $40\,\mathrm{ml\,g^{-1}}$. This says that the glycoprotein will extend throughout the whole solution at a concentration of $25\,\mathrm{g\,l^{-1}}$. At this concentration all the solvent is intramolecular.

(ii). The onset of intermolecular interaction, as indicated by the change in viscous behaviour, takes place in the $20–30\,\mathrm{g\,l^{-1}}$ range of concentration.

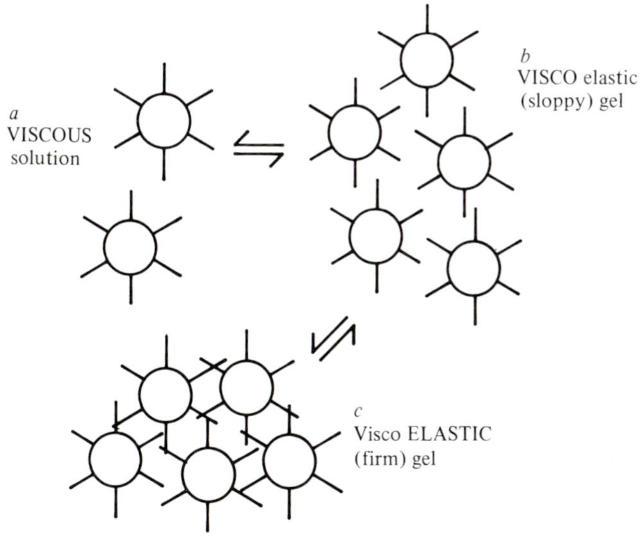

Fig. 5. A model illustrating the transition from, *a*, a purely Newtonian viscous solution of glycoprotein at lower concentrations to, *c*, a gel with highly shear-dependent viscosity and high stiffness at low shear. Interactions are indicated diagrammatically by interdigitation between oligosaccharide chains, without any implications as to the nature of the intermolecular interactions. Each circle represents one glycoprotein unit.

(iii). The average glycoprotein concentration in gastric mucous gel in vivo is $30\,g\,l^{-1}$.

From these findings, it is evident that gel formation takes place as the molecules fill the solution and their surfaces are brought into contact. In reality, of course, the molecules do not have even, spherical, surfaces but present many oligosaccharide chains reaching out towards other molecules. Thus, interaction is most likely to occur when there is a degree of interdigitation between vicinal molecules and, further, it should be possible for molecules to invade each other's domains in this way to a variable extent depending on the water content of the gel (Fig. 5). The greater the degree of interpenetration, the greater the opportunity for interactive forces to be stabilized and the slower the kinetics of solubilization of the gel, as is found experimentally.

The precise nature of the interactions is not known. Possibilities include multiple hydrogen bonding whose potential for cooperativity would allow effective competition with hydrogen bonds to water, or multiple hydrophobic interactions. In relation to the latter, there is no evidence for a negative temperature coefficient of interaction, either for freely interacting molecules or for gel stability, which usually accompanies hydrophobic interactions. A further possibility, which the author favours, is that interaction is driven by reducing the water content below the minimum amount required for full hydration of other hydrophobic groups on the glycoprotein. This limit will occur when chains are brought close to each other and therefore compete for water. Bonding will then occur between chains to minimize the enthalpy. In this way one can explain the low cooperativity of a sloppy gel like gastric mucus, the relative ease of solubilization of the gel, and the large dependence of its stability on water content, which is far greater than that for the much more cooperative agar type of gel, for instance. The shear resistance of the gel will thus increase continuously with decreasing water content leading to the unusual viscoelastic properties of mucus.

Physiological aspects of mucus

Based on the above chemical and physical approaches to the glycoproteins which compose the mucous gel, it is possible to start relating the structure of the gel to some of its physiological functions.

Gastric mucus

Gastric mucus must somehow prevent access of protons and macromolecular proteases to the surface of the wall, while allowing small molecules to diffuse through (Florey, 1962). Given the large content of non-bound water in the gel, the latter is unlikely to offer a significant barrier to the diffusion of protons. The gel can, however, provide a region of relative calm where turbulence and mixing are reduced to a minimum. This would allow a pH gradient to be set up, given the secretion of bicarbonate from the wall and the diffusion of protons towards the wall from the stomach contents (cf. Heatley, 1959).

Protection of the wall against protease may be viewed as the gel restricting diffusion of the protease or as a phase exclusion phenomenon (Edwards, 1978). In either instance, the glycoprotein and the protease may be seen as competing for the same solvent volume in a way which is not so in an agar gel, where the diffusion of proteins is relatively unhindered. The structural model described above accounts for this difference in that the glycoprotein is distributed throughout the gel space whereas the polysaccharide chains of agar are gathered together locally in rather densely packed strands.

It is significant to note that the gel obtained from the stomach wall is heterogeneous in terms of viscoelasticity while that formed by removing water from the isolated, soluble components is homogeneous. This may reflect two factors. The first is that the gel can be broken down by proteases, which separate the subunits (see Fig. 1). Thus, the surface of the gel will be attacked by pepsin and removed, while fresh mucus is secreted on the underside. This will augment the general solubilization which will be going on at the surface, so that there will be a gradient of stability and water content in a plane normal to the stomach wall. Second, we have seen that the gel 'texture' will depend on water content. It is likely that this may vary from place to place on the wall, depending on the environment at the point of secretion, the rate of secretion of glycoprotein into the mucus and the concentration at which it is finally secreted. While little is known of these details, this discussion serves to show that the function of a sample of mucus will be dependent, through its water content, on a whole variety of physical, biochemical, structural and environmental factors associated with the processes and location of secretion.

Cervical mucus

Cervical mucus changes its rheological properties dramatically at the time of ovulation in response to hormonal control. At this point the total secretion of glycoprotein and its water content increase, as does the ability to transport sperm; some chemical changes, including a reduction in sialic acid content, also occur. This mucous secretion is anisotropic and filamentous and can be shown *in vitro* to orientate into fibres on stretching. According to the observation of Odeblad (1977) the glycoprotein at this time is extruded rapidly from the crypts to form filaments which lie along the wall of cervix. These are accompanied by some free blobs of mucus while the remainder of the fluid is viscous. It is tempting to speculate that this unusual mode of secretion keeps the glycoprotein more concentrated in the thick filaments which, according to the arguments advanced above, would be expected to persist longer. The localization of glycoprotein in this way allows a lower viscosity medium in which the sperm can swim, their noses being kept in the right direction under the constraint of the filaments.

At other times during the menstrual cycle the mucus is more homogeneous, has a lower water content and is more viscous throughout and in this state can be compared more with the structure of the gastric mucus. It would be expected, therefore, on this basis to offer an impenetrable barrier to sperm transport.

Salivary mucus

According to the literature, sheep salivary mucus appears to possess a structure which is qualitatively different from those of gastric, cervical and tracheal origin. The structure as described earlier is that of a long protein core glycosylated overall with short disaccharide side-chains. It is tempting to speculate that this long, flexible and rather highly-charged molecule will interact with and coat the relatively more basic protein and small food particles, thus easing their flow down to the stomach. It could also bind amylase which would then be kept in closer contact with food particles during this journey. At the acid pH of the stomach, the carboxyls being neutralized, the mucus would dissociate and leave the food open to attack by proteases and lipases.

Fundamental questions

This brief and, at points, speculative review has pointed to some of the areas in which further work is needed before we can fully understand the relation between these unique biological materials and their function. They provide adhesive, flexible and highly variable media which may act as barriers to or facilitators of diffusion. Very little is known, however, about the nature of the interactions between the glycoproteins at the molecular level. Nor is there much knowledge concerning the specific effects, if any, of the remarkable variety of carbohydrate side-chains on the conformation, interactions and rheology of the glycoproteins and the mucus. Last, and perhaps most important for any complete understanding of their function, it is necessary to know more about the processes and the environment of secretion *in vivo*, for this is what would appear to be the major determining factor in the function of mucus.

I wish to acknowledge many interesting and fruitful discussions with my colleague Dr A. Allen, in collaboration with whom much of the work on which this paper is based was done.

References

ALLEN, A. (1978). Structure of gastrointestinal mucus glycoproteins and the viscous and gel forming properties of mucus. *British Medical Bulletin,* **34,** 28–33.

ALLEN, A., PAIN, R. H. & ROBSON, T. R. (1976). Model for the structure of the gastric mucus gel. *Nature, London,* **264,** 88–9.

CLAMP, J. R. (1977). Mucus in health and disease. In *Mucus in Health and Disease*, ed. M. Elstein & D. V. Parke, pp. 1–18. New York and London: Plenum Press.

CREETH, J. M. (1978). Constituents of mucus and their separation. *British Medical Bulletin,* **34,** 17–24.

CREETH, J. M. BHASKAR, K. R., HORTON, J. R., DAS, I., LOPEZ-VIDRIERO, M.-T. & REID, L. (1977). The separation and characterisation of bronchial glycoproteins by density gradient methods. *Biochemical Journal,* **167,** 557–69.

CREETH, J. M. & HORTON, J. R. (1977). Macromolecular distribution near the limits of density-gradient columns. Some applications to the separation and fractionation of glycoproteins. *Biochemical Journal,* **161,** 449–63.

EDWARDS, P. A. W. (1978). Is mucus a selective barrier to macromolecules? *British Medical Bulletin,* **34,** 55–6.

ELSTEIN, M. (1978). Functions and physical properties of mucus in the female genital tract. *British Medical Bulletin,* **34,** 83–8.

FLOREY, H. W. (1962). The secretion and function of intestinal mucus. *Gastroenterology,* **43,** 326–9.

GIBBONS, R. A. (1978). Mucus of the mammalian genital tract. *British Medical Bulletin,* **34,** 34–8.

GOTTSCHALK, A., BHARGAVA, A. S. & MURTY, V. L. N. (1972). Submaxillary gland glycoproteins. In *Glycoproteins: their composition, structure and function,* ed. A. Gottschalk, revised 2nd edn. pp. 810–29. Amsterdam & London: Elsevier.

HEATLEY, N. G. (1959). Mucosubstance as a barrier to diffusion. *Gastroenterology,* **37,** 313–17.

HERMANS, J. J. (1953). *Flow properties of disperse systems.* New York: Interscience.

HOLDEN, K. G., YIM, N. G. F., GRIGGS, L. J. & WEISBACH, J. A. (1971). Gel electrophoresis of mucous glycoproteins. II Effect of physical deaggregation and disulphide-bond cleavage. *Biochemistry,* **10,** 3110–13.

HUNT, S. & OATES, K. (1977). Intracellular cationic counterion composition of an acid mucopolysaccharide. *Nature, London,* **268,** 370–2.

KUNTZ, I. D. & KAUZMAN, W. (1974). Hydration of proteins and polypeptides. *Advances in Protein Chemistry,* **28,** 239–45.

LITT, M., KHAN, M. A., SHIH, C. K. & WOLF, D. P. (1977). The role of sialic acid in determining rheological and transport properties of mucus secretions. *Biorheology,* **14,** 127–32.

LITT, M., KHAN, M. A. & WOLF, D. (1976). Mucus rheology: relation to structure and function. *Biorheology,* **13,** 37–48.

MEYER, F. A. & SILBERBERG, A. (1978). Structure and function of mucus. In *Respiratory Tract Mucus.* Ciba Foundation Symposia, No. 54 (new series), pp. 203–18. Amsterdam: Elsevier.

MORRIS, E. R. & REES, D. A. (1978). Principles of polymer gelation. *British Medical Bulletin,* **34,** 49–54.

ODEBLAD, E. (1977). Physical properties of cervical mucus. In *Mucus in Health and Disease,* ed. M. Elstein & D. V. Parke, pp. 217–25. New York & London: Plenum Press.

PAIN, R. H. (1980). Gastric mucus gel: a challenge for biophysics. In *Biomolecular Structure, Conformation, Function and Evolution,* ed. R. Srinivasan, London: Pergamon.

PHELPS, C. F. & YOUNG, A. M. (1977). The control of submaxillary gland mucin production. In *Mucus in Health and Disease,* ed. M. Elstein & D. V. Parke, pp. 143–54. New York & London: Plenum Press.

REES, D. A. (1969). Structure, conformation and mechanism in the formation of polysaccharide gels and networks. *Advances in Carbohydrate Chemistry and Biochemistry,* **24,** 267–332.

REID, L. & CLAMP, J. R. (1978). The biochemical and histochemical nomenclature of mucus. *British Medical Bulletin,* **34,** 5–8.

ROBERTS, G. P. (1978). Chemical aspects of respiratory mucus. *British Medical Bulletin,* **34,** 39–42.

ROBSON, T., ALLEN, A. & PAIN, R. H. (1975). Non-covalent forces hold glycoprotein molecules together in mucous gel. *Biochemical Society Transactions,* **3,** 1105–7.

SAKETKHOO, K., YERGIN, B. M., JANUSZKIEWICZ, A., KOVITZ, K. & SACKNER, M. A.

(1978). The effect of nasal decongestants on nasal mucous velocity. *American Reviews of Respiratory Diseases,* **118,** 251–4.

Shih, C. K., Litt, M., Khan, M. A. & Wolf, D. P. (1977). Effect of non-dialysable solids, concentration and viscoelasticity on ciliary transport of tracheal mucus. *American Reviews of Respiratory Diseases,* **115,** 989–95.

Silberberg, A. (1977). Basic rheological concepts. In *Mucus in Health and Disease*, ed. M. Elstein & D. V. Parke, pp. 181–90. New York & London: Plenum Press.

Silberberg, A., Meyer, F. A., Gilboa, A. & Gelman, R. A. (1977). Function and properties of epithelial mucus. In *Mucus in Health and Disease*, ed. M. Elstein & D. V. Parke, pp. 171–80. New York & London: Plenum Press.

Snary, D., Allen, A. & Pain, R. H. (1970). Structural studies on gastric mucoproteins: lowering of molecular weight after reduction with 2-mercaptoethanol. *Biochemical Biophysical Research Communications,* **40,** 844–51.

Snary, D., Allen, A. & Pain, R. H. (1974). Conformational changes in gastric mucoproteins induced by caesium chloride and guanidinium chloride. *Biochemical Journal,* **144,** 641–6.

Wolf, D. P., Sokoloski, J., Khan, M. A. & Litt, M. (1977). Human cervical mucus. III. Isolation and characterisation of rheologically active mucin. *Fertility and Sterility,* **28,** 53–8.

ARTICULAR CARTILAGE

S. A. V. SWANSON

Department of Mechanical Engineering,
Imperial College, London SW7 2BX, UK

Introduction

Mechanical functions of articular cartilage

Articular cartilage is the tissue which forms the bearing surfaces in the synovial joints of mammalian skeletons. The usual arrangement is that the tubular shafts of long bones merge into enlarged ends having a cancellous internal structure (so named from the *cancelli*, the open lattice screens behind which a Roman judge sat) supporting thin cortices of generally rounded shape. The cortices of the two bone ends forming a joint are generally complementary in shape (e.g., one is convex and the other concave), but the shapes are not always exactly matched. Articular cartilage covers at least as much of each cortex as can make contact with the other within the range of motion permitted and thus, in an undamaged joint, contact is never between bone and bone or bone and articular cartilage but always between articular cartilage and articular cartilage, unless the cartilage surfaces are separated by a fluid film.

It is thus apparent that articular cartilage could perform two mechanical functions: it could so distribute the forces transmitted through the joint as to limit the stresses in the underlying bone, and it could act as a better bearing material than bone. That articular cartilage is orders of magnitude more deformable than bone makes the first function inherently probable. To ascribe the second function also to articular cartilage is tempting if one supposes that (whatever the development process) the mammalian musculo-skeletal system as we now see it can legitimately be viewed as a designed mechanical system; from this point of view a smooth, slippery layer of tissue is there for a purpose. It is hardly necessary to pause to have doubts about such a teleological view, because there is abundant evidence that when articular cartilage is absent from parts of the joint surfaces, as in advanced osteoarthrosis, the performance of the joint as a bearing is markedly inferior to that of a sound joint.

Scope of this paper

This paper is concerned only with adult human articular cartilage (except when results are quoted from experiments performed on cartilage of other mammals). In common with others in the Symposium, it will relate mechanical properties to composition and structure and to mechanical function; but it will be more towards the 'function' end of the spectrum than many. This is because two motives lead to the investigation of the mechanical properties of human articular cartilage: the common scientific interest in understanding how the properties are related to the composition and structure and the human, social, or economic interest in understanding the pathogenesis of osteoarthrosis.

The consequences of severe osteoarthrosis for the individual sufferers and their families are obvious; but the magnitude of the economic consequences is not always appreciated. If three indicators of burden on the economy are taken to be (1) General Practioner consultations, (2) out-patient referrals and (3) sickness and invalidity payments, then in recent years in Britain musculo-skeletal disorders ranked third, first and second respectively; although this category includes conditions other than osteoarthrosis, there can be no doubt that osteoarthrosis was a significant contributor. Another measure is that in 1974 an estimated 30 000 hips were replaced in 24 000 patients in Britain, nearly all of which would have been because of osteoarthrosis or rheumatoid arthritis. Taylor (1976) estimated the cost of these procedures to be £15 000 000; his estimate of the cost of maintaining the same patients if they had not been treated by joint replacement is necessarily less accurate, but is a total of £84 000 000, at 1974 prices, spread over the future lives of this cohort of patients. Even allowing for the uncertainties in all such estimates, three things are clear: joint replacement as now practised saves the community money, improvements in the practice of joint replacement might save more money, and means of preventing or arresting osteoarthrosis or rheumatoid arthritis so that joint replacement was unnecessary would save much more money (as well as suffering). Thus much of the work on the mechanical properties of articular cartilage has been undertaken and supported with a view to describing the function of the tissue in health and in the early stages of the process which, when further advanced, is called osteoarthrosis.

The various joint diseases are described in the appropriate books. Here it should be mentioned that numerically the two most important joint diseases in developed countries are rheumatoid arthritis, which commonly affects several joints in the same person and is believed to result from malfunction in the body's immune response system, and osteoarthrosis, which is classified as primary and secondary. Secondary osteoarthrosis follows, for example, injuries which disrupt the joint surfaces. Primary osteoarthrosis occurs with no history of obvious insult to the joint, and is regarded as idiopathic. It often affects joints in the same body to very different extents. The name 'osteoarthrosis' is derived from the fact that in advanced stages significant remodelling of bone occurs under and around the articular cartilage.

All aspects of the composition, structure and properties of adult human articular cartilage have been fully discussed in a recent volume (Freeman, 1979); therefore this paper will not attempt to review all the relevant literature, but will instead outline the present state of knowledge of the structure and mechanical properties in relation to the mechanical functions of the tissue. Though forms of cartilage other than articular do exist, only articular cartilage is considered here and for convenience it will be referred to simply as 'cartilage'.

Composition and structure

Size and disposition of cartilage in the body

Cartilage is found on the rounded ends of bones wherever bones are connected by synovial joints. The size of the sheet of cartilage is therefore determined by that of the joint, and ranges from a few millimetres in each direction at the temporo-mandibular joint to several centimetres at the knee joint. Most joint surfaces are doubly curved, i.e. few are simply cylindrical or conical in shape. The thickness of cartilage varies with the size of the joint, from one or two millimetres in the smallest joints to three, four or occasionally five millimetres in the biggest joints. Cartilage is firmly attached to bone, but a fairly definite junction exists, as is shown by the smooth surface of dried dead bone from which cartilage has been removed.

A cross-sectional view of a notional synovial joint, with its principal components, is shown in Fig. 1. In passing it may be noted that the name 'synovial' is believed to have been given by an anatomist in antiquity, who likened the fluid to the white of an (uncooked) egg.

Major constituents

Cartilage contains roughly 70–75% water, 2–10% proteoglycans and 15–20% collagen, and probably a small quantity of a glycoprotein. The

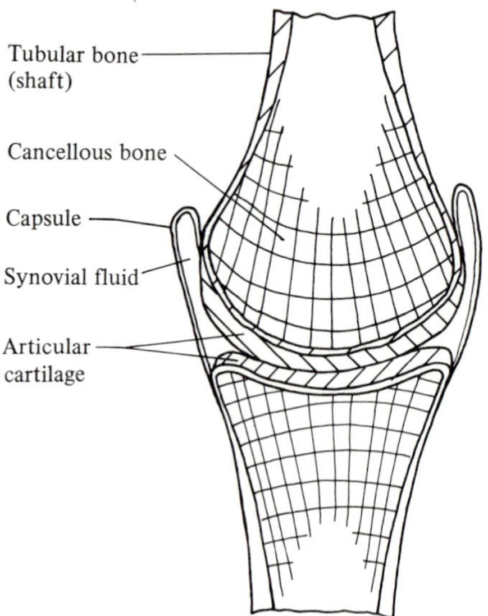

Fig. 1. A cross-sectional view of a notional synovial joint connecting two tubular bones.

exact composition varies within one sheet of cartilage, with age and condition, and no doubt with other factors.

Some of the water is free, in the sense that it can be expelled by the application of mechanical pressure, but much of it is combined with the proteoglycans to form a gel.

The proteoglycans are present as a two-stage assembly. The smallest units are glycosaminoglycans, principally chondroitin sulphate, with a relative molecular mass (RMM) of typically about 20 000, and keratan sulphate, with a rather smaller molecule, both being unbranched chains of disaccharide units. About 100 molecules of chondroitin sulphate and about 50 of keratan sulphate are attached, each by one end, along the length of a protein itself having an RMM of about 20 000. This assembly, with an RMM in the range 1 to 4×10^6, is called a proteoglycan. A few tens of these proteoglycans are in turn attached, each by one end, to a hyaluronic acid molecule about 1.2 μm long, forming an aggregate with an RMM of about 50×10^6.

Cells (chondrocytes) are of course present, about 15 000 of them per cubic millimetre (Stockwell, 1967), but from the point of view of mechanical function it will be assumed that they form part of the matrix of gel and collagen fibres.

The above bald statements represent the present state of belief, and have omitted all the many questions which are incompletely resolved; full descriptions and discussions are given by Meachim & Stockwell (1979) and Muir (1979).

Spatial arrangement of constituents

It is apparently certain that the layer of cartilage next to the articular surface contains proportionately more collagen than the deeper layers, that in this superficial zone (up to about 200 μm deep) the dominant direction of the collagen fibres is parallel to the surface, that in the deepest zone the dominant direction is perpendicular to the surface (and therefore to the bone–cartilage junction), and that in the intermediate zone the collagen fibres run obliquely. This certainty is described as apparent because the effect of a sequence of scanning electron micrographs of cartilage produced by many observers has been to show collagen lying in almost every direction in any region chosen for examination, and thereby to confuse the simpler picture previously deduced from polarised light micrographs; but the above description is compatible with the bulk of the observations so far made. What is more clearly certain is that when the articular surface is pricked with a sharp circular needle the result is not a circular hole but a split, and that on any one sheet of cartilage a series of such splits forms a regular pattern which is repeated on examples of the same joint from different bodies. For a survey of such observations and the interpretations which have been placed on them, the reader is referred to Weightman & Kempson (1979). It seems legitimate to assume that the split or prick pattern shows the dominant direction of the collagen fibres in the superficial zone and that the collagen fibres have a generally curved course, from being mainly perpendicular to the subchondral bone in the deepest zone of cartilage to being mainly parallel to the surface when close to it. Whether any one fibre has a roughly semicircular arched shape, starting and finishing in the deep zone, so that a series of fibres forms an arcade, as proposed by Benninghoff (1925), cannot be determined on the basis of present knowledge. Fig. 2 shows a highly simplified cross-sectional view of a representative piece of cartilage with the adjacent cancellous bone.

The above, grossly simplified, account presents the tissue in a form which can potentially be related to its mechanical properties and functions: an hydrated gel, stabilised by collagen fibres passing through it in definite ways, with cells capable of renewing at least some of the constituents of the matrix.

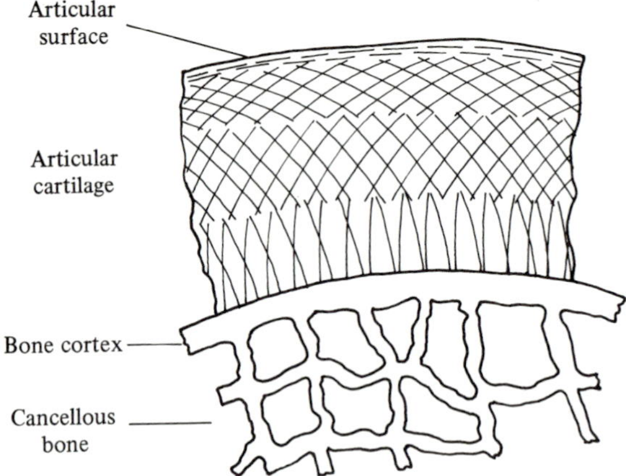

Fig. 2. A cross-sectional view of a portion of articular cartilage in place on cancellous bone, showing in a greatly simplified form the dominant directions of the collagen fibres in the different zones of cartilage.

Mechanical properties

Preliminaries

The two assumed mechanical functions of cartilage, to distribute load and to provide a bearing surface, lead naturally to the investigation of its deformability and its lubricating properties respectively. The strongly directional nature of the structure of cartilage suggests that the mechanical properties will vary with direction, i.e. that the material is anisotropic; the known variations in composition suggest that the mechanical properties will be far from uniform either within a sheet or between sheets; and the presence of an hydrated gel and free water suggests that the mechanical properties will be time-dependent. When the lubricating properties come to be considered, account must obviously be taken of the presence in healthy joints of synovial fluid.

Compression and indentation tests

Choice of type of test

The forces applied to cartilage in service are primarily compressive, and this is therefore an obvious mode to use in laboratory tests. Because cartilage does not naturally exist in pieces of convenient shape (e.g. circular cylinders with plane ends) there is no possibility of performing the type of test that yields the most understandable results on cartilage as it is found in the body. If it is preferred to interfere as little as possible with the cartilage and yet to explore spatial variations in properties, only one small region at a time must be compressed and the test is effectively an indentation test. If, instead, a more controlled test is wanted, many cylindrical specimens can be removed from a sheet of cartilage and tested either attached to, or removed from, the adjacent bone. The former type of test has the difficulties associated with indentation tests on thin layers; the latter type produces more straightforward results but at the expense of having disrupted the structure of the cartilage and, in particular, having considerably changed the conditions controlling the flow of water into or out of the specimen.

Origin of specimens

Most laboratory testing has been done on cartilage of the femoral head, partly because this is more readily available as operative or *post-mortem* specimens than components of most other joints, partly because its part-spherical shape lends itself to tests over the whole area, partly because the clinical features of osteoarthrosis are better documented for the hip joint than for probably any other joint, and partly because 10 or 15 years ago more was known of the forces at the hip joint than of those at any other joint.

Testing techniques

These have been fully discussed recently by Kempson (1979); it will suffice here to say that precautions are taken by most investigators to eliminate the more serious artefacts of both biological and mechanical origin, and that, though the accuracy of the results is less good than in mechanical testing of more consistent man-made materials, it is sufficient to enable major variations to be detected.

Results

From indentation tests on femoral head cartilage, the following facts have been established.

(1) On any one femoral head, the cartilage stiffness is not constant; a range of four to one is typical.

(2) On any one femoral head, the stiffness varies systematically, the stiffest cartilage lying in a band passing from the anterior region over the supero-lateral to the posterior region.

(3) On a femoral head with local disruption of the cartilage surface (fibrillation) the unfibrillated cartilage is everywhere less stiff than that on a head with no fibrillation.

(4) The general level of stiffness of unfibrillated cartilage is roughly

inversely related to the severity and extent of local fibrillation; the average stiffness of unfibrillated cartilage on a femoral head with severe local fibrillation is about one third that of the cartilage on a healthy femoral head.

(5) When indentations are made on cartilage which itself shows the mildest degrees of disruption, the stiffness is inversely correlated with the degree of disruption.

(6) Stiffness is positively correlated with proteoglycan content.

(7) Stiffness does not seem to be directly correlated with age, although when the results from many specimens are considered an indirect correlation is seen because fibrillation occurs more frequently in older specimens.

Results obtained from uniaxial compression tests on cylindrical specimens are consistent with the above.

Tensile tests

Significance

Cartilage is not subjected to tensile forces in service, although tensile strains may be developed as a consequence of non-uniform deformations. Thus tensile tests are less directly related to service conditions than are compressive or indentation tests but have the advantage of giving, in addition to a measurement of stiffness, an unambiguous measurement of strength. A further advantage is that tensile tests can be performed on specimens taken from different depths and in different directions, so as to explore spatial variations in properties.

Techniques

What was said about compressive and indentation tests is relevant here also. Because the preparation of a specimen of predetermined shape and size is of the essence of a tensile test, the difficulties attending the preparation of such specimens are inevitable, and in particular the extent to which the collagen fibres are disrupted may affect the recorded strength of the specimen. Obviously all fibres crossing the boundaries of the specimen are cut and this is an unavoidable source of artefacts; but the effects of cutting them cleanly may be expected to differ from those of tearing them out of the matrix by the use of inadequate techniques.

Origin of specimens

Mainly from a desire to use specimens no smaller than necessary, most tensile testing has been done with specimens from the femoral condyles (i.e. the rounded portions of the femur which form part of the knee joint), which offer larger regions of smaller curvature than the femoral head. Even so, the parallel length of specimens is limited to about 10 mm, with a width of about 2 mm and a thickness of about 0.2 mm if several specimens are to be tested from different depths.

Results

Allowing for the inaccuracies of the tests, certain points have become clear.

(1) In the surface zone of cartilage, the tensile stiffness and strength are higher when tensile stress is applied parallel to the dominant collagen direction rather than perpendicular to it.

(2) This difference is less marked in the other zones; representative tensile strengths, parallel and perpendicular respectively to the surface prick pattern, are 25 and 10 MNm^{-2} in the surface zone and 15–20 and 6–7 MNm^{-2} in the middle and deep zones.

(3) Tensile stiffness and strength are not significantly correlated with glycosaminoglycan content.

(4) Tensile stiffness and strength are positively correlated with collagen content.

Fatigue tests

Significance

In service, cartilage is loaded repetitively as are all components of the locomotor system; the number of paces taken per year obviously varies widely with the individual's way of life, but is usually of the order of a few million, with perhaps one million as the lower limit for people in normal health. Man-made materials are well known to fail in fatigue, i.e. after repeated applications of a stress too small to cause failure when applied once, and it is at least possible that biological materials also can fail in fatigue. Fatigue fracture of bone is a well-documented clinical phenomenon (Devas, 1958; Orava, Puranen & Alaketola, 1978), but laboratory fatigue tests on biological materials are open to two obvious sources of error. For a component to break in fatigue a crack must be started and it must propagate. The means by which a fatigue crack is started are not fully understood, but involve the accumulation of stress cycles applied to a piece of material; the natural turnover of biological materials can in principle prevent the accumulation of a sufficient number of cycles. If a crack can be started, its propagation is resisted in many biological materials (conspicuously in bone) by the excitation of a healing response and whether the crack propagates or not depends on the balance between the agents tending to make it propagate (stress amplitude and frequency) and the agents tending to reorganise the damaged material. It follows that a laboratory fatigue test on a biological material is valid only if it uses a number of stress cycles small enough to be applied in life in too short a time for turnover to be effective in replacing parts of the tissue before a crack could be started.

Estimates of the rate of turnover of the constituents of cartilage cannot be closer than orders of magnitude, because of the natural variations within and between individuals. Recent figures (Maroudas, 1979) are 600 to 1000 days for the mean life of proteoglycans and about 400 years for collagen. Therefore on a pessimistic assumption (that the fatigue process starts in the proteoglycans) fatigue tests using up to 10^5 or 10^6 stress cycles are valid, whereas if the fatigue process involves essentially only the collagen in cartilage, there is effectively no limit to the number of stress cycles that can properly be used (other limits will apply in practice).

Methods

Fatigue tests have used broadly the same types of specimen and loading as the static indentation and tensile tests mentioned above. Because the interactions between the gel and the collagen fibres are likely to be at least as important in fatigue tests as in static tests, indentation tests on intact sheets of cartilage attached to bone are preferable to compressive tests on isolated specimens. Tensile tests again have the advantage of a clearly-defined end-point but the drawback of using a specimen of which the structure has already been disrupted to some extent. In order to allow complete recovery of deformation between successive stress cycles, the time under no load may have to be longer than the time under load; otherwise a creep test would be superimposed on a fatigue test. Full details are given by Weightman, Chappell & Jenkins (1978) and Weightman & Chappell (1980).

Results

Indentation fatigue tests have shown surface damage, visually similar to that described as fibrillation when seen in life or at post-mortem examination, after about 400 000 applications of a nominal stress (i.e. the average stress under the indenter) of 4 MNm^{-2} to femoral head cartilage from 55 year old cadavers, and after about 90 000 cycles at a nominal stress of about 2 MNm^{-2} applied to femoral head cartilage from a 47 year old cadaver (Weightman, Freeman & Swanson, 1973; Weightman & Chappell, 1980). Surface splitting occurred before detectable loss of proteoglycans. A separate series (Johnson, Dowson & Wright,

1978) in which cylindrical specimens removed from the cartilage of the femoral condyles were subjected to a stress range of 0 to 5.6 MNm^{-2} gave similar results; also considerable scatter is found in the fatigue testing of any material and the results from a small number of specimens should be interpreted with caution.

Tensile fatigue tests (Weightman, 1976) have shown that

(1) the susceptibility of cartilage to fatigue damage increases with increasing age;

(2) the reduction in fatigue strength with increasing age is greater than can be accounted for merely by the accumulation of stress cycles on a given piece of cartilage;

(3) at a given age there seems to be more variation in the fatigue strength of male cartilage than in that of female cartilage.

Friction and wear

Cartilage and synovial fluid

Although this paper is concerned with articular cartilage, in considering the function of this tissue as a bearing material attention must be paid to the fact that it always appears in association with synovial fluid. Synovial fluid is essentially a dialysate of blood plasma to which is added hyaluronic acid, a polymeric disaccharide similar in many respects to the chondroitin sulphate and keratan sulphate present in cartilage. The hyaluronic acid is generally believed to be present as a hyaluronate–protein complex which assumes a roughly spherical shape about 0.5 μm in effective diameter when unconstrained in relatively large volumes of aqueous solution (Ogston & Stanier, 1951). There is evidence (Radin, Swann & Weisser, 1970) for the presence of a protein attached to hyaluronate either loosely or not at all.

In passing it should be mentioned that the characteristic constituents of synovial fluid are produced by the cells lining the capsule which encloses the joint and that one important function of synovial fluid is to convey nutrients to the cartilage; but the present concern is with the lubricating properties of the fluid, i.e. its effects on the friction and wear of cartilage.

As would be expected from its composition, synovial fluid is markedly non-Newtonian; its viscosity decreases with increasing shear rate. Reported values have ranged between 10^{-1} and 10^{2} N s m^{-2} at a shear rate of 10^{-1} s^{-1} and between 10^{-2} and 10^{-1} N s m^{-2} at a shear rate of 10^{3} s^{-1}.

Experimental observations

Methods

Frictional tests on joints in a living body suffer from the ineluctable defect that, even if the body is anaesthetised, the passive forces exerted by ligaments and muscles cannot be distinguished from the forces of friction between the joint surfaces. Therefore almost all the results quoted below have been obtained from tests on specimens removed from the body, either both bones connected by a joint (with the ligaments removed) or small specimens of cartilage, with or without bone attached, made to slide against glass. For a full discussion of the various types of test, the reader is referred to a recent review by the present author (Swanson, 1979).

Results

(1) The general level of friction is low by engineering standards, particularly for sliding bearings which have to function at rubbing speeds down to zero. Values of the coefficient of friction (the ratio of tangential force to normal force) for cartilage on cartilage are between 0.003 and 0.04, compared to 0.07 for polytetrafluorethylene against itself or 0.2 for a steel shaft in an oiled bronze bush.

(2) The variation of frictional force with rubbing speed suggests that much of the force arises from the sliding of solid surfaces, but that part of the force arises from the shearing of a layer of fluid between the surfaces.

(3) If synovial fluid is removed and replaced with saline solution, the friction in a given joint increases, but the range of values of coefficients of friction obtained in several series of experiments with saline solution (0.005–0.022) lies within the range found with synovial fluid present (0.003–0.04).

(4) If synovial fluid is treated so as to depolymerise the hyaluronic acid, little change is seen in the coefficient of friction.

(5) If synovial fluid is treated so as to detach hyaluronic acid from protein (which is held to impair its adsorption to cartilage), the friction increases noticeably.

(6) The general level of frictional forces with synovial fluid present, if assumed to result entirely from the shearing of a fluid layer, requires that layer to have a viscosity two or three orders of magnitude higher than any measured with synovial fluid at a relevant shear rate.

(7) Compression of a layer of synovial fluid between two layers of cartilage results in the expulsion of water and small solutes from the most highly loaded region while the large molecules of hyaluronate–protein are trapped in the most highly loaded region; thus the concentration of the distinctive component of the fluid is increased where presumably it is most needed.

(8) Repeated massive dilution of synovial fluid in rabbits' knees led to fibrillation during periods of up to 85 days of apparently normal activity.

(9) A few tests under exaggerated conditions have shown that cartilage tested *in vitro* wears at a greater rate when lubricated by saline solution than when lubricated by synovial fluid and that the wear-limiting property of synovial fluid is impaired by interfering with the attachment of the hyaluronate to protein, and thereby with its ability to be adsorbed on to cartilage, but not by depolymerising it.

Service requirements

Stresses

The forces transmitted through several of the joints of the human skeleton during representative activities have been measured or calculated by many workers using various means. One series of such results has been produced at the University of Strathclyde using non-invasive measurements of force and position to calculate the forces transmitted through the hip and knee joints during level walking, climbing and descending ramps and climbing and descending stairs (Paul, 1967; Morrison, 1970). Similar work has been done in various places on other joints.

From a knowledge of the force transmitted through a joint and the possible contact area of the cartilage surfaces, a lower limit for the average stress can easily be calculated. But there is good reason to expect that at any instant the stress will vary widely over the contact area, and attention has been turned to measuring pressures on small regions within an articular surface. The most complete set of such results so far published is that of Adams, Kempson & Swanson (1978), who inserted 11 pressure transducers at the bone-cartilage junction in the acetabulum of each of nine cadaveric hips, and subjected each hip to loads representing six instants in a representative walking cycle according to Paul (1967). For the present purpose, it is sufficient to note that

(1) the highest pressure in a given hip under a given load was up to three times the average pressure;

(2) when the load was the highest during the walking cycle (four times body weight) the highest local pressure ranged from 5 to 8.6 MNm^{-2} in the nine hips tested;

(3) in every hip tested, the region which was subjected to the highest pressure when the load was the highest was subjected to a pressure of less than 1 MNm^{-2} at some other instant in the cycle.

The above stresses correspond to a walking cycle typical of a middle-aged person, in which the resultant force transmitted through the hip joint rises to about four times body weight; this was chosen because it matched the ages of most of the cadavers from which the joints were taken, which is also the age range during which the symptoms of early osteoarthrosis commonly appear. Younger people in general walk more vigorously, and at any age some walk more vigorously than others. Hughes, Paul & Kenedi (1970) found hip forces during level walking of up to eight or nine times body weight, while Smith (1975) calculated a knee force of up to 24 times body weight during drop-landing from a height of 1 m on a hard surface.

Rubbing speeds

The highest rubbing speed in the human hip during normal level walking was estimated by Paul (1967) to be about 80 $mm\,s^{-1}$. Higher speeds presumably occur during running and some athletic activities and games, and many joints of the human skeleton are required to move after periods of no motion, often under load.

Possible modes of failure of cartilage

Fibrillation and osteoarthrosis

Fibrillation, in the sense of some disruption of the articular surface of cartilage, is seen so commonly in post-mortem specimens from middle-aged and elderly cadavers as to be regarded as a usual (but not invariable) accompaniment to ageing. Primary osteoarthrosis, meaning severe disruption or loss of cartilage and associated changes in the adjacent bone, with no history of trauma or other insult to the joint, is much less common than fibrillation. A precise figure for its incidence is unobtainable for many reasons, including the difficulty of defining the condition precisely; but the figure, quoted above, of about 24 000 patients undergoing hip replacement in one year in Britain suggests an order of magnitude. Allowance must be made for the facts that not all those replacements were of osteoarthrosic hips (though it is probable that most were), that in 1974 there were probably many people with osteoarthrosic hips which merited replacement but which were not replaced, that many more probably suffered from osteoarthrosic hips to an extent which affected their way of life but was not sufficient to justify replacement (even if the facilities for replacement had been available), and that joints other than the hip are affected by osteoarthrosis. On this very uncertain basis, an incidence for the whole population of between 0.1% and 1% per annum of significant osteoarthrosis can be suggested. Of course the incidence is not uniform at all ages, and the incidence for the age groups principally affected is therefore considerably higher.

The possible connection between the commonly observed mild fibrillation and the less commonly observed osteoarthrosis has been discussed by Freeman & Meachim (1979). Here it is sufficient to note that, whatever the connection, one purpose of such investigations as those surveyed above is to attempt to describe how, in a small proportion of joints, cartilage damage progresses at a rate which leads to the phenomena of osteoarthrosis. (It is not difficult to see how, once damage to the cartilage has started, continued use of the joint will lead to the removal of cartilage from some regions, the exposure of bone as a bearing material, and an increase in the maximum stresses in the bone, probably causing microfractures in the cancellous bone structure with consequent pain; the present point is to see what light can be shed on the earliest stages.)

Fatigue failure of cartilage

Disruption of the surface of cartilage has been produced in the laboratory by repeated applications of stress at an amplitude within the range corresponding to normal walking and for a number of cycles too small for effective turnover to have occurred in life. Therefore fatigue failure in life is certainly a possibility. Both the range of maximum stresses measured in hip joints from different cadavers and the range of rates of decrease in fatigue strength with increasing age, suggest that the probability of fatigue failure should vary widely between individuals.

Three main causative factors can be identified:

(1) the individual's level of physical activity; one whose work and leisure occupations involve only mild exertion will in general impose lower forces on his joints than one whose work or leisure activities are more strenuous; a clerk who spends his leisure time

playing squash or practising high jumps may from this point of view be in the same category as a miner who spends his leisure time playing chess;

(2) the distribution of stresses in the joints; it seems, from the limited number of observations so far made, that a similar load will produce markedly different maximum stresses in different hip joints, and the differences must result from differences in the configuration or properties of the bone and cartilage, i.e. in the detailed anatomy of the joint (and it is probable that similar variations will in due course be found at other joints);

(3) the rate of decrease in the fatigue strength of cartilage with increasing age; the work briefly referred to above showed that this rate of decrease varies between individuals, more widely in males than in females.

If the hypothesis now being discussed, that some osteoarthrosis results from fatigue failure of cartilage, is correct, then one must assume that the many people who do not suffer from osteoarthrosis have favourable values of each of the above three variables, or that unfavourable values of at least two of the variables are needed for osteoarthrosis to develop. The information does not exist to enable the question to be pursued quantitatively, but such information as does exist is compatible with the hypothesis. Largely anecdotal evidence exists suggesting that people engaging in activities which load certain joints abnormally highly are likely to develop osteoarthrosis in the joints concerned, which is relevant to the first of the three factors listed above. The epidemiological studies now in progress would need to be supplemented by mechanical tests on post-mortem specimens from at least some of the populations studied if light were to be shed on the practical importance of the second and third factors.

Failure of lubrication

From the experimental findings surveyed above it can be inferred that the mechanical function of synovial fluid (as distinct from its nutritional function) is primarily to protect cartilage against wear and only secondarily to reduce the frictional forces. Consideration of the structure of the relevant substances suggests that this is probable. The matrix of cartilage is an hydrated gel with collagen fibres in it; in normal cartilage the uppermost 10 μm or thereabouts contains keratan sulphate or a similar substance (Stockwell, 1970); thus the surface presented by cartilage to the joint cavity consists of a polymeric disaccharide. If two cartilage surfaces are rubbed together, the frictional force will be that due to the rubbing of two sets of these polymeric disaccharides. If synovial fluid is present and if the surfaces are completely covered with hyaluronate–protein, the friction will again be that of polymeric disaccharides against each other (though now different disaccharides). It is therefore not surprising that the frictional forces are found to be not greatly different, but the important difference is that with synovial fluid present one layer of disaccharides can be continuously renewed, and it is also not surprising that if it is not renewed or even present the cartilage surfaces wear at a greater rate. Whether the hyaluronate–protein layer is simply adsorbed on to the cartilage surfaces to form a boundary lubricant, or whether the synovial fluid is so concentrated by the process mentioned above as to form a gel with the same properties as an adsorbed layer, is irrelevant to the question of the frictional forces but may be highly relevant to the question of protection against wear. The experimental observations showing that depolymerisation of hyaluronate has little effect on friction or wear, but that interference with its ability to be adsorbed on to cartilage leads to increased friction and wear, suggest that the wear-protecting function is performed by adsorption of hyaluronate–protein on to cartilage. If this adsorption is hindered or prevented in life, then significant wear of the cartilage surfaces might be expected. How the adsorption could be hindered or prevented is a biochemical question outside the scope of this paper.

Conclusion

This brief and simplified survey has shown that the broad outlines of the mechanical properties of articular cartilage are now fairly well established, and that what is known of them is compatible with what is known of the structure of the tissue. From this survey, and even more from a study of the works referred to, it will be obvious that many questions await answers. In the present state of knowledge, it is possible to hypothesise that primary osteoarthrosis can result either from unfavourable combinations of mechanical factors leading to fatigue failure of cartilage under repeated stress or from some unexplained failure of synovial fluid to protect the cartilage surfaces against wear. The two causal chains postulated are not proven but are compatible with the evidence now available; they could exist together and do not preclude the existence of other causal chains.

References

ADAMS, D., KEMPSON, G. E. & SWANSON, S. A. V. (1978). Direct measurement of local pressures in the cadaveric human hip joint. *Journal of Physiology,* **278,** 33–34*P*.

BENNINGHOFF, A. (1925). Form und Bau der Gelenkknorpel in ihren Beziehungen zur Funktion. I: Die modellierenden und formerholtenden Factoren des Knorpelreliefs. *Zeitschrift für Anatomie und Entwicklungsgeschichte,* **76,** 43–63.

DEVAS, M. B. (1958). Stress fractures of the tibia in athletes or 'Shin Soreness'. *Journal of Bone and Joint Surgery,* **40B,** 227.

FREEMAN, M. A. R. (ed.) (1979). *Adult Articular Cartilage*, 2nd edn. Tunbridge Wells: Pitman Medical.

FREEMAN, M. A. R. & MEACHIM, G. (1979). Ageing and Degeneration. In *Adult articular cartilage*, ed. M. A. R. Freeman, 2nd edn, pp. 487–544. Tunbridge Wells: Pitman Medical.

HUGHES, J., PAUL, J. P. & KENEDI, R. M. (1970). Control and movement of the lower limbs. In *Modern Trends in Biomechanics,* **1,** ed. D. C. Simpson, pp. 147–79. London: Butterworth.

JOHNSON, G. R., DOWSON, D. & WRIGHT, V. (1978). The fracture of articular cartilage under impact loading. In *The wear of non-metallic materials*, ed. D. Dowson, M. Godet & C. M. Taylor, pp. 113–15. London: Mechanical Engineering Publications Ltd.

KEMPSON, G. E. (1979). Mechanical properties of articular cartilage. In *Adult articular cartilage*, ed. M. A. R. Freeman, 2nd edn, pp. 333–414. Tunbridge Wells: Pitman Medical.

MAROUDAS, A. (1979). Physicochemical properties of articular cartilage. In *Adult articular cartilage*, ed. M. A. R. Freeman, 2nd edn, pp. 215–90. Tunbridge Wells: Pitman Medical.

MEACHIM, G. & STOCKWELL, R. A. (1979). The Matrix. In *Adult articular cartilage*, ed. M. A. R. Freeman, 2nd edn, pp. 1–67. Tunbridge Wells: Pitman Medical.

MORRISON, J. B. (1970). Biomechanics of the knee joint in relation to normal walking. *Journal of Biomechanics,* **3,** 51–61.

MUIR, I. H. M. (1979). Biochemistry. In *Adult articular cartilage*, ed. M. A. R. Freeman, 2nd edn, pp. 145–214. Tunbridge Wells: Pitman Medical.

OGSTON, A. G. & STANIER, J. E. (1951). The dimensions of the particle of the hyaluronic acid complex in synovial fluid. *Biochemical Journal,* **49,** 585.

ORAVA, S., PURANEN, J. & ALAKETOLA, L. (1978). Stress fractures caused by physical exercise. *Acta Orthopaedica Scandinavica,* **49,** 19–27.

PAUL, J. P. (1967). Forces transmitted by joints in the human body. *Proceedings, Institution of Mechanical Engineers*, **181** (3J), 8.

RADIN, E. L., SWANN, D. A. & WEISSER, P. A. (1970). Separation of a hyaluronate-free lubricating fraction from synovial fluid. *Nature, London,* **228,** 377.

SMITH, A. J. (1975). Estimates of muscle and joint forces at the knee and ankle during a jumping activity. *Journal of Human Movement Studies,* **1,** 78–86.

STOCKWELL, R. A. (1967). The cell density of human articular and costal cartilage. *Journal of Anatomy,* **101,** 753–63.

STOCKWELL, R. A. (1970). Changes in the acid glycosaminoglycan content of the matrix of ageing human articular cartilage. *Annals of the Rheumatic Diseases,* **29,** 509.

SWANSON, S. A. V. (1979). Friction, wear and lubrication. In *Adult articular cartilage*, ed. M. A. R. Freeman, 2nd edn, pp. 415–60. Tunbridge Wells: Pitman Medical.

TAYLOR, D. G. (1976). The costs of arthritis and the benefits of joint replacement surgery. *Proceedings of the Royal Society of London, Series B,* **192,** 145–55.

WEIGHTMAN, R. (1976). Tensile fatigue of human articular cartilage. *Journal of Biomechanics,* **9,** 193–200.

WEIGHTMAN, B. & CHAPPELL, D. J. (1980). A collagen fatigue model for osteoarthrosis. In *Models of osteoarthrosis*, ed. G. Nuki. Tunbridge Wells: Pitman Medical.

WEIGHTMAN, B., CHAPPELL, D. J. & JENKINS, E. A. (1978). A second study of the tensile fatigue properties of human articular cartilage. *Annals of the Rheumatic Diseases,* **37,** 58–63.

WEIGHTMAN, B., FREEMAN, M. A. R. & SWANSON, S. A. V. (1973). Fatigue of articular cartilage. *Nature, London,* **244,** 303.

WEIGHTMAN, B. & KEMPSON, G. E. (1979). Load carriage. In *Adult articular cartilage*, ed. M. A. R. Freeman, 2nd edn, pp. 291–332. Tunbridge Wells: Pitman Medical.

DEFORMATION IN TENDON COLLAGEN

JOHN KASTELIC and ERIC BAER

Macromolecular Science Dept, Case Western Reserve University, Cleveland, Ohio 44106, USA

Introduction

Large-scale mechanical structures in higher organisms are apparently achieved through hierarchical assembly of cellularly produced macromolecules along with other constituents such as ions, water and minerals. Hierarchically organized structures are widespread and the size, scale and organizational character at each level are dictated by the particular mechanisms of assembly and growth. Little is known concerning these mechanisms so, for the present, we can deal only with their consequences.

Fibrous proteins are known to possess the property of self-assembly (Crane, 1950) though the *in vivo* nucleation and formation mechanisms are not at all known. Such proteins are frequently found organized into tensile load-bearing tissues. Muscle (Huxley, 1972), mammalian keratin (Sikorski, 1975; Fraser & MacRae, 1980) and tendon are prime examples. All exhibit a uniaxial fibrous character at virtually all levels of organization from molecular to macroscopic.

The hierarchical organization of tendon collagen has been widely studied and reviewed (Bear, 1952; Verzar, 1964; Elliott, 1965; Viidik, 1972; Baer, Gathercole & Keller, 1975; Lees & Davidson, 1977). Present knowledge of the structural units involved is summarized in Fig. 1 (Kastelic, Galeski & Baer, 1978). Starting from the smallest and proceeding upward in size they are: the single protein macromolecule, the tropocollagen triple helix, the microfibril of five tropocollagen units appropriately staggered lengthwise, and finally the tetragonal lattice of microfibrils possibly forming subfibrils and certainly the collagen fibril. At this level other macromolecules are incorporated: it is the current view that proteoglycans in association with much water come into play as a matrix binding the fibrils together (Wassermann, 1956; Sobel, 1967). Proceeding toward macroscopic dimensions the fascicles comprising crimped collagen fibrils are embedded in the proteoglycan-water gel

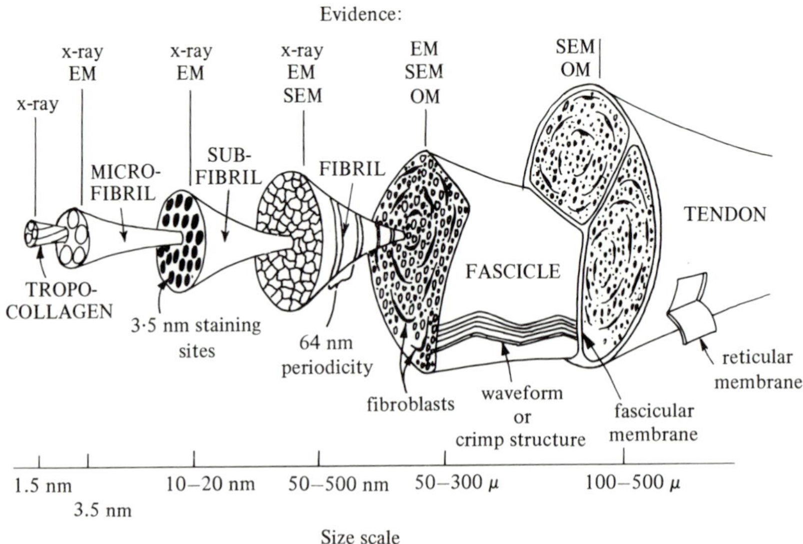

Fig. 1. Hierarchical organization of tendon.

with several fascicles in turn making up the functioning tendon. It is tendon, with this complex structure, which serves as the mechanical link connecting muscle with skeleton.

The tendinous structure bears a strong resemblance to a common type of engineering composite. Specifically, this tissue belongs to the family of uniaxial composites, that is, uniaxially aligned fibres embedded in a binding matrix. Since the engineering model applies at several levels of organization the term 'uniaxial multicomposite' aptly describes the hierarchical organization of tendon. Clearly the hierarchical organization of most tissues can be classed into the various composite families, perhaps with different categories applying at the various levels of organization.

The significant aspect of a composite is its unique set of mechanical properties. One property may be superior or even spectacular without sacrificing other important physical characteristics. This is achieved through the synergistic mechanical interplay between structural elements (Broutman & Krock, 1967). With proper design, mechanically inferior or undesirable properties of one element are compensated through sophisticated mechanical interactions with other elements of the composite. Generally the interaction mechanisms control significantly the deformation and fracture processes within the composite.

Adhesion effects are foremost in importance. In the tendon multi-composite the levels and types of possible interactions are many. Bonding forces, such as chemical or physical cross-links, water bridges, hydrogen bonding and ionic or van der Waals' attractions are undoubtedly involved. These are all topics of present-day research and little definitive information exists (Verzar, 1964; Wassermann, 1956; Yannas, 1972; Traub & Piez, 1971; Hulmes *et al.*, 1973). Qualitative changes in these bonding interactions are thought to occur with aging. It is known that the solubility and the rate of digestion by enzymes decrease with age (Bonfield, 1972; Hamlin & Kohn, 1971, 1972; Jolma & Hruza, 1972). This suggests that the bonding interactions at the lower levels of the hierarchy are intensifying with age (Verzar, 1964). Chemical cross-linking between tropocollagen chains at lysine or hydroxylysine derivatives is thought to occur progressively with maturation and aging (Tanzer, 1973; Robins, Shimokomaki & Bailey, 1973). One may speculate that the known, age-related changes in tendon deformation behaviour arise as a consequence of these systematic changes in bonding interactions.

When tendon is deformed beyond its elastic limit, permanent elongation occurs. While this can easily be followed externally on the testing machine chart, it is not known what internal events take place within the structural hierarchy. Slippage or fracture may be occurring at various levels of organization but this has not been thoroughly investigated. Individual elongated collagen fibrils have been observed. Historically, the first observations were accidental (Schmitt, Hall & Jakus, 1942). Beam damage caused the film supporting unfixed dispersed collagen fibrils under observation in the electron microscope to fracture and peel back and by chance extended some of them. Elongations of up to 900% were reported. The fibril banding was proportionally extended, though above about 200% extension certain band regions extended slightly more than others. Two other studies have appeared, one concerning reconstituted and the other both reconstituted and mechanically dispersed fibrils (Schwartz, Geil & Walton, 1967; Barenberg, Filisko & Geil, 1978). In each, fibrils were elongated or compressed, depending upon their orientation on the substrates, and certain regions of the banding pattern were identified as deforming more than other regions. Unfortunately, the results of these two investigations are in conflict. Schwartz *et al.* (1967) concluded that for a limited range of deformation the stained (presumably polar) regions underwent greater deformation than the unstained regions, while Barenberg *et al.* (1978) found just the opposite. It should be noted that the staining methods employed were different. The first used phosphotungstic acid while the latter used uranyl acetate. At much higher elongations of the collagen Barenberg *et al*. (1978) reported highly localized deformation of the bands followed by regular transverse splitting. There was complete absence of age effects. A far more serious drawback of these two studies and the preceeding one is that, like much of the early work on bone, specimens were air dried during extension.

In sharp contrast to the subtle effects observed with dry collagen, severe damage to fibrils has been reported due to deformation *in vitro* (Torp, Baer & Friedman, 1975). In rat-tail tendons which were mechanically cycled or fractured while wet, collagen fibrils were observed to dissociate into numerous fibrous fragments of 15.0 nm and 3.5 nm diameter. Since banding was absent in the shreds, slippage of subfibrils and microfibrils was suggested. Damage was less with older specimens. It is not known at what point in the stress–strain curve this damage takes place or if it progresses systematically with deformation. Also, it is not known if a sequence of micro-deformation events precedes this irreversible damage.

Here we have attempted a systematic investigation of deformation in tendon collagen. Rat tail tendons were deformed *in vitro* into certain regions of their stress–strain curve and then examined for structural damage with ultra-thin section electron microscopy. Small angle x-ray diffraction was used to monitor the net crystallographic condition of fibrils in the deformed specimens. In addition polarizing optical microscopy was performed to record the effects of deformation in the tendon crimp structure (W. C. Dale, *Ph.D. thesis*, 1974; S. Torp, *Ph.D. thesis*, 1974; Torp, Arridge, Armeniades & Baer, 1975). Observations were carried out over a wide age range from immature to old age.

Experimental methods

The tail tendons of male Sprague Dawley rats 1.7, 3, 6, 12, 24 and 37 months old were used in this investigation. Methods for tendon removal and mechanical testing have been described previously (Torp *et al*., 1975; Galeski, Kastelic & Baer, 1978). Tensile tests on single fascicles were performed in neutral 0.9% saline at 21–23°C and completed within 4 h of death. The strain rate was 8% min^{-1}. Extension was stopped at 1%, 2.5%, 4% or 8.5% elongation representing respectively deformation into the elastic region, into the first yield region (twice), and into the second yield region. Specimens were subsequently fixed using one of two fixation procedures, which are described below.

Specimens which underwent 'taut fixation' were fixed while still under tension in the testing machine. Upon achieving a given strain the crosshead was halted and the saline specimen bath was quickly replaced with a bath containing fixative solution. Specimens which underwent 'relaxed fixation' were removed from the testing apparatus and fixed in their relaxed unloaded state. In both procedures 6.25% glutaraldehyde, buffered with sodium cacodylate to pH 7.3–7.4, was used for 1 h at room temperature. Fixation was always followed by two soakings, lasting at least 45 min each, in a buffered sucrose rinse solution to remove unreacted fixative. The specimens were fixed, relaxed and subsequently dehydrated in hexylene glycol and then examined in the optical microscope to assess the effects of mechanical deformation on crimping. 'Taut fixed' specimens look like birefringent but featureless glass rods. Photomicrographs were taken with an Olympus Photomax microscope equipped with polarizers and automatic camera.

After either taut or relaxed fixation, specimens to be viewed in the electron microscope were post-fixed for 4 h at 0–4°C in 4% OsO_4 in collidine buffer (pH 7.3–7.4), dehydrated in a hexylene glycol series (approx. 1 h each in 0, 20, 40, 60, 80, 100 and 100% hexylene glycol, aqueous solutions), infiltrated with Spurr epoxy (24 h each in 30, 60, 100 and 100% epoxy in hexylene glycol) and cured in flat moulds at 60°C. Final dehydration and all infiltration was done within a partly evacuated desiccator. Longitudinal sections of the embedded tendon specimens were made on a Richert OM-2 ultra microtome with DuPont diamond knife. The sectioning direction was approximately parallel to the collagen fibrils. All sections were stained with uranyl acetate (Watson, 1958) and lead citrate (Reynolds, 1963) prior to examination in a Hitachi-Hu-12 electron microscope run at 50 kV.

Small angle x-ray diffraction patterns (Bear, 1952) were obtained on a Rigaku Denki rotating anode unit equipped with an evacuated small angle camera using pin-hole collimators. Deformed, taut-fixed, hexylene-glycol-dehydrated specimens were sealed in powder pattern tubes. These were each carefully attached in place over the final collimator pin-hole using a microscope to view the alignment. Exposures ranged from 12 to 92 h with a camera-to-film distance of 18.2 cm. Microdensitometer scans were performed with a Joyce Lobel MK-III. Widths at half height were measured from the traces and converted to lattice size using the Scherrer equation (Kakudo & Kasai, 1972).

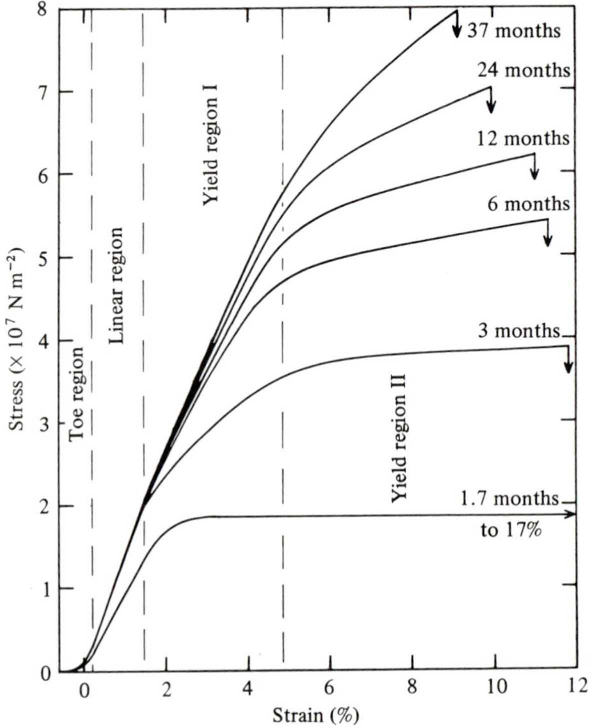

Fig. 2. Stress–strain behavior of rat tail tendon as a function of age. Neutral saline, 8% per minute strain rate.

Results

Tensile stress–strain behavior of tendon

The tensile stress–strain curve for rat tail tendon (RTT) can be divided into four regions (Fig. 2). The first is the 'toe region' where the change in stress with respect to strain, $d\sigma/d\varepsilon$, steadily increases until it reaches a constant value, E_1. This behavior has long been attributed to the straightening of the crimped fibrils, and the following linear region, the second region, to the elastic extension of the fibrils. The length of the toe region decreases with age due to a systematic decrease in crimp angle with age (Torp *et al.*, 1975). The elastic modulus, E_1, within the linear region, increases up to maturity (3–4 months for RTT) and then remains constant (Torp *et al.*, 1975).

Up to maturity of the tendon the linear region is followed by a single yield region in which irreversible elongation and structural damage takes place. A near-zero apparent modulus is observed in this yield plateau. Thus, at the plateau stress level, the rate of structural deformation of the tendon is sufficiently high to match the rate of applied straining.

After maturity this single yield plateau is replaced by two distinct yield regions, ('yield region one' and 'yield region two'). The apparent modulus, E_2, within the first yield region (YR–I) is measurably higher than that, E_3, of the second (YR–II). The values of both E_2 and E_3 increase systematically with aging, with the value of E_2 approaching E_1, the modulus in the linear elastic region. Occasionally a failure region is observed at the end of the stress–strain curve where, rather than a sudden break at fracture, the stress–strain curve rounds over to some degree prior to actual failure. The above results are in accord with those reported previously (Torp *et al.*, 1975; Galeski *et al.*, 1978). Within both YR–I and YR–II, elongation is irreversible so we must conclude that structural damage takes place. The fact that two distinct yield regions are observed with mature RTT suggests that two different yield processes are operating in sequence. In composites the deformation mechanism is determined by what can fail or slip at the lowest applied stress, i.e. the weakest 'link' in the mechanical structure. In the complex organization of tendon many mechanisms are possible and those actually taking place must be identified. With increasing age higher and higher forces are required for the yield mechanisms to occur, suggesting that bonding interactions, whatever their exact nature, are strengthening with age. In immature RTT, when only a single yield plateau is observed, presumably a single yield mechanism predominates. In order to investigate these mechanisms clearly one cannot examine only tendon which has been deformed all the way to failure. In mature tendon, with the likelihood of at least two yield processes and possibly a fracture process occurring in sequence, the cumulative total structural damage might be too complicated to interpret, or structural damage caused by one yield mechanism may be obliterated by that of subsequent processes. To determine the sequence of deformation events, a series of specimens which have been extended to systematically greater elongations must be examined. Furthermore, specimens which have been fixed while still under tension must be studied because, as will be seen, additional structural alteration can take place as the tensile load is released. 'Taut fixed' specimens must be prepared, regardless, for x-ray diffraction studies.

Small angle x-ray diffraction of deformed tendon

Clear small angle x-ray diffraction (SAX) patterns could be obtained only from 'taut fixed' specimens. Crimping, which naturally occurs in tendon and is modified by deformation, causes the collagen SAX fibril pattern to be only poorly oriented. The broad arcing which results (Kastelic *et al.*, 1978) completely overshadows the more subtle effects which arise from yield damage. Taut fixing 'freezes out' crimps and aligns the collagen fibrils parallel to one another. The resulting SAX pattern thus shows good orientation (Fig. 3) and is suitable for semiquantitative assessment of crystallographic size or perfection as well as measurement of repeat period, all of which could be affected by deformation.

Representative SAX patterns for 1.7, 6 and 24 month RTT are given in Fig. 4. Patterns obtained for all ages after only 1% deformation are similar to those appearing in the literature for normal undeformed collagen (Bear & Boldman, 1951; Rougvie & Bear, 1953; Miller & Wray, 1971). Typically these patterns are obtained with the specimen clamped under mild tension to remove crimping. Here the specimens are under no tension but instead have been fixed while under tension to achieve the same effect. This method has the advantage that stress relaxation effects are avoided. This is an important consideration when patterns are required from specimens loaded to high stresses. With immature tendon (1.7 month series) the patterns become decidedly weaker as the amount of deformation is increased. The onset of detectable weakening of the pattern occurs with very little deformation, early in the yield plateau region. With 8.5% deformation the SAX exposure time had to be increased by a factor of four to obtain a detectable trace of 64.0 nm fiber pattern. Apparently the crystalline packing within collagen fibrils is destroyed as deformation proceeds along the yield plateau. Diffuse scattering does build up at the center of the patterns. For mature tendon the situation is more complex. In addition to a general weakening in diffraction intensity, there is a very pronounced line broadening effect. With deformation into YR–I the diffraction pattern becomes progressively wider, but without a serious drop in intensity. Furthermore, the broadening occurs only in the equatorial direction. As deformation proceeds into YR–II there is a definite decrease in intensity and still further broadening. The loss in intensity of diffraction is slightly less in the 24 month pattern than in the 6 month, but neither approaches the dramatic loss apparent in the immature 1.7 month specimens. Nevertheless, the general effect of loss of intensity of pattern during deformation of mature RTT in YR–II and deformation of immature RTT in the yield plateau are quite similar. This suggests that the respective deformation mechanisms may be

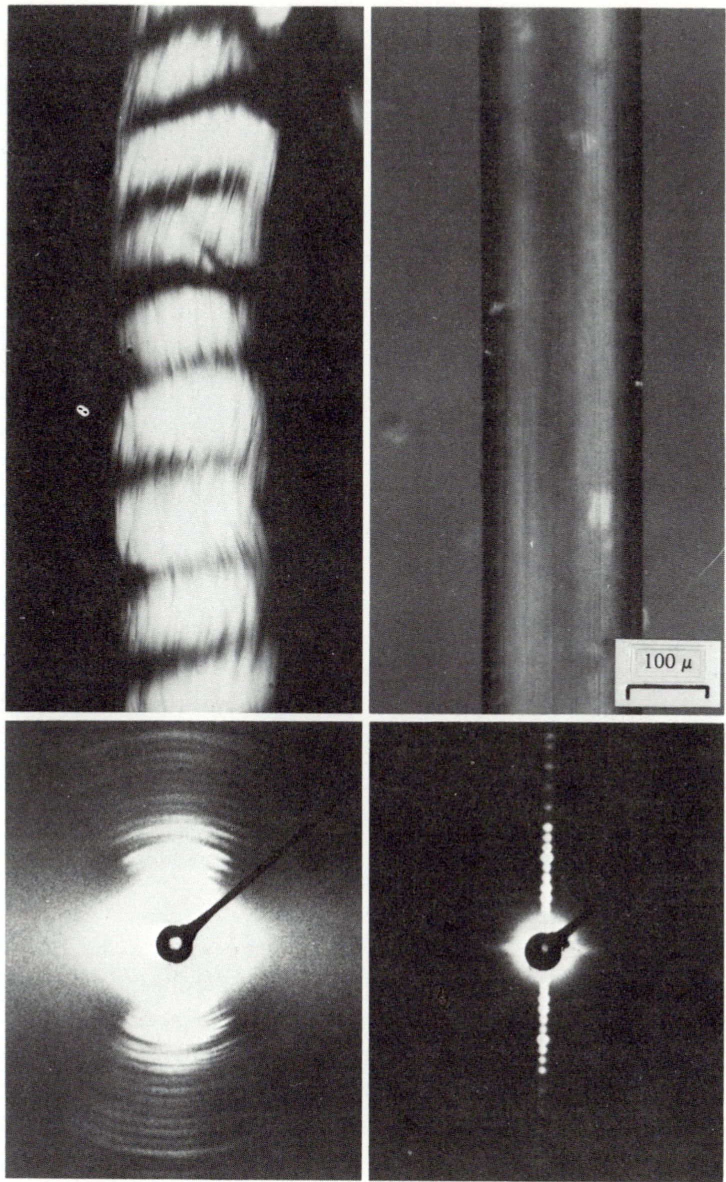

Fig. 3. Orientation of the small angle x-ray diffraction pattern of tendon achieved by fixation while under tension. Top left: Optical micrograph of specimen fixed without applied tension showing crimping. Bottom left: Corresponding arced pattern. Top right: Optical micrograph of specimen fixed while extended to 1% strain to straighten our crimping. Bottom right: Corresponding well aligned pattern.

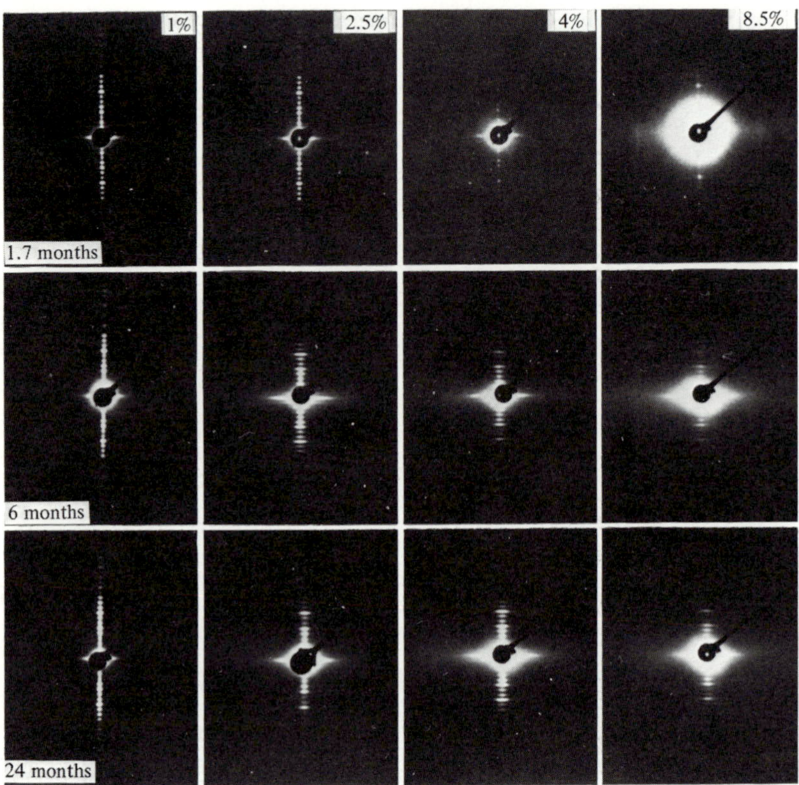

Fig. 4. The effect of age and mechanical deformation on the small angle x-ray diffraction pattern of rat tail tendon. Age and percent extension are indicated. Approximate exposure factors: Top Row: 1, $1\frac{1}{2}$, 3 and 6. Middle and Bottom Row: 1, 1, $1\frac{1}{2}$ and 2.

related. Diffraction broadening can arise either from loss in crystallographic spatial coherence (loss of 'crystalline perfection') or decrease in the size of the scattering (reflecting) lattices. Both can occur simultaneously and unfortunately the x-ray diffraction here does not discriminate between the two. Here the streaking and broadening are only in the equatorial direction. This indicates breakup and/or loss of coherence in a direction transverse to the collagen fibril (Boldman & Bear, 1951; Rougvie & Bear, 1953; Kakudo, 1972).

Microdensitometer scans

It is possible to gain more insight into the nature of the microdeformation events by examining the diffraction patterns in more detail. A semiquantitative line-broadening method is to measure the diffraction width at half height and then use the Scherrer equation to calculate the

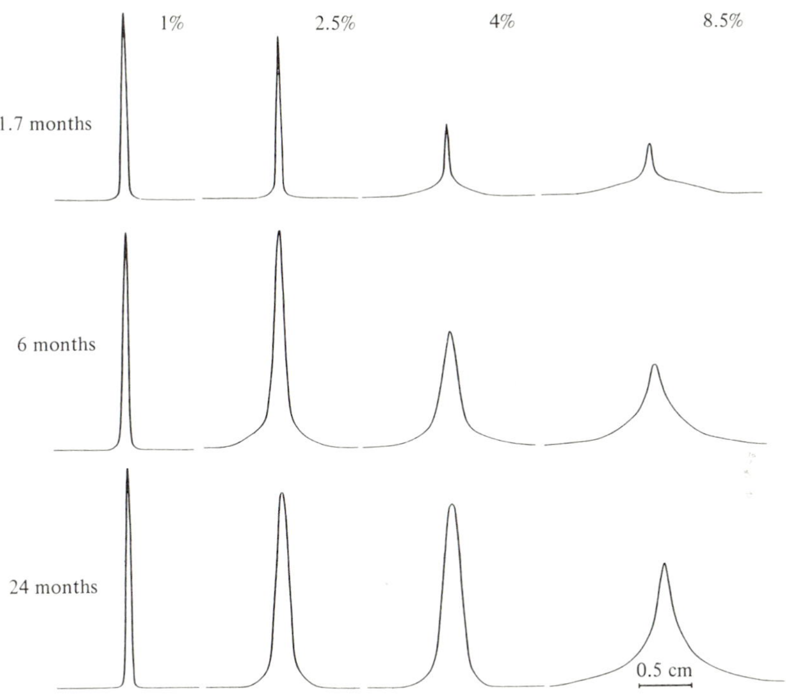

Fig. 5. Microdensitometer scans of SAX patterns from Fig. 4.

size of the diffracting lattice. The calculation assumes that crystalline perfection is not degraded by deformation. In this case the assumption is unjustified so such calculated dimensions have limited validity. But the line-broadening profiles and rough numbers do assist interpretation. Microdensitometer scans of the SAX patterns of Fig. 4 are given in Fig. 5. Slit scans were done across the 9th order reflections (most intense), in the equatorial direction. Naively calculated lattice dimensions are given in Table 1.

In 1.7 month RTT, streaking can be detected with the microdensitometer in the 4% and 8.5% deformation SAX patterns. This is not readily apparent in the photographic reproductions of the patterns themselves. Each reflection can be resolved into two superimposed peaks, a strong narrow central peak poised on a second broad weak background peak. In this case, the central peak does not increase noticeably in width even after extensive deformation. Only its intensity falls. This indicates that, at least in the area sampled by the x-ray beam, which should be representative of the entire sample, a contingent of

Table 1. *The effect of age and mechanical deformation on the transverse size of the diffracting unit in RTT. Calculations were made with the Scherrer equation (Galeski* et al., *1978). Data were not corrected for instrument broadening and ideal crystallinity was assumed*

Age (months)	Deformation (%)	Microdensitometer half width (mm ×20 scan)	Crystallite Dimension (nm)
1	1.7	5	100
	2.5	5	100
	4	5/45	100/12
	8.5	6/~70–80	83/~5.5–6.5
6	1	4	125
	2.5	14	36
	4	~17	~29
	8.5	~26/~70–80	~19/~5.5–6.5
24	1	4	125
	2.5	15	33
	4	~17	~29
	8.5	~22/~70–80	~23/~5.5–6.5

undamaged fibrils exists but declines in number as deformation proceeds. We probably should not be talking about a population of undamaged whole fibrils, but rather a population of undamaged *portions*. A fibril need not lose its ability to diffract x-rays uniformly along its length, it is more likely, and even suggested in electron micrographs (Fig. 10) that damage is not uniform along the length of a given fibril. It is even possible that a region of damage propagates in each fibril much as a draw band can propagate in a synthetic polymeric material. This requires a strain hardening mechanism whereby a deformed region is actually stronger than the undeformed material.

The lattice dimension calculated for the central peak half width (Table 1) is 100 nm which is in reasonable correspondence with the average collagen fibril diameter expected for this age, about 120 nm (Torp, Baer & Friedman, 1975). Perfect agreement can not be expected because instrument broadening has not been taken into account and because fibril crystallinity may well be less than perfect. For the 8.5% pattern, which has a broad background peak of a sufficient intensity for half width measurement, a transverse crystallite dimension of 5.5–6.5 nm has been calculated. If crystalline perfection persists this might be viewed as a limiting size for lateral breakdown, at least for 8.5%

deformed immature RTT. If perfection does not persist, then the actual lattice size could be much larger (Kakudo & Kasai, 1972). Possibly this is merely the limiting size for SAX scattering by collagen and further breakdown might be occurring. This pattern also shows a faint equatorial reflection at a position corresponding to a spacing of about 5.4 nm. It is not known what gives rise to this reflection. This could be a spurious reflection caused by lipid (Miller & Wray, 1971). The spacing is suspiciously close to that calculated for the background streaking in the meridional SAX fibre pattern. Fibril fragments could be packing into some sort of imperfect lattice. This and the background peak transverse spacing are slightly larger than the 4–5 nm cited for the microfibril (Smith, 1968) and smaller than that for the proposed sub-fibril, about 15 nm (see Kastelic, 1978; Wassermann, 1956; Torp, Baer & Friedman, 1975).

In mature specimens line-broadening shows up clearly in the microdensitometer scans (Fig. 5, middle and bottom rows). In contrast to immature material the central reflection itself is prominently broadened (line broadening vs. streaking). This indicates that a great proportion of the fibril regions has been structurally damaged by deformation, again by loss of coherence and/or breakdown in size, but only in the transverse direction. It appears unrealistic to discuss this broadening profile in terms simply of two superimposed peaks as with young RTT (although we have dutifully calculated lattice dimensions for Table 1 doing the best we can). It is safest just to comment that the equatorial diffraction profile, in general, becomes increasingly broad with increasing deformation and can be fitted only by a wide distribution of Gaussian distributions. The latter implies that damage, whatever its exact nature, is highly non-uniform in the collection of fibrils sampled by the x-ray beam. Since intensity of diffraction is not as readily destroyed by deformation, some collagen banding should persist in the mature specimens; microscopic evidence shows this (below, p. 420). When a fibril is 'damaged' it no longer contributes to the narrow x-ray diffraction pattern. It can be safely surmized that damaged fibril regions give rise to both diffraction broadening effects and the central diffuse scattering around the beam stop. The intensity of both increases systematically with deformation as the height of the central peak diminishes. It must be pointed out that these patterns were all obtained from specimens which had not failed and, indeed, were under substantial tensile loads when fixed. Thus, even though there is definite crystallographic damage within the collagen fibrils, they are still entirely capable of bearing load. Their strength is not dependent on crystallographic perfection. This has implications concerning any strain-hardening mechanism within the fibril.

Surprisingly no permanent elongation of the 67 nm periodicity (Cowan, North & Randall, 1955; Viidik & Ekholm, 1968) is discernible in the SAX patterns, not even when longitudinal microdensitometer scans are studied. If elongation of the period does take place it is not preserved by the chemical fixation employed here and recovers after tensile loading is released. Longitudinal broadening or streaking also are not observed, indicating that the longitudinal 67 nm lattice structure is extensive and more highly resistant to mechanical damage than is the lateral structure. This might be expected since the tropocollagen molecules are aligned in the fibril direction.

Small angle x-ray diffraction gives useful information about the structural perfection of deformed collagen. The technique samples a relatively large volume of material and thus gives an average result rather than details of a few specific fibrils. Still, the latter is quite desirable. X-ray diffraction must also be corroborated. Furthermore, the technique fails to yield information about non-crystalline structure. As a consequence ultra-thin section electron microscopy is necessary both to make up for the inadequacies of x-ray diffraction and to verify its interpretation.

Electron microscopy of deformed immature tendon

Electron microscopy of immature RTT shows that during deformation within the yield plateau the banding pattern of the native collagen fibril is progressively destroyed. In 'taut fixed' specimens (Fig. 6) less and less normal banding is apparent as the extension is increased. First, with only 2.5% extension, a degree of waviness shows up in the banding (Fig. 6*a*). That is, the banding no longer is straight across the fibril, but appears disjointedly tilted. With further elongation the amount of tilting increases and band contrast fades and in some fibrils even vanishes (Fig. 6*b*). With 'taut fixation' after 8.5% elongation most fibrils appear as grainy, featureless columns when observed by electron microscopy (Fig. 6*c*). Structural damage is highly non-uniform from fibril to fibril and along the length of a given fibril. Also, no broken fibril ends can be found even after 8.5% elongation. It is thus concluded that highly damaged regions do not immediately fracture and apparently continue supporting loads, allowing damage to proceed in other regions of the same fibril. In this way great volumes of fibrils can be transformed into a non-banded structure, as observed in the electron microscope, and as predicted by the drop in SAX meridional pattern intensity. This implies

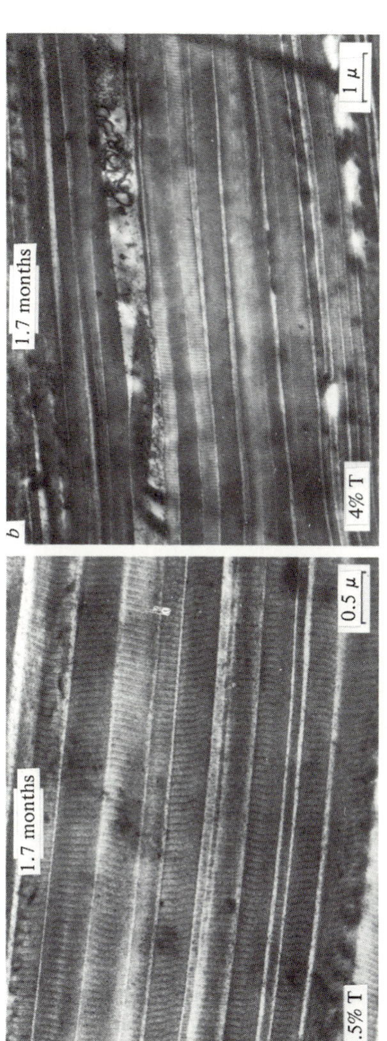

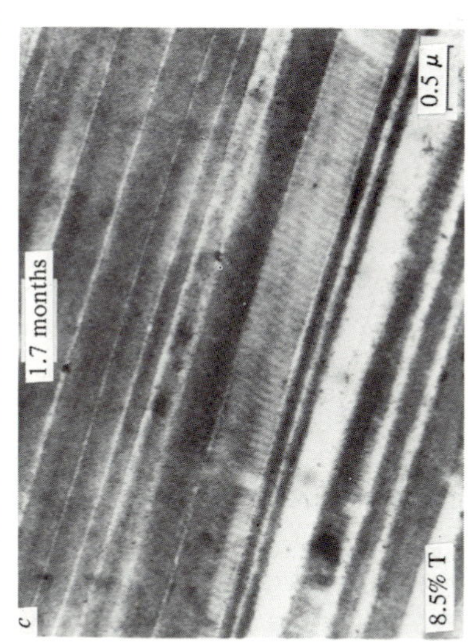

Fig. 6. Effect of deformation on the electron microscope appearance of collagen fibrils in immature RTT. Age: 1.7 months. Taut fixation (T), percent extension indicated.

a work hardening deformation mechanism for each fibril. Some type of shear-slip process must be taking place within the fibrils. The mechanism apparently allows small structural units to slip past one another and become first slightly sheared, to cause tilting of the bands, and then completely misregistered to cause loss of banding. The slipping units should be no wider than about 10 nm, for anything exceeding this would be resolved in the micrographs. The possibilities include tropocollagen molecules, microfibrils, or an aggregate of the latter. The x-ray diffraction results suggest that the last two possibilities can not be possible exclusive of the first. If only intact microfibrils or microfibril clusters were slipping the SAX diffraction intensity should not fall as dramatically as is observed. In addition, without slip of the tropocollagen and consequent extension of microfibrils, these two possibilities require profuse fracture of the microfibril or natural discontinuity of the microfibril. Both are inconsistent with high tensile strength and a strain hardening mechanism.

To impart tensile strength in the collagen fibril, which is constructed of discrete structural units, the tropocollagen macromolecules, some type of lateral bonding force is required where tropocollagen units overlap. End-to-end bonding across the 'hole' or 'gap' zone (Smith, 1968; Lees & Davidson, 1977) of the current microfibril model seems unlikely because of the great distance unless other, as yet undocumented, macromolecules are involved. The slippage mechanism by which a collagen fibril elongates involves, to the best of our reasoning, slippage of tropocollagen units within the microfibril. This requires overcoming the lateral bonding forces but apparently does not result in these forces being completely and permanently destroyed. Load bearing ability is maintained, or even improved, when intrafibrillar strain hardening occurs. This would be difficult to explain if we were arguing that only larger units were sliding. However, we are advocating tropocollagen slip and simply require that the lateral bonding interactions between tropocollagen units are labile and/or persistent (frictional) and, to account for any strain hardening, possibly strengthen somehow once slip has occurred (Rigby, 1964). A contributing factor in strain hardening may be that the more weakly bonded regions slip first so that further slip requires higher loading. Unfortunately, the bonding situation becomes even more complex when immature specimens *fixed after loading* are considered.

Micrographs of the deformed immature specimens which were fixed while under no load are presented in Fig. 7*a, b*. Interestingly in the 'relaxed fixed' specimens shearing of the bands is not prevalent. This

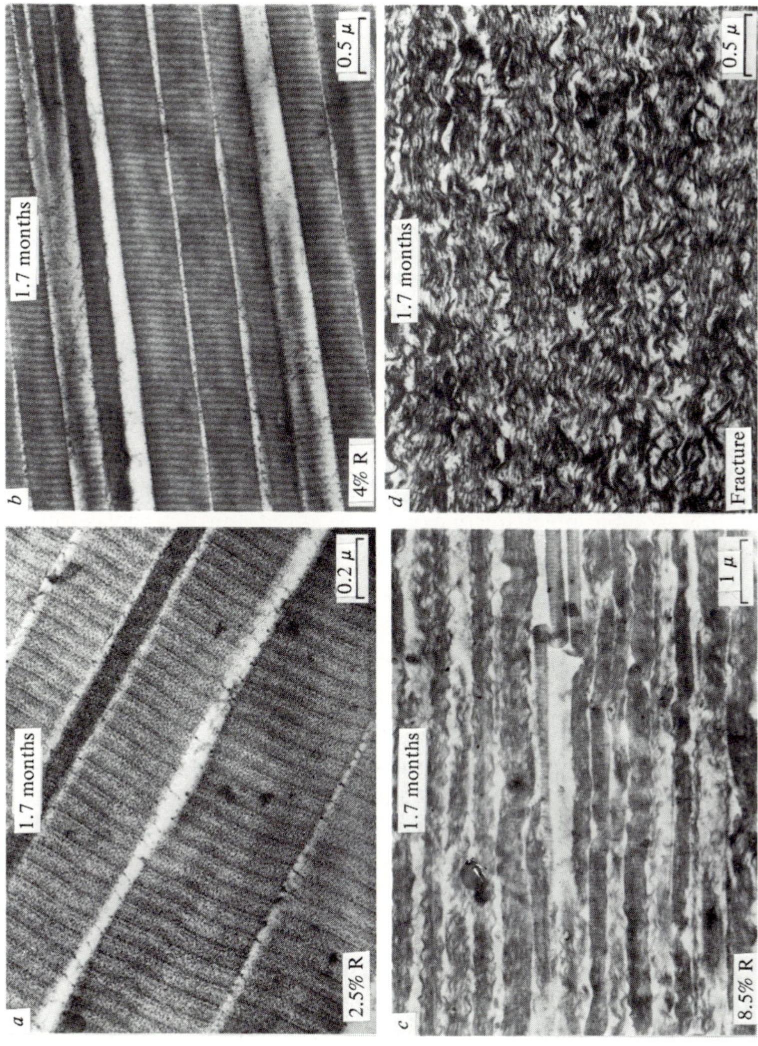

Fig. 7. Effect of deformation on the electron microscope appearance of collagen fibrils in immature RTT. Age: 1.7 months. Relaxed fixation (R), percent extension indicated. Note MPS filaments in *a* and *b*.

may indicate that the fibril structure is more mobile and dynamic than previously thought and that a degree of shear recovery can take place within the fibril. Also, non-banded regions are not observed in these samples. Instead, shredded regions are found. Upon release of the tensile load, areas of the fibril which during elongation had lost periodicity now appear disintegrated into numerous long undulating shreds (Fig. 7*c*). In specimens strained to fracture, damage is so complete and uniform that virtually every fibril is found to be disintegrated (Fig. 7*d*). The filamentous shreds, the smallest of which have a diameter of 7–10 nm and a wavelength of 200–400 nm, are much like those observed previously in cyclicly loaded tendon (Torp, Baer & Friedman, 1975). Their waviness could be a property intrinsic to the shreds, perhaps caused within each shred by a differential recovery mechanism, or it could actually be buckling caused by a compressive force, though there seems to be no plausible origin for such a force except perhaps in the mucopolysaccharide matrix. In any event specimens end up shorter after deformation than before. This is thought to be due to the high crimp angle apparent in the shreds. The shreds appear to have no periodicity, suggesting that slip and consequent misregistry have occurred between organizational units within them. The question is: what mechanism caused the shredding? Clearly certain bonding forces holding the fibril together in the transverse direction must be in large part permanently destroyed during the creation of the non-banded region in order for the fibrils to shred apart. It seems possible that there are two categories of lateral interaction to consider within the collagen fibril. We suggest that mechanical strength is maintained by persistent or quickly re-formed lateral bonding forces at a low level of organization, i.e. between tropocollagen molecules within the microfibril, while transverse failure occurs after unloading by the permanent failure of lateral bonds at some higher level of organization, such as between microfibrils or clusters of microfibrils. This permanent failure may be a consequence of excessive slip between microfibrils and/or clusters thereof.

Although it is evident that the principal deformation events take place within the collagen fibril, the mucopolysaccharide matrix may play some secondary role. A given region in a fibril appears to elongate and lose the banding independently of neighboring fibrils. This suggests the possibility of shear forces and a small amount of relative slippage between fibrils. The forces would be borne only by the hydrated mucopolysaccharide matrix (Wassermann, 1956; Sobel, 1967; Wolff *et al.*, 1971; Anderson & Jackson, 1972), which is often found concentrated at regular intervals along the fibrils and usually displays a filamentous morphology (Myers, Highton & Rayns, 1969, 1973; Nakao & Bashey, 1972; Merker & Gunther, 1973; Reed, 1973; Kischer & Shetlar, 1974; R. H. Lam & P. H. Geil, personal communication). In Figs. 7*a*, *b* and perhaps in Fig. 6*c* small filaments can be seen which interconnect the collagen fibrils. They are observable without ruthinium red staining (Viidik & Ekholm, 1968) and appear connected periodically in register with the 67.0 nm band pattern. In areas where the fibrils are widely separated they are broken and retracted. This, together with the fact that they often run at an angle acute to the fibril surfaces, suggests that one fibril has moved relative to its neighbor. The influence of these filaments on deformation and the mechanical characteristics of tendon collagen is probably small.

Electron microscopy of deformed mature tendon

With maturation, two distinct deformation regions develop in the stress–strain behavior of RTT. Damage observed for these two regions is different from that observed in the yield plateau of immature RTT. Damage caused by deformation into YR–I is very slight. A small amount of tilting of the bands, suggesting a limited degree of intrafibrillar slip, is observable in 'taut fixed' 6-month-old tendon (Fig. 8*a*). As age increases, damage resulting from YR–I deformation seems to decrease. Nothing unusual can be seen in the micrograph of 24 month tendon (Fig. 8*b*). Apparently, resistance to intrafibrillar shear and slip starts at maturation concurrent with the appearance of YR–I behavior. This resistance increases with age as does the apparent modulus of the region.

Deformation of mature tendon in YR–II, however, has definite effects on fibril structure much like those observed for immature RTT in the yield plateau region. With 'taut fixed' 6 month specimens, waviness and tilting develop in the banding pattern, of the fibril, but the effects are quite variable (Fig. 9*a*). Tilting can go to very high angles and often chevron patterns (Bruns, Trelstad & Gross, 1973) are discernible. This is accompanied by loss of band contrast and some fibrils lose their banding altogether. As with immature RTT, when fixation is performed without applied tension the nonbanded regions appear shredded (Fig. 9*b*). After deformation to fracture all fibrils are completely shredded (Fig. 9*c*), with the implication that lateral cohesion has been destroyed, at least at the 10 nm level. The shreds have no visible periodicity and exhibit undulations of generally lower wavelength and lower amplitude than in immature RTT. It is interesting that we observe no evidence of

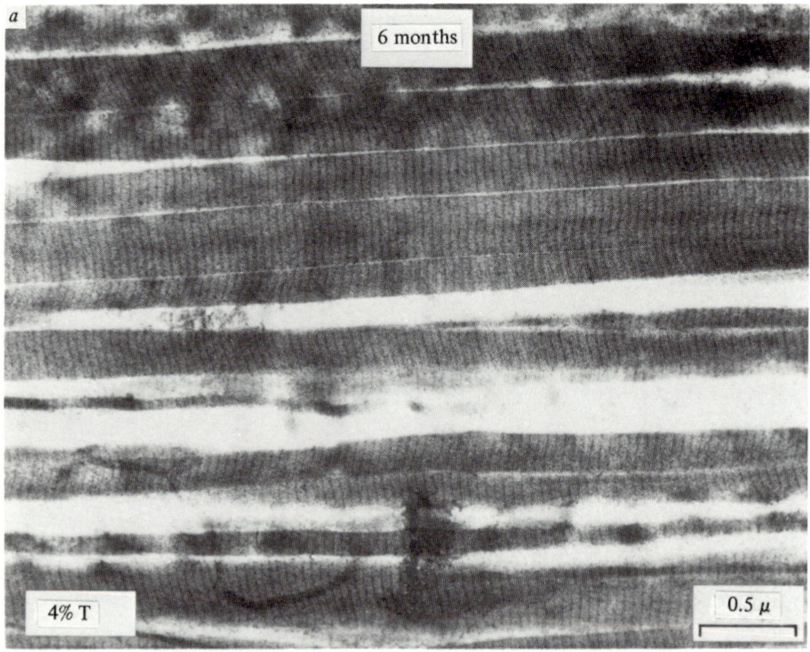

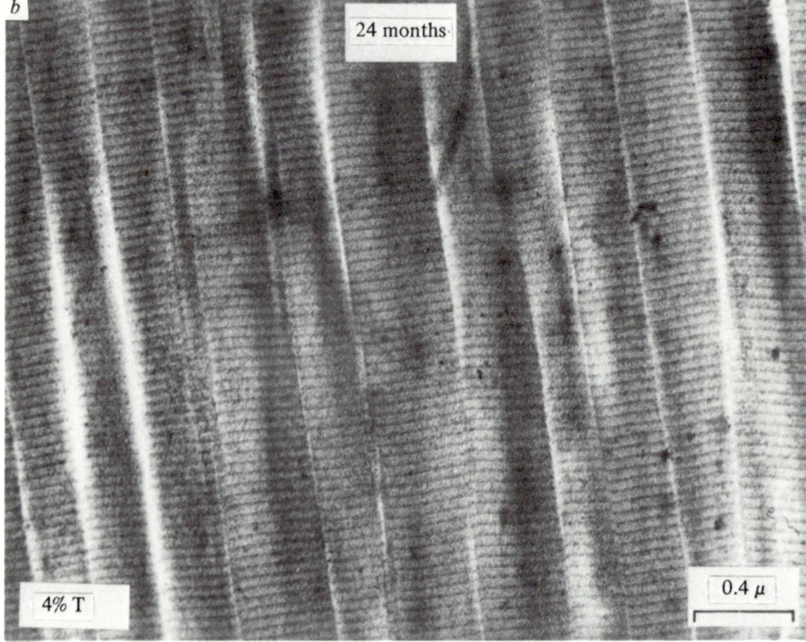

Fig. 8. Electron micrographs of mature RTT deformed into Yield Region I. *a*, 6 month, taut fixed; *b*, 24 month taut fixed. Extensions indicated.

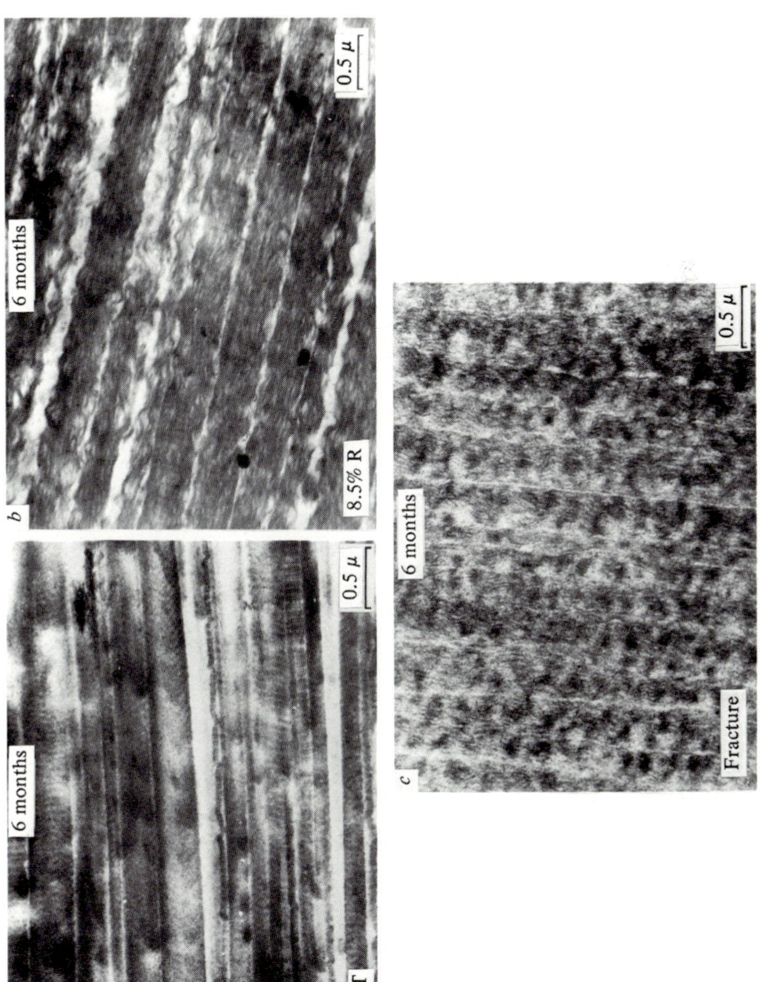

Fig. 9. Electron micrographs of 6 month RTT deformed into Yield Region II. *a*, 8.5% deformation, taut fixed; *b*, 8.5% deformation, relaxed fixed; *c*, deformed until fractured, relaxed fixed.

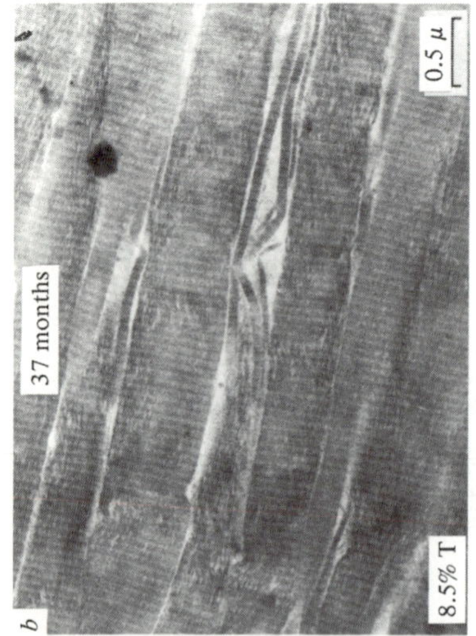
b
37 months
0.5 μ
8.5% T

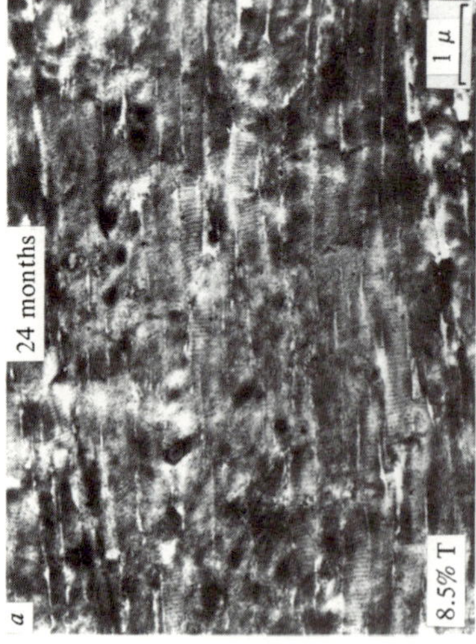
a
24 months
1 μ
8.5% T

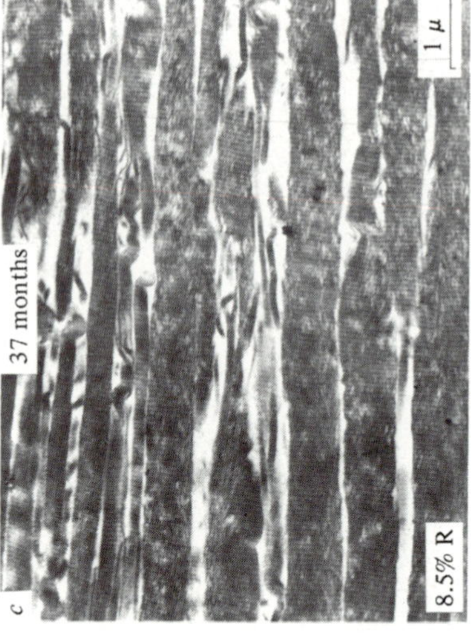
c
37 months
1 μ
8.5% R

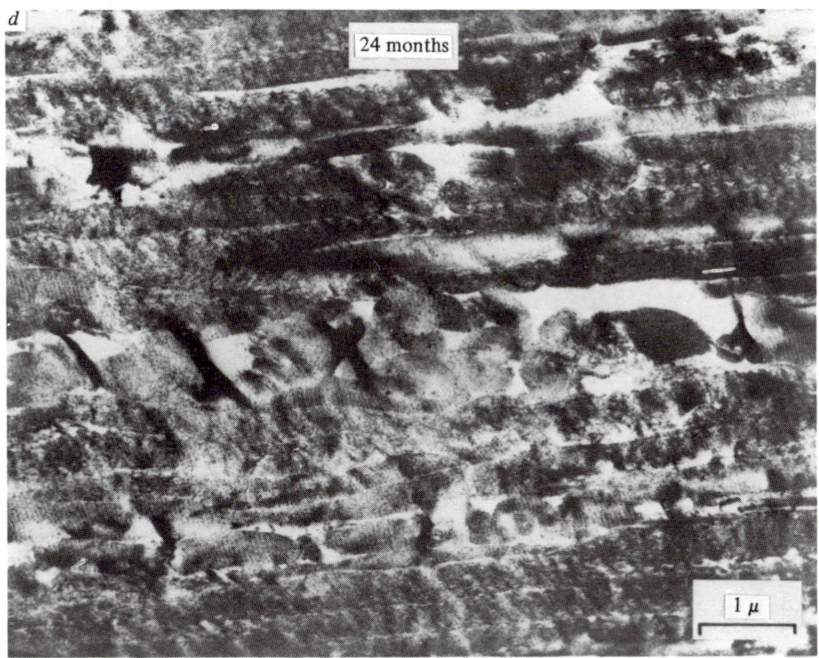

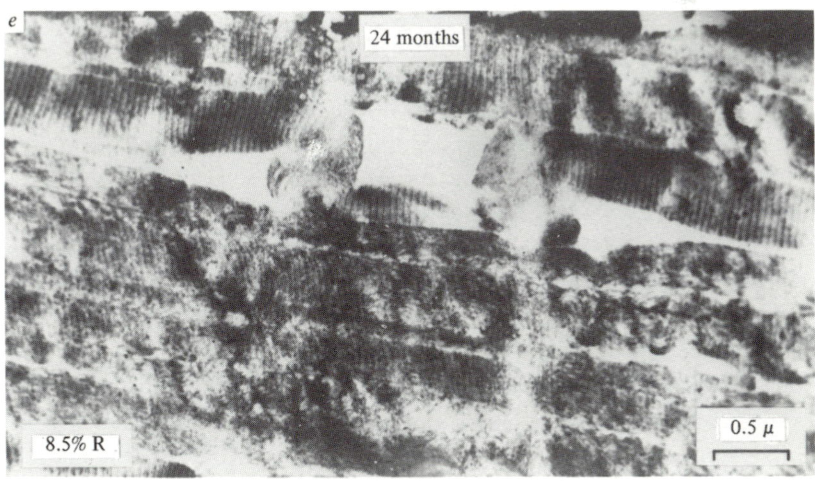

Fig. 10. Electron micrographs of tail tendons from old rats deformed into Yield Region II. *a*, 24 month, taut fixed; *b*, 37 month, taut fixed; *c*, 37 month, relaxed; *d*, *e*, 24 month, relaxed fixed. Fields showing isolated fractured fibrils.

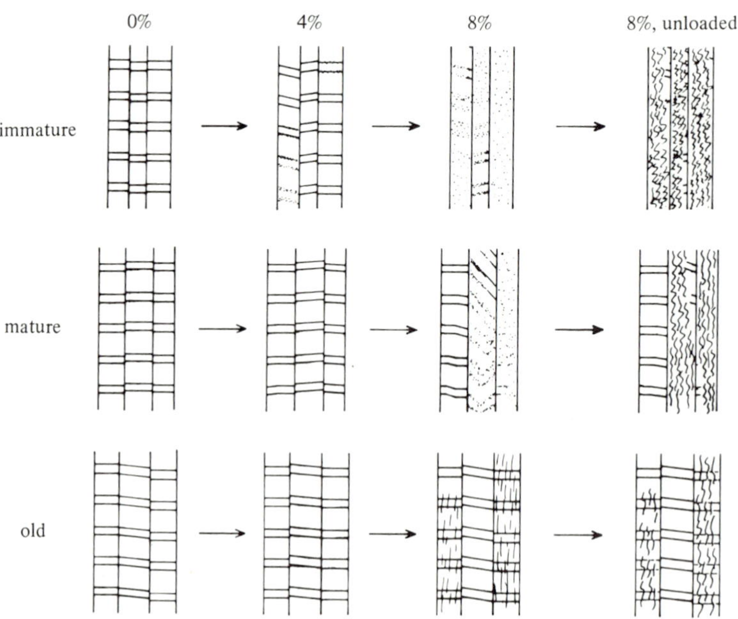

Fig. 11. Schematization for immature, mature and old RTT of fibril damage incurred during deformation. Extensions indicated at top.

any helical organization such as has been described for chemically damaged material (Braun-Falco & Rupec, 1963; Bouteille & Pease, 1971; Rayns, 1974; Lillie, MaCallum, Scaletta & Occhino, 1977). Significant changes in YR–II damage occur with increasing age. Shear-slip effects within the fibrils seem greatly reduced and banding more persistent (Fig. 10*a*). Damage seems to be confined to more localized regions along the length of the collagen fibril with numerous small regions maintaining prominent banding. Even when the specimens are fixed under tension, shredding can be observed. Periodicity is often apparent in the shreds (Fig. 10*b*) which tend toward larger diameters in old material. They do not appear displaced very much when fixed in the relaxed condition (Fig. 10*c*). In addition, isolated fractured fibrils are observed on rare occasions. These usually appear to have recoiled violently upon fracture since they are buckled up on themselves and must crowd their neighbors (Fig. 10*d*, *e*): they are not observed in younger specimens.

The above observations on deformed tendon have been schematized in Fig. 11. Proceeding from left to right the pictures are of tendons unstrained, deformed 4% and held taut, deformed 8.5% and held taut and deformed 8.5% but with load now removed.

With immature material, effects on banding are apparent even after only low extension. With mature and old specimens, effects of low extension are minimal in electron micrographs, yet line broadening in SAX indicates that the width of the lattice and/or perfection are significantly reduced. Limited microfibril slip is implicated as responsible for these effects. With greater strain, prominent deformation effects occur at all ages. For immature tendon, high extension eradicates banding and, after 'taut fixation', fibrils appear as grainy but otherwise featureless columns. SAX diffraction intensity drops off in correlation with the loss of banding seen in electron micrographs: slip between tropocollagen units within the microfibril is probably responsible. When the tensile stress is removed prior to fixation the fibrils with highly damaged or destroyed banding fall apart laterally into shreds. This indicates failure of cohesion between microfibrils or clusters of these, probably due to excessive shear-slip between microfibrils. With maturation and aging, resistance to such extreme damage develops progressively. Deformed tissues appear not so thoroughly destroyed and the tensile forces required to achieve a given extension progressively increase. As with young material, upon release of stress, damaged fibrils appear to disintegrate laterally into shreds. Fibrillar damage and subsequent shredding is the least in the oldest specimens. We infer that in older material transverse bonding forces within the fibril are stronger. The collagen fibril is then better able to resist the internal deformation processes which in younger material would proceed, thereby obliterating banding and allowing profuse shredding. Unfortunately, we were not able to identify the nature of these bonding forces. The literature is replete with possibilities, particularly with those advocating chemical cross-linking with aging. There appear to be at least two types of bonds to consider: those that have failed completely and allow the profuse shredding and those internal to the shreds which maintain load bearing ability yet which allow a degree of slippage and elongation together with the consequent misregistry effects on banding. The suspected hierarchical locations of these two types of bonding interactions are, respectively, between microfibrils (or small clusters of microfibrils) and between tropocollagen units within the microfibril.

Recrimping

Deformation causes additional effects on a much larger scale. It is known that the crimp structure on tendon is altered by tensile deformation (Dale, 1974; Torp, 1974; Torp *et al.*, 1975). The original crimp waveform is perturbed by the development of a new small period crimp which appears superimposed on the original. Polarizing optical micrographs were taken of relaxed fixed deformed specimens (Fig. 12). For all ages it can be seen that after low deformation the native crimping takes on a less prominent and less regular appearance. The crimps are somewhat straightened out with bends becoming less distinct. After higher deformations, small crimps appear which are generally sharper, irregularly spaced and much shorter. This has been termed recrimping (Dale, 1974; Torp, 1974; Torp *et al.*, 1975; Dale & Baer, 1975). As the amount of deformation is increased the extent of recrimping increases. Some subtle age effects can be identified. The new crimps are slightly sharper in the tendons from older animals. Also, at very low elongation recrimping seems to be more highly developed in the immature specimens. Details of the recrimping process show up more clearly in highly magnified optical microscope views (Fig. 13). It can be seen that in mature RTT deformation into YR–I removes the crimp (Fig. 13*b*). Whatever locks the native crimp into the fibril array, probably the mucopolysaccharide matrix, must give way mechanically. This could involve yield, fracture, or creep. Consequently, in the 'relaxed-fixed' specimens, the native crimp angle appears reduced. With the deformation into YR–II new crimps are formed upon removal of the load (Fig. 13*c*, *d*). These are clearly of a much finer nature and are more numerous than native crimps. These new crimps do not appear to traverse the entire cross-section of the fascicle as do the native crimps. Instead, they seem to be more or less a surface feature. In fact, after high extensions the recrimping pattern (Fig. 13*d*) is reminiscent of the numerous buckles which occur in a thin tubular column which has collapsed in axial compression.

In electron micrographs the recrimping can be seen to involve fibrils within the outer circumference of the fascicle (Fig. 14*a*). The actual bend in the fibrils seems to occur preferentially at the more highly damaged regions which might be expected to be structurally less stable (Fig. 14*b*). Collagen fibrils within the interior of the tendon do not exhibit this same severe bending. Instead they remain straight (Fig. 14*d*) yet appear more highly damaged. Those at an intermediate position run more or less straight except for a few intact fibrils which seem to be buckled as if under compression (Fig. 14*c*).

It has been previously proposed that recrimping involves fibrils slipping in an elastic matrix (Dale, 1974; Dale & Baer, 1974). Upon unloading the matrix retracts to its original undeformed length and in doing so buckles the fibrils. This explanation is difficult to accept in the

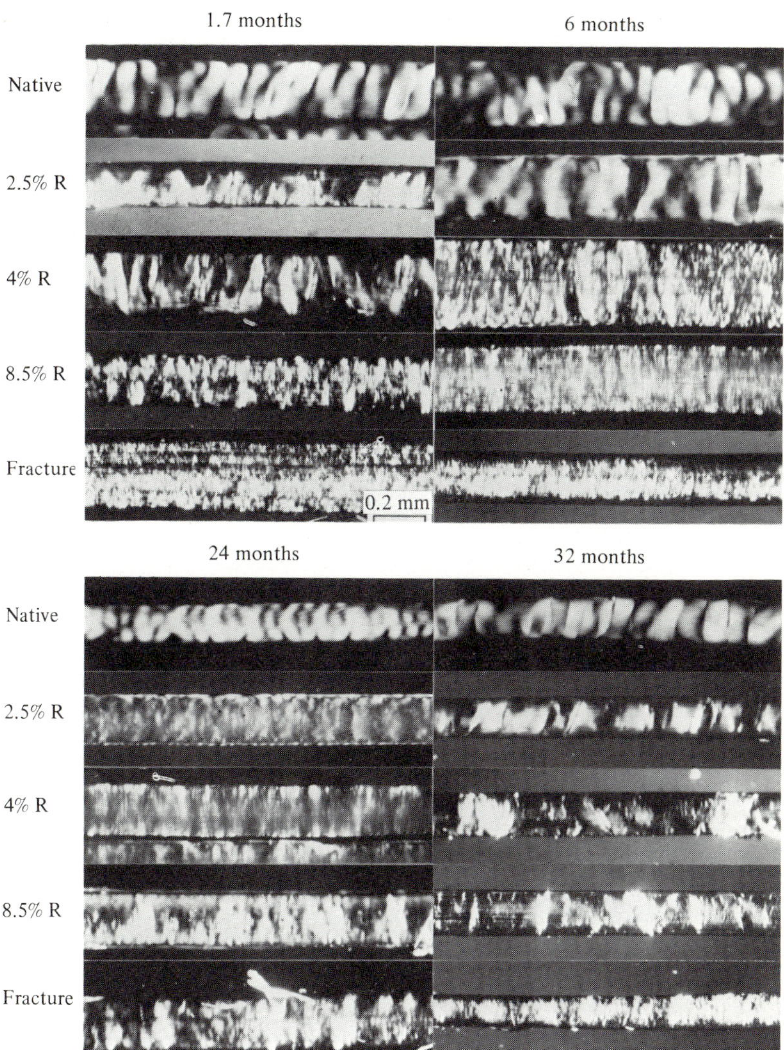

Fig. 12. Effect of mechanical deformation on the crimp structure of RTT. Ages and deformations indicated. Polarizing optical micrographs.

case of RTT since a strong resilient matrix is required. The initial effects on the original crimping may indeed be due to fibrils slipping in the matrix or, more importantly, with respect to one another. This could account for the apparent reduction in crimp angle with low strain deformation. But any contraction of the matrix seems to be giving way even at these low strains. At higher deformations an alternative mechanism must be put forward to explain the recrimping. An attractive concept is that fibrils at the center of the fascicle are older

a

Native

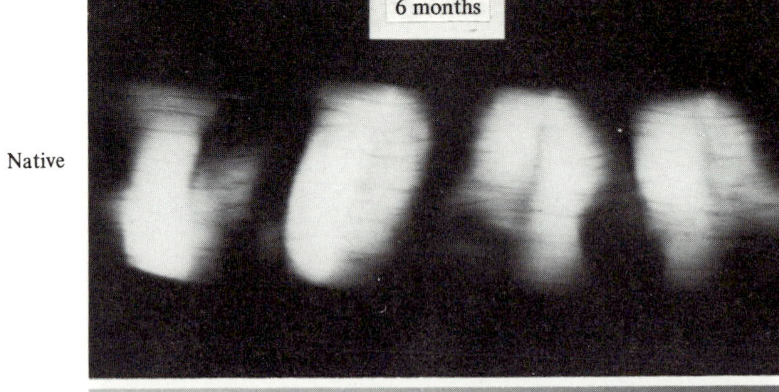

b

2.5% R

c

4% R

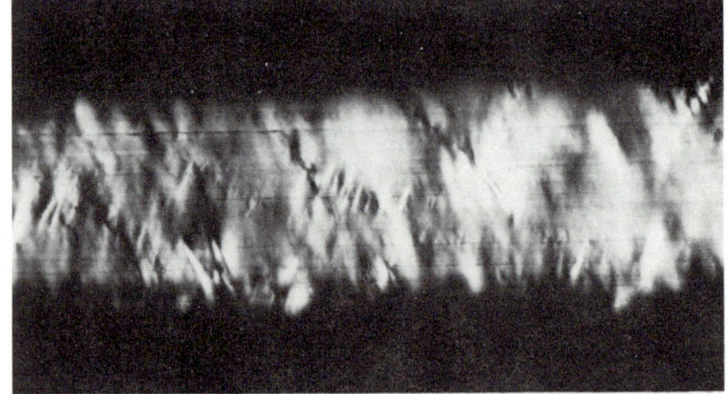

d

8.5% R

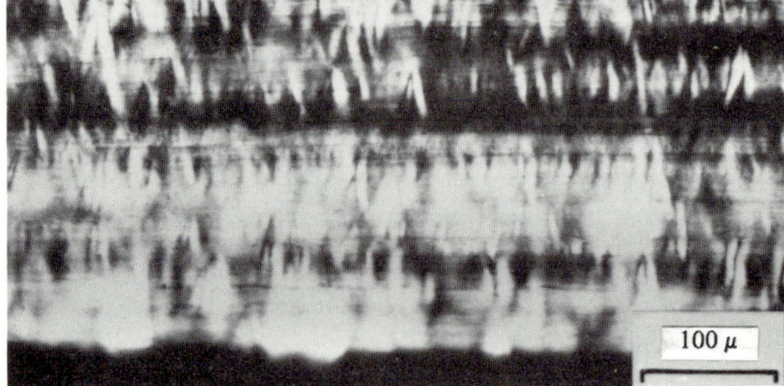

than those at the surface (Jackson & Bentley, 1960; Hayes & Allen, 1967). If this were so, the deformation characteristics of fibrils at the two locations could be different: upon extension the outer and thus 'younger' fibrils would suffer greater permanent elongation than the inner 'older' fibrils. Upon unloading, the greater elastic recovery of the inner fibrils could buckle the outer, more highly deformed, ones. Unfortunately, this model fails to account for, and even violates, the phenomenon of length contraction that is observed with RTT. Yielding results in defibrillation, which causes shortening rather than permanent lengthening.

An interesting consequence of shredding which was mentioned earlier is that highly deformed specimens end up with a shortened unloaded length. It seems that the great waviness of the shreds actually causes the tendon to contract to a length shorter than its undeformed initial length. For 8.5% tensile extension a 2–6% contraction of the gauge length occurs. Immature tendon contracts the most and with age the amount declines. It is not known whether this contraction serves any physiological purpose, such as triggering cellular response or affording additional protection against shock. This phenomenon of shredding and shortening could explain recrimping. If the collagen fibrils were more highly damaged at the tendon centre than at the surface, then contraction, arising from the undulation of the consequent shreds, should be greatest at the center. This could then cause the outer, less contracted, fibrils to crimp. It is interesting to note that in Fig. 14*c* the buckled fibrils are generally not severely damaged but are each surrounded by fibrils that are. Also, damage at the center of the tendon (Fig. 14*d*) is somewhat greater than at the edges (Fig. 14*b*). This distribution of damage would be expected if the native crimp angle decreased at the center of the tendon, as has been proposed elsewhere (Kastelic *et al.*, 1978). The straighter, central fibrils would come under damaging loads before the outer more highly crimped fibrils. It was noted earlier that immature specimens suffer yield damage and subsequently 'defibrillate' at lower strains than do mature specimens. This fact could explain why immature specimens are able to recrimp after only small extension, while mature ones cannot.

Some problems remain. It must be noted that this mechanism invokes a particular fascicle structure (Kastelic *et al.*, 1978) which has not been thoroughly verified (J. Gathercole, A. Keller & X. Shah, personal

Fig. 13. Higher magnification views of tendon recrimping, seen in an optical microscope. Relaxed fixation after indicated deformations. Six month RTT.

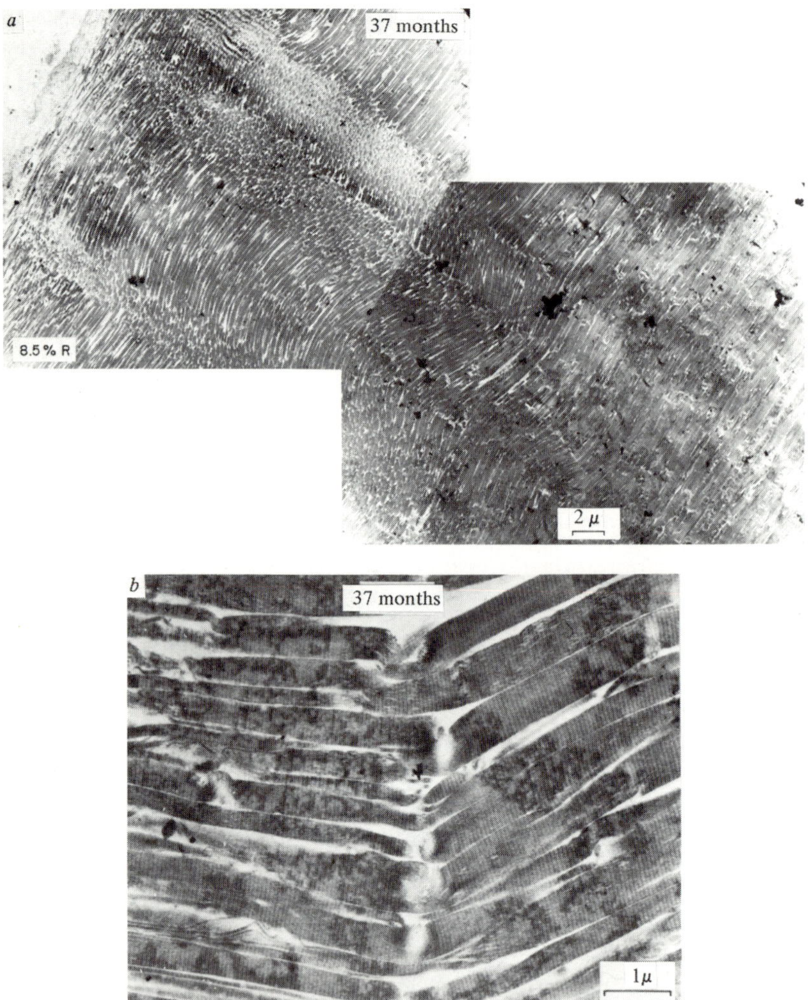

Fig. 14. Recrimping viewed at the electron microscope level. 37 month RTT deformed 8.5% and prepared for EM in the free unloaded condition. *a*, low magnification view showing 'recrimped' fibrils at tendon surface and straight fibrils deep inside tendon; *b*, higher magnification view of 'recrimped' fibrils near surface; *c*, fibrils at intermediate depth; *d*, fibrils deep inside tendon.

c
1 μ

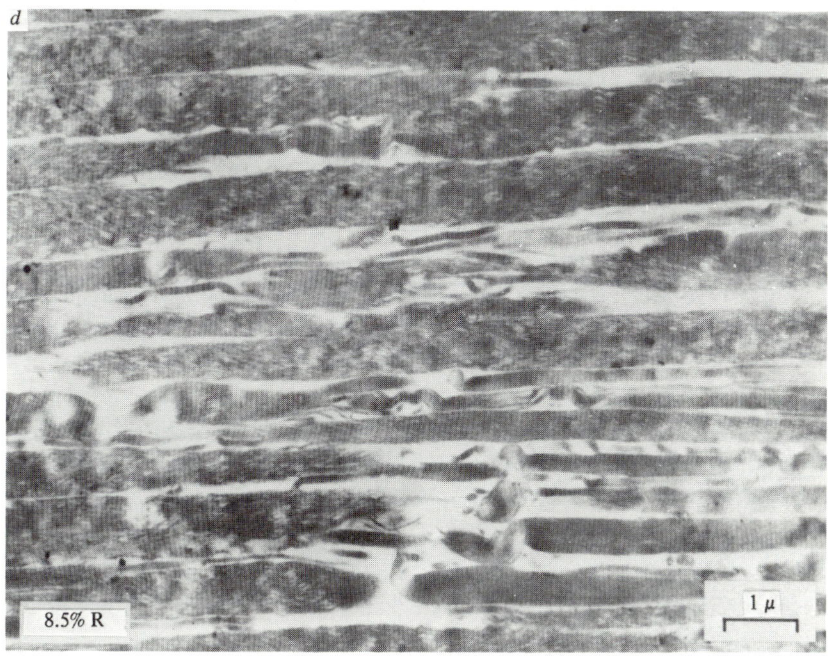
d
8.5% R
1 μ

communication). In addition we have not proved mathematically that the extent of differential contraction can actually be sufficient to cause the observed recrimping. There is also the possibility that a combination of effects causes recrimping. Finally, nothing is known about possible recrimping in tendon from other sources.

Functions of the MPS matrix in the tendon composite

The matrix in a conventional composite serves only a mechanical purpose. In tendon the situation is markedly different. The mucopolysaccharide–water gel matrix performs many non-mechanical functions. It is implicated in the control of the distribution of diameters of the collagen fibrils and the spatial distribution of fibrils (Hart & Farrell, 1969; Cox, Farrell, Hart & Langham, 1970; Borcherding *et al.*, 1975), though the mechanisms of control are not yet known. It serves as the medium for the diffusion of nutrient and waste to and from cells embedded in the dense collagen fibril array. It probably serves to regulate the ionic and tonic environment of the cells and fibrils. These non-mechanical functions are clearly very important. But the mechanical functions do not seem to be important.

What are the mechanical functions of the matrix? The most obvious is to bind the collagen fibrils into a functional sliding cord. However, in the rat tail each tendon is contained within a very strong tissue sheath. Since the size of the tendon and the tensile loads are small the matrix need not be very strong to perform this minor function. A skein of unbound fibrils might serve as well. The case may be quite different for other tendons, particularly those from much larger animals where higher loads, absence of a strong sheath and complex bifurcating structures are encountered. These same arguments apply for the binding of fascicles within tendon (Kastelic, 1978). In conventional composites the matrix serves the key function of transferring mechanical loads from fiber to fiber. This is important because fibers are of finite length and/or break due to flaws and brittleness. Yielding can often be attributed to fiber slip which can be allowed by shear in the matrix or loss of cohesion between matrix and fiber. However, collagen fibrils in tendon seem to be endless. Naturally occurring ends have never been reported in the course of the countless electron microscope studies of this tissue. Even after deformation into YR–II, broken fibrils are not observed except when deformation approaches the fracture strain. In the absence of broken fibrils and fibril ends, load transfer cannot be regarded as a significant function of the matrix in RTT.

When a RTT fascicle fractures while being extended in the testing machine, a break is not immediately apparent. The only indication of failure is the abrupt drop in tensile load. With continued movement of the crosshead, long tapering ends are eventually produced. These are not the product of a ductile drawing process, since negligible load is required for their generation. They appear to grow out from what must be the central region of fibril failure. When the specimen is pieced back together an overlap of about a centimeter of the broken tapering ends is required. These observations indicate that fracture in RTT is highly interleaved. Neighboring fibrils do not fracture on the same plane; instead the fracture points are dispersed over a relatively long region of the specimen. The limp broken fibril ends, of what must be widely varying length, cling together into a long tapering mass. This high degree of what could be called fibril 'pull out' is characteristic of uncoupled fracture which predominates in composites with an especially weak or poorly adhering matrix (McCullough, 1971). The concept, in this discussion, of a strong influential matrix is untenable.

The stress–strain curve of tendon is remarkably similar to that of a uniaxial composite of strong-ductile fibers embedded in a ductile matrix. The matrix elastically bears loading up to its own yield at which point one yield region begins. A second yield region begins at the fiber yield point which in this case must occur at higher elongation. Such a specialized composite consequently exhibits two yield regions (Weeton & Signorelli, 1966). It is tempting to use this composite as a model for mature tendon. Age changes in YR–I could be attributed to an increase in the yield strength of the matrix. It is known that, with development and aging, changes in the saccharide side-chains *do* occur and the matrix *does* lose a small amount of water (Torp *et al.*, 1975). It is conceivable that these changes could alter matrix properties. The age changes in YR–II would be attributed to an alteration in the yield behavior of the fibrils themselves.

However, there are several arguments which rule out this model for yield behavior. In order for the matrix to contribute measurably in bearing stresses within the composite it must either possess adequate yield strength or comprise a large volume fraction. But the matrix volume fraction is relatively small and is known to decline with age (Torp, Baer & Friedman, 1975). In addition, as we have seen, there are indications that RTT matrix is very weak. An analysis of restoring forces in the crimp, undertaken to explain mechanical behavior in the toe region, suggests that the shear modulus of the matrix of RTT is extremely low (Kastelic, Arridge, Palley & Baer, 1980). In this model composite, deformation into the first yield region does not cause any type of fiber damage. This is not consistent with the marked line broadening of the fibril SAX pattern observed in YR–I with tendon. Crystallographic packing within the fibril is definitely modified, suggesting non-elastic deformation. If any changes in the physical properties of the matrix occur with aging the fracture process should also be influenced. Observing only casually, we have not noticed any effect of age on fracture. For RTT we suggest that great changes in the mechanical properties of the matrix do not occur with development and aging and that analogies cannot be drawn between the specialized model composite and tendon. The yield behavior observed for the whole tendon must be attributed to the properties of the isolated collagen fibril. Changes in mechanical deformation behavior occurring with age must be ascribed to an intrafibrillar origin. The matrix has been implicated in creating and controlling fibril crimping within tendon (Dale & Baer, 1974). There have been suggestions that the matrix also influences mechanical behavior in the toe region. These phenomena may be affected by the matrix but will be treated in more detail elsewhere and will not be dealt with here (Kastelic *et al.*, 1980).

Some evidence exists which does at first glance seem to elevate the mechanical role played by the mucopolysaccharide–water matrix in tendon. When the matrix is removed with chemical treatments, usually involving enzymes, the tendon is considerably weakened (Minns, Soden & Jackson, 1973). But these reported severe effects might be attributed to proteolytic impurities and subsequent damage to the collagen fibrils (Partington & Wood, 1963; Elden, 1968). The MPS matrix, in its control of the collagen fibril environment, may well have yet unknown effects on intrafibrillar cohesion. The well known weakening effects of salt and acid solutions can be ascribed to influences on intrafibrillar bonding interactions rather than weakening and/or removal of the interfibrillar matrix. There are indications that the fibrils themselves are affected. Tendons often become transparent and enlarged during such treatments, probably because the fibrils swell (Wassermann, 1956; Kohn & Rollerton, 1958; Borysko, 1963). Until stronger evidence to the contrary appears, the role of the mucopolysaccharide binding matrix in mechanically determining the composite behaviour of RTT must be regarded as minor.

Our stand in this controversial area and our supporting arguments hold only for RTT or other tendons with a weak matrix. For those tendons with much stronger interfibrillar bonding, such as are frequently found in larger animals and humans, our comments are invalid.

However, it is our hope that this discussion will serve to direct and organize future thought on this subject.

Other aspects of the composite design of tendon

Tendon serves as the mechanical link connecting muscle with skeleton and so must possess high tensile modulus, high toughness and good resistance to tensile creep, fatigue and shock. Yet it must be flexible enough to bend at joints and absorb slack when muscle tone is relaxed. These physical requirements are all met by the particular crimped fiber composite structure of tendon with its ductile yielding fibrils. It can be justifiably surmized that the multicomposite structure of tendon is optimized principally for its mechanical function in the body. The structure also must fit constraints imposed by the available mechanisms for assembly and growth in life. Through means which we do not yet understand, cellularly produced macromolecules assemble into the functioning structure. During growth the tendon lengthens though all the while bearing loads. Undoubtedly it is these processes that to a large degree dictate the details of the ultrastructural levels on the hierarchy.

There is an additional, much more subtle, design requirement which could influence micromechanical yield processes within the hierarchy. Tendon must maintain its function throughout life and consequently must heal when damaged by mechanical overload. The matter is not as simple as cellular ingrowth, resumption of macromolecule production and subsequent repair.

Tendon is generally nonvascular and instead relies on diffusion for nutrient supply (Elliott, 1965). Any yield mechanism involving highly localized damage would require a localized healing process which conceivably could greatly exceed the capacity for transport of nutrient and waste of the synovial fluid–tissue pathway. Consequently, we speculate that the yield mechanisms of tendon are neatly adapted to disperse mechanical yield damage over a broad volume of material. In this way hot spots of intense repair activity are avoided. An interesting thought is that shredding, which occurs after a damaging overload, may additionally serve to enhance diffusion in the tissue and enable increased cellular metabolism.

Any damage mechanism which leaves intact suitable building material which could subsequently re-self-assemble with minimum cellular intervention would also seem desirable. Slippage of microfibrils and tropocollagen units, which we have proposed as the prime mechanisms of yielding, in addition to being consistent with all experimental observations, does not involve damage to the tropocollagen. If this view is correct the possibility exists for reassembly after yield damage. Furthermore, this allows a healing process in which function can be maintained during repair provided fracture has not occurred. Of course, it is the great ability of tendon to yield in response to mechanical overload which affords protection from the latter.

Conclusions

(a) The stress–strain curve of tail tendon from the mature rat exhibits two yield regions with the apparent modulus (or strain hardening) of each region increasing with age. Immature RTT exhibits a single low stress yield plateau.

(b) After deformation into YR–I the small angle x-ray diffraction pattern of the collagen is broadened in the direction transverse to the fibril axis. This suggests that either the fibril crystallographic perfection and/or lattice size is decreased, but only in the transverse direction. The latter possibility is weakly supported by evidence from electron microscopy of shearing and tilting of the 67.0 nm fibril banding.

(c) Deformation into the second yield region causes a lessening of diffraction intensity as well as further transverse line broadening. This is accompanied by pronounced shearing and even loss of the 67.0 nm band pattern. Damage is less severe in specimens from old animals.

(d) In immature RTT, the SAX weakens during deformation in the yield plateau. Line broadening is decidedly less than in mature RTT. Loss of fibril banding occurs progressively with deformation but without extreme shearing and tilting.

(e) Upon unloading, severely damaged collagen fibrils disintegrate laterally into numerous, long, highly undulating shreds. Specimens contract to a length actually a few percent shorter than that before deformation. A mechanism based on this length contraction is put forth to explain the phenomenon of 'recrimping'.

(f) For YR–I, a mechanism involving slight shear-slip between microfibrils or clusters thereof is suggested to explain the x-ray line broadening and the limited tilting in fibril banding. For both YR–II of mature and the yield plateau of immature a two-part mechanism is suggested. This involves: (1) slippage of tropocollagen units within the microfibril (to account for the SAX pattern weakening and the loss of fibril banding in electron micrographs), and (2) permanent loss of cohesion between microfibrils or clusters thereof, caused by excessive shear-slip (to account for subsequent deformation upon unloading). Age effects on deformation are attributed to the strengthening of intrafibrillar bonding interactions.

The generous financial support of the US National Institute of Health, under Grant AG–00361–19A1, is gratefully acknowledged.

References

ANDERSON, J. C. & JACKSON, D. S. (1972). The isolation of glycoproteins from bovine Achilles tendon and their interaction with collagen. *Biochemical Journal*, **127,** 179–86.

BAER, E., GATHERCOLE, L. J. & KELLER, A. (1975). Structural hierarchies in tendon collagen: an interim summary. In *Structure of Fibrous Biopolymers*, ed. E. D. T. Atkins & A. Keller, pp. 189–95. London: Butterworth.

BANFIELD, W. G. (1952). The solubility and swelling of collagen in dilute acid with age variations in man. *The Anatomical Record*, **114,** 157–71.

BARENBERG, S. A., FILISKO, F. E. & GEIL, P. H. (1978). Ultrastructural deformation of collagen. *Connective Tissue Research*, **6,** 25–35.

BEAR, R. S. (1952). The structure of collagen. *Advances in Protein Chemistry*, **7,** 69–160.

BEAR, R. S. & BOLDMAN, O. E. A. (1951). Periodic statistical distortion of unidirectionally ordered diffractors, with application to collagen. *Journal of Applied Physics*, **22,** 191–8.

BORCHERDING, M. S., BLACIK, L. J., SITTIG, R. A., BIZZELL, J. W., BREEN, M. & WEINSTEIN, H. G. (1975). Proteoglycans and collagen fibre organization in human corneoscleral tissue. *Experimental Eye Research*, **21,** 59–70.

BORYSKO, E. (1963). Collagen. In *Ultrastructure of Protein Fibres*, ed. R. Borasky, pp. 19–37. London: Academic Press.

BOUTEILLE, M. & PEASE, D. C. (1971). The tridimensional structure of native collagenous fibrils, their proteinaceous filaments. *Journal of Ultrastructure Research*, **35,** 314–38.

BRAUN-FALCO, O. & RUPEC, M. (1963). Some observations on dermal collagen fibrils in ultra-thin sections. *Journal of Investigative Dermatology*, **42,** 15–19.

BROUTMAN, L. J. & KROCK, R. H. (1967). *Modern Composite Materials*. Reading, Mass: Addison-Wesley.

BRUNS, R. R., TRELSTAD, R. L. & GROSS, J. (1973). Cartilage collagen: a staggered substructure in reconstituted fibrils. *Science*, **181,** 269–71.

COWAN, P. M., NORTH, A. C. T. & RANDALL, J. T. (1955). X-ray diffraction studies of collagen fibres. In *Fibrous Proteins and their biological significance*, (9th Symposium of the Society for Experimental Biology) pp. 115–26. Cambridge University Press.

COX, J. L., FARRELL, R. A., HART, R. W. & LANGHAM, M. E. (1970). The transparency of the mammalian cornea. *Journal of Physiology*, **210,** 601–16.

CRANE, H. R. (1950). Principles and problems of scientific growth. *Scientific Monthly*, **70,** 376–89.

DALE, W. C. & BAER, E. (1974). Fibre-buckling in composite systems: a model for the ultrastructure of uncalcified collagen tissues. *Journal of Materials Science*, **9,** 369–82.

ELDEN, H. R. (1968). Physical properties of collagen fibres. *International Review of Connective Tissue Research*, **4,** 283–348.

ELLIOTT, D. H. (1965). Structure and function of mammalian tendon. *Biological Reviews*, **40,** 392–421.

FRASER, R. D. B. & MACRAE, T. P. (1980). Molecular structure and mechanical properties of keratins. In *The Mechanical Properties of Biological Materials* (34th symposium of the Society for Experimental Biology) pp. 211–46. Cambridge University Press.

GALESKI, A., KASTELIC, J., BAER, E. & KOHN, R. R. (1978). Mechanical and structural changes in rat tail tendon induced by alloxan diabetes and aging. *Journal of Biomechanics*, **10,** 775–82.

HAMLIN, C. R. & KOHN, R. R. (1971). Evidence for progressive age-related structural changes in post-mature human collagen. *Biochimica et Biophysica Acta*, **236,** 458–67.

HAMLIN, C. R. & KOHN, R. R. (1972). Determination of human chronological age by study of a collagen sample. *Experimental Gerontology*, **7**, 377–9.

HART, R. W. & FARRELL, R. A. (1969). Light scattering in the cornea. *Journal of the Optical Society of America*, **59**, 766–74.

HAYES, R. L. & ALLEN, E. R. (1967). Electron-microscopic studies on a double-stranded beaded filament of embryonic collagen. *Journal of Cell Science*, **2**, 419–34.

HULME, D. J. S., MILLER, A., PARRY, D. A. D., PIEZ, K. A. & WOODHEAD-GALLOWAY, J. (1973). Analysis of the primary structure of collagen for the origins of molecular packing. *Journal of molecular Biology*, **79**, 137–48.

HUXLEY, H. E. (1972). Molecular basis of contraction in cross-striated muscles. In *The Structure and Function of Muscle*, 2nd edn, vol. 1, ed. G. H. Bourne, pp. 301–87. New York: Academic Press.

JACKSON, D. S. & BENTLEY, J. P. (1960). On the significance of the extractable collagens. *Journal of Biophysical and Biochemical Cytology*, **7**, 37–42.

JOLMA, V. H. & HRUZA, Z. (1972). Differences in properties of newly formed collagen during aging and parabiosis. *Journal of Gerontology*, **27**, 178–82.

KAKUDO, M. & KASAI, N. (1972). *X-ray Diffraction by Polymers*, New York: Elsevier.

KASTELIC, J., GALESKI, A. & BAER, E. (1978). The multicomposite structure of tendon. *Connective Tissue Research*, **6**, 11–23.

KASTELIC, J., ARRIDGE, R. G. C., PALLEY, I. & BAER, E. (1980). Structural and mechanical models for tendon crimping. *Proceedings of the Royal Society of London, Series B,* (in press).

KISCHER, C. W. & SHETLAR, M. R. (1974). Collagen and mucopolysaccharides in the hypertrophic scar. *Connective Tissue Research*, **2**, 205–13.

KOHN, R. R. & ROLLERTON, E. (1960). Aging of human collagen in relation to susceptibility to the action of collagenase. *Journal of Gerontology*, **15**, 10–14.

LEES, S. & DAVIDSON, C. L. (1977). The role of collagen in the elastic properties of calcified tissues. *Journal of Biomechanics*, **10**, 473–86.

LILLIE, J. H., MACCALLUM, D. K., SCALETTA, L. J. & OCCHINO, J. C. (1977). Collagen structure: evidence for a helical organization of the collagen fibril. *Journal of Ultrastructure Research*, **58**, 134–43.

MCCULLOGH, R. L. (1971). *Concepts of Fiber-Resin Composites*, New York: Marcel Dekker.

MERKER, H. J. & GUNTHER, TH. (1973). Die elektronmikroscopische Darstellung von Glykosaminoglykanen im gewebe mit Rutheniumrot. *Histochemie*, **34**, 293–303.

MILLER, A. & WRAY, J. S. (1971). Molecular packing in collagen. *Nature, London*, **230**, 437–9.

MINNS, R., SODEN, P. & JACKSON, D. (1973). The role of the fibrous components and ground substance in the mechanical properties of biological tissues: A preliminary investigation. *Journal of Biomechanics*, **6**, 153–00.

MYERS, D. B., HIGHTON, T. C. & RAYNS, D. G. (1969). Acid mucopolysaccharides closely associated with collagen fibrils in normal human synovium. *Journal of Ultrastructure Research*, **28**, 203–13.

MYERS, D. B., HIGHTON, T. C. & RAYNS, D. G. (1973). Ruthenium red-positive filaments interconnecting collagen fibrils. *Journal of Ultrastructure Research*, **42**, 87–92.

NAKAO, K. & BASHEY, R. I. (1972). Fine structure of collagen fibrils as revealed by ruthenium red. *Experimental and Molecular Pathology*, **17**, 6–13.

PARTINGTON, F. R. & WOOD, G. C. (1963). The role of non-collagen components in the mechanical behaviour of tendon fibres. *Biochimica et Biophysica Acta*, **69**, 485–95.

RAYNS, D. G. (1974). Collagen from frozen fractured glycerinated beef heart. *Journal of Ultrastructure Research*, **48**, 59–66.

REED, R. (1973). Freeze-etched connective tissue. *International Review of Connective Tissue Research*, **6**, 257–305.

REYNOLDS, E. S. (1963). The use of lead citrate at high pH as an electron-opaque stain in electron microscopy. *Journal of Cell Biology*, **17**, 208–12.

Rigby, B. J. (1964). Effect of cyclic extension on the physical properties of tendon collagen and its possible relation to biological ageing in collagen. *Nature, London*, **202,** 1072–4.

Robins, S. P., Shimokomaki, M. & Bailey, A. J. (1973). The chemistry of the collagen cross-links. Age-related changes in the reducible components of intact bovine collagen fibres. *Biochemical Journal*, **131,** 771–80.

Rougvie, M. A. & Bear, R. S. (1953). An x-ray diffraction investigation of swelling by collagen. *Journal of the American Leather Chemists' Association*, **48,** 735–51.

Schmitt, F. O., Hall, C. E. & Jakus, M. A. (1942). Electron microscope investigations of the structure of collagen. *Journal of Cellular and Comparative Physiology*, **20,** 11–33.

Schwartz, A., Geil, P. H. & Walton, A. G. (1969). Ultrastructural deformation of reconstituted collagen. *Biochimica et Biophysica Acta*, **194,** 130–7.

Sikorski, J. (1975). Structural studies of mammalian keratin. In *Structure of Fibrous Biopolymers*, ed. E. D. T. Atkins & A. Keller, pp. 271–87, London: Butterworth.

Smith, J. W. (1968). Molecular pattern in native collagen. *Nature, London*, **219,** 157–8.

Sobel, H. (1967). Aging of ground substance in connective tissue. In *Advances in Gerontological Research*, **2,** pp. 205–83. New York: Academic Press.

Tanzer, M. L. (1973). Cross-linking of collagen. *Science*, **180,** 561–6.

Torp, S., Arridge, R. G. C., Armeniades, C. D. & Baer, E. (1975). Structure–property relationships in tendon as a function of age. In *Structure of Fibrous Biopolymers*, ed. E. D. T. Atkins & A. Keller, pp. 197–221, London: Butterworth.

Torp, S., Baer, E. & Friedman, B. (1975). Effects of age and of mechanical deformation on the ultrastructure of tendon. In *Structure of Fibrous Biopolymers*, ed. E. D. T. Atkins & A. Keller, pp. 223–50. London: Butterworth.

Traub, W. & Piez, K. A. (1971). The chemistry and structure of collagen. *Advances in Protein Chemistry*, **25,** 243–352.

Verzar, F. (1964). Aging of the collagen fibre. *International Review of Connective Tissue Research*, **2,** 243–300.

Viidik, A. (1972). Simultaneous mechanical and light microscopic studies of collagen fibres. *Zeitschrift für Anatomie und Entwicklungsgeschichte*, **136,** 204–12.

Viidik, A. & Ekholm, R. (1968). Light and electron microscopic studies of collagen fibres under strain. *Zeitschrift für Anatomie und Entwicklungsgeschichte*, **127,** 154–64.

Wassermann, F. (1956). The intercellular components of connective tissue: origin, structure and interrelationship of fibers and ground substance. *Ergebnisse der Anatomie und Entwicklungsgeschichte*, **35,** 240–333.

Watson, M. L. (1958). Staining of tissue sections for electron microscopy with heavy metals. *Journal of Biophysical and Biochemical Cytology*, **4,** 475–8.

Weeton, J. W. & Signorelli, R. A. (1966). Fibre-metal composites. In *Strengthening Mechanisms, Metals and Ceramics*, ed. J. J. Burke, N. C. Reed & V. Weiss, pp. 477–00. New York: Syracuse University Press.

Wolff, I., Fuchswans, W., Weiser, M., Furthmayer, H. & Timpl, R. (1971). Acidic structural proteins of connective tissue. Characterization of their heterogeneous nature. *European Journal of Biochemistry*, **20,** 426–31.

Yannas, I. V. (1972). Collagen and gelatin in the solid state. *Journal of Macromolecular Science – Reviews in Macromolecular Chemistry*, **C7,** 49–104*B*.

ADAPTIVE MATERIALS: A VIEW FROM THE ORGANISM

STEPHEN A. WAINWRIGHT

Zoology Department, Duke University,
Durham, North Carolina, USA

In this symposium we have stood within the world of materials science and have concerned ourselves with a material's view of materials. I want now to consider materials from the point of view of the plants and animals that produce, contain and use them. First, I will deal with the mechanisms that control changes in material properties in organisms and then I will think out loud about the question 'What are the most important mechanical properties of structural materials in organisms?'

Changes in mechanical properties

In evolutionary time, stiff wood in trees has come from gelatinous algal precursors and the crisp cuticle of insects has evolved from some pliable and much less complex worm skin. Fraser and MacRae have shown us that keratins of existing mammals are more complex than, and have different properties from, those of reptiles and birds, and we do believe they all had common ancestors. In developmental time, we are familiar with the bending bones of bouncing babies and how they pass through a maximum toughness and strength at maturity and become the brittle bones of the aged. My focal point is the very much shorter time of the duration of a wave on intertidal seaweeds and barnacles or the time in the cyclic movements in running between foot strike and lift-off or the fraction of a second it takes a startled shark to accelerate from cruising to escape speed. The changes in material properties over times of seconds to minutes are not based on changes in the kinds and relative amounts of macromolecules as are the changes over developmental and evolutionary times. Rather, these quicker changes are due to changes in ionic composition of the ambient fluid or to the changes in stress or rate of strain that arise from the animal's behavior.

Control by the central nervous system

When a larva of the bug *Rhodnius prolixus* takes a blood meal, its abdominal wall is stretched greatly in accommodating the large increase in body volume. Between meals the abdominal cuticle is flexible but has the relative inextensibility we associate with insect cuticle. Maddrell (1966) has shown that, with the onset of the blood meal, a stimulus is delivered to the abdominal body wall by a nerve from the central nervous system. The stimulus is followed by a dramatic increase in the extensibility of the cuticle that persists throughout the meal. Following the meal, extensibility is gradually lost within 30 min.

While we do not know the mechanism of the change in extensibility, it is important to know that such changes do occur and that they are, in *Rhodnius*' larval abdominal cuticle, controlled by the central nervous system. Since the central nervous system receives stimuli from all other systems, one may imagine the number and complexity of control mechanisms of material properties that we are bound to discover in other systems in the next few years.

Control by hormones

Immediately following emergence from the puparium, the cuticle of the adult blowfly *Calliphora vicina* is relatively inextensible. Shortly following emergence, at a time that is affected by environmental conditions, the animal begins to swallow air. As it does so, the cuticle of the presumptive sclerites becomes extensible, thus allowing the animal to increase in length and girth as a result of the pneumatic pumping. After air-swallowing stops, the cuticle once again becomes inextensible (Cottrell, 1962). Reynolds (1976) showed that the extensibility of the cuticle was brought about by the presence of a blood-borne factor indistinguishable from the tanning hormone, bursicon.

Whereas neural stimuli originate in the central nervous system, endocrines are secreted by a variety of tissues: indeed, some are neurosecretory. Each tissue secretes according to its developmental and physiological state in a particular chemical ambience. The diversity of mechanisms for the control of changes in mechanical properties of connective tissues is obviously great and we have knowledge of only a tiny fraction of it. This view, more than any other, shows that material properties are interdependent with other structural and functional features of an organism. Our understanding of the organization of organisms will be seriously flawed if mechanical properties and their control are left out.

Control by the ionic nature of ambient fluid

In sea urchins (Takahashi, 1967a) and brittle-stars (Wilkie, 1978*a*) muscles that connect the calcitic ossicles in the movable parts of the skeleton lie parallel and adjacent to collagenous ligaments connecting the same ossicles. These ligaments are unique in that their extensibility at constant load and breaking strength are alterable. Some are reported to shorten beyond elastic recoil to resting length.

If an arm of the brittle-star *Ophiocomina nigra* is pinched distally, the muscles between ossicles at a more proximal point will detach from the ossicles and the ligaments will quickly undergo an increase in the rate of extension under constant load of the arm itself (Wilkie, 1978*a, b*). This is accompanied by dissolution of the ligaments until they are just microscopic fibers that either pull out or actually break as the arm falls off. Wilkie (1978*b*) found that the rate of extension under constant load is increased when the preparation is flooded with isosmotic potassium and that it can be retarded and even stopped by being subsequently flooded with excess calcium. Because the effect of the potassium was blocked in the presence of four common anesthetics and was depressed by magnesium chloride, Wilkie concluded that the potassium acts via a nervous mechanism. However, he found that solutions of 10^{-8}–10^{-2} M adrenalin, noradrenaline, 5-hydroxytryptamine, γ-aminobutyric acid, Na-L-glutamate and glycine had no effect on the normal extension rate of the ligament.

Wilkie views the collagenenous ligament as a polyelectrolyte whose highly hydrated molecules may be cross-linked by divalent cations and that may have their charges neutralized by monovalent cations. He notes that excess calcium or pH below 7 renders the ligament extensible, whereas chelation of calcium from the ligament, pH between 7 and 9 or absence of magnesium from the ambient medium all cause an increased rate of extension under constant load.

Takahashi (1967*b*) showed that the rate of extension of the ligament supporting the spines of the urchin *Anthocidaris crassispina* under constant load increases when the preparation is exposed to 10^{-6} M adrenalin and decreases when it is exposed to 10^{-4} M acetylcholine, 10^{-6} M 5-hydroxytryptamine and excess potassium. These results indicate that there may be significant differences between collagenous ligaments in echinoids and ophiuroids, but they do show that properties of echinoderm ligaments can be changed by changes in their ionic milieu.

The changes in properties in the brittle-star ligament are more interesting because there are juxtaligamental cells (Wilkie, 1979) of unique ultrastructural characteristics surrounding the ligaments. These cells contain droplets of electron-dense material. Extensions of the cells containing these droplets penetrate the ligament and lie between the large bundles of collagen fibrils. In four pairs of sites per ligament, the juxtaligamental cells occur in aggregates and each aggregate receives a nerve from the radial nerve of the central nervous system. Synapses have not yet been seen.

An unpublished hypothesis of J. P. Eylers holds that the range of magnitudes of the values of tensile properties that an individual echinoderm can control, by flooding its collagenous connective tissues with appropriate ions, is a paradigm of the range of collagenous connective tissue properties we see represented by the evolutionary spread of cnidarians, annelids, nematodes, molluscs and chordates. Echinoderms maintain the range of magnitudes and its control, whereas the rest of us settle for much narrower ranges of magnitudes of properties of our collagenous connective tissues. When these tissues are experimentally exposed to excessive monovalent and divalent cations, they too have been shown to change properties beyond those values found in life. The hypothesis sees the collagen and its attendant glycosaminoglycans as being highly hydrated gel-like aggregates that are polyanionic. If the anionic sites are satiated with monovalent cations, the ability of the macromolecules to form temporary links with each other is reduced and extensibility and creep rate will increase. If the anionic sites are satiated with divalent cations, the divalent cations may act as cross-links between macromolecules and would decrease their ability to slide past one another and thus decrease extensibility and creep rate. Anyone who can know about all these things and not rush off to initiate an enormous research project on the ionic control of tensile properties in collagenous connective tissues in development, evolution and disease must be either cynical or tired.

Control by body shape and behavior

It has been difficult to shake the notion that mechanical properties of a piece of material have constant, unique values. The great tables of elastic moduli in reference texts give a false sense of security to the person who learns later that a few percent change in the rate at which a strain is imposed on an algal cell wall or a piece of eel skin can cause the elastic modulus to change markedly.

In fluid-filled thin-walled vessels, the stress in the skin is equal to the pressure multiplied by the radius of the vessel divided by the thickness of the wall. When a shark accelerates from slow cruising to a burst of fast swimming, the pressure under the skin increases (Wainwright, Vosburgh & Hebrank, 1978) causing a concomitant increase in the stiffness of the skin. This allows the otherwise stretchy skin to become a relatively inextensible tendon that can transmit the force and the shortening (the work) of the contraction of anteriorly placed muscles to the tail to effect the thrust-producing bend in the fish. Here, indeed, is a modulatable spring that releases us from concepts that would prevent our finding elastic function in fish swimming (McCutcheon, 1970).

The mechanical design and performance of two species of giant sea anemones allow us to see clearly the interdependence of body shape and the material properties of its constituent materials (Koehl, 1977*a*). *Anthopleura xanthogrammica* is 6–7 cm tall, thick-walled (2–3 mm) and lives in pockets in the bottom of surge channels scoured by waves every time the tide changes. *Metridium senile* (Fig. 1A) is commonly 40 cm tall, thinner walled (1–2 mm) and lives on vertical rock faces far below the effects of surface waves where tidal currents have one tenth the velocity of wave surge. Both anemones have an outer diameter of 9–10 cm.

Because *Anthopleura* is short, it sees much less than midstream velocity of the surge channel, and both anemones experience the same maximum force on a nonstormy day – about 1 Newton. However, the stress in the connective tissue of their body wall depends also on the length of their body as it projects into the flow. Thus the maximum tensile stress in *Anthopleura* is only 950 Nm^{-2} whereas in *Metridium* it is 39 000 Nm^{-2}. Because *Anthopleura*'s body wall is thicker, its critical stress for kinking or local buckling is 4500 Nm^{-2} whereas for *Metridium* it is only 790 Nm^{-2}. So we should not be surprised to learn that *Metridium* normally bends (crown displacement 8 cm) in tidal currents while *Anthopleura* doesn't do so measurably (calculated crown displacement <0.001 cm). And, finally, the strain rate of *Metridium* at home is 10 times that of *Anthopleura*, so that the highly strain-rate-dependent tensile modulus is the same for both (0.2 Nm^{-2}).

The point here is that things simply aren't as they might seem. By that I mean that because water flow in surge channels is 10–20 times faster than tidal currents, large anemones in surge are not necessarily exposed to greater forces than are those in tidal flows: one needs to know body dimensions and the habit of *Anthopleura* to sit in pockets and thus avoid the fast flow. Then by knowing that both animals experience the same force and that the effective stiffness of their body walls is the same, the stress in *Metridium* connective tissue is seen to be 40 times that in *Anthopleura* and produces bending in the former and not in the latter.

The biological importance of these unexpected interrelationships is that *Metridium* is an effective filter feeder because it bends and exposes

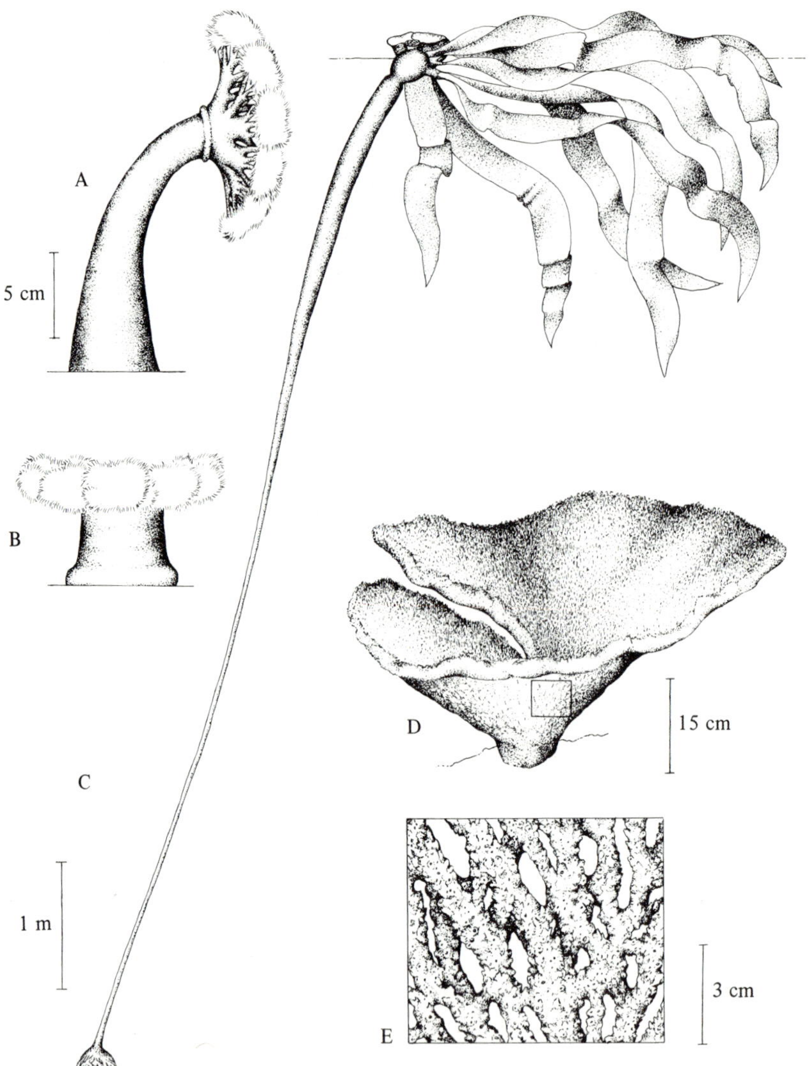

Fig. 1.A. Sea anemone *Metridium senile fimbriatum* fully expanded in filter feeding posture (redrawn from Koehl, 1977*a*). B. *Metridium* partially expanded. C. Bull kelp, *Nereocystis luetkeana*. D. Colony of the stony coral *Acropora reticulata*. E. Magnified view of inset box in D.

more plankton-catching surface to the flow while *Anthopleura* is effectively an open dustbin, facing always upward, where it can catch mussels and sea urchins knocked off the rocks by the waves.

Mechanical properties of biological materials do change. What makes this concept even more interesting is the notion that these changes are controlled by physiological or behavioral mechanisms. The point of the present exercise is to see material properties as just one more physiological function of organisms that we must fit into our larger understanding of organismal development, function and evolution.

Critical mechanical properties

Structural materials perform diverse functions in diverse plants and animals. Our knowledge of how whole plant and animal bodies generate and accommodate forces is so scant that each time such a study is made, we find surprises. Our notions of the mechanical conditions of whole-organism function are sometimes qualitatively far off base. We have been known to find that a mechanical property not previously considered turns out to be the critical one for our study.

One situation we are only just beginning to learn about in mechanical terms is that of organisms living attached to the bottom of the sea. Some examples of recent revelations are given here to show how naive judgments about which properties may be important are being corrected by observation. They will also show that very different properties are critical to organisms in seemingly similar conditions.

Breaking strain

The bull kelp *Nereocystis luetkeana* (Fig. 1C) attains a length in excess of 20 m. It grows fastened to the sea floor by an elaborate holdfast, but its long slender stipe is hollow and ends in a gas-filled float that carries the photosynthesizing blades at or near the sea surface. It grows in great beds within a few hundred meters of the wave-beaten shores of northwestern America. The stipe is subjected to a continuous forced oscillation by buoyant forces caused by waves. Dr Koehl and I (1977) predicted from this information that the stipe of the bull kelp must have a high tensile strength. We were surprised to find that it had the lowest tensile strength ever published for a biological structural material (Table 1*a*). However, the toughness or the work required to break a specimen has two components: breaking stress and breaking strain. Scrutiny of the stress–strain curves showed that in order to break the stipe in tension, we had to stretch it by 30–45% of its resting length. The work to break bull kelp stipe was 0.1 MJ m^{-3}, which is an entirely respectable value for toughness compared to that of other materials (Table 1*b*).

Two aspects of this high breaking strain merit further note. First, it takes a wave applying its drag force to a kelp a finite amount of time to stretch the stipe of many meters length by 30%. Apparently the duration of the pulling phase of even large storm waves – rarely as much

Table 1. *Breaking strengths and breaking energies of some materials*

a. Breaking strengths (MN m^{-2})	*b. Breaking energies* (MJ m^{-3})
0.001	0.001
sponge with spicules[a]	sponge with spicules[a]
0.01	0.01
–	stony coral shell crab carapace
0.1	0.1
sea anemone mesoglea[a]	bull kelp stipe[a] sea urchin spine bone insect cuticle oak timber sea anemone mesoglea[a]
1.0	1.0
bull kelp stipe[a]	resilin chitin steel
10	10
stony coral crab carapace oak timber sea urchin spine shell resilin	tendon silk cellulose
100	100
bone stiff insect cuticle tendon chitin cellulose silk steel	—

[a] Koehl, personal communication. All other values from Wainwright, Biggs, Currey & Gosline (1976).

as 8 s – is simply not long enough for the kelp to reach its breaking strain. The critical property is therefore the breaking strain rather than the more popular breaking stress.

Second, the cell walls that provide mechanical support for most plants are cylindrical structures wound with fibrils of cellulose or other straight-chain polysaccharides of high tensile strength. This basic structure allows plants to bend with the wind and waves, but only in the seaweeds, including the bull kelp, does it allow tensile extensions greater than 10%. Dr Koehl and I suggested a mechanism that may contribute to this: the angle of preferred orientation of the cell wall polysaccharides was measured to be 60° with respect to the cell's long axis. If a cylinder with such a fiber winding angle is stretched by 30%, the fiber angle will change to about 35°, and there need be no breaking of fibrils. Unfortunately, at present there are no measurements of the fiber winding angle of kelp cells at high extensions.

Seaweeds are very good at what they do that no other plants do, namely, attain large size and live in violent surf. It behoves us to discover the morphological mechanism in the material of the plant body that allows this great extensibility.

Material properties, shape and direction of force

Enewetak is a coral atoll in the Marshall Islands about half-way between Hawaii and Japan. When there is a storm at sea, wind generates huge waves that move out from the storm center. At sea, such a wave may be hundreds of meters long and only a meter high and it travels long distances at great speed with remarkably little energy loss. Enewetak, like most islands in midocean, receives waves every day from storms that took place days before in places all over the Pacific Ocean. When such a wave comes into shallow water, it becomes compressed in length and consequently greater in height. This is why surf waves are much higher than waves further seaward.

This situation renders the outer reefs of coral atolls some of the most mechanically violent habitats on earth. We have no useful measurements of flow velocities taken down where these organisms live, but we can estimate the flow velocity at any given depth below surface waves of known height.

Vosburgh (1977) studied the coral *Acropora reticulata* (Fig. 1D) in order to learn which mechanical properties and other design features allow this delicate and brittle coral to be one of the most abundant large corals on the outer reefs of Enewetak. Wind velocity data from the weather station on the atoll suitably processed through equations of ocean engineering allow the estimate that the height of the largest storm waves of the year at Enewetak is 4.88 m. Vosburgh calculated the velocity of water driven by such a wave just above the sea floor at various depths. From tests measuring the drag on a coral colony as a function of flow velocity, he converted the flow velocities of the storm waves to values of drag at different depths. He measured breaking strength in four-point bending of machined samples of the coral skeletal material and estimated the breaking strength for the stalk of the colony. From this he learned that any colony of this species growing in water shallower than its known distribution (7 m) will be broken by the worst storm of the year.

This may seem to be an exercise in proving the obvious, and it certainly contains no surprise, but it is a step towards linking the lives of organisms to mechanical factors in the environment via their material properties. It is noteworthy that the mechanisms used by this coral and the bull kelp – both large organisms – to survive in a mechanically rigorous habitat are vastly different.

Another aspect of this study addresses a subject that I believe will, in future, make good use of our knowledge of mechanical properties of biological materials. This subject is the interaction of material properties with the shape of the skeletal element. An example of this interaction is the flexural stiffness, EI, of a cylinder where E is Young's modulus of elasticity, a material property, and I is the second moment of area of the cross-section of the cylinder, a property of the shape of the section. EI is useful in assessing the design for bending of tree trunks, limbs, stems, long bones, arthropod leg segments and the stalk of *A. reticulata*.

What can be said about the broad, open cone-shaped top of the colony of this coral? Given a respite from storm waves and surge-driven chunks of broken coral, these colonies grow into tables 1.5 to 2 m in diameter whose thickness is 2–5 cm and whose structure is anything but solid (Fig. 1E). The solid part of the material is more than 98% by volume aragonite with a diaphanous organic matrix of unidentified protein, polysaccharide and lipid moieties. Coral skeletal material is extremely brittle and at the outer, growing, edge of the colony, it can be broken by ungloved fingers. A light hammer blow anywhere on the colony produces a ring like a bronze gong and may produce breakage at the edges.

It is reasonable to suppose that a large surface area would allow a coral colony to have a correspondingly large number of feeding polyps and to have a great expanse of its photosynthesizing symbiotic algae exposed to direct sunlight. But how can such a large colony made of such porous and brittle material survive in this habitat long enough to serve functions that depend on a large, exposed area? Vosburgh measured the breaking strength of the colony to forces normal to the stem and to forces normal to the table top. The colony is twice as strong to horizontal forces directed more or less parallel to the delicate table and normal to the very rigid and strong stem as it is to vertically directed forces. Also, the estimated vertical forces due to surface waves where this coral grows are only half as great as the horizontal forces. The colony shape is thus appropriate both for resisting the strongest directional force and for achieving a large surface area vital to the capture of food and light. This is the first case where an analysis of colony form has been made taking the direction and magnitude of environmental forces, material properties and ecological function into account.

The staghorn coral, *Acropora cervicornis*, is a closely related species that occurs in large tracts on Caribbean reefs. In contrast to *A. reticulata* just discussed, the staghorn grows as an isolated, branched twig 2–4 cm in diameter and up to a half meter in length. These branched twigs start life attached to solid substrata. Tunnicliffe (1979) has shown that due to boring activities of various organisms, particularly sponges, the twigs break readily at the base and fall to lie flat on the often sandy sea floor where they continue to grow and break and grow and break. Their linear growth rate averages 12 cm a year; they are much the fastest growing coral on crowded reefs, and the grow–break–topple–grow habit allows them to overgrow other corals and seaweeds and to spread faster than any other species on the reef.

Remembering that the solid coral skeletal material has perhaps the highest volume percent of mineral and is the most brittle of any animal support material, we might predict (indeed some of us have) that high breaking strength would be very important to reef-building corals. Now we see that while that is true for *A. reticulata*, the porous nature of the material facilitates weakening by borers and the brittleness encourages frequent breakage in the staghorn coral. *A. reticulata* is thus allowed to attain a large colony size in a high-flow-rate habitat while the staghorn is allowed to out-compete its monolithic neighbors for vital space in calmer seas.

Creep recovery rate

The large plumose anemone *Metridium senile* var. *fimbriatum* (Fig. 1A, B), whose behavioral control of connective tissue properties we noted above, lives at depths where wave action is so attenuated that it creates insignificant drag forces on the anemone. *Anthopleura xanthogrammica* is a large anemone which attaches to surf-beaten intertidal rocks on the American northwest coast. It is shorter and has a thicker body wall than *Metridium*, but the connective tissue of both consists of collagen fibrils in a gel matrix (Koehl, 1977*b*). When a piece of the body-wall mesoglea

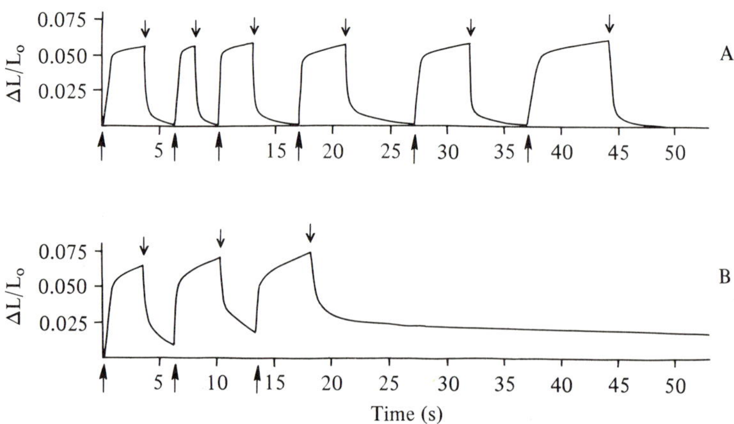

Fig. 2. Results of creep tests performed on samples of body wall from sea anemones at wave-like forces and time intervals. A. *Anthopleura xanthogrammica* body wall recovers completely after each 'wave'. B. *Metridium senile* body wall, exposed to wave conditions which prevail in the habitat of *Anthopleura*, does not recover. Redrawn from Koehl (1977*b*).

of each species was dynamically extended at stresses of the magnitude they experience in their respective habitats but at durations and frequencies of the waves in *Anthopleura*'s habitat, an interesting result was observed.

Figure 2 shows that the mesoglea of *Anthopleura* always recovered its deformation in time to receive the next wave, while the mesoglea of *Metridium* was unable to recover from any wave before the next one hit. Without the help of energetically expensive, high endurance muscle, an anemone with mesoglea like that of *Metridium* could not survive in the habitat of *Anthopleura*. Given the mesoglea of *Anthopleura*, an anemone can resist unremitting wave action without recourse to muscular activity. Before this study was made, there was no reason to believe that the differences between species of squashy, 'lower' animals like sea anemones would include differences in dynamic properties of their connective tissues. As we have seen in all the attached marine plants and animals mentioned here, not only does each species have a unique set of material properties, but the properties are 'tuned', presumably by natural selection, and they allow the organism to respond appropriately to the direction, magnitude and frequency of the forces that impinge on the organism in nature.

Properties which aid locomotion

It is not enough that structural materials bear the loads and suffer the strains that they do: they are also found performing functions that are co-ordinated with muscular activity in the cyclic movements of terrestrial locomotion.

Elastic energy storage and release

Just as a bouncing ball stores elastic strain energy on impact with the ground and releases it in regaining altitude, hopping kangaroos (Alexander & Vernon, 1975) and running men (Cavagna, Saibene & Margaria, 1964) store elastic strain energy in the long tendons of the lower leg on foot strike and use the energy at lift off to accelerate the body upward and forward. In a discussion of several examples of elastic energy being used in locomotion, Alexander & Bennet–Clark (1977) have pointed out that the ability of an element to store elastic energy is proportional to its length. They also note that muscle stores little elastic energy, and the best design for this is that of pennate muscles with short fibers in combination with long tendons. This accurately describes the foot extensor systems of mammals and locusts. Evidence from measurements of the forces and kinematics of hopping in kangaroos (Alexander & Vernon, 1975) indicates that as much as 40% of the energy necessary for this locomotion may be saved by the storage of elastic strain energy in tendons.

Reversible transformation from solid to fluid

When a slug crawls over the ground, it first lays down a layer of mucus about 100 μm thick. Crawling is accomplished by causing a wave of longitudinal contraction of the epidermis to pass forward along the sole of the foot. Denny (1979) has shown that the mucus is an elastic solid that attaches the slug to the ground at the frictional point of each wave. As the wave moves forward, it shears the mucus up to 1000%. Denny showed that at shear strains of about 400%, the mucus undergoes a change from an elastic solid to a viscous fluid (Fig. 3) that allows the overlying bit of skin to move relative to the ground and not to be pulled back by elastic recoil to square one.

This surprising phenomenon provides further encouragement not to be satisfied with measuring only breaking strength and relaxation time when we meet a material in a new situation. It also supports Wainwright's maxim that aside from nucleic acids, mucus is the most widely distributed macromolecular substance with the broadest range of vital functions in biology. There is a truly exciting future in research into functions of mucus, their mechanisms and control.

There are many roads into the future from here. Speakers at this symposium, their associates and their students will continue to reveal

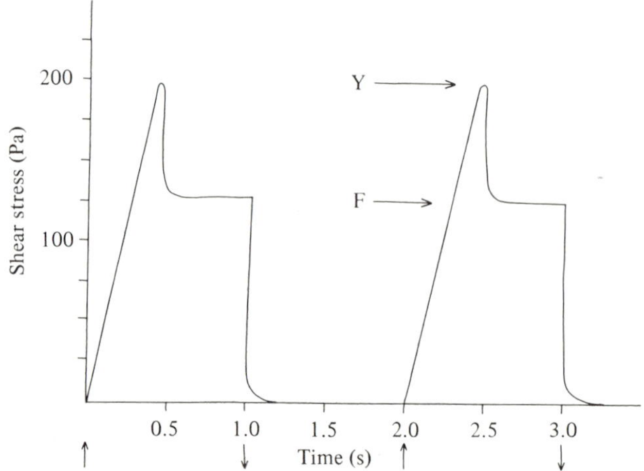

Fig. 3. Response of pedal mucus from the slug *Ariolimax columbianus* to shearing forces under conditions analogous to those in life. When the force is applied (upward arrows), stress is proportional to strain indicating that the mucus is an elastic solid. At a strain of 500–600% the mucus yields (Y). With further strain, the stress is proportional to strain rate and the mucus behaves as a liquid (F). If shearing is stopped (downward arrows), the mucus will heal and repeated tests give nearly identical results. Redrawn from Denny (1979).

the mechanisms underlying mechanical properties of biological materials. Others will see to the interaction of materials in more complex systems. I offer two subjects wherein these groups of researchers may profitably combine their efforts.

Structural modulus

The question has been asked here 'When does a material become a system?' The answer depends on your point of view. An organic chemist certainly would see insect cuticle, bone, wood or even collagen as complex systems. Nevertheless, we treat these items as materials and talk about each one as though it were a continuum. My own vision is defective at molecular level, so I spend time fitting the structures I see in various microscopes into my concept of materials. Coral skeleton and shark cartilage are awkwardly but finally stuffed into the box labelled materials, but the skeletal organization of the alcyonarian or soft corals has me stumped. For example, the tropical Pacific sea-fan *Melithaea ochracea* has a colony formed of tapering, branched branches that bend a little in tidal currents. The branches are like our fingers in that they have rigid shafts connected by flexible joints (Muzik & Wainwright, 1977).

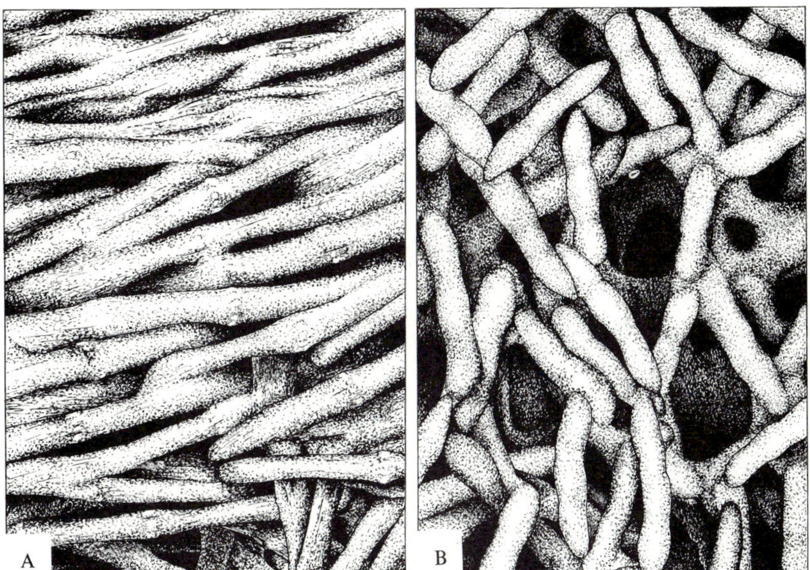

Fig. 4. Size, shape, spacing, orientation and adhesion of sclerites in the sea-fan *Melithaea ochraces*. A, Sclerite array in a rigid segment of the axis. Sclerites are joined by calcium carbonate. B, Sclerite array in a flexible joint. Sclerites are joined by a pliant polymer. Drawn from scanning electron micrographs in Muzik & Wainwright (1977).

The basic unit of microscopic structure in this skeleton is the sclerite. Each sclerite is an aggregate of many calcite crystals and each sclerite is about 10 μm in diameter and 100 μm long (Fig. 4). In the stiff segments, the sclerites are tightly packed, aligned parallel to the skeletal axis and glued together with more calcium carbonate (Fig. 4A). In the flexible joints the sclerites have the same shape and diameter but are only about 60 μm long (Fig. 4B). They are randomly oriented with respect to each other and they touch one another only at single points where they are glued together by an unidentified pliant polymeric substance. The considerable spaces between the sclerites are filled with a collagenous mesoglea. The reasons why the skeleton flexes where it does and is rigid where it is are obvious. Is it a material or a system?

I would like to be able to compare the axial skeleton of *Melithaea* to bone with its highly ordered complexity (Katz, 1980) or to a skeletal system whose elements are much larger. What is required is a structural descriptor that has terms for any number of material or morphological components and the size, shape, orientation, volume fraction, density, modulus and spacing of each one. The area of adhesion between items and the strength of the adhesive should be included. Perhaps then we could compare sea urchins with geodesic domes and stick-insects with giraffes. No one contemplating this structural modulus could avoid thinking about structure and function in biological systems over several levels of organization.

Workmanship

Among the many species of British barnacles, two have been studied with respect to their abilities to withstand mechanical forces applied to them in testing rigs in a laboratory (Murdock & Currey, 1978). Testing data showed that the entire shell of *Balanus balanus* withstood crushing more effectively than did the shell of *Semibalanus balanoides*. Individual skeletal plates from the two species were found to have very similar material properties. The difference in properties of whole barnacles between the two species was ascribed to 'workmanship'. That is, the plates of *B. balanus* (the stronger one) overlapped each other more and were joined together laterally tongue-in-groove. They were also joined by an interdigitating suture to a rigid calcareous basis that was attached to the substratum. *S. balanoides* (the weaker one) had no basis and its plates overlapped little laterally without tongue-in-groove joints.

Many phenomena are involved here. Each one might be studied in detail and the mechanisms of crushability might one day be fully understood. I believe we need the concept of workmanship in thinking about how biological materials are involved in structures. Workmanship should be discussed and a definition sought that would include such engineering wisdoms as 'compatibility of materials' in our thinking as we ponder adjacent materials in complex systems. Workmanship and the structural modulus may not be susceptible to operational definition, but they can keep our thinking out where the winds of change can get at it.

The conclusion to be drawn from all the foregoing is that mechanical properties are functional characteristics of materials. Materials serve in formed elements as parts of systems at more complex levels of organization up to and including the organism in its environment. We ignore material material properties at our peril. No, that is not the right emphasis. We know about them and how important they are. We are not in danger of ignoring them. It is the molecular biologists, cell biologists, developmental biologists, physiologists, physicians, anatomists, ecologists, behaviorists and evolutionary biologists who need to know about material properties. We are their only source of information. Let's take it to them. Now there's a project for the future.

References

Alexander, R. M. & Bennet-Clark, H. C. (1977). Storage of elastic strain energy in muscle and other tissues. *Nature, London*, **265,** 114–17.

Alexander, R. M. & Vernon, A. (1975). The mechanics of hopping by Kangaroos. *Journal of Zoology*, **177,** 265–303.

Cavagna, G. A., Saibene, F. P. & Margaria, R. (1964). Mechanical work in running. *Journal of applied Physiology,* **19,** 249–56.

Cottrell, C. B. (1962). Imaginal ecdysis of blowflies. Evidence for a change in the mechanical properties of the cuticle at expansion. *Journal of experimental Biology,* **39,** 449–58.

Denny, M. W. (1979). The role of mucus in slug locomotion. Ph.D. Thesis, in Zoology, University of British Columbia, Vancouver.

Fraser, R. D. B. & MacRae, T. P. (1980). Molecular structure and mechanical properties of keratins. In *The Mechanical Properties of Biological Materials* (34th Symposium of the Society for Experimental Biology) ed. J. F. V. Vincent & J. D. Currey, pp. 211–46. Cambridge University Press.

Katz, J. L. (1980). The structure and biomechanics of bone. In *The Mechanical Properties of Biological Materials* (34th Symposium of the Society for Experimental Biology) ed. J. F. V. Vincent & J. D. Currey, pp. 137–68. Cambridge University Press.

Koehl, M. A. R. (1977*a*). Effects of sea anemones on the flow forces they encounter. *Journal of experimental Biology*, **69,** 87–105.

Koehl, M. A. R. (1977*b*). Mechanical diversity of connective tissue of the body wall of sea anemones. *Journal of experimental Biology,* **69,** 107–25.

Koehl, M. A. R. & Wainwright, S. A. (1977). Mechanical adaptations of a giant kelp. *Limnology and Oceanography*, **22,** 1067–71.

Maddrell, S. H. P. (1966). Nervous control of the mechanical properties of the abdominal wall at feeding in *Rhodnius*. *Journal of experimental Biology*, **44,** 59–68.

McCutchen, C. W. (1970). The trout tail fin: a self-cambering hydrofoil. *Journal of Biomechanics,* **3,** 271–81.

Murdock, G. R. & Currey, J. D. (1978). Strength and design of shells of the two ecologically distinct barnacles, *Balanus balanus* and *Semibalanus balanoides*. *Biological Bulletin,* **155,** 169–92.

Muzik, K. & Wainwright, S. (1977). Morphology and habitat of five Fijian sea fans. *Bulletin of marine Science,* **27,** 308–37.

Reynolds, S. E. (1976). Hormonal regulation of cuticle extensibility in newly emerged adult blowflies. *Journal of Insect Physiology,* **22,** 529–34.

Takahashi, K. (1967a). The catch apparatus of the sea urchin spine. I. Gross histology. *Journal of the faculty of Science, Tokyo University,* **4,** 109–20.

Takahashi, K. (1967b). The catch apparatus of the sea urchin spine. II. Responses to stimuli. *Journal of the faculty of Science, Tokyo University,* **4,** 121.

Tunnicliffe, V. (1979). The Ecology of the Coral *Acropora cervicornis*. Ph. D. Thesis, Biology Department, Yale University, New Haven.

Vosburgh, F. (1977). Response to drag of the reef coral *Acropora reticulata*. *Proc. Third Int. Coral Reef Symp.* Miami: University of Miami Press.

Wainwright, S. A., Vosburgh, F. & Hebrank, J. H. (1978). Shark skin: function in locomotion. *Science,* **202,** 747–9.

Wainwright, S. A., Biggs, W. D., Currey, J. D. & Gosline, J. M. (1976). *Mechanical Design in Organisms*. London: Edward Arnold.

Wilkie, I. C. (1978*a*). Arm autotomy in brittle stars. *Journal of Zoology,* **186,** 311–30. 311–30.

Wilkie, I. C. (1978*b*). Nervously mediated change in the mechanical properties of a brittlestar ligament. *Mar. Behav. Physiol.,* **5,** 289–306.

Wilkie, I. C. (1979). Juxtaligamental cells of *Ophiocomina nigra* and their possible role in mechano-effector function of collagenous tissue. *Cell Tissue Research*, **197,** 515–30.

POSTERS DISPLAYED AT THE SYMPOSIUM

(4–6 September 1979)

An electron microscope study of the calcareous network in bone
by I. G. Turner (*p*. 457)

Correlation of hatching techniques in some avian species with the mechanical properties of their eggs
by G. M. Bond, V. D. Scott, R. G. Cooke and R. G.Board (*p*. 459)

Biomaterials as foods
by P. W. Lucas (*p*. 463)

Structure and mechanics of tendon
by J. H. Evans, J. C. Barbenel, T. R. Steel and A. M. Ashby (*p*. 465)

Deformation of slender filaments with planar crimp: general theory and applications to tendon collagen
by D. W. Lloyd and C. P. Buckley (*p*. 471)

The Cuverian tubules of Holothuria: design for successful failure in a collagenous system
by L. J. Gathercole, A. J. Bailey, J. Dlugosz and A. Keller (*p*. 475)

Creep testing of isolated cervix from pregnant rats
by M. Hollingsworth, S. Gallimore and L. M. Williams (*p*. 477)

Biological thixotropy – the unifying factor in basement membrane function
by L. O. Simpson (*p*. 479)

The hardness of locust incisors
by J. E. Hillerton (*p*. 483)

Design and materials of feather shafts: very light, rigid structures
by D. G. Crenshaw (*p*. 489)

Small-scale tensile tests
by R. F. Ker (*p*. 487)

AN ELECTRON MICROSCOPE STUDY OF THE CALCAREOUS NETWORK IN BONE

I. G. TURNER

Department of Metallurgy & Materials Technology,
University College of Swansea, Singleton Park,
Swansea SA2 8PP, UK

Bone consists of two main components, the organic collagen fibre network and the inorganic mineral part which is mainly hydroxyapatite. Despite many years of research the actual nature of the mineral component in bone and its spatial arrangement are still in dispute (Pautard, 1978).

Bone does not lend itself readily to standard preparation techniques for electron microscopy, such as electro-polishing or chemical etching. Microtoming, the most generally adapted method for biological materials, can introduce undesirable artefacts (Thorogood & Craig-Gray, 1975). Accordingly the technique of ion bombardment, a method which can be used where other processes present practical difficulties, has been used to etch the surface of transverse sections of compact bovine and human long bone. The prepared surfaces were then coated with gold and viewed in a JEOL 120C electron microscope operating in the scanning mode.

The organic material is preferentially removed (de Nee, 1976), revealing a dense mineral fibre network. Fibres are segmented into spheroidal and cylindrical components with a diameter of $\sim 0.1\ \mu m$. A network is achieved by frequent bifurcation. The dimensions correlate with similar less well defined spherical particles reported by others (Pautard, 1978). Similar morphologies have been reported in equivalent human and bovine compact long bone.

From electron and x-ray diffraction studies we are led to conclude that the mineral fibres consist of spherical segments which are made up of a defective form of hydroxyapatite.

Early work suggests that bone consists of discrete particles of mineral embedded in or around the collagen fibres. The arrangement we propose is one of interpenetrating networks of collagen and the mineral fibres we observe. Such a model can be more easily reconciled with the high stiffness, >20 GPa, parallel to the long axis and high strain rate, $\sim 2\%$, achieved in bone.

References

DE NEE, P. B. (1976). Identification and analysis of particles in biological tissue using SEM and related techniques. In *Proceedings of the Ninth Annual Symposium of Scanning Electron Microscopy*, Chicago, pp. 461–468.

PAUTARD, F. G. E. (1978). Phosphorus and bone. In *New Trends in Bio-inorganic Chemistry*, ed. R. J. P. Williams and J. P. R. du Silva, pp. 261–353, Academic Press.

THOROGOOD, R. V. & CRAIG-GRAY, J. (1975). Demineralization of bone matrix: observations from electron microscope and electron probe analysis. *Calcified Tissue Research,* **19,** 17–26.

CORRELATION OF HATCHING TECHNIQUES IN SOME AVIAN SPECIES WITH THE MECHANICAL PROPERTIES OF THEIR EGGS

G. M. BOND, V. D. SCOTT, R. G. COOKE and R. G. BOARD

The School of Material Science and School of Biological Sciences, University of Bath, Claverton Down, Bath BA2 7AY, UK

Eggs from the domestic hen, domestic duck, Japanese quail and feral pigeon were selected for study, since earlier work (Oppenheim, 1972) indicated that they exhibited differences in hatching characteristics.

Unincubated eggs were tested in compression between flat plates. Pipping and hatching simulations, in which loading was on the inner surface of the shell, were performed on half-shells of incubated eggs mounted on annuli of resin in an Instron compressive test rig. With hen and duck eggs, the onset of fracture is followed by a marked and rapid decrease in load (Fig. 1*a*) corresponding to the development of a crack which, with brief arrest periods, spreads across the shell. In contrast, quail and pigeon eggs show no sharp peak in the load–displacement curve (Fig. 1*b*) but continue to support substantial loads as cracks develop. Hen eggs, which have relatively low energy requirements for crack propagation, produce comparatively few large acoustic emissions during fracture (Fig. 2*a*). Quail eggs, however, give greater numbers of large acoustic emissions (Fig. 2*b*), associated with many short, energy-consuming steps in crack growth.

We can thus make a clear distinction between eggs of different species based upon their mechanical characteristics. The rapid and catastrophic propagation of a single crack together with the accompanying loss of strength of the whole structure once fracture has been initiated, as observed in the hen and duck eggs, are typical of brittle failure. With the quail and pigeon eggs, however, cracks formed at the onset of fracture remain short and can be extended only by further application of fairly substantial loads; crack propagation occurs in short bursts and more cracks are likely to be generated. Such behaviour is characteristic of a tough, more flexible material. It should be noted that the above results refer to eggshells with membranes present. Removal of the shell

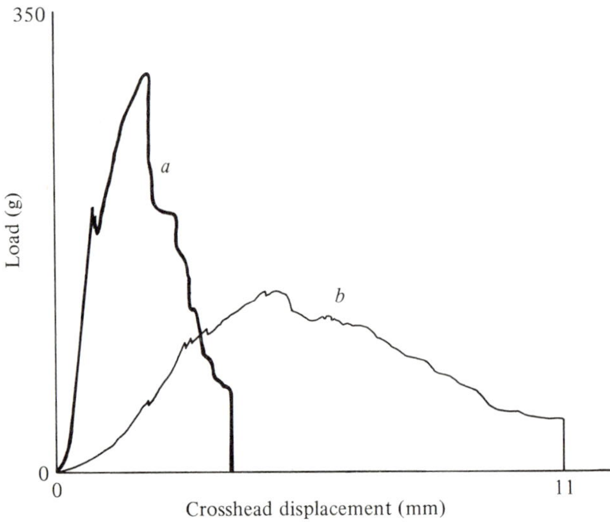

Fig. 1. Hatching simulation with egg halves on an Instron tensometer: *a*, hen; *b*, quail (see text for details).

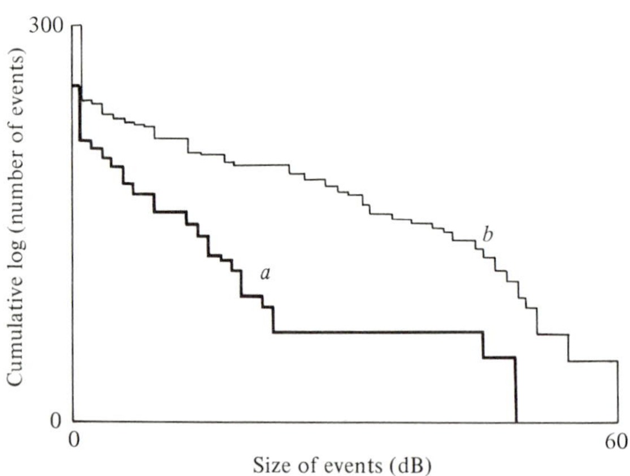

Fig. 2. Acoustic–emission data from compression of a whole, blown, eggs. *a*, hen; *b*, quail.

membranes prior to pipping or hatching simulations lessens but does not eliminate the distinction between brittle and tough behaviour.

We have studied the hatching techniques of a range of species in the laboratory, at bird-breeding establishments and in the field, and have looked at films of hatching. It appears that two main types of hatching techniques are exhibited. For example, hen and duck neonates, after pipping when the initial fracture occurs, may chip away at the shell to produce a small hole. Further cracking is induced until the damage extends for approximately two thirds of the circumference of the shell, and then the cap is pushed off to allow the chick to emerge. Quail and pigeon neonates, however, chip virtually the whole way round the egg, perforating shell and membranes at quite small intervals, before pushing off the cap; the quail may take more than one revolution to achieve adequate perforation (Oppenheim, 1972).

These two broad categories of hatching technique correlate with the classification of the four species based upon the mechanical test and acoustic emission data from their eggs. With hen and duck eggs a crack, once produced, can be propagated fairly easily and the hatching technique of the chick consists of making relatively few initial cracks before pushing off the cap. By contrast the strength of the tougher eggs of quail and pigeon does not decrease greatly after the initial damage and, consequently, chicks have to assist crack propagation by making an extensive series of cracks or holes around the circumference before they can push off the cap.

These results form the basis for a wider spectrum of behaviour. At one extreme would lie species such as the quail, pigeon and little bittern (Zwergrohrdommel, 1962), whose eggs display toughness, while at the other would lie species such as the rhea and ostrich whose eggs are very hard and brittle and whose chicks chip, at most, a quarter of the way round the egg before pushing away a cap (Bruning, 1974; BBC, 1977).

We thank Dr J. Kear and Professor G. V. T. Matthews of the Wildfowl Trust, Slimbridge, and the British Egg Marketing Board Research Trust for their assistance, and Mr M. Kendall of the British Broadcasting Corporation Natural History Unit for access to copyright material. The support of the SRC for one of us (GMB) is also gratefully acknowledged.

References

BBC, (1977). *Namib – strange creatures of the skeleton coast.* 'World About Us' series, film no. 15222.

Bruning, D. F. (1975). Social structure and reproductive behaviour in the greater rhea. *The Living Bird, Thirteenth Annual, 1974*, pp. 251–94.

Oppenheim, R. W. (1972). Prehatching and hatching behaviour in birds: a comparative study of altricial and precocial species. *Animal Behaviour,* **20,** 644–55.

Zwergrohrdommel, X. Y. (1962). *Ixobrychus minutus (Ardeïdae) – Schlüpfen, erste Lebensstunde* [Hatching, first hour of life. Little bittern]. From *Encyclopeadia Cinematographica,* film no. E275. Göttingen: Institut für der Wissenschaftlichen Film.

BIOMATERIALS AS FOODS

P. W. LUCAS
Unit of Anatomy in relation to Dentistry,
Guy's Hospital Medical School,
London, E1, UK

The teeth of mammals are designed mainly for the reduction of particle size of foods and thus their shapes are dependent upon the physical properties of foodstuffs. Given that mammals need to chew their food and that chewing is efficient (i.e. maximised for energy gain to the animal), the shapes of teeth should be predictable from an analysis of solid foods.

The simplest mode of loading is to press the tooth on to the food. The area of contact at fracture can be derived from Hertz's equations (Hertz, 1896) for harder foods. Minimal areas of contact are expected for greatest efficiency. This is compromised by the necessity to grip the food, and not to weaken sharp teeth due to high stress concentrations. Provided that the strength of foods is greater in foods of higher elastic modulus then a relationship between the sharpness (radius of tip curvature) of teeth and the elastic modulus of foods is expected. Blunt 'pestle-and-mortar' and 'double-pestled' tooth shapes, dependent on the degree of comminution required, are derived for foods of high modulus which shatter, and sharp 'double bladed' teeth are best with foods of low modulus in which crack propagation speed is slow. It is likely that many soft foods, for which blades are also necessary, will show plastic behaviour. The bluntness of man's cheek teeth (Table 1) suggests adaptation to a diet of hard brittle foods.

Biomaterials are viscoelastic and differ greatly in structure. The viscous component will be less at higher loading speeds and fast chewing may therefore save energy in the fracture of foods (e.g. white potato, Table 2) as well as easing other (theoretical) problems of food behaviour. It is concluded that detailed studies of the mechanical behaviour of animal diets may provide a test of the theory. Portable testers, such as used by food scientists (Mohsenin, 1970) may be well-suited for use in ecological studies. Some understanding of the shape and evolution of simple hand tools may also be provided by this analysis.

Table 1. *The bluntness of human cheek teeth*

Tooth	No. of cusps measured	Mean radius of curvature of cusps (mm)	±Standard deviation
Upper First Premolar	16	0.67	0.21
Upper Second Premolar	16	0.72	0.21
Lower First Premolar	13	0.67	0.20
Lower Second Premolar	8	0.64	0.13
Upper First Molar	17	0.69	0.21
Upper Second Molar	12	0.59	0.15
Lower First Molar	20	0.60	0.24
Lower Second Molar	13	0.52	0.24

The teeth were measured from camera lucida tracings of the teeth at ×32 magnification and fitted (using least-squares) to an equation for conic section curves.

Table 2. *Energy to fracture foods at different loading rates*

Loading rate ($cm\ min^{-1}$)	Energy to fracture unit volume of food ($kN\ m^{-2}$)
0.25	0.203
0.5	0.220
1.0	0.177
2.5	0.158
5.0	0.162
10.0	0.165
12.5	0.165
20.0	0.148
25.0	0.141

Uniaxial compression tests on an Instron testing machine.

References

Hertz, H. R. (1896). *Miscellaneous Papers.* London: Macmillan.

Mohsenin, N. N. (1970). *Physical Properties of Plant and Animal Materials.* New York: Gordon & Breach.

STRUCTURE AND MECHANICS OF TENDON

J. H. EVANS, J. C. BARBENEL, T. R. STEEL and A. M. ASHBY

University of Strathclyde, Glasgow, UK

In mechanical terms, tendon serves primarily as a passive transmitter of uniaxial force. This is reflected in its relatively simple microstructure (Fig. 1) which is essentially a unidirectional arrangement of fibres. The arrangement of the tendon relative to the muscle is dictated by the geometry of the joint on which the muscle acts: muscular tension may be redirected by the ability of the tendon to transmit the tension around bony prominences or under fibrous bands. In consequence normal forces are developed between tendon and the supporting structures. Tendon is vascularised except in regions repeatedly subject to these normal compressive forces where the structural adaptations may be likened to articular cartilage. Structural specialisations also occur at the aponeurosis and insertion.

In the relaxed state, tendon shows regular undulations of the surface, just visible to the naked eye, of wavelength 70 μm. These undulations disappear progressively with the application of tension. Additionally when the collagen fibres are free of stress they show a secondary crimping (Fig. 2) which gives rise to long bands, visible on the surface, which lie at approximately 45° to the long axis and are spaced at intervals of millimetres.

Pertinent structure

To incorporate all these structural phenomena into a model of tendon would be exceedingly difficult and of doubtful value. It is also necessary to neglect the specialised structures which occur over joint 'pulleys' and at the ends. Consequently a first-stage model is being considered which relates to the behaviour of tendon in normal function. As tendons are normally pretensioned to some, as yet unknown, degree, the occurrence of secondary crimping which disappears at extremely low loads may be considered to be functionally insignificant. The structure which has been considered is that of an essentially parallel array of protein fibres,

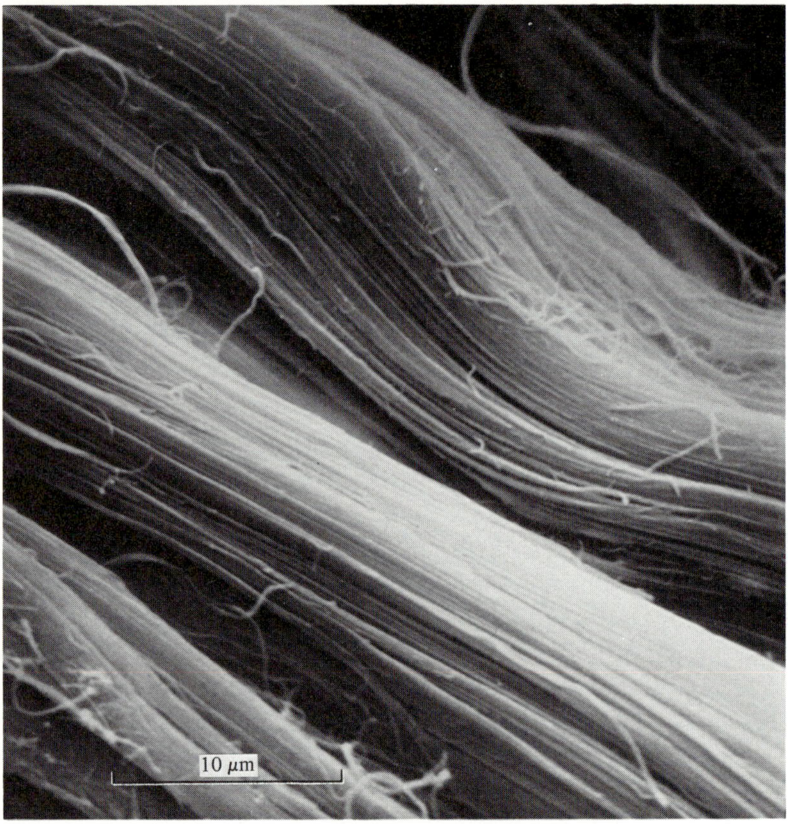

Fig. 1. Collagen fibrils in an essentially parallel array. Scanning electron micrograph.

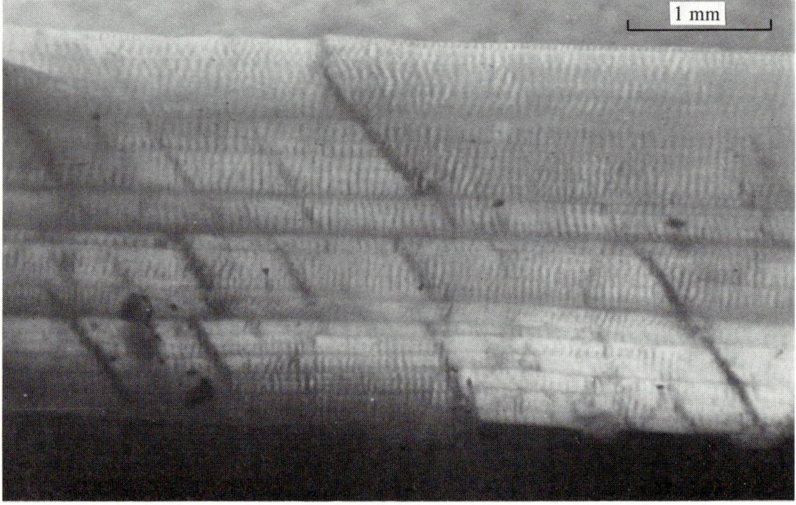

Fig. 2. Surface waveform as observed in unstrained *palmaris longus* tendon. Adjacent collagen coils released by enzymolysis with hyaluronidase. Scanning electron micrograph ×300.

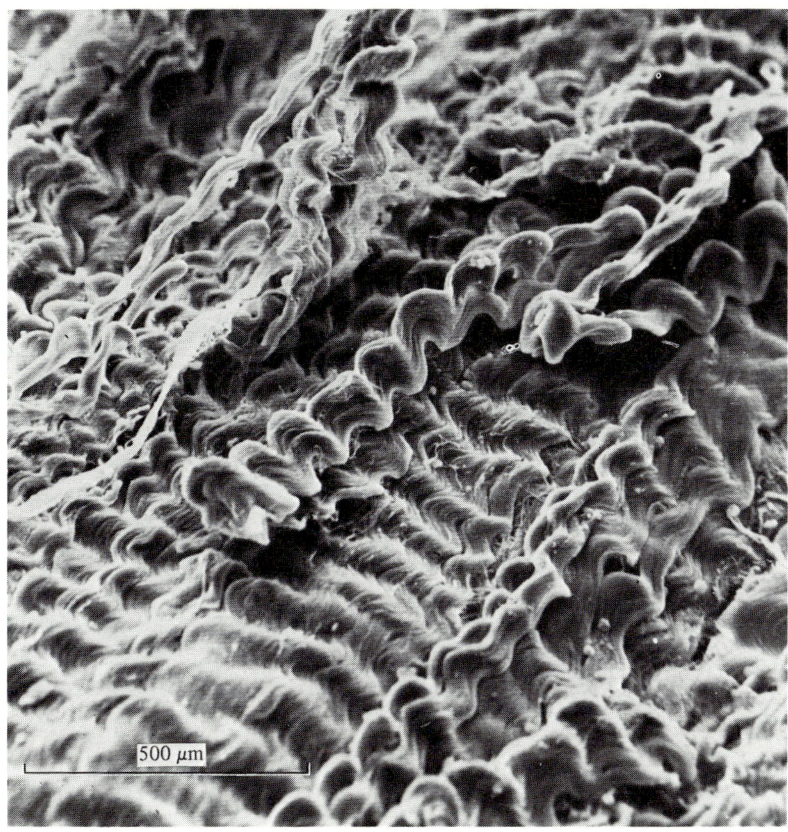

Fig. 3. Secondary crimping at 45° to tendon long axis in the unstrained condition. Primary crimping, with a 70 μm wavelength, can also be seen at right angles to the long axis. Light micrograph.

principally collagen, which are regularly contorted to produce the familiar 70 μm surface waveform in the tension-free state. There is some contention as to the nature of these contortions but in a range of mammalian tendons the fibres have been observed to follow an approximately helical path, possibly with an elliptical cross-section (Fig. 3). Within the fibres the collagen fibrils, with their regular structure reflecting the internal molecular packing, are also in sensibly parallel array.

Mechanics

The *in vitro* mechanics of tendon are typical of most soft connective tissues, non-linear and time-dependent (Elliott, 1965). On extension

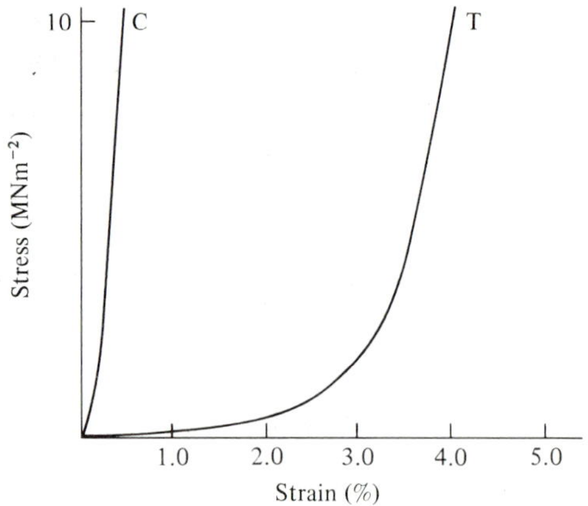

Fig. 4. Stress–strain curves for tendon (T) and collagen (C).

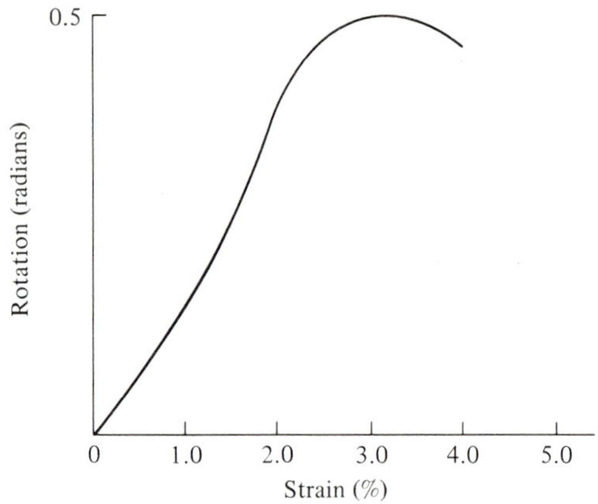

Fig. 5. Rotation – axial strain curve for tendon in tension.

from the relaxed condition the tendon initially offers little resistance but rapidly stiffens at approximately 3% strain (the toe region) (Beskos & Jenkins, 1975; Lanir, 1978) to exhibit a nearly linear response similar to that of single collagen fibres (Fig. 4). The stress is time-dependent over this strain range, the initial 'toe' region being the most nearly elastic. Failure can occur abruptly on rapid extension but frequently progressive tearing leads to large extensions prior to splitting.

The model developed thus far incorporates a parallel array of helically contorted collagen fibres which are assumed to be elasticas (Barbenel, Evans & Gibson, 1971; Kastelic, Galeski & Baer, 1978). In addition to describing the non-linear behaviour of tendon in tension the model predicts torsion/tension coupling in the 'toe' region of the response. This behaviour has been confirmed qualitatively and the current research programme is dedicated to obtaining a quantitative correlation (Fig. 5) (Evans & Barbenel, 1974). Axial force–extension–rotation tests are being conducted with conventional mechanical testing equipment incorporating an essentially frictionless air bearing and a rotational differential transformer. The test environment is air at 37°C saturated with water vapour.

References

Barbenel, J. C., Evans, J. H. & Gibson, T. (1971). Quantitative relationships between structure and mechanical properties of tendon. *Digest, 9th International Conference on Medical and Biological Engineering, Melbourne.*

Beskos, D. E. & Jenkins, J. T. (1975). A mechanical model for mammalian tendon. *Journal of Applied Mechanics,* **42,** 755–8.

Elliott, D. H. (1965). Structure and function of mammalian tendon. *Biological Reviews,* **40,** 392–421.

Evans, J. H. & Barbenel, J. C. (1974). Structure and mechanical properties related to function. *Equine Veterinary Journal*, **7,** 1–7.

Kastelic, J., Galeski, A. & Baer, E. (1978). The multicomposite structure of tendon. *Connective Tissue Research,* **6,** 11–23.

Lanir, Y. (1978). Structure-function relations in mammalian tendon. *Journal of Bioengineering,* **2,** 119–28.

DEFORMATION OF SLENDER FILAMENTS WITH PLANAR CRIMP: GENERAL THEORY AND APPLICATIONS TO TENDON COLLAGEN

*D. W. LLOYD and *C. P. BUCKLEY*
Department of Textile Industries, University of Leeds, and
*Department of Textile Technology, University of Manchester Institute of Science and Technology

Tensile deformation of a slender filament crimped into the form of a plane wave has been analysed, using the theory of extensible planar elasticas (Fig. 1). The load–extension relation can be expressed in terms of the dimensionless quantities: crimp level, D_0; slenderness (ratio of thickness to original contour length), R; and shape function (curvature normalised to its peak value) only. A particular family of shapes has been examined where the shape function is specified through a single parameter q. All shapes of practical interest lie between the limits of a wave of circular arcs ($q = -\infty$) and a planar zig-zag ($q = +\infty$). Previous solutions by other authors can be shown to be special cases within this family of shapes. The effects of varying crimp level, slenderness and shape were all examined by repeated numerical solution, avoiding the restrictive assumptions of previous work. Normalised forms of load and extension were found which bring together the load–extension curves for wide ranges of crimp, slenderness and shape. The parameters used, φ and ξ, are illustrated in Fig. 2. Replotting in log–log form leads to a procedure for fitting experimental load–extension curves to those calculated by a simple shifting of plots. This permits the experimental error in the zero strain to be calculated, and a corrected value for crimp level to be deduced. The procedure also yields a value for slenderness ratio, and hence the filament thickness.

The method has been applied to data for tendon collagen, satisfactory agreement being obtained between the theoretical and experimental results (Fig. 3). The values of thickness obtained from three independent sets of data (Partington & Wood, 1963; Elden, 1968; Yannas & Huang, 1972) were mutually consistent, but exceeded by an order of magnitude the dimensions of observed collagen fibrils. There are three possible explanations of this paradox: first, adjacent fibrils might be mechanically coupled (Torp, Baer & Friedman, 1975) causing groups of

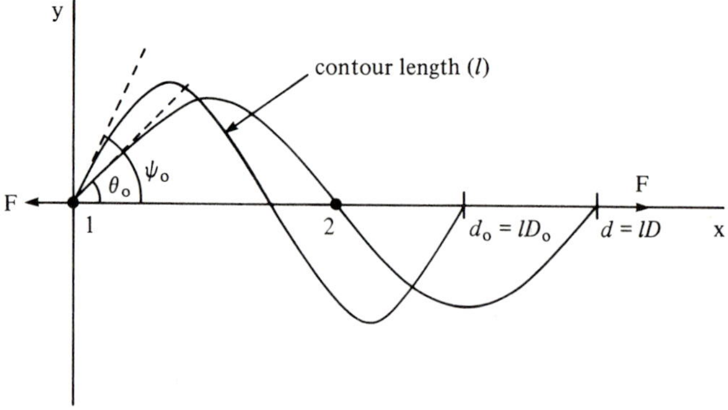

Fig. 1. One wavelength of the crimped filament in stress-free and stretched states.

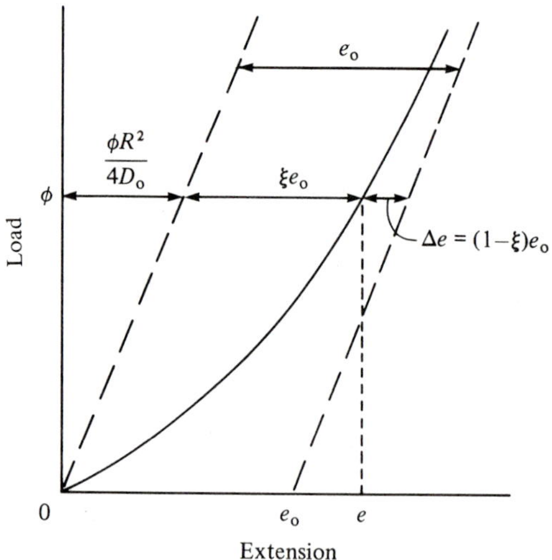

Fig. 2. Load–extension curve showing definition of ξ.

a few fibrils to deform together; second, the possible existence of a range of crimp levels within tendon (Viidik & Ekholm, 1968; Kastelic, Galeski & Baer, 1978) which could be modelled using the present results by assuming parallel coupling of fibrils of different crimp level; third, that tendon is viscoelastic. If this originates in the fibrils, the present results would apply only to stress–relaxation experiments, whereas the data are from constant strain–rate tests. Alternatively, if the viscoelasticity originates in the amorphous ground substance (Cohen, Hooley & McCrum, 1976) the present results would apply only to stress–relaxation tests after long times.

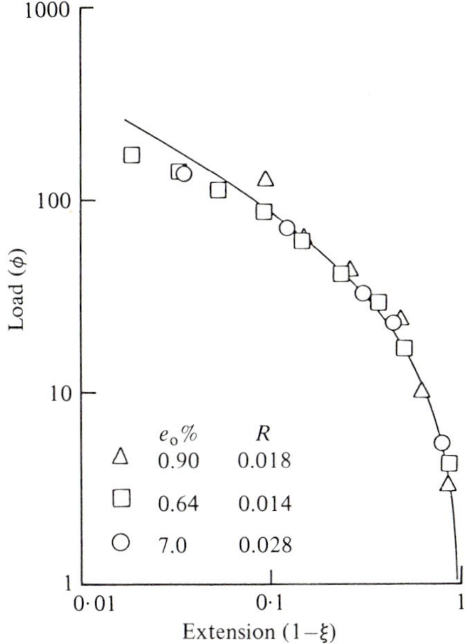

Fig. 3. Fit of load–extension data for tendon by eye to the calculated curve for $q=0$, $R=0.01$, $D_0=0.97$. Triangles, data of Partington & Wood (1963); squares, Elden (1968); circles, Yannas & Huang (1972).

References

COHEN, R. E., HOOLEY, C. J. & MCCRUM, N. G. (1976). Viscoelastic creep of collagenous tissue. *Journal of Biomechanics,* **9,** 175–84.

ELDEN, H. R. (1968). Physical properties of collagen fibres. *International Review of Connective Tissue Research,* **4,** 283–348.

KASTELIC, J., GALESKI, A. & BAER, E. (1978). The multicomposite structure of tendon. *Connective Tissue Research,* **6,** 11–23.

PARTINGTON, F. R. & WOOD, G. C. (1963). The role of non-collagen components in the mechanical behaviour of tendon fibres. *Biochimica et Biophysica Acta,* **69,** 485–95.

TORP, S., BAER, E. & FRIEDMAN, B. (1975). Effects of age and of mechanical deformation on the ultrastructure of tendon. In *Structure of Fibrous Biopolymers*, ed. E. D. T. Atkins & A. Keller, pp. 223–50. London: Butterworth.

VIIDIK, A. & EKHOLM, R. (1968). Light and electron microscopic studies of collagen fibres under strain. *Zeitschrift für Anatomie und Entwicklungsgeschichte,* **127,** 154–64.

YANNAS, I. V. & HUANG, C. (1972). Fracture of tendon collagen. *Journal of Polymer Science A-2*, **10,** 577–84.

THE CUVERIAN TUBULES OF HOLOTHURIA: DESIGN FOR SUCCESSFUL FAILURE IN A COLLAGENOUS SYSTEM

*L. J. GATHERCOLE, A. J. BAILEY, *J. DLUGOSZ and A. KELLER*

H. H. Wills Physics Laboratory, University of Bristol, Bristol BS8 1TL, UK and
*Meat Research Institute, Langford, Nr. Bristol BS18 7DY, UK

Many species of Holothurians possess Cuvierian tubules or organs. These can be everted to form a sticky highly extensible mass of uniaxial collagen fibrils, used for food capture or entanglement of predators. The organization of the collagen fibrils and polysaccharide matrix is unusual. While the intermolecular crosslinking of the collagen is typical of a strong connective tissue, in this structure it is associated with extreme tensile weakness. Rupture loads are very low, of the order of 0.5 g. Fig. 1 shows a comparison of stress–strain curves for tubules of *Holothuria forskäli* with the data of Diamant, Keller, Baer, Litt & Arridge (1972) for rat tail tendons (RTT) of comparable diameters.

The failure stress of the tubules is only 10^{-4} times the stress at the limit of the 'toe' region for RTT. The fibre has a long failure zone before final rupture. In contrast, the collagen is highly insoluble and contains the acid- and heat-stable crosslink hydroxylysino-S-keto-norleucine in quantities comparable to young cartilage or bone.

The likely explanation is found in the morphology of the tubules. In the intact state the collagen fibrils are in convoluted sheets, packed in a way reminiscent of chitin (e.g. Neville, 1975): an analogue on the fibril level of molecular cholesteric liquid crystals (Dlugosz, Gathercole & Keller, 1979). Transmission electron microscopy (TEM) and low-angle electron diffraction reveal fibrils of small diameters and of irregular indented outline in transverse section, well separated by matrix. TEM of sections and dispersed tubules, stained with uranyl acetate, ruthenium red and cationized ferritin show large amounts of staining material between the fibrils, often stranded, with bands of overlying material round the fibrils at intervals of 68 nm. The tubules contain large amounts of uronic acid (probably derived from chondroitin sulphate), about 32 times the content in RTT.

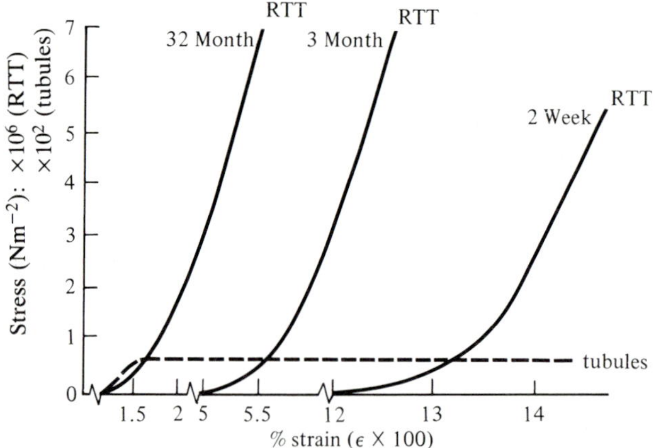

Fig. 1. Stress–strain curves for *H. forskäli* (broken line) and RTT (full lines), the latter from Diamant *et al.* (1972).

A model for the failure might be a local shearing and rotation of the liquid crystalline sheets, leading to a neck of uniaxial fibrils which then flow past each other, lubricated rather than restrained by the abundant matrix.

References

Diamant, J., Keller, A., Baer, E., Litt, M. & Arridge, R. G. C. (1972). Collagen: ultrastructure and its relation to mechanical properties as a function of ageing. *Proceedings of the Royal Society of London, Series B,* **180,** 293–315.

Dlugosz, J., Gathercole, L. J. & Keller, A. (1979). Cholesteric analogue packing of collagen fibrils in the Cuverian tubules of *Holothuria forskäli* (Holothuroidea, Echinodermata). *Micron,* **10,** 81–7.

Neville, A. C. (1975). *Biology of the Arthropod Cuticle* pp. 171–5. Berlin–Heidelberg: Springer.

CREEP TESTING OF ISOLATED CERVIX FROM PREGNANT RATS

M. HOLLINGSWORTH, S. GALLIMORE, and L. M. WILLIAMS

Departments of Pharmacology, Materia Medica and Therapeutics and Child Health, Medical School, University of Manchester, Manchester M13 9PT

The isolated cervix of the non-pregnant rat extends only slightly when a radial load is applied. By contrast the cervix of the term pregnant rat must extend considerably to allow foetal delivery. The mechanical properties of the isolated rat cervix have been measured using a creep test (Hollingsworth & Isherwood, 1977; Hollingsworth, Isherwood & Foster, 1979). Inner circumference increased at a linear rate with time after an initial curve following application of the load. Cervical creep rate was defined as the fractional increase in circumference per min, with units of min^{-1}, during the linear rate of extension. The initial curve was not accurately measured by this technique. The cervix of the non-pregnant rat was extended only slightly during 100 min under load, creep rate had increased eightfold by day 22 of pregnancy (term) but was similar to values from non-pregnant rats by day 1 *post partum*. At term the cervix exhibited strains of greater than 100% before rupture. It is likely that the measured mechanical properties do not involve active tension developed by the small amount of smooth muscle within the tissue as creep rate measured on day 18 or 22 was unaltered in the presence of isoprenaline (10^{-4}M) following 30 min incubation. Measurements of uterine motility in conscious rats by means of intra-uterine latex balloons preceding and during delivery show that contractions occurred with a frequency of about one per minute. This intermittent stress applied to the cervix has been mimicked *in vitro* by application of the load in a 'one min on, one min off' cycle (Fig. 1), now using a linear variable differential transducer (Sangamo Transducers) and a potentiometric pen-recorder (Rikadenki Ltd.) to record cervical extension. First application of the load produced an immediate extension followed by a non-linear rate of extension, successive loads produced linear rates of extension as described above. Following removal of the load there was a decrease in circumference, initially rapid, and then a second, slower, phase. The results demonstrate that the mechanical properties

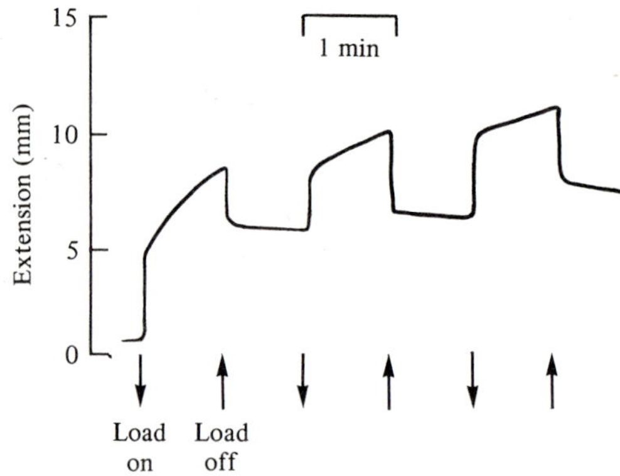

Fig. 1. Changes in extension of a cervix from a day 22 pregnant rat following one minute loading and unloading cycles.

of the rat cervix can be described by elastic and viscoelastic components and their magnitudes increase in late pregnancy.

The support of the Lalor Foundation, USA is gratefully acknowledged. L.M.W. is an SRC (CASE) student with ICI Pharmaceuticals.

References

HOLLINGSWORTH, M. & ISHERWOOD, C. N. M. (1977). Changes in the extensibility of the cervix of the rat in late pregnancy produced by prostaglandin $F_{2\alpha}$, ovariectomy and steroid replacement. *British Journal of Pharmacology*, **61,** 501–502*P*.

HOLLINGSWORTH, M., ISHERWOOD, C. N. M. & FOSTER, R. W. (1979). The effects of oestradiol benzoate, progesterone, relaxin and ovariectomy on cervical extensibility in the late pregnant rat. *Journal of Reproduction and Fertility,* **56,** 471–7.

BIOLOGICAL THIXOTROPY – THE UNIFYING FACTOR IN BASEMENT MEMBRANE FUNCTION

L. O. SIMPSON

Pathology Department, University of Otago, New Zealand

On the basis of observations on the protoplasm of sea-urchin larvae, Peterfi (1927) applied the term 'thixotropy' to the sol–gel–sol transformation which he observed during cell movement. Thus the term was defined as a reversible, pressure-induced, isothermal reduction in viscosity. Although Peterfi expressed the view that thixotropy was a phenomenon of widespread biological significance, it was not until 1964 that thixotropic properties were ascribed to the basement membranes of glomerular capillaries (Menefee, Mueller, Bell & Myers, 1964). With the development of biorheology as a separate discipline, various components of the mammalian body, such as whole human blood, living human skin and synovial fluid have been shown to satisfy biophysical criteria of thixotropy.

Basement membranes, probably as fully hydrated gels, subtend epithelial and endothelial cell layers forming boundaries through which, in the case of epithelia, all substances must pass when entering or leaving the cell layer via the blood stream. Basement membranes may be produced by cells derived from any of the three primary tissue layers. Despite this diversity of origin there is general agreement that basement membranes are composed of Type IV collagen and glycoproteins, and in general rely upon the Type I and III collagen in adjacent or investing connective tissue for their tensile strength. Amino-acid analysis reveals differences in composition of the same basement membrane in different species, and differences in the proportion of amino acids in different basement membranes of the same species.

Despite this diversity of origin and chemical composition, basement membranes share similar functional characteristics. Thus they act as selectively permeable, self-repairing tissue boundaries capable of being pierced by cell processes and traversed by substances up to approximately 68 000 daltons molecular weight under normal conditions. In abnormal circumstances when larger than normal molecules cross basement membranes, the functional efficiency of the membrane is unchanged when the system returns to normal, without apparent change in ultrastructure. Motile cells are able to pass through basement membranes without leaving ultrastructural evidence of their passage. It is proposed that large molecules, cell processes, motile cells and motile organisms which either by virtue of increased blood pressure or as a result of intracellular metabolic processes are able to exert sufficient energy to induce localised deformation and flow in the basement membrane gel, would be able to pass through the membrane. With the removal of the source of external energy the gel would reform by the random approximation of displaced lattice units through the agency of Brownian movement. Thus in contrast to biophysical concepts of thixotropy in which viscosity is reduced by shearing, biological thixotropic systems undergo localised reduction in viscosity under normal stress with penetration being the most common feature. It has been proposed that renal glomerular basement membranes are thixotropic systems (Simpson & Batten, 1975). In a mouse model of spontaneous immune complex glomerulonephritis it has been shown (Simpson, 1980) that there is a time-related association of increasing systemic blood pressure, increasing total urinary protein levels and increasing molecular size of urinary proteins. These relationships had been predicted on the basis that the renal basement membrane was a biological thixotropic system. The migration of duodenal epithelial cells with basal processes which penetrate basement membrane, so also explicable on the basis of biological thixotropy of the basement membrane (Partridge & Simpson, 1980). A hypothesis that all basement membranes are biological thixotropic has been formulated (Simpson, 1980*b*).

It is apparent (Peterfi, 1927; Bauer & Collins, 1967) that current usage of the term 'thixotropy' is rheologically oriented and much broader than the original concept, although it is not possible to apply biophysical criteria to living systems. For this reason it seems prudent to distinguish between biological and non-biological thixotropic systems.

References

BAUER, W. H. & COLLINS, E. A. (1967). Thixotropy and dilatancy. In *Rheology theory and applications*, vol. 4, ed. F. R. Eirich, pp. 423–60. N.Y. and London: Academic Press.

MENEFEE, M. G., MUELLER, C. B., BELL, A. L. & MYERS, J. K. (1964). Transport of globulin by the renal glomerulus. *Journal of experimental Medicine,* **120,** 1129–38.

PARTRIDGE, B. T. & SIMPSON, L. O. (1980). Duodenal epithelial cell migration and loss in NZB mice. *Micron*, **11,** 63–72.

PETERFI, T. (1927). Die Abhebung der Befructungsmembran bei Seeigeleiern. *Archives für Entwicklungsmechanik,* **112,** 660–95.

Simpson, L. O. (1980*a*). A mouse model of spontaneous renal hypertension. Blood pressure, heartweight, kidneyweight and proteinuria relationships in NZB X OUW F_1 hybrid female mice. *Pathology*, **12** (in press).

Simpson, L. O. (1980*b*). Basement membranes and biological thixotropy. *Pathology*, **12** (in press).

Simpson, L. O. & Batten, E. H. (1975). A functional re-appraisal of glomerular ultrastructure (Abstr). *Journal of Anatomy*, **120**, 613.

THE HARDNESS OF LOCUST INCISORS

J. E. HILLERTON

Biomechanics group, Department of Zoology, The University, Whiteknights, Reading RG6 2AJ, UK

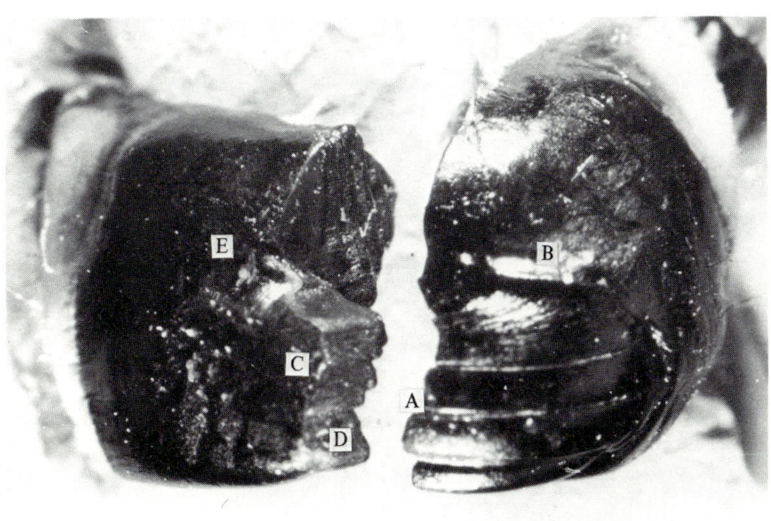

Fig. 1. Mandibles from adult *Locusta migratoria migratorioides* viewed from the exterior in cutting position. This shows the wear of the right mandible resulting in its angled shape.

The mandible of the locust has two functions, biting and chewing, and separate parts of the structure are adapted for these.

In the biting motion the left mandible shears across the ventral surface of the right mandible rather like the action of a pair of scissors. This results in the inside of the left incisor and the outside of the right incisor being worn away (Figs. 1, 2). The resultant wear leaves a prominent edge to the mandibles. The hardnesses of the surfaces involved in cutting and of adjacent cuticle have been determined. The cutting surfaces have a hardness twice that of the sheared surfaces (Table 1). Thus the shearing action wears away the softer parts of the incisor and produces a self-sharpening pair of scissors.

Fig. 2. Locust mandibles viewed as from within the mouth. The wear on the underside of the left mandible is shown.

Table 1. *Hardness of locust mandibles*

Region of mandible		Identification	Hardness ($kg\ mm^{-2}$) Mean	s.e.
Left	Cutting edge	Fig. 1 A	36.4	1.3
	Sheared face	Fig. 2 F	Cannot be measured because of its shape	
	Base	Fig. 1 B	19.7	0.9
Right	Cutting edge	Fig. 2 G	32.2	0.8
	Trailing edge	Fig. 1 C	20.2	0.9
	Sheared face	Fig. 1 D	18.2	0.8
	Base	Fig. 1 E	19.7	0.8

The extra hardness of the cutting edges appears to result from a different form of sclerotization from that of the rest of the mandible. The cutting edges are very heavily sclerotized caps resistant to boiling methanol:chloroform for 24 h, boiling M NaOH for 5 h and boiling M HCl for 2 h. Also they are partly sclerotized before the adult locust moults, unlike the rest of the mandible.

The hardness measured varies with applied load and surface architecture of the specimen. This variation has been eliminated as far as possible in these measurements.

DESIGN AND MATERIALS OF FEATHER SHAFTS: VERY LIGHT, RIGID STRUCTURES

D. G. CRENSHAW

Department of Zoology, Duke University, Durham, North Carolina 27706, USA

The shafts of primary feathers resist the aerodynamic loads imposed by flight. The broad design criteria for feather shafts are: (1) Do not bend excessively. (2) Do not break. (3) Be as light as possible. Feathers meet these criteria by means of 'foam sandwich' construction. The cortex of the shaft forms a tube of compact keratin with approximately rectangular cross section. The medulla is filled with spongy keratin.

Several material properties of the shafts of pigeon primary feathers were determined by tensile tests at slow strain rates of intact sections of the shaft, of sections of the shaft with the medulla removed, and of strips of compact keratin from either the dorsal or ventral surfaces. Tensile stress–strain curves show an initial linear region followed by a region of lower slope, followed by a steeply ascending region. Mean strain at failure was 12.5%. Mean stress at failure was 226 $\mathrm{MNm^{-2}}$. The mean elastic modulus in the early part of the stress–strain curve for compact keratin was 2.52 $\mathrm{GNm^{-2}}$. Whole shafts showed lower modulus and stress at failure than did altered shafts. If the cross-sectional area of the medulla is discounted in calculating the stress on the whole shaft, the values for the modulus and stress at failure rise to about the values for compact keratin. These data indicate that the spongy keratin contributes little to the tensile strength of the shaft.

To examine the role of spongy keratin in bending, intact shafts and shafts with the medulla removed were tested to failure in three-point bending with the dorsal surface in compression, thus partially simulating the direction of in-flight loads. Failure was always by local buckling of the compression surface. Mean buckling stress at failure was 0.24 $\mathrm{GNm^{-2}}$ for both whole and altered shafts.

Among rigid biological materials, compact keratin has a very high stress at failure but a rather low modulus (Table 5.3, Wainwright, Biggs, Currey & Gosline, 1976). A feather shaft is very light, given its bending strength, as shown by a low ratio of density of the material to stress at failure. For intact primary feather shafts, this value is 1.77 $\mu\mathrm{s^2\,m^{-2}}$. For shafts with the medulla removed, this value rises to 3.65 $\mu s^2\ m^{-2}$. The spongy keratin therefore makes a slight but discernable contribution to the performance of the shaft. The lowest previously reported value for this density : stress ratio was 5.1 $\mu s^2\ m^{-2}$ for Swedish Pine. The column buckling parameter (material density divided by the square root of the elastic modulus) for compact keratin is 15.0 ms $kg^{1/2}\ m^{-5/2}$. This value is unexceptional compared to other biological materials (Wainwright *et al.*, 1976).

The material properties of keratin show it to be a strong, light substance. The arrangement of hard and soft keratins in the shaft provides superior weight to strength ratio. Buckling and tensile failure occur at similar stresses, indicating that the shaft is not 'overbuilt'. Purslow & Vincent (1978) have shown that the removal of the medulla reduces the flexural stiffness in dorso-ventral bending by about 16%. This may or may nor represent excessive bending for a flying pigeon. Failure of the shaft in flight may not be the onset of local buckling but the deformation of the shaft beyond the elastic region of the stress–strain curve. Detailed study of the aerodynamics of bird wings will allow better estimates of the forces acting on feathers and hence sharpening of the design criteria.

References

Purslow, P. P. & Vincent, J. F. V. (1978). Mechanical properties of primary feathers from the pigeon. *Journal of experimental Biology,* **72,** 251–60.

Wainwright, S. A., Biggs, W. D., Currey, J. D. & Gosline, J. M. (1976). *Mechanical Design in Organisms.* New York: Halstead Press.

SMALL-SCALE TENSILE TESTS

ROBERT F. KER

A.R.C. Unit, Dept. of Zoology, Oxford University
(Present address: Dept. of Zoology, The University, Leeds, LS2 9JT, UK)

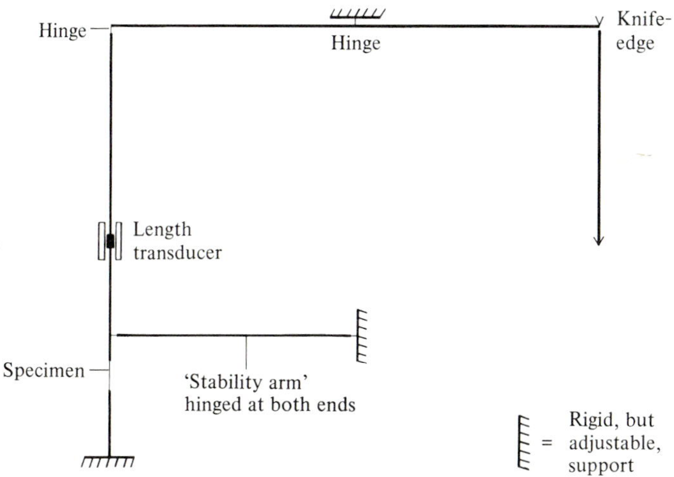

Fig. 1. Principle of the tensile testing machine. The inclusion of the 'stability arm' and the use of steel-strip hinges defines the alignment.

Biological materials of uniform structure are often available only as small samples. Even with large samples (e.g. vertebrate tendon) it may be useful to test a small part (e.g. a fascicle). I have designed a tensile creep testing machine, appropriate for small polymer samples (down to about 1 mm $\times$ 10^{-3} mm^2). The load, up to about 20 N, can be applied smoothly in <0.1 s. The measurement of extension is sufficiently reliable to allow work with the more rigid biomaterials.

Figure 1 shows the principle of the machine, which follows that used, on a larger scale, by Dr N. G. McCrum and others in the Engineering Department of Oxford University. My machine includes features to allow the precise alignment required for small samples. Alignment can be checked by viewing the specimen through an optical microscope from a range of directions. The specimen is contained in a temperature-controlled bath in an appropriate fluid. It is held between clamps (Fig.

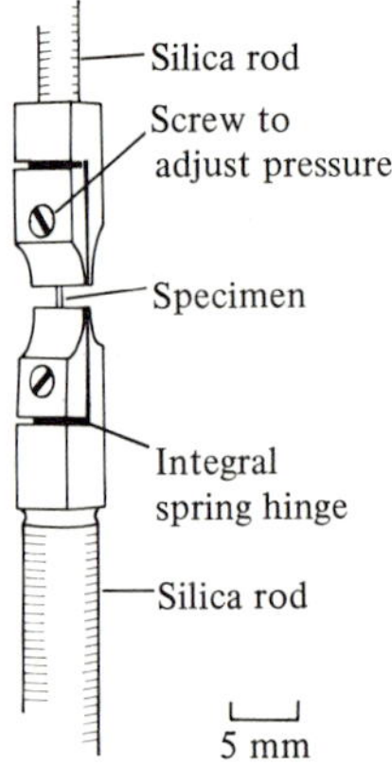

Fig. 2. Clamps.

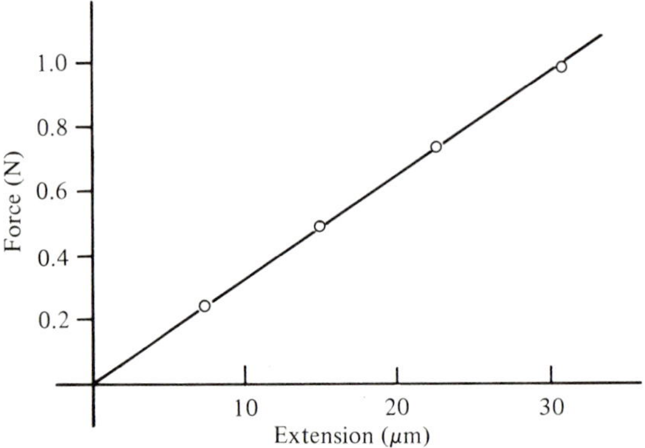

Fig. 3. Force v. extension for an adult, female locust tibial flexor apodeme. Length, 4.78 mm; cross-sectional area, 0.0090 mm^2.

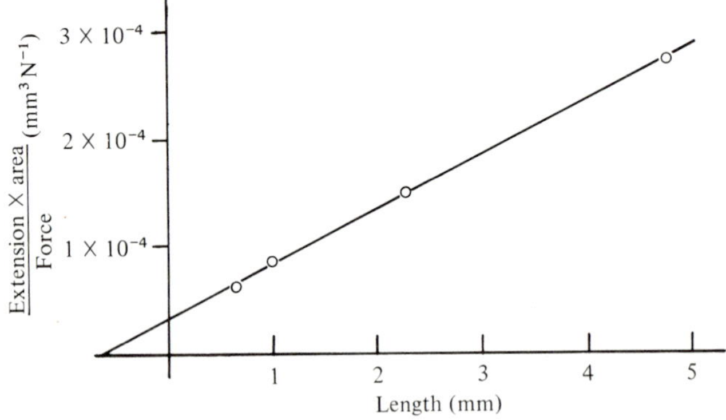

Fig. 4. Correction for end-effects. The intercept on the length axis is the 'end-effect', which is acceptably small and well-defined using these clamps and this material. Young's modulus (given by the reciprocal of the slope), 20 GN m^{-2}.

2) made of phosphor-bronze, each having one jaw shaped to allow bending, as with fine forceps. The integral construction of the spring hinge ensures that the jaws are always in register.

Figures 3 and 4 show an example of the results obtained. The length transducer output gives change in length as a function of time for a fixed load. Each extension in Fig. 3 was measured 5 s after loading. To check that the system is operating correctly, it is important to make measurements at more than one length with the same specimen. This also allows the 'end-effect' to be eliminated in the way shown by Fig. 4.

The machine was made in Oxford by Mr R. Cheney, Mr A. Price and Mr A. R. W. Savage, whose skill I gratefully acknowledge.

AUTHOR INDEX

Figures in bold type indicate pages on which references are listed

Abendschein, W., 155, **163**
Abrahams, M., 58, **72**
Adams, D., 389, **394**
Adams, D. F., 46, 47, **74**
Adler, L., 157, **163**
Alaketola, L., 385, **394**
Alderman, B., 42, **72**
Alexander, N. J., 213, 220, **244**
Alexander, R. McN., 292, 293, 299, **312;** 341, 349, **354;** 449, **453**
Alfrey, T., 301, **312**
Allen, A., 360, 361, 366, 369, 370, **374, 375, 376**
Allen, E. R., 425, **434**
Allen, T. D., 230, **240**
Alley, V. L., 52, **72**
Ambardar, A., 156, **163**
Amberson, W. R., 131, **134**
Andersen, S. O., 183, 184, 198, 199, 201, **207;** 259, **270;** 333, 335, 336, 341, 343, 349, **354, 355, 357**
Anderson, J. C., 414, **433**
Andreasen, J. O., 124, 125, **133**
Andrews, E. H., 13–35; 18, 19, 21, 22, 25, 26, 27, 32, 33, 34, **34, 35**
Ansell, M. P., 71n, **72;** 176, **180**
Apter, J. T., 307, **312**
Arends, J., 120, **133**
Armeniades, C. D., 400, 402, 403, 429, **435**
Arrhenius, S., 310, **312**
Arridge, R. G. C., 400, 402, 403, 429, 430, **434, 435;** 475, **476**
Ascenzi, A., 141, 144, 150, **163**
Ashby, A. M., 465–9
Astbury, W. T., 213, 218, 220, 221, 222, 233, 236, **240, 241**
Atack, D., 175, **181**
Atkins, A. G., 202, **208**
Aveston, J., 69, **72**
Azzi, V. D., 59, **72**

Baden, H. F., 213, 220, 227, 228, 235, 236, **241**
Bader, M., 61, **72**
Baer, E., 41, **72;** 332, **356;** 397–435; 397, 400, 402, 403, 404, 408, 409, 414, 422, 425, 429, 430, **433, 434, 435;** 468, **469;** 471, 472, **473;** 475, 476, **476**
Bailey, A. J., 399, **435;** 475–6
Banfield, W. G., 399, **433**
Bang, S., 149, **164**
Banky, E. C., 223, **241**
Barbenel, J. C., 465–9; 468, 469, **469**
Barber, N. E., 318, **329**
Barenberg, S. A., 399, 400, **433**
Bargren, J. H., 149, 152, **163**
Barth, F. G., 183, 184, **207**
Bashey, R. I., 415, **434**
Bassett, C. A., 149, 152, **163**
Batten, E. H., 480, **481**
Bauer, W. H., 480, **480**
BBC, 461, **461**
Bell, A. L., 479, **480**
Bear, R. S., 397, 401, 404, 406, **433, 435**
Becker, H., 150, **163**
Beedham, G. E., 93, **96**
Beighton, E., 261, 262, **270**
Bendit, E. G., 220, 221, 224, 227, 228, 233, 234, 235, 236, 238, **241**
Bennet-Clark, H. C., 205, 206, **207;** 333, 349, **355;** 449, **453**
Bennett, C. P., 177, **181**
Benninghoff, A., 381, **394**
Bentley, J. P., 425, **434**
Bergel, D. H., 307, **312;** 351, 352, 353, **355**
Beskos, D. E., 468, **469**
Bevan, B., 149, **163**
Bhargava, A. S., 362, **375**
Bhaskar, K. R., 361, **374**
Bhimasenachar, J., 96, **96**
Bigelow, C. C., 199, **207**
Biggs, W. D., 42, 72, **74;** 189, 207, **210;** 220, 221, 222, 234, **245;** 249, 250, 251, 253, 265, **271;** 334, **357;** 444, 453; 485, **486**
Billington, E. W., 18, 19, 21, **35**
Birbeck, M. S. C., 223, **241**
Bird, F., 150, **163**
Bizzell, J. W., 428, **433**
Blacik, L. J., 428, **433**

Black, G. V., 109, 110, **133**
Black, J., 152, **163**
Black, M. M., 42, **72**
Blackwell, J., 183, **209**
Blakey, P. R., 235, 238, **241, 242**
Blank, I. H., 226, **241**
Board, R. G., 459–61
Boas, W., 155, **163**
Bøggild, O. B., 76, **96**
Boldman, O. E. A., 404, 406, **433**
Boltzmann, L., 294, 296, **312**
Bonar, L. C., 220, **241**
Bond, G. M., 459–61
Bonfield, W., 150, 151, 153, 155, 157, 161, 162, **164**
Bonucci, E., 141, 144, 150, **163**
Borcherding, M. S., 428, **433**
Borysko, E., 430, **433**
Bothwell, J. W., 227, **245**
Bouteille, M., 420, **433**
Bowen, R., 112, 113, **133**
Bowyer, W. H., 61, **72**
Boyd, J. D., 174, **181**
Boyde, A., 124, **135;** 141, **164, 165**
Bradbury, L. H., 220, **241**
Braden, M., 114, 115, 116, 118, 119, **133, 135**
Braun, F., 289, **312**
Braun-Falco, O., 420, **433**
Breen, M., 428, **433**
Brear, K., 71, **72**
Brenden, B. B., 117, **135**
Bridgen, J., 219, **245**
Brody, I., 213, 226, 230, **241**
Broutman, L. J., 398, **433**
Brunet, P. C. J., 196, **208**
Bruning, D. F., 461, **461**
Bruns, R. R., 415, **433**
Brezinski, J. P., 176, **181**
Brown, C. R., 57, **73**
Buckley, C. P., 471–3
Bueche, F., 348, **355**
Buggy, J., 318, **328**
Burstein, A. H., 146, 149, 151, **164, 167**
Burton, A., 353, **356**

Caldwell, R., 121, **133**
Capaldi, R., 199, **208**
Carriker, M. R., 95, **96**
Cavagna, G., 449, **453**
Cave, I. D., 170, 172, **181**
Centeno, J. P., 175, **182**
Chandler, H. D., 185, 186, 188, 189, 190, 191, 192, 203, **208;** 249, 250, 252, 253, 254, 256, 264, **271**
Chaplin, C. R., 207
Chapman, B. M., 220, 224, 233, 234, **241, 242**
Chappell, D. J., 386, **395**
Charrier, J. M., 281, **287**
Chen, P. E., 49, **73**
Clamp, J. R., 361, **374, 375**
Clark, E. A., 150, 151, **164**
Cleland, R. E., 315, 318, **328**
Cobbold, R. S. C., 157, **165**
Cohen, C., 265, **271**
Cohen, J., 140, **164**
Cohen, R. E., 295, 310, **312, 313;** 472, **473**
Coles, B. C., 196, **208**
Collins, E. A., 480, **480**
Cook, J., 8, **11;** 66, **72;** 84, **96;** 130, **133**
Cooke, E. W., 71, **72**
Cooke, R. G., 459–61
Cooper, R. R., 140, **164**
Cooper, W. E. G., 115, 116, **133, 135**
Corey, R. B., 258, 260, **271**
Corten, H. T., 54, 55, **72**
Cottrell, A. H., 172, **181**
Cottrell, C. B., 197, **208;** 438, **453**
Cowan, P. M., 410, **433**
Cowdrey, D. R., 174, **181;** 319, **328**
Cox, H. L., 4, **11;** 52, 59, **72;** 155, **164**
Cox, J. L., 428, **433**
Craig, R. G., 109, 110, 111, 112, 121, 125, 132, **133, 135;** 152, **164**
Craig-Gray, J., 457, **458**
Crane, H. R., 397, **433**
Craven, J. D., 155, **165**
Creath, J. M., 361, 364, **374, 375**
Crenshaw, D. G., 485–6
Crewther, W. G., 216, 225, **242, 243**
Crick, F. H. C., 215, **242**
Crowninshield, R. D., 151, **164**
Cummins, W. R., 318, **328**
Currey, J. D., vii–ix; 42, 71, 72, **72, 74;** 75–97; 78, 81, 86, **96;** 149, 150, 151, 152, 154, 155, 157, 159, 162, **164;** 189, 207, **210;** 220, 221, 222, 234, **245;** 249, 250, 251, 253, 265, **271;** 334, **357;** 444, 452, **453;** 485, **486**

Dale, B. A., 215, **242**
Dale, W. C., 400, 421, 422, 430, **433**
D'Arcy, R. L., 200, **207**
Darnbrough, G., 176, **182**
Das, I., 361, **374**
Datta, P. K., 151, **164**
Davidoff, M. R., 249, 250, 252, 253, 254, 256, 264, **271**
Davidson, C. L., 120, **133;** 397, 412, **434**
Davies, G. J., 61, **73**
Davis, W. L., 146, 151, **167**
De Nee, P. B., 457, **458**
Debaise, G., 175, **181**
Debevoise, N. T., 155, **165**
Deinse, A. van, 260, **271**
Dempster, W. T., 149, **164**
Dennell, R., 188, 197, **208**

Denny, M. W., 247–71; 249, 250, 251, 252, 253, 254, 256, 266, 269, **270;** 334, **355;** 359, 361, 364; 449, 450, **453**
Denton, E. J., 93, **96**
Devas, M. B., 385, **394**
Diamant, J., 475, 476, **476**
Dibdin, G. H., 131, **133**
Dimmock, J., 58, **72**
Dinwoodie, J. M., 173, 174, **181**
DiSalvo, N. A., 109, **134**
Dlugosz, J., 475–6; 475, **476**
Dobb, M. G., 216, **242;** 260, **270**
Doner, D. R., 46, 47, **74**
Donkelaar, A. van, 233
Doolittle, A. K., 310, **312**
Dopheide, T. A. A., 217, **242**
Dorrington, K. L., 117–18, **133;** 152, **164;** 201, 204, 205, **208;** 289–314; 293, 311, **312, 313;** 316, 321, 323, **328;** 346, 347, **355**
Doty, P., 301, **312**
Douglass, A. B., 227, **245**
Dowling, L. M., 216, 225, **242**
Dowson, D., 386, **394**
Druhala, M., 227, 228, 232, **242**
Duncanson, M. G., 125, 126, **133, 134**
Dunn, K. L., 130, **134;** 156, **165**

Earland, C., 238, **242**
Eberstein, A., 301, **314**
Ebner, V. von, 141, **168**
Eckold, G. C., 59, **72, 74**
Edwards, D. C., 96, **97**
Edwards, P. A. W., 372, **375**
Eguchi, G., 332, **356**
Ekholm, R., 410, 415, **435;** 472, **473**
Elden, H. R., 430, **433;** 471, 473, **473**
Elder, H. Y., 332, **355**
El-Hosseiny, F., 172, 176, **181**
Elias, P. M., 226, **242**
Elleman, T. C., 217, **242**
Elliott, D. H., 431, **433;** 467, **469**
Ellis, C. D., 69, **72**
Ellis, G. E., 338, **355**
Elöd, E., 222, **242**
Elstein, M., 360, **375**
Encyclopaedia Britannica, 266, **270**
Enamoto, S., 221, **242**
Erickson, R. O., 318, **328**
Evans, F. G., 149, 150, **164, 165, 168**
Evans, J. H., 465–9, 468, 469, **469**
Evert, R., 151, **167**
Eylers, J. P., 440
Eyring, J., 325, **329**

Fairhurst, C. W., 119, **135**
Faison, R. W., 52, **72**
Farrell, R. A., 428, **433, 434**
Favre, J.-P., 69, **72**
Fehr, F. R. von der, 120, **134**
Feltham, P. F., 301, **312**
Ferran, E. M. de, 62, **73**
Ferri, C., 341, **356**
Ferris, C. D., 156, **163**
Ferry, J. D., 126, **134;** 309, 310, **314;** 349, **355**
Feughelman, M., 220, 221, 223, 224, 227, 228, 232, 233, 234, **241, 242, 243**
Filisko, F. E., 399, 400, **433**
Filshie, B. K., 218, **242**
Fisher, F. M., 333, 336, **356**
Flioriani, L P., 155, **165**
Florey, H. W., 372, **375**
Flory, P. J., 287, **287;** 334, 339, 340, 341, 343, 346, **355**
Foley, D. E., 119, **135**
Forslind, B., 236, **242**
Foster, J. A., 336, 337, 338, 341, **355**
Foster, R. W., 477, **478**
Fox, F. R., 184, **209**
Foye, R. L., 62, **73**
Fraenkel, G., 184, 188, 191, 193, 196, 197, **208, 210**
Frankel, V. H., 149, 151, **164, 167**
Frankel, W. H., 146, 151, **167**
Frasca, P., 141, 142, 143, 144, 145, 146, 147, 156, 159, **165**
Fraser, R. A. W., 29, **35**
Fraser, R. D. B., 211–45; 213, 215, 216, 217, 218, 219, 220, 223, 224, 225, 235, **242, 243, 244;** 260, 262, **270;** 397, **433;** 437, **453**
Freeman, M. A. R., 379, 386, 391, **394, 395**
Frei, E., 316, 321, 322, **328**
French, C. J., 305, 306, 307, 310, **313;** 347, 351, **355**
Frenkel, M. J., 219, 220, 225, 234, **243**
Friedman, B., 400, 402, 403, 408, 409, 414, 421, 422, 429, **435;** 471, **473**
Friedman, R., 155, **165**
Friend, D. S., 226, **242**
Fristrom, J. W., 184, **208**
Fruhauf, V., 81, **97**
Fuchswans, W., 414, **435**
Fukahori, Y., 19, 22, 26, 27, **35**
Furthmayer, H., 414, **435**
Fushitani, M., 301, **313**

Galeski, A., 397, 400, 403, 404, 425, **433, 434;** 468, **469;** 472, **473**
Galileo, G., 237, **243**
Gallimore, S., 477–8
Garcia, B. J., 157, **165**
Gardiner, B. C., 207, **208**
Gathercole, L. J., 397, 425, **433;** 475–6; 475, **476**
Gebhardt, W., 141, **165**

Geddes, A. J., 261, 262, **270**
Gee, G., 279, **288**
Geil, P. H., 399, 400, 415, **433, 435**
Gelman, R. A., 359, **376**
Gent, A. N., 281, **287**
Geoghegan, T., 189, 192, 196, **209**
Gibbons, R. A., 360, 362, 364, **375**
Gibson, T., 468, **469**
Gilboa, A., 359, **376**
Gillam, E., 81, **97**
Gillespie, J. M., 219, 220, 225, 234, 235, **241, 243**
Gillis, P. P., 171, **181**
Gilman, J. J., 82, **97**
Gilmore, R. S., 156, **165**
Gilmore, R. W., 117, 121, 130, **133, 134**
Gilpin-Brown, J. B., 93, **96**
Giroud, A., 213, **243**
Gjelsvik, A., 149, 152, **163**
Glagov, S., 353, **357**
Glaser, A. E., 200, **208**
Glimcher, M. J., 138, 158, **165**
Goatham, J., 49, **73**
Goerke, J., 226, **242**
Goldberg, M., 184, 195, 196, **208**
Goldsmith, L. A., 227, 228, 236, **241**
Goldsmith, W., 151, **166**
Gordon, J. E., 1–11; 4, 8, **11;** 66, **72;** 84, **90;** 130, **138;** 170, 177, **181;** 206, **208;** 270, **270;** 332, **355**
Gosline, J. M., 42, 72, **74;** 185, 189, 196, 207, **208, 210;** 220, 221, 222, 234, **245;** 249, 250, 251, 253, 265, **271;** 305, 306, 307, 310, 311, **312, 313;** 331–57; 332, 334, 335, 341, 342, 346, 347, 351, **355, 357**
Gotte, L., 307, 309, 310, **313;** 347, **355**
Gottschalk, A., 362, **375**
Gough, K. H., 216, **243**
Granit, R., 290, **313**
Gray, W. R., 332, 335, 336, 337, 338, 341, **355, 356**
Green, P. B., 318, **328**
Greenfield, M. A., 155, **165**
Greenslade, F. C., 184, **209**
Grégoire, C., 75, **97**
Grenoble, D. E., 130, **134;** 156, **165**
Griffith, A. A., 17, 18, **35;** 175, **181**
Griggs, L. J., 364, **375**
Griggs, P. R., 59, **72, 74**
Grogg, B., 301, **313**
Gross, J., 415, **433**
Gruning, W., 341, **357**
Grut, W., 311, **312;** 343, **355**
Grynpas, M. D., 155, 161, 162, **164**
Guild, F. J., 57, **73;** 202, **208**
Gumbrell, S. M., 279, **288**
Gunther, Th., 415, **434**
Guth, E., 282, **288**
Guy, R., 235, **241**
Hackman, R. H., 183, 184, 185, 186, 195, 196, **208**
Hagebö, T., 119, **134**
Haggith, J. W., 221, **240**
Haines, D. J., 109, 117, **134**
Hall, A., 76, 81, 93, **97**
Hall, C. E., 399, **435**
Halpin, J. C., 177, **181**
Haly, A. R., 223, **243**
Hamlin, C., 399, **433, 434**
Hancock, D. A., 96, **97**
Handa, S., 332, **356**
Hannah, C. McD., 112, 113, 115, **134**
Happey, F., 235, **241**
Hardacker, K. W., 176, **181**
Harper, R. A., 141, 142, 143, 144, 146, 156, **165, 167**
Harrup, B. S., 216, 218, 238, **243**
Harris, B., 37–74; 57, 62, 66, 67, 69, **72, 73;** 176, 177, **182;** 185, 189, 192, 200, 201, **208;** 222, **243**
Harris, W., 140, **164**
Harrison, A., 107, **134**
Hart, R. W., 428, **433, 434**
Harwood, J. A. C., 285, 286, **288**
Hashin, Z., 130, **134;** 159, **165**
Haughton, P. M., 301, 302, **313;** 321, 322, 323, 324, 325, 326, 327, **328**
Hayashi, T., 62, **73**
Hayes, R. L., 425, **434**
Hayes, W. C., 151, **168;** 311, **313**
Healer, J., 150, **163**
Hearle, J. W. S., 52, **73;** 220, 232, **244**
Hearmon, R. F. S., 82, 90, **97**
Heatley, N. G., 372, **375**
Hebrank, J., 440, **453**
Hepburn, H. R., 185, 187, 188, 189, 190, 191, 192, 202, 203, **208, 209;** 249, 250, 252, 253, 254, 256, 264, **271**
Hegdahl, T., 119, **134**
Helfet, J., 67, **73**
Helmcke, J. G., 124, **134**
Hener, A. H., 120, 122, 123, 124, **134**
Hermans, J. J., 370, **375**
Hershanov, B., 109, **134**
Hertz, H. R., 463, **464**
Highton, T. C., 415, **434**
Hill, R., 46, **73**
Hill, R. J., 184, **208**
Hillerton, E. J., 184, 185, 187, 188, 190, 191, 192, 193, 194, 195, 196, 197, 199, 200, 204, 205, 207, **208, 209, 210;** 483–4
Hobdell, M. H., 141, **164, 165**
Hoekstra, I. S., 120, **133**
Hoeve, C. A. J., 310, **313;** 341, 343, 346, 350, **355, 356**
Hoffman, A. S., 341, **355**
Holden, K. G., 364, **375**
Hollingsworth, M., 477–8; 477, **478**

Holt, C., 323, **328**
Hooley, C. J., 295, 310, **312, 313;** 472, **473**
Horton, J. R., 361, 364, **374, 375**
Huang, C., 471, 473, **473**
Huber, L., 141, **167**
Hubbard, S. J., 312, **313**
Huddleston, A. L., 155, **165**
Huebner, J. D., 96, **97**
Huggins, M. L., 280, **288**
Hughes, J., 390, **394**
Hruza, Z., 399, **434**
Hulmes, D. J. S., 399, **434**
Hulsen, K. K., 149, **165**
Hunt, S., 336, **356;** 361, **375**
Huxley, H. E., 397, **434**
Hyatt, G. W., 155, **163, 165**

Idler, W. W., 213, **245**
Iizuka, E., 251, 257, 260, **271**
Iler, R. K., 81, **97**
Inglis, A. S., 216, 218, **242, 243, 244**
Isherwood, C. N. M., 477–8; 477, **478**

Jackson, D., 430, **434**
Jackson, D. S., 414, 425, **433, 434**
Jakus, M. A., 399, **435**
Januszkiewicz, A., 369, **375**
Jenkins, E. A., 386, **395**
Jenkins, J. T., 468, **469**
Jensen, M., 186, 188, 190, 195, 197, 198, 202, 203, 205, **209;** 349, **355**
Jeronimidis, G., 64, **73;** 169–82; 170, 176, 177, **181;** 188
Joffe, J., 188, 202, **208, 209**
Johnson, D. W., 110, 111, 112, **133**
Johnson, G. R., 386, **394**
Jolma, V. H., 399, **434**
Jones, L. N., 215, 216, 224, **243**
Jones, R. M., 48, **73**
Jones, S. J., 124, **135**
Jones, W. A., 218, 220, **245**
June, R. R., 62, **73**

Kaburagi, J., 301, **313**
Kahler, G. E., 333, 336, **356**
Kakivaya, S. R., 310, **313;** 350, **356**
Kakudo, M., 257, **271;** 401, 406, 409, **434**
Kamiya, N., 318, **328**
Kampula, J. W., 109, 110, 111, **135**
Kasai, N., 257, **271;** 401, 409, **434**
Kastelic, J., 332, **356;** 397–435; 397, 400, 403, 404, 409, 425, 428, 429, 430, **433, 434;** 468, **469;** 472, **473**
Katz, J., 41, **73;** 117, 118, 128, 130, **134;** 137–68; 141, 142, 143, 144, 146, 149, 152, 153, 154, 156, 157, 158, 160, 162, **165, 166, 167, 168;** 451, **453**
Kauzman, W., 368, **375**
Kay, L. M., 259, **271**
Keiper, D. A., 152, **167**
Kellenberger, E., 141, **166, 167**
Keller, A., 397, 425, **433;** 475–6; 475, 476, **476**
Kelly, A., 37, 50, 51, 61, **73;** 176, **181;** 221, **243**
Kelly, M., 235, 236, **241**
Kelvin, W. T., 230, **243, 244**
Kempson, G. E., 381, 383, 389, **394, 395**
Kendall, K., 81, 84, **97**
Kenedi, R. M., 390, **394**
Kennedy, W. J., 76, 81, 93, **97**
Ker, R. F., 187, 188, 189, 190, 192, 193, 198, 202, 203, 204, 205, 206, **209;** 487–9
Khan, M. A., 360, 361, 362, 366, 369, **375, 376**
Khan, M. F., 207, **208**
Kimura, S., 332, **356**
King, N. E., 26, 34, **35**
Kingsbury, H. B., 152, **166**
Kinloch, A. J., 32, 33, 34, **35**
Kischer, C. W., 415, **434**
Klein, H., 131, **134**
Klier, K., 200, **210**
Kligman, A. M., 226, 227, **244**
Knott, J., 16, **35**
Ko, R., 152, **166**
Koch, J. C., 149, **166**
Koehl, M. A. R., 441, 442, 443, 444, 445, 447, 448, **453**
Koehler, A., 179, **181**
Kohn, A. J., 86, **96**
Kohn, R. R., 399, 430, **433, 434**
Komiya, T. von, 120, **134**
Koratky, G., 260, **271**
Korostoff, E., 125, 126, **133, 134;** 151, 152, **163, 166**
Kovitz, K., 369, **375**
Koyama, K., 62, **73**
Krane, S. M., 138, 158, **165**
Krenchel, H., 59, **73;** 222, **244**
Krimm, S., 221, 222, **242, 245**
Krock, R. H., 398, **433**
Kröncke, A., 120, **134**
Kubala, A. L., 150, **168**
Kuntz, I. D., 368, **375**

Lager, J. R., 62, **73**
Laird, G. W., 152, **166**
Lake, G. J., 26, 27, **35**
Lakes, R. S., 148, 151–2, 153, 154, **166**
Lam, R. H., 415
Lancaster, A. P. S., 172, **181**
Landel, R. F., 309, 310, **314**
Lang, S. B., 155, **166**
Langham, M. E., 428, **433**
Lanir, Y., 468, **469**
Layman, M., 94, **97**
Leadbetter, D., 59, **72, 74**

Learoyd, B. M., 307, **313**
Leblond, C. P., 213, **243**
Lee, L., 227, 228, 236, **241**
Lees, S., 112, 117, 118, **134;** 156, 157, **166;** 397, 412, **434**
Lehman, M. L., 112, 113, 114, **134**
Levitt, M., 199, **209**
Lewis, J. L., 151, **166**
Lewis, T. T., 107, **134**
Li, C. H., 153, 157, **164**
Liddicoat, R. T., 149, **164**
Lillie, J. H., 420, **434**
Lindley, H., 217, **242**
Lindley, P. B., 27, **35**
LingaMurty, K., 130, **134**
Lipke, H., 189, 192, 196, **209**
Litt, M., 360, 361, 362, 366, **375, 376;** 399; 475, 476, **476**
Liu, T. S., 58, **74**
Lloyd, D. W., 52, **73;** 471–3
Lockwood, P., 235, **241**
Long, M. M., 338, 341, **356**
Lopez-Vidriero, M. T., 361, **374**
Lubin, Y. D., 267, **271**
Lucas, F., 248, 249, 250, 251, 253, 255, 256, 259, 260, 262, 264, 266, **271**
Lucas, P. W., 463–4
Lucey, E. C. A., 333, 349, **355**
Lugassy, A. A., 151, 161, 162, **166**

MacCallum, D. K., 420, **434**
McCrum, N. G., 293, 295, 307, 310, 311, **312, 313;** 343, **355;** 472, **473**
McCullogh, R. L., 429, **434**
McDonald, D. A., 351, **356**
McElhaney, J. H., 151, **167**
McCutchen, C. W., 440, **453**
Mackenzie, I. C., 230, **244**
MacKenzie, J. K., 155, **163**
McKern, N. M., 216, **242**
Mackie, W., 322, 323, **328, 329**
McLachlan, A. D., 216, **244**
McLeish, R. D., 58, **74**
McNeill, K. G., 157, **165**
MacRae, T. P., 211–45; 213, 215, 216, 217, 218, 219, 220, 223, 224, 225, 235, **242, 243, 244;** 260, 262, **270;** 397, **433;** 437, **453**
Maddrell, S. H. P., 438, **453**
Magoshi, J., 257, **271**
Magoshi, Y., 257, **271**
Mahler, D. B., 109, **134**
Majno, G., 141, **167**
Makinson, K. R., 232, 235, 236, **244**
Mammi, M., 307, 309, 310, **313,** 347, **355**
Mantle, D., 365, 367
Margaria, R., 449, **453**
Mark, R. E., 170, 171, 172, **181**
Maroudas, A., 386, **394**
Marquez, E., 307, **312**
Marsh, R. E., 258, 260, **271**
Maruyama, K., 332, **356**
Marwick, T. C., 213, 218, 220, 222, **240**
Mather, B. S., 150, **167**
Matoltsy, A. G., 213, 215, **244, 245**
Matoltsy, M. N., 213, **245**
Matsubara, S., 332, **356**
Maxwell, J. C., 298, **313**
May, W. D., 175, **181**
Meachim, G., 381, 391, **394**
Melcher, D., 42, **72**
Melms, D., 301, **313**
Melville, H. A. H., 42, **72**
Menefee, M. G., 479, **480**
Menton, D. N., 229, **244**
Mercer, E. H., 213, 223, **241, 244**
Meredith, R., 228, **244**
Merker, H. J., 415, **434**
Messer, M., 150, **163**
Metraux, J. P., 317, 318, **329**
Meyer, E., 119, **134**
Meyer, F. A., 359, 363–4, **375, 376**
Meyer, K. H., 274, **288;** 341, **356**
Meyer, O. E., 289, **313**
Micale, F. J., 200, **210**
Milgram, J. W., 140, **164**
Miller, A., 399, 404, 409, **434**
Mills, R. R., 184, **209**
Minke, R., 183, **209**
Minns, R., 430, **434**
Mockras, L. F., 311, **313**
Mohsenin, N. N., 463, **464**
Moncrieff, R. W., 232, **244**
Montague, W., 222, 230, **244**
Moreno, E. C., 131, **134**
Moriizumi, S., 301, **313**
Morley, J. G., 64, 66, 67, **73**
Morris, E. L., 175, **181;** 307, **313**
Morris, E. R., 362, **375**
Morrison, J. B., 389, **394**
Morton, W. E., 220, 232, **244**
Mow, V. C., 149, **166**
Mueller, C. B., 479, **480**
Muir, I. H. M., 381, **394**
Mullins, L., 203, 204, **209;** 252, **271;** 273–88; 279, 282, 283, 284, 285, 286, **288;** 333, 345, **356**
Muntz, M. L., 121, **133**
Murakami, F., 332, **356**
Murdock, G. R., 452, **453**
Murphy, M. C., 66, 71, **73**
Murty, K. L., 156, **165**
Murty, V. L. N., 362, **375**
Muzie, K., 450, 451, **453**
Myers, D. B., 415, **434**
Myers, J. K., 479, **480**

Nadia, A., 114, **134**

Nah, S. H., 283, **288**
Nakamura, S., 257, **271**
Nakao, K., 415, **434**
Natori, R., 332, **356**
Nemethy, G., 341, **356**
Neville, A. C., 183, 206, 207, **209;** 475, **476**
Newman, H. H., 109, **134**
Newnham, R. E., 118, **135**
Newton, I., 297, **313**
Nielsen, D. J., 184, **209**
Nielsen, L. E., 49, **73**
Niranjan, V., 151, **167**
Nonomura, Y., 332, **356**
North, A. C. T., 410, **433**
Nygaard, V., 102, **135**

Oates, K., 361, **375**
Occhino, J. C., 420, **434**
Odeblad, E., 373, **375**
O'Donnell, I. J., 218, 219, 224, **244**
Ogston, A. G., 387, **394**
Ohashi, K., 332, **356**
Olivo, O. M., 150, **167**
Oppenheim, R. W., 459, 461, **461**
Orava, S., 385, **394**
Outwater, J. O., 66, **73**
Owen, G., 93, **96;** 332, **355**

Packer, K. J., 338, **355**
Paffenberger, G. C., 109, 110, 111, 112, **135**
Pagano, N. J., 177, **181**
Page, D. H., 172, **181**
Pain, R. H., 359–76; 360, 361, 365, 366, 367, 369, 370, **374, 375, 376**
Palley, L., 429, 430, **434**
Panjabi, M. M., 151, **167**
Parakkal, P. F., 213, 220, 222, 230, **244**
Parker, K. D., 261, 262, **270, 271**
Parry, D. A. D., 217, 218, 225, **243, 244;** 399, **434**
Partington, F. R., 430, **434;** 471, 473, **473**
Partridge, B. T., 480, **480**
Partridge, S. M., 341, **356**
Patchin, R. E., 120, 122, 123, 124, **134**
Patel, D. J., 351, **356**
Paul, J. P., 389, 390, **394**
Pauling, L., 258, 260, **271**
Pautard, F. G. E., 235, **244;** 457, **458**
Payne, A. R., 285, 286, 287, **288**
Peakall, D. B., 248, 259, 267, **271**
Pease, D. C., 420, **433**
Pentoney, R. E., 175, **181**
Peterfi, T., 479, **480**
Peyton, F. A., 109, 110, 111, 112, 121, 125, **133, 134;** 152, **164**
Pezzin, G., 307, 309, 310, **313;** 347, **355**
Phelps, C. F., 364, **375**
Phillips, D. C., 66, 67, **73**

Pickard, H. M., 123, 128, 129, **134, 135**
Piekarski, K., 122, **134;** 148, 157, **167**
Piez, K. A., 338, **356;** 399, **434, 435**
Piggott, M. R., 62, **73**
Pigman, W., 121, **133**
Pily, L., 260, **271**
Pollack, R. P., 117, 126, **134**
Pope, M. H., 71, **73;** 151, **164**
Porter, A. W., 175, **181**
Porter, N. S., 117, **135**
Potten, C. S., 230, **240, 244**
Powers, J. M., 132, **135**
Pozniak, R. A., 175, **182**
Prentice, J. H., 205, **210**
Preston, R. D., 170, 171, 174, 176, **181, 182;** 301, 302, **313;** 315, 316, 317, 318, 319, 321, 322, 324, **328, 329**
Price, N. R., 336, **356**
Probine, M. C., 315, 316, 317, 318, 319, 320, 322, **329**
Pryor, M. G. M., 192, 194, 197, 198, **209**
Puranen, J., 385, **394**
Purslow, P. P., 9; 206; 236, 238, **244;** 486, **486**

Radin, E. L., 387, **394**
Randall, J., 301, **314**
Randall, J. T., 410, **433**
Ranvier, L., 141, **167**
Rao, H., 141, **165**
Rasmussen, S. T., 120, 122, 123, 124, **134**
Rauber, A. A., 149, 153, **167**
Rawcliffe, D. J., 81, **97**
Rayns, D. G., 415, 420, **434**
Reed, R., 415, **434**
Rees, D. A., 362, 370, **375**
Reich, F. R., 117, **135**
Reid, L., 361, **374, 375**
Reilly, D. T., 149, **164, 167**
Reis, D., 316, **329**
Reis, P., 228, **243**
Renson, C. E., 114, 115, 116, 118, 119, 124, **135**
Reuss, A., 157, **167**
Reynolds, E. S., 401, **434**
Reynolds, S. E., 186, 188, 190, 197, 204, 205, 207, **209;** 438, **453**
Riblin, R. S., 17, 18, **35**
Rigby, B. J., 234, **244;** 412, **435**
Rivlin, R. S., 279, **288**
Roach, M., 353, **356**
Robert, A. M., 341, **356**
Robert, B., 341, **356**
Robert, L., 341, **356**
Roberts, G. F., 361, **375**
Robins, S. B., 399, **435**
Robinson, R. A., 140, **164**
Robson, T. R., 361, 370, **374, 375**
Rodriguez, M. S., 112, 113, **133**
Roelofson, P. A., 315, **329**

Rogers, G. E., 213, 215, 218, 220, 223, **242, 243, 245**
Rohrbach, R., 341, **357**
Roland, J. C., 316, **329**
Rölla, G., 130, 131, **135**
Rollerton, E., 430, **434**
Rollins, F. R., 112, 118, **134**
Rollins, F. R., jun., 156, **166**
Romer, A. S., 212, **245**
Rootare, H. M., 132, **135**
Rosen, B. W., 54, 61, 62, **73, 74;** 159, **165**
Roth, S. I., 218, 220, **241, 245**
Rougvie, M. A., 404, 406, **435**
Rouiller, C., 141, 144, **166, 167**
Rowlands, R. J., 217, 235, **242, 243**
Roy, C. S., 289, **313**
Royal Society, London: *Proceedings of Discussion Meeting,* 39, **73**
Roydhouse, R. H., 116, **135**
Rudall, K. M., 184, 188, 191, 193, 197, **208;** 213, 215, 218, 220, 230, 238, **245;** 255, 259, 260, 261, 262, **271**
Rupec, M., 420, **433**
Ruth, E. B., 141, **167**
Rutishauser, E., 141, **167**
Ryder, M. L., 231, 232, **245**
Ryge, G., 119, **135**

Sackner, M. A., 369, **375**
Sage, E. H., 332, 335, 336, **356**
Saha, S., 148, 154, **166**
Saibene, F. P., 449, **453**
Saketkhoo, K., 369, **375**
Sammarco, G. J., 146, **167**
Sandberg, L. B., 336, 338, 341, **355, 356**
Sass, R. L., 333, 336, **356**
Scaletta, L. J., 420, **434**
Scandola, M., 307, **313**
Scheitele, S. J., 71, **72**
Scheraga, H. A., 341, **356**
Schmidt, F. O., 399, **435**
Schneider, J. W., 340, **357**
Schniewind, A. P., 175, 176, **182**
Schor, R., 222, **245**
Schroeder, W., 259, **271**
Schwartz, A., 399, **435**
Scott, D. B., 102, 120, 122, 123, 124, **134, 135**
Scott, V. D., 459–61
Scurfield, G., 178, **182**
Sedlin, E. D., 152, **167**
Sekora, A., 260, **271**
Sellen, D. B., 301, 302, **313;** 315–29; 321, 322, 323, 324, 325, 326, 327, **328, 329**
Shah, X., 425
Shanahan, W. J., 52, **73**
Shaw, J. T. B., 249, 250, 251, 255, 256, 259, 264, **271**
Shetlar, M. R., 415, **434**
Shih, C. K., 362, 366, 369, **375, 376**
Shimokomaki, M., 399, **435**
Shtrikman, S., 130, **134**
Signorelli, R. A., 429, **435**
Sikorski, J., 397, **435**
Silberberg, A., 359, 360, 363–4, **375, 376**
Simkin, A., 144, **163**
Simkiss, K., 75, **97**
Simmelink, J. W., 102, **135**
Simmonds, D. H., 225, **243**
Simpson, L. O., 478–81; 480, **480, 481**
Sisson, W., 236, **241**
Sittig, R. A., 428, **433**
Skerrow, D., 213, **245**
Slayter, G., 221, **245**
Slen, S. B., 223, **241**
Smith, A. J., 390, **394**
Smith, D. C., 115, 116, **133, 135**
Smith, J. W., 141, 149, 152, **167, 168;** 409, 412, **435**
Smith, R. W., 152, **167**
Smith, S. G., 249, 250, 251, 255, 256, 259, 264, **271**
Snary, D., 366, 369, **376**
Sobel, H., 397, 414, **435**
Soden, P., 430, **434**
Soden, P. D., 58, 59, **72, 74**
Sokoloski, J., 361, **376**
Southwick, W. O., 151, **167**
Spark, L. C., 176, **182**
Speakman, J. H., 224, 228, **245**
Spearman, R. I. C., 230, 231, 235, 236, 237, **245**
Spei, M., 235, **245**
Sproule, R. N., 175, **181**
Stacy, R. W., 301, **314**
Stanford, J. W., 109, 110, 111, 112, **135**
Stanier, J. E., 387, **394**
Steel, T. R., 465–9
Steinert, P. M., 213, **245**
Stell, J. G. P., 238, **242**
Stern, R., 155, **165**
Sternstein, S. S., 153, 154, **166**
Stevenson, A., 34, **35**
Stewart, F. H. C., 260, **270**
Stewart, M., 219, 238, **245**
Stockwell, R. A., 381, 392, **394**
Storer, T. I., 237, **245**
Stowell, E. Z., 58, **74**
Street, A., 213, 233, **241**
Strout, V., 189, 192, 196, **209**
Susich, G. von, 274, **288**
Suzuki, E., 216, 218, 219, 225, 231, **243, 245**
Swann, D. A., 387, **394**
Swanson, S. A. V., 41, **74;** 377–95; 386, 388, 389, **394, 395**
Swart, L. S., 216, **245**

Swedlow, D. B., 141, 144, 156, 158, **166, 167**
Sweeney, A. B., 109, 110, 111, **135**
Sweeney, W. T., 109, 110, 111, 112, **135**
Synge, J. L., 127, **135**
Szent-Gyorgi, A. G., 265, **271**

Taiz, Z., 317, 318, **329**
Takahashi, K., 438, 439, **453**
Takata, T., 318, **328**
Tanzer, M. L., 399, **435**
Tappin, G., 121, **135;** 176, **182**
Tattersal, H. G., 121, **135;** 176, **182**
Taylor, D. G., 378, **394**
Taylor, J. D., 76, 77, 81, 91, 93, 94, **97**
Taylor, M. G., 307, **313**
Tazawa, M., 318, **328**
Tennyson, R. C., 151, **167**
Thomas, A. G., 17, 18, 26, **35;** 283, **288**
Thompson, G., 152, 153, **167**
Thompson, P. R., 187, 188, 191, **209**
Thompson, W. A., 158, **166**
Thorogood, R. V., 457, **458**
Timpl, R., 414, **435**
Tischendorff, F., **167**
Tobin, N., 282, 283, **288**
Tobolsky, A., 325, **329**
Topping, A. D., 52, **74**
Torchia, D., 338, **356**
Torp, S., 400, 402, 403, 408, 409, 414, 421, 422, 429, **435;** 471, **473**
Traub, W., 399, **435**
Treloar, L. R. G., 276, 278, 280, **288;** 339, 344, **356**
Trelstad, R. L., 415, **433**
Trueman, E. R., 333, **356**
Tsai, S. W., 44, 47, 59, **72, 74**
Tsuda, K., 152, **167**
Tulloch, P. A., 217, 235, **243**
Tunnicliffe, V., 447, **453**
Turner, I. G., 457–8
Tychsen, P. H., 187, 190, **209**
Tyldesley, W. R., 114, 115, **135**

Ukraincik, K., 118, **134;** 156, 158, **166, 168**
Urry, D. W., 338, 341, **356**

Vaishnav, R. N., 351, **356**
Valko, P., 274, **288**
Vanderkool, G., 199, **208**
Vernon, A., 449, **453**
Verzar, F., 399, **435**
Vian, B., 316, **329**
Viidik, A., 125, **135;** 295, **313;** 410, 415, **435;** 472, **473**
Vincent, J. F. V., vii–ix; 183–210; 184, 185, 186, 187, 190, 191, 192, 193, 194, 195, 196, 197, 200, 203, 204, 205, 206, **208, 209, 210;** 236, 238, **244;** 332, **356;** 486, **486**
Vincentelli, R., 150, **165, 168**
Voigt, W., 157, **168**
Vosburgh, F., 440, 445, **453**
Vose, G. P., 150, **168**

Wainwright, S. A., 42, 72, **74;** 169, **182;** 189, 207, **210;** 220, 221, 222, 234, **245;** 249, 250, 251, 253, 265, **271;** 334, **357;** 437–53; 440, 444, 450, 451, **453;** 485, **486**
Walker, I. D., 219, **245**
Walmsley, R., 149, 152, **167, 168**
Walton, A. G., 399, **435**
Warburton, F. L., 236, **245**
Ward, I. M., 23, **29, 35;** 305, **313**
Warwicker, J. O., 258, 260, 261, **271**
Wassermann, F., 397, 399, 409, 414, 430, **435**
Waters, N. E., 99–135; 119, 127, 131, **135**
Watson, M. L., 401, **435**
Watt, F., 184, 200, **208**
Watt, L. C., 200, **207**
Wawra, H., 260, **271**
Weber, W., 289, 290, 291, **313, 314**
Weeton, J. W., 429, **435**
Weibull, W., 54, **74**
Weidenreich, F., 141, **168**
Weigel, K. V., 109, 110, 111, 112, **135**
Weightman, R., 381, 386, 387, **394, 395**
Weinstein, H. G., 428, **433**
Weis-Fogh, T., 185, 186, 188, 190, 191, 195, 197, 198, 202, 203, 205, **209, 210;** 311, **314;** 333, 341, 343, 344, 345, 347, 349, **355, 357**
Weisbach, J. A., 364, **375**
Weiser, M., 414, **435**
Weisser, P. A., 387, **394**
Weitnauer, H., 341, **357**
Welinder, B. S., 198, **210**
Wertheim, M. G., 149, 152, **168**
White, A. A., 151, **167**
Whiting, J. F., 159, **168**
Whitney, J. M., 177, **181**
Wiechert, E., 301, **314**
Wiegand, W. B., 340, **357**
Wilbur, K. M., 75, **97**
Wilde, J. de, 249, 253, 254, 256, **271**
Wildman, A. B., 231, **245**
Wildnauer, R. H., 227, **245**
Wilkie, I. C., 438, 439, **453**
Williams, J. G., 23, 24
Williams, L. G., 95, **96**
Williams, L. M., 477–8
Williams, M. L., 309, 310, **314**
Winkler, K., 172, 176, **181**
Wisko, D. S., 155, **165**
Witt, P. N., 267, **271**
Wöhlisch, E., 341, **357**

Wolf, D. F., 360, 361, 362, 366, 369, **375, 376**
Wolff, L., 414, **435**
Wolinsky, H., 353, **357**
Wood, G. C., 430, **434;** 471, 473, **473**
Wood, J. L., 151, **168**
Wood, S. D. E., 186, 190, 192, 198, 204, 205, **210**
Woodhead-Galloway, J., 399, **434**
Woodin, A. M., 218, **245**
Woods, E. F., 218, 238, **243**
Woods, H. J., 217, 221, 222, 233, **241, 245, 246**
Work, R. W., 251, 256, 257, 266, **271**
Wray, J. S., 404, 409, **434**
Wright, K. W. J., 128, 129, **135**
Wright, T. M., 151, **168**
Wright, V., 386, **394**

Yamada, H., 149, **168**
Yannas, I. V., 399, 435; 471, 473, **473**
Ycas, M., 216, **246**
Yergin, B. M., 369, **375**
Yettram, A. L., 128, 129, **135**
Yim, N. G. F., 364, **375**
Yokoo, S., 149, 152, **168**
Yonge, C. M., 96, **97**
Yoon, H. S., 118, **135;** 156, 157, 160, **168**
Young, A. M., 364, **375**

Zahn, H., 220, 222, 235, **242, 245, 246**
Zahradnik, R. T., 131, **134**
Zatzman, M., 301, **314**
Zdarek, J., 196, **210**
Zeidman, H., 71, **72**
Zener, C., 298, **314**
Zettlemoyer, A. C., 200, **210**
Ziegler, D., 141, **168**
Zimmerman, S. B., 213, **245**
Zweben, C., 54, **74**
Zwergrohrdommel, X. Y., 461, **461**

SUBJECT INDEX

abduction, protein rubber of scallop ligaments, 273, 333
amino-acid composition of, 334, 335–6
elasticity of, 346, 348, 349; elastic energy stored in, as change in entropy, 341
abrasion, resistance of different types of mollusc shell to, 95
Acetabularia crenulata (alga), stress-relaxation in stipe of, 322
acetylgalactosamine, acetylglucosamine: in glycoproteins of mucus, 362
acid, resistance of different types of mollusc shell to, 95
Acropora cervicornis, A. reticulata (corals), 442
mechanical properties of, in relation to wave forces acting on, 445–7
activation energy
of fracture of chemical bonds, 25–6
of stress-relaxation in *Nitella* and *Penicillum*, 321, 325–7
adhesion
failure of, 30–4
of filaments to matrix, 225–6
thermodynamic work of, 33
age of animal
and crimp angle of collagen in tendon, 402
and rate of enzyme digestion of tendon, 399
and stress-strain behaviour of tendon, 402–3, 432
and susceptibility to fatigue of articular cartilage, 387
alanine content
of glycoproteins of saliva, 362–3
of protein rubbers, 334, 335
of silks, 258, 259, 260
amino acids
contents of: in insect cuticle, 198–9; in protein rubbers, 334–6; in silks, 259
sequence of, in protein rubbers, 336–9
amniotic membrane, stress–strain curve of, 9–10
Anaphe moloneyi (moth), cocoon silk of, 249, 250, 256, 259
Anaphe infracta, cocoon silk of, 256
Antheraea pernyi (moth), cocoon silk of, 259, 261
Antheraea spp., cocoon silk of, 256
Apidae (bees), larval silk of, 259
Apis mellifera, larval silk of, 253, 254, 256, 262
Araneus diadematus (spider), producing five types of silk, 248
cocoon silk of, 252, 253, 256
dragline silk of, 251, 256, 259, 266–7
viscid silk of, 253, 254, 256, 259
Araneus sericatus (spider)
dragline silk of, 249, 251, 252, 267–8
viscid silk of, 253, 256, 270
arterial walls, 351–5
collagen fibrils in, 351, 353
elastin in, 311, 350, 351, 353
mechanical properties of: elasticity, 273; stress-relaxation times, 292, 293, 299, 301; viscoelasticity, 289
atherosclerosis, 311

Balanus (barnacle), resistance to crushing of shell of, 452
barnacles, resistance to crushing of shells of, 452
basement membranes, as thixotropic systems, 479–80
beak of birds, keratins of, 214
bending fracture, in nacre, 81–5
bending strength
of feather shafts, 485
of laminates, 202, 203
of mollusc shell, 79, 80
bending stress, on feathers, 237–8
bending structures, 6–7
birds, 'hard' keratins of, 213, 214, 221
in feathers, 218–19, 236–9; *see also* feathers
in scale, beak, and claw, 219–20, 239–40
bivalve molluscs, shells of, 76, 79, 94–5
evolution of, 91

Boltzmann principle for viscoelasticity
assumption of linearity, 294–5
assumption of linear superposition, 296–7
Bombyx mori (silk moth), 247
cocoon silk of, 248, 249, 256, 259, 261
length of silk per cocoon of, 265–6
structure of silk of, 257–8, 260
bone, 137–40, 162–3
articular cartilage and, 377, 379
collagen as matrix for hydroxyapatite in, 40, 41
composition of, 100–1
mechanical properties of, 149–50; compressive and tensile strengths, 63; elasticity, 118; fracture behaviour, 69, 71; shear strength, 63; viscoelasticity, 150–4
modelling of, as hierarchical composite, 157–62
structure of: electron microscopy of calcareous network, 457; ultra- and micro, 140–4
ultrasonic wave propagation in, 154–7
see also osteons
boron fibres, 38, 40, 54
plastics reinforced with, 56, 63
breaking strain
of chitin, 189
of different fibres, 40
of *Nereocystis* stipe, 443–5
of silks, 264
of wood, 177
breaking strength
of chitin, 189
of dentine, 124
of different materials, 444
of silks, 254
of stratum corneum, 227
see also fracture, tensile strength
breaking stress, 2–3, 443, 444
'brittleheart' of trees, 174
brittlestars: ions of ambient fluid, and expansibility of ligaments of, 438–40
Buccinum undatum (whelk), sulphated cellulose in mucus of, 361
bulk modulus, from ultrasonic wave propagation
of bone, 155, 160
of dentine and enamel, 117–18
bursicon (tanning hormone of *Calliphora*), controls extensibility of cuticle, 438

calcium carbonate, in mollusc shell, 75
passage of cracks between blocks of, rarely through, 82
Calligula japonica, cocoon silk of, 156
Calliphora (blowfly): tanning hormone of, and extensibility of cuticle, 438
calluses, 230
carbon fibres, 38, 40, 54
plastics reinforced with, 38, 40, 56, 58
cartilage, 41, 311
cartilage, human articular, 377–9
constituents of, 379–81; spatial arrangement of, 381–2
mechanical properties of, 382; in compression and indentation tests, 382–4; in fatigue tests, 385–7; in tensile tests, 384–5
possible modes of failure of, 390–3
service requirements of, 389–90
synovial fluid and, 387–9, 393
cell walls of plants, *see* plant cell walls
cellulose, 40
mechanical properties of: breaking energy, 251; extensibility, 250; *see also* plant cell walls
microfibrils of: in plant cell walls, 315–16, 322, 323; in wood, 170–1
sulphated, in mucus of *Buccinum*, 361
central nervous system, changes in mechanical properties controlled by, 437–8
cervix, uterine
mucous secretion of, 360, 363, 364
pregnancy: and creep rate of, 477–8; and deformability of, 42
Chaetomorpha (alga), cell walls of, 316, 321
chemical bonds
activation of fracture of, 25–6
between matrix and filaments of keratin, 225–6
in composite materials, 37
fracture of, in fibrous composites, 66–7, 71
primary or secondary, in adhesive systems, 34
see also disulphide bonds, hydrogen bonds
chitin, 40, 250, 251
microfibrils of, in insect cuticle, 183; lengths of, 189, 192; orientations of, 180–1, 183, 184; percentage of, in different cuticles, 183
chondrocytes, 380, 381
chondroitin sulphate, in matrix of cartilage, 380
Chrysopa carnea (green lacewing), egg-stalk silk of, 253, 254, 256, 261, 262
Chrysopa flava (lacewing), amino-acid composition of silk of, 259
Cladophora (alga), structure of cell wall of, 316, 321
claws, keratins of
in birds and reptiles, 239–40
in mammals, 225, 236

Codium fragile (alga), mannans in cell wall of, 323
cohesion, forces of (primary and secondary), 24–6
collagen fibres, 40, 250, 251
  in arterial walls, 351, 355
  in articular cartilage, 385, 386
  in basement membrane, 479
  in bone, 41, 137, 138, 141; orientation of, 141–2, 143, 150
  in cartilage, 41, 379
  in dermis, 211
  in Holothuria, 475–6
  in pre-dentine surface, 102, 103
  in skin, 42
  in tendon, 397; events in deformation of, 410, 412, 414–15, 420–1; shredding of (on deformation), 414, 415, 420, 421, 432
compliance, of cell wall of *Nitella*
  creep proportional to, 321, 323
  varies with direction of stress, 319, 320
composite materials, in engineering, 37–9
  see also fibrous composites
composite strength efficiency factors, 59, 60
compression crease, in wood, 173, 174
compression and indentation tests, on articular cartilage, 382–4
compression wood, in soft woods on side of trunk in compression, 179
compressive strength
  of bone, 151
  of dentine, 109–10
  of enamel, 110–12; sintered, 132
  of fibrous composites, 61–3; fibre diameter and, 63, 207
  of keratins, 234
  of mollusc shell, 79, 80; crossed-lamellar type, 88, 89,
  of wood, 172–3, 174
compressive stress, in tree trunk, 174–5
connective tissue: dermal, of birds and mammals, 42
contiguity factor, in estimate of elastic modulus of fibrous composites, 44
Conus spp. (snails), mechanical properties of shells of, 79, 85–6
  fracture tests on, 85–9
Cook–Gordon mechanism (stopping crack propagation by opening of weak interfaces), 8, 66, 71
corals
  mechanical properties and shape of, in relation to wave forces acting on, 445–7
  soft (Alcyonaria), skeletal organization of, 450–2
crack propagation
  in adhesive failure, at interfaces, 30–4
  in articular cartilage, 385–6
  in bone, in ground substance rather than in osteons, 148
  conditions for, 15–18, 23, 24
  Cook–Gordon mechanism limiting, 8, 66, 71
  in eggshells of different species, and hatching behaviour of chicks, 459–61
  in enamel, 130
  in insect cuticle, energy storage and, 205–6
  in laminates, 202
  in mollusc shell, 82, 84, 85–9
  in wood, 175–7
crazing, in a glassy polymer, 29
creep
  in bone, 152
  in insect cuticle, 186–7, 203–4
  in isolated uterine cervix, pregnancy and, 477–8
  in *Nitella* cell walls, 317; proportional to compliance, 321, 323
  plant cell wall growth as, biologically controlled, 318, 321, 327
  and stress–relaxation, 292–3
crimp
  in collagen molecule, 295
  and elastic behaviour of pliant composites, 51
  in hair, 232
  in rat tail tendon: effect of deformation on, 401; recrimping after deformation, 421–8, 432; and stress–strain curves, 402
cross-linking of molecules
  in keratins, 211, 223, 230
  in matrix of insect cuticle, in sclerotization, 192–3, 194–5
  in protein rubbers, 333, 334, 336; elastin, 336, 337, 338–9; resilin, 196, 199, 200, 336, 337
  in swelling of rubbers, 281
cross-β structure, in silks, 261, 262
crossed-lamellar structure, in mollusc shell, 76, 77, 94
  mechanical properties of, 79, 80; fracture behaviour, 85–9, 90; resistance to various treatments, 95
crushability of barnacle shells, and 'workmanship', 452
cuticle
  of insects, *see* insect cuticle
  of mammalian hairs, 217–18, 220, 232
cysteine residues, keratins rich in, 214
  and compressive strength, 234
  in cuticle of hairs, 218
  in filaments of 'hard' keratins, 216, 219

cysteine residues, keratins rich in; (*contd.*)
in matrix of 'hard' keratins, 214, 215, 216–17; of 'soft' keratins, 224–5

deflection, in coping with excessive loads, 2
deformation
loading/unloading cycle in, with loss of energy (hysteresis), 14
of tendon, *see under* tendon
tests of, related to load, on materials with edge cracks, 21–2
density, of enamel and hydroxyapatite, 132
dentinal tubules, 101, 102
dentine, 99, 102
composition of, 100–1, 130–1
junction between enamel and, 103
mechanical properties of, 108–9; compressive strength, 109–10; elasticity, 117–18; fracture behaviour, 120–5; hardness, 119–20, 121; shear strength, 115–17; tensile strength, 112–15; viscoelasticity, 125–7
dermis, 212
extracellular collagen and elastin of, 211
desmosine, isodesmosine (lysine condensates), provide cross-links in elastin, 336, 337, 338–9
disulphide bonds
in glycoproteins of mucus, 363, 364
in keratins, 211, 223, 230
dopa-decarboxylase, in sclerotization of insect cuticle, 196

earthworms, layer of mucus on, 359
Echidna quill, protein of, 225
EDTA, resistance of various types of mollusc shell to, 95
efficiency of reinforcement by fibres, orientation of fibres and, 222–3
egg shells
of different species: correlation of crack propagation in, with hatching behaviour of chicks, 459–61
stress–strain curves of membranes of, 9–10
elastic modulus (Young's)
of bone, 118, 149–50, 150–1, 160, 161–2; calculated from hierarchical model, 158–9; determined by ultrasonic wave propagation, 155; of osteons, 146–8
of cellulose, 40, 171, 172
of chitin, 40, 189
of dentine, 109, 113, 114, 117–18
of enamel, 111, 114–15, 117–18
of fibres, natural and artificial, 40
of fibrous composites, 47–51; Voigt and Reuss estimates for, 43–5
of insect cuticle, 189, 192, 199; matrix, 192, 193, 198
of keratins, 221, 228
of mollusc shell, 79, 80, 82, 90–1, 93
of protein rubbers, 344–6
of rubbers, filled and unfilled, 282–3
of tendon, 402, 403
of wood, longitudinal and transverse, 172–3; of wood cells and bast fibres, 172
of wool, effect of drying on, 223
elasticity
of foods, and shape of teeth, 463–4
of protein rubbers: dynamic, 346–50; static, swelling solvent in, 339–44
rubber-like, 273; biological examples of, 273; of composites, 281–7; general conditions for, 274–6; kinetic theory of, 273–4, 339; phenomenological theory of, for large deformations, 278–80; statistical theory of, 276–8, (applied to swelling) 280–1
true linear, 14
elastin, protein rubber, 41, 42, 332–3
amino-acid composition of, 334, 335
in arterial wall, 311, 350, 351, 353
cross-linking in, 336, 337
in dermis, 211
elasticity of, 341–4, 346; dynamic, 346–50
evolution of, 311–12, 351
force–extension curve of, 346
glass transition in, 310
hydrophobicity of, 334–5
plasticizing effect of water on, 310–11
response of, to forced vibration, 306
stress–relaxation of, 292, 293, 299, 301
electron microscopy
of calcareous network in bone, 457
of deformed tendon, 401, 432; immature, 410–15; mature, 415–21
scanning, for elucidating structure of bone, 141–2, 143, 145
enamel, 99, 103
composition of, 100–1, 130–1
Hunter–Schreger bands in, 103, 105
junction between dentine and, 103
mechanical properties of, 108–9, 121; compressive strength, 110–12; elasticity, 117–18; fracture behaviour, 120–5; hardness, 119–20, 121; shear strength, 115–17; of sintered, compared with hydroxyapatite, 132; tensile strength, 112–15; viscoelasticity, 126–7
prism structures in, 103, 104; keyhole model for interlocking of, 103, 105; orientation of successive layers of, 106
energy
of activation, *see* activation energy
available to create crack surface, 19–21

rate of release of: and crack propagation, 18–19; and fracture resistance, 18–19, 34; temperature and, 24
stored elastic: in arterial walls, 352; in insect cuticle, 202, 205–6; in laminates, 202, 203; in protein rubbers, 341–2; in tendons of foot extensor systems, 449
epidermis, 212
keratins produced in, 211
see also stratum corneum
evolution
of different types of mollusc shell, 78, 81, 91
of elastin, 311–12, 351
of insect cuticle proteins, 201
extensibility
of insect cuticle, water content and, 196, 197
of keratins, 238
maximum force supported by a thread as function of, 268–9
of *Nereocystis* stipe, 443, 445
of silks, 250, 255, 256; Group I, 252, 269; Group II, 252, 253; Group III, 253, 254, 262, 269; compared with other materials, 250; molecular structure and, 265; water content and, 264
extension, molecular events in silk fibre during, 263–5

fabrics, woven: elastic behaviour of, coated, 52, and uncoated, 51–2
failure stress, of collagen in Cuvierian tubules of Holothuria, 475–6
fatigue failure, in articular cartilage, 391–2
fatigue fracture, 24
fatigue tests, on articular cartilage, 385–7
feathers, 7, 236–7
bending stress on, 237–8
design and materials of shaft of, 485–6
keratins of, 214, 218–19; amino-acid sequence of, 218; filament orientation in, 238; non-β part of molecule of, acts as matrix component? 218–19, 225, 238
stress–strain curves of rachis of, 227
fibres
in composite materials, 38, 39–42, 184–7; in crossed-fibre pattern or random, 331–2
diameter of, in composities, and compressive strength, 63, 207
properties of natural and artificial, 40
pull-out of, in fracture of composites, 67–8; in bone, 71; in tendon, 429
slender, with planar crimp: theory of deformation of, 471–3
see also filaments
fibrillation, in articular cartilage, 383–4, 390, 391
caused by dilution of synovial fluid, 389
in fatigue tests, 386
fibroblasts, in dermis, 211
fibrous composites, 38, 39–42, 331–2
arterial wall as, 351–5
bone as, 138, 140
dental hard tissues as, 130–3
with discontinuous fibres, 332
insect cuticle as, 183–9
keratins as, 213, 215, 220–1, 240
mechanical properties of: compressive and shear strengths, 61–3; elasticity, 43–52, 282–3; stress–strain curves, 53; tensile strength, 52–61; toughness and impact resistance, 63–71
plant cell walls as, 315
teeth as, 130–3
tendon as, 41, 431–2
wood as, 41, 180
filaments of keratin
adhesion of matrix to, 225–6
diameters of, 213, 215, 218
lengths of, 222
mechanical properties of, 221–2
orientation and packing of, 222–3, 235; in scales of birds and reptiles, 239; in stratum corneum, 227, 229
regulation of distance between, 235
structure of: α-helix in 'soft' keratins, and 'hard' keratins of mammals, 213, 215, 221; β-sheet in 'hard' keratins of birds and reptiles, 213, 218, 221
fish (especially eels), layer of mucus covering, 359
fleas
energy-storing pads in legs of, 273
release of energy in jump of, 349
flexural properties, of dentine and enamel, 114
fluorapatite, elastic constants of, 118
foliated structure, in mollusc shell, 76, 77, 94
mechanical properties of, 79, 80
resistance of, to various treatments, 95
food
elastic modulus of, and shape of teeth, 463–4
energy required to fracture, at different loading rates, 464
force–extension curves, of protein rubbers, 344–5, 346
fractographs, of dentine, 122, and enamel, 123
fracture
of bone, 69, 71
of dentine and enamel, 120–5
of nacre, 85–9, 90

fracture, (*contd.*)
torsional, of osteons, 146
of wood, along and across the grain, 175–7, 178, 180
fracture energy (work of fracture), 9, 19–29
of bone, 71
of brittle solids, 63
of dentine and enamel, 120–2
of different materials, 251, 444
of fibrous composites, 64; orientation dependence of, 67–8, 69
of nacre, 84–5
per unit volume (resilience modulus), 132
of silks, 250–1, 252, 254, 266–7
temperature and, 177
of wood, 71; along and across the grain, 175–6
fracture mechanics, 3–4, 13–15
engineering, 15–18
generalized, 18–19
friction
directional effect of, on hair, 232
tests of, on articular cartilage, 388–9
frog, layer of mucus on palate of, 359
fucose, methyl-, in glycoproteins of mucus, 362

galactose, in glycoproteins of mucus, 362
Galleria mellonella (waxmoth), cocoon silk of, 252, 253, 256, 259
gastropod molluscs
evolution of shell type in, 91
mechanical properties of shells of, 79
gel
of glycoprotein of mucus, 368–72
of hydrated proteoglycan, matrix for collagen in cartilage, 381
of mucus, model for, 370–2
glass fibres, 38, 40, 54
plastics reinforced with, 38, 39, 40; failure of, in compression, 62; orientation dependence of elasticity of, 48; optical micrograph of transverse cracking in, 56–7; tensile strength of, 56
glass transition temperature, 308–10
of matrix proteins of insect cuticle, water content and, 201, 205
of protein rubbers, 346–7, 350
glycine residues, proteins rich in
in glycoproteins of saliva, 362–3
in non-helical parts of filaments in 'hard' mammalian keratins, 216
in protein rubbers, 334, 335
in silks, 258, 259, 260
glycine and tyrosine residues, proteins rich in: in matrix of 'hard' mammalian keratin, 214, 216, 217, 224–5
content of, and compressive strength, 234
glycoproteins of mucus, 361–6
flexibility of, 366–7
interactions between, 369–70, and with water, 368–9
isolation of, from mucous gels, 368
model for gel of, 370–2
glycosaminoglycans, in matrices of pliant biological composites, 332
in cartilage, 380
in tendon, 41

hairs, 231–5
cuticle of, 217–18, 220, 232
elasticity of, 228
high-sulphur and high-glycine–tyrosine proteins in, 225
tactile whiskers of lions and tigers, filament orientation in, 235
hardness
of dentine and enamel, 119–20, 121
of different parts of locust incisors, 483–4
of enamel and hydroxyapatite, 132
of insect cuticle, 206–7
Haversian canals, in bone, 140
healing, in damaged tendon, 431–2
α-helical structure
in keratins, 213, 215, 221
in silks, 259, 262
hemicellulose, in matrix for cellulose fibrils in wood, 41
Holothuria, failure stress of collagen in Cuvierian tubules of, 475–6
homogeneous structure, in mollusc shell, 76, 77
in burrowing bivalves, 94
mechanical properties of, 79, 80, 90
resistance of, to various treatments, 95
Hooke's law, assumed by engineers, 1, but not obeyed by animal membranes, 9
horns
anisotropic properties of, 236
high-sulphur and high-glycine–tyrosine proteins in, 225
hyaluronic acid
in cartilage, 380
in synovial fluid, combined with protein, 387, 388, 389, 393
hydrogen bonds
between filaments and matrix of keratins, 225
between glycoproteins of mucus, and water, 366
in chitin, 183
in dry protein rubbers, 347
in silk, 258, 264
temperature of immobilization of, 323

hydrophobicity, of proteins of insect cuticle, 198, 199, 200
  and amount of water bound by unsclerotized cuticle, 201
  of elastin, 347, 351
  less in later-evolved insects, 201
hydroxyapatite
  in bone, 101, 137, 138; interpenetration of networks of collagen and, 457; orientation of crystallites of, 138, 143
  in claws of some animals, 236
  in dentine, 101; mechanical properties of, 118, 121, 132; phosphoproteins in enamel with affinity for, 130–1
hyperkeratosis, 230
hysteresis, 14
  in dentine, 125
  in different extensible materials, 22
  in silks: Group I, 251–2, 270; Group II, 252; Group III, 254, 270
hysteresis ratio, 19, 20, 22–4, 34
  reflects internal friction of solid, 23

impact resistance, of fibrous composites, 63–71
indentation and compression tests, on articular cartilage, 382–4
  on locust mandible, 483–4
insect cuticle
  as fibrous composite, containing chitin microfibrils, 183–9
  as layered structure (exo- and endo-cuticle), 201–3
  hydrophobicity of, 198, 199, 200
  mechanical properties of, 190–1; elasticity, 273; extensibility, 437–8; energy storage and toughness, 202, 205–6; hardness, 206–7; viscoelasticity, 203–5
  sclerotization of, 192–201
ions of ambient fluid, and extensibility of ligaments in brittlestars and sea urchins, 438–40

jellyfish, creep and stress relaxation experiments on mesogloea of, 292, 293, 299
joints
  between bones, structure of, 379, 380, 382
  in rigid structures, 10–11
  stress at hip joint in walking, 389–90
  stress at knee joint in drop-landing, 390

keratan sulphate, in matrix of cartilage, 380, 392
keratinocytes, 211
keratins, 211–12
  classification of, 213, 214
  compact and spongy, in feather shafts, 485–6
  'hard', of birds and reptiles, with β-sheet structure, 213, 218; in bird beak, claw, and scale, 219–20; in feathers, 218–19; in reptiles, 220
  'hard', of mammals, with α-helix structure, 214; filaments of, 215–16; in hair cuticle, 217–18; matrix of, 216–17
  matrices of, *see under* matrices
  mechanical properties of, 220–1; breaking energy, 251; extensibility, 238, 250; of filaments, 221–4; of 'hard' of birds and reptiles, 236–40, and of mammals, 231–6; of matrix, 224–6; of stratum corneum, 226–31
  'soft', with α-helical structure, 213, 214, 215, 216; between scales of reptiles, 239
  see also filaments of keratin
Kevlar fibres (aromatic polyamide), 38, 40, 49
  plastics reinforced with, 38, 39, 56

lamellae
  in bone: collagen-rich, in ground substance, 144; interstitial, 140
  in plant cell walls, 315–16
laminates
  in fibrous composites, toughness arising from, 68–71
  insect cuticle as, 201–3
lateral connection of subdivided structure, to add strength in compression, 7–8
ligaments
  elasticity of, 273
  of brittlestars and sea urchins, ions of ambient fluid and extensibility of, 438–40
lignin, in matrix for cellulose fibrils in woods, 41
lipids, in stratum corneum, 226
load–deformation curves
  of crossed-lamellar shell, 86–8
  of nacre, 81, 88; with and without prisms, 85
  of teeth, 127
load–extension curves, *see* stress–strain curves
locusts,
  hardness of different parts of incisors, of, 483–4
  mechanical properties of cuticle of, 190–1
  proteins in cuticle of, 186
loss modulus, 305
  defines dissipation of energy in protein rubbers, 348
lubrication failure, in articular cartilage, 392–3

lysine condensates, *see* desmosine
lysinonorleucine, provides cross-links in elastin, 336, 337

mannan, structural polysaccharide of *Acetabularia* and *Codium*, 322–3
mastication, 107–8, 129
matrices
  adhesion of, to filaments, 225–6
  of articular cartilage, 381, 392
  of fibrous composites, 39–42, 332; parallel and series arrangements of filler in, 43–4
  of insect cuticle: elasticity of, 192, 193, 198; loss of water from, in sclerotization, 198; protein of, 183–5, cross-linked in sclerotization, 192–3, 194–5
  of keratins: 'hard', 214, 216–17, 224–5, and 'soft', 214, 215; mechanical properties of, 223–4; percentage content of, 224–5
  protein rubbers in, 332
  of tendon, 397, 428–31
mechanical properties of biological materials, changes in, 437–43
  in marine organisms, 443; in aids to locomotion, 448–9; in breaking strain, 443–5; in creep recovery rate, 447–8; in elastic energy storage and release, 449; in reversible change from solid to liquid, 449–50; in shape and other properties, in relation to direction of wave force, 445–7; in structural modulus, 450–2; in 'workmanship', 452
Melithaea ochracea (sea fan), sclerite arrays in rigid segments and flexible joints of, 450–2
membranes, safety of continuous, 8–10
Meta reticulata (spider), viscid silk of, 253, 254, 256
microdensitometer scans, of deformed tendon, 401, 406–10
models
  of bone as hierarchical composite, 157–62
  for explaining fracture behaviour of wood, 176–7
  for mechanical properties of laminates, 202
  for prediction of mechanical behaviour of teeth, 128–30
  to relate structure to function in bone, 140
mollusc shell
  functions of different materials in, 91–6
  mechanical properties of, 78–81; elasticity, 90–1, 93; fracture behaviour, in crossed-lamellar structure, 85–9, and in other types, 89–90; tensile and bending fracture, 81–5
  structural types of (crossed-lamellar, foliate, homogeneous, nacre, prisms), 75–8
mucopolysaccharides
  in dentine, 103
  in matrix of tendon, 428
  viscoelasticity of, 41
mucus, 359–60
  chemical composition of, 361–4
  flexibility of, 366–7
  formation of gel of, 368–72
  homogeneity of, 364–5
  molecular conformation of, 365–6
  molecular structure of, 361
  reversible viscosity of, in slug, 449, 450
muscle, protein rubber in, 332

nacre (mother-of-pearl) structure, in mollusc shells, 76, 77
  mechanical properties of, 79, 80, 81–5
  with prisms, 94
  resistance of, to various treatments, 95
nails, human
  growth direction and elasticity of, 236
  orientation of filaments of, 235–6
Nautilus (Cephalopoda)
  mechanical properties of shell of, 79, 96
  shell of, as buoyancy tank, 93–4
Nephila madagascarensis, dragline silk of, 256
Nereocystis luetkeana (bull kelp), 442; breaking strength of stipe of 443–5
Nitella opaca, N. translucens (algae)
  cell wall of: anisotropy of, 319–20; compliance of, 320; growth of, 315, 316–18; relaxation times of, 302

optical effects, in plastic deformation of nacre, 82, 83
orientation of fibres, in fibrous composites, 58–9
  and fracture energy, 67–8, 69
  in plant cell walls, 315–16, 318–19
  and tensile strength (keratins), 222–3, 230
orientation factor, 61
osteoarthrosis, 378, 379, 390–1
  possible causes of, 393
osteons, 138, 139, 140
  in hierarchical model, 158–62; hexagonal array of, 159, 161
  mechanical properties of, 144–6; shear and elastic moduli, 146–8
  structure of, 140–3

Ostrea edulis (oyster): resistance of shell of, to various treatments, 95–6

Pacinian corpuscles, mechanical properties of, 312
Penicillus dumetosus
  stress–relaxation in single cells of, 322, 323, 326, 327
  stress-strain curves of, 324
periodontal membrane, 99, 101, 127
periostracum of mollusc shell, mainly protein, 75
  in locomotion of *Solemya*, 93
phenols, in insect cuticle
  binding of water by? 200
  question of cross-linking of proteins by, in sclerotization, 192–3, 194–5, 196, 201
Pinctada margaritifera (pearl oyster)
  crack path in nacre of shell of, 84–5
  mechanical properties of shell of, 79
Pinna muricata (bivalve), 93
  mechanical properties of shell of, 79, 80
plant cell walls
  mechanical properties of: and growth, 315–18, 327; and structure, 318–21
  viscoelasticity of, 321–7, 327–8
plant cells, twisting of
  in growth, 316
  in return to normal turgor pressure, 321
plastic flow, 14, 114
plexiform bone, 138, 139
Poisson's ratio, for dentine and enamel: determined by ultrasonic wave propagation, 117, 118
polarity, of proteins from different insect cuticles, 199
porcupine quill, 232
  elasticity of, 238
  proteins in, 225
  stress–strain curves of, 227; ratchet profile in, 232, 233
  water content of, 223
pressure-radius curves, for arteries, 352–3
prestressing
  of keratin, from assembly and cross-linking in hydrated form, followed by drying, 224
  of periphery of tree trunk in tension, 174–5
prismatic structure, in mollusc shell, 76, 77
  mechanical properties of, 79, 80; elasticity, 91; fracture behaviour, 89–90
  with nacre, 94
  resistance of, to various treatments, 95
  in *Solemya,* 93
proline
  in silks, 259, 265
  in some glycoproteins of mucus, 362
protease
  effects of, on glycoproteins of mucus, 364, 366
  gastric mucus in protection of stomach wall against, 372–3
  resistance of different types of mollusc shell to, 95
protein rubbers, 332–3
  in arterial walls, 351–5
  chemistry of, 333–4; amino-acid composition, 334–6, and sequences, 336–9; cross-linking, 336
  dry, are glasses at normal temperatures, 347
  elasticity of, 346; dynamic, 346–50; static, swelling solvent in, 339–44
  network properties of, 344–6
  only rubber-like when swollen with water, 339
  see also abductin, elastin, resilin
proteins
  of enamel, 106; phospho-, with affinity for hydroxyapatite, 130–1
  fibrous, self-assembly of, 397
  of insect cuticle (arthropodins), matrix of chitin, 183–4; amino-acid composition of, 198–9; cross-linked in sclerotization? 192–3; polarity of, 199; from stiff and pliant cuticles, hydrophobicity of, 198, 199, 200
  of mollusc shells, 75; failure of, in fracture, 82
  of protein rubbers, kinetically free random polymers, forming random network with permanent cross-links, 333–4, 354
  of proteoglycans, 380
  of silks: amino-acid composition of, 259; β-pleated sheet structure in, 257, 258; in solution in silk gland of *Bombyx*, 257; structural change in, on extraction from gland, 257
  in synovial fluid, attached to hyaluronic acid, 387, 388, 389, 393
proteoglycans
  hydrated, in matrix of cartilage, 41, 379, 397
  mean life of, 386
  protein of, 380
  stiffness of cartilage correlated with content of, 384

quills, 232
  of Echidna, 225
  filament orientation in, 236
  of porcupine, 223, 225, 228; stress–strain curves of, 227, 232, 233
  water content, and elasticity of, 236

reaction wood, 170, 174, 178–9
reinforcement factor
  elastic modulus as, 47
  shear modulus as, 46
relaxation modulus, 294–5, 297–9
relaxation time, 298
  for bone in torsion, 153
  in theory of viscoelasticity, 299–304
  see also stress–relaxation
reptiles, 'hard' keratins of, 213, 214, 218, 220, 221
  mechanical properties of, 239–40
resilience, of protein rubbers, 348, 350–1
resilience modulus (fracture energy per unit volume), of enamel and hydroxyapatite, 132
resilin, protein rubber of insects, 273, 333
  amino-acid composition of, 334, 335
  cross-linking in, 196, 199, 200, 336, 337
  elasticity of, 348, 349
  force–extension curve of, 344–5
  hydrophobicity of, 198, 199
  storage of elastic energy in, 205, 311; as change in entropy, 341
respiratory system (bronchus and trachea), moving belt of mucus in, 359
  glycoprotein of, 363, 364
  water content and function of, 369
Reuss estimate, for elastic modulus of fibrous composites, 43–5, 157
rheumatoid arthritis, 378–9
Rhodnius prolixus (bug), nervous control of extensibility of abdominal cuticle of, 437–8
rigidity modulus, of wool: effect of drying on, 223
rubbers
  elasticity of: general conditions for, 274–6; kinetic theory of, 273–4, 339; statistical theory of, 276–7, (applied to swelling) 200–1
  filled: elasticity of, 282–3; strain amplification in, 283–4; stress softening in, 284–7
  phenomenological theory of large deformations in, 278–80
  structural formulae for natural and synthetic, 275
rubbing speed, on articular cartilage, 390

safety factor
  in arterial walls, 353
  in engineering, 2
saliva, glycoproteins of mucus of, 362–3, 364, 366
  function of, 373–4
salts, in insect cuticle, 183
  in control of water content? 203
scales
  of birds, 238–9
  of reptiles, 239; snakes, 227, 228
scallops, ligaments of hinges of, 273
sclerotization (tanning), of matrix of insect cuticle at ecdysis, 192
  of cutting edges of locust incisors, 484
  loss of water in, 196–8, 201, 202, 203, 205
  question of cross-linking by phenols in, 192–3, 194–5, 196, 201
sea anemones
  conditions in habitats of *Anthopleura* and *Metridium:* and creep recovery rates, 447–8; and design and performance, 441–2
  connective tissues of, 41
  viscoelasticity of body wall of, 299
sea urchins: ions of ambient fluid, and extensibility of ligaments of, 438–40
Semibalanus balanoides (barnacle), resistance to crushing of shell of, 452
serine
  in glycoproteins of mucus, 262
  in silks, 259, 260
shark: alteration in stiffness of skin of, at onset of rapid movement, 440–1
shear modulus
  of bone, 155, 160
  cross-lamination and, 49
  of dentine and enamel, 115–17; determined by ultrasonic wave propagation, 117–18
  of fibrous composites (unidirectional), 45–6
  of insect cuticle, 198, 220
  temperature and, 309
shear storage modulus, of osteons, 146–8
shear strength
  of dentine and enamel, 115, 116–17
  of fibre composites, 63
  interlaminar, 63
shear stress
  resistance to: of composite materials, 42, 50–1; of glycoprotein gel, 372; of pegged and riveted joints, 10; of tough soft tissues, 9
  viscoelasticity of bone in, 153–4
β-sheet structure
  in 'hard' keratins of birds and reptiles, 213, 218, 221
  in silks, 257, 258
ships
  pegged, riveted, and welded joints in, 10
  sailing, tension structures in, 5–6
sialic acids, in glycoproteins of mucus, 362, 366
silks, 247–8
  chemical structure of, 255–65
  effect of tanning, 196

mechanical properties of: Group I, 248–52; Group II, 252–3; Group III, 253–5; relation to function, 265–70; stress–relaxation, 299; viscoelasticity, 290–1
skin
of shark: alteration in stiffness of, at onset of rapid movement, 440–1
structure of (collagen and elastin fibres in protein-polysaccharide matrix), 42, 211–12
skull, sutures in, 10–11
slugs, pedal mucus of, 359
change of viscosity of, with change of shear strain, 449, 450
glycoproteins of, 364
snails
mucus trail left by, 368
shells of, 76
Solemya (mollusc), flexible shell of, 92–3
solid, standard linear
relaxation modulus of, 297–9
response of, to forced vibration, 306
steel fibres (high-carbon), 40
sterols, in stratum corneum, 226
stiffness
of articular cartilage, 383–4
of chitin and matrix from insect cuticle, 189
of different insect cuticles, 190–1
see also elastic modulus
stiffness coefficients, of textiles, 52
stiffness ratio, of fibre composites, 46, 47
stomach, mucous secretion of, 360, 374–5
glycoproteins of, 361, 364, 365; glycosylation of, 363; structure of, 361, 364, 365; viscosity of, 367, 370
storage modulus, 305
of protein rubbers, 348
strain
dependence of hysteresis ratio on, 23
engineering, and 'true', 292–3
strain energy, 2, 3–4
and fracture mechanics, 4
not transmitted between subdivided tension members, 4–5
strain hardening, in tendon, 412, 432
stratum corneum of epidermis, 212
cell columns in, 229, 230
mechanical properties of, 226–7, 230; elasticity, 228
'soft' keratins of, 213, 214
stress intensity factor, 16
stress–relaxation, 292–3
activation energies for, 324–5, 327–8
Boltzmann's principle applied to, 294–7
creep and, 292–3
in insect cuticle, 186–7, 204–5
in plant cell walls, 321–3, 327
of silks, 299
stress softening, in filled and unfilled rubbers, 284–7
stress–strain curves, 9–10
of algae: cells of *Nitella*, 323–4; stipe of *Nereocystis*, 443
of extensible solids, 22
of fibre, matrix and two composites, 53
of insect cuticle, 186–7; after tanning with different concentrations of catechol, 194
of keratins, 227, 234; feather rachis, 238, 485; porcupine quill, 227; snake scale, 239
linear and non-linear,
of nacre, 81
of osteons, 147
of rubbers (artificial), 22; (filled), 8, 283–4, 285, 286, 287
of silks, 248, 249; Groups I, 249, 250; Group II, 252–3; Group III, 270
before and after tanning, 196
of teeth, 109; for dentine and enamel, 114
of tendon, 399–400, 429, 432, 468; for tendon matrix, 397, 428–31
of wobbly joint, 10
of wood, 177
of wool, 227
yield regions in: keratins, 234, 238, 239; tendon, 402–3, 432
sulphate, in glycoproteins of mucus, 362
surface energy of solids (energy required to overcome forces of cohesion), 24–6, 34
measurement of, 26–9
significance of, in a glassy polymer, 29
theoretical and experimental values for, 28
swelling
of glycoproteins of mucus, 362
of rubbers, statistical theory of, 280–1
of protein rubbers, solvent and, 339–46
synovial fluid, in joints, 379, 380, 382
and friction, 387–9
wear-preventing function of, 392–3

teeth
composite nature of, 130–3
conditions of operation of, 107–8
mechanical properties of, 127–8; *see also* dentine, enamel
problems of testing, 108
shape of, related to elasticity of food, 463–4
structure of, 99–107

temperature
  and distribution of relaxation times, 303
  and elasticity of rubbers, 270–4, 275–6
  and fracture energy, 24, 177
  and hysteresis ratio, 23
  of immobilization of hydrogen bonds, 323
  and relaxation modulus: equivalance of time and, 307–8; isochronal, 308–9
  and resilience of elastin, 351
  and shear modulus, 309
  and surface energy, 25
  see also glass transition temperature
tendon
  as fibrous composite, 41, 431–2
  models: for mechanics of, 466–9; for structure of, 465–6
  of rat tail, under deformation, 399–400; electron microscopy of, immature, 410–15, and mature, 415–21; experimental methods, 400–1; function of matrix in, 428–31; microdensitometer scans of, 406–10; recrimping in, 421–8; stress–strain behaviour of, 402–3; tensile tests on, 400–2; X-ray diffraction study of, 403–6
  structure of, 41; hierarchical, 397–9
tensile modulus, of wool; effect of drying on, 223
tensile strength
  of articular cartilage, 385
  of cellulose, 171, 172
  of dentine and enamel, 112–15, 132
  of fibrous composites, 52–61
  of fibres, natural and artificial, 40
  of keratins: filament orientation and, 222–3, 230; matrix and, 222; water content and, 226, 227
  of mollusc shell, 79, 80, 88–9
  of *Nereocystis* stipe, 443
  of silks, 250, 255, 256; Group I, 251–2; Group II, 252, 253; Group III, 253–4, 254–5
  tests of: on articular cartilage, 384–5; on rat-tail tendon, 400–2; small-scale, 487–9
  of wood, 172, 174; of bast fibres and wood cells, 172
tensile stress
  in fibrous composites, 50–1
  in nacre, fracture under, 81–5
  in periphery of prestressed tree trunk, 174–5
tension structures, 4–6
tension wood, in hardwoods on side of trunk in tension, 178–9
thermoelastic measurements on elastic solid, to determine energy and entropy contributions to elastic force, 340
thixotropy, in basement membranes, 479–80
threonine, in glycoprotein of mucus, 362
time: equivalence of temperature and, in determining relaxation modulus, 307–8
time dependence
  of hysteresis ratio, 23
  of surface energy, 25
toughness, of fibrous composites, 63–4
  from fibre–matrix interaction, 64–6
  from laminate effects, 68–71, 202
tracheids of wood, 170, 171
  in reaction wood, 179
tropocollagen macromolecules, 398
  slipping between, in deformation of tendon, 412, 413, 431, 432
turgor pressure
  in cell, and growth rate of cell wall, 316, 318, 328
  experiments on reduction of, in *Nitella*, 321
tyrosine, di- and ter-, providing cross-links in resilin, 336, 337
  proteins rich in, *see* glycine and tyrosine

ultrasonic waves
  elastic, shear, and bulk moduli, and Poisson's ratio, determined for dentine and enamel by specific impedence to, 117–18
  propagation of, in cortical bone: bulk waves, 155–6; extensional waves, 154–5; surface waves, 156–7
uterine cervix, *see* cervix, uterine

Vespidae (wasps), amino-acid composition of silk of, 259
vibration, forced: for study of elasticity, 304–7
Victory, H.M.S., tension structures in, 5–6
viscoelasticity (elasticity with loss of work energy in form of heat), 311
  of bone, 140, 150–4
  of dentine and enamel, 125–7
  early experiments on, 289–92
  illustrated by experiments on creep and stress–relaxation, 292–4
  of insect cuticle, 203–5
  mechanism of: glass and other transitions in, 308–10; plasticizing action of water in, 310–11; time–temperature equivalence in, 307–8
  of mucus, water content and, 361
  of plant cell walls, 321–7, 327–8
  relaxation modulus and, 297–9;

distribution of relaxation times and, 299–304
studied by forced vibration, 304–7
theory of, in biological materials, 311–12; Boltzmann's principle of superposition in, 294–7
viscosity
of glycoproteins of mucus, 367, 370; effect of ions on, 365; water content and, 369
of slug pedal mucus, changes with shear strain, 449, 450
of synovial fluid, decreases with increasing shear rate, 387
Voigt estimate, for elastic modulus of fibrous composites, 43–5, 157
Volutocorona nobilis (snail), tensile strength of shell of, 89

water
in bone, 101, 119, 147
in cartilage, 379, 380
in dentine, 101, and enamel, 101, 130–1
in elastin, 350, 351
in insect cuticle, 185, 188, 190–1; bound by phenols? 200; and extensibility, 196, 197; lost in sclerotization, 196–8, 201, 202, 203, 205
in keratins, 223, 236, 240
in mucus, 359–60, 361, 369, 373; interaction of, with glycoproteins, 366, 368–9, 372
plasticizing action of, in tissues, 310–11
in silks, 251, 252, 254, 264
wood, 169–70, 178
as fibrous composite, 41, 180
mechanical properties of; compressive and tensile strengths, 61; fracture behaviour, 69, 70, 71, 175–7; prestressing and, 174–5; shear strength, 63; structure and, 170–4
reaction, 170, 174, 178–9
soft, weight-bearing tracheids of, 40
wool
effects of drying on matrix of, 223–4
elasticity of, 228
electron micrograph of cross-section of fibre of, 215
high-sulphur and high-glycine-tyrosine proteins in, 225
keratins of, 214
stress–strain curve of, 227

X-ray diffraction studies
of deformed tendon, 401, 403–6, 410, 432
of keratins, 213–14, 215; feathers, 218, 238; other 'hard' keratins, 219, 220, 223; stretching of keratins, 221, 222
of silk, 257, 260
of single osteons, 142–4
xylan, structural polysaccharide in *Penicillus*, 322, 323

yield regions, in stress–strain curves
of keratins, 234, 238, 239
of tendon, 402–3, 432